Manufacturing Processes for Textile and Fashion Design Professionals

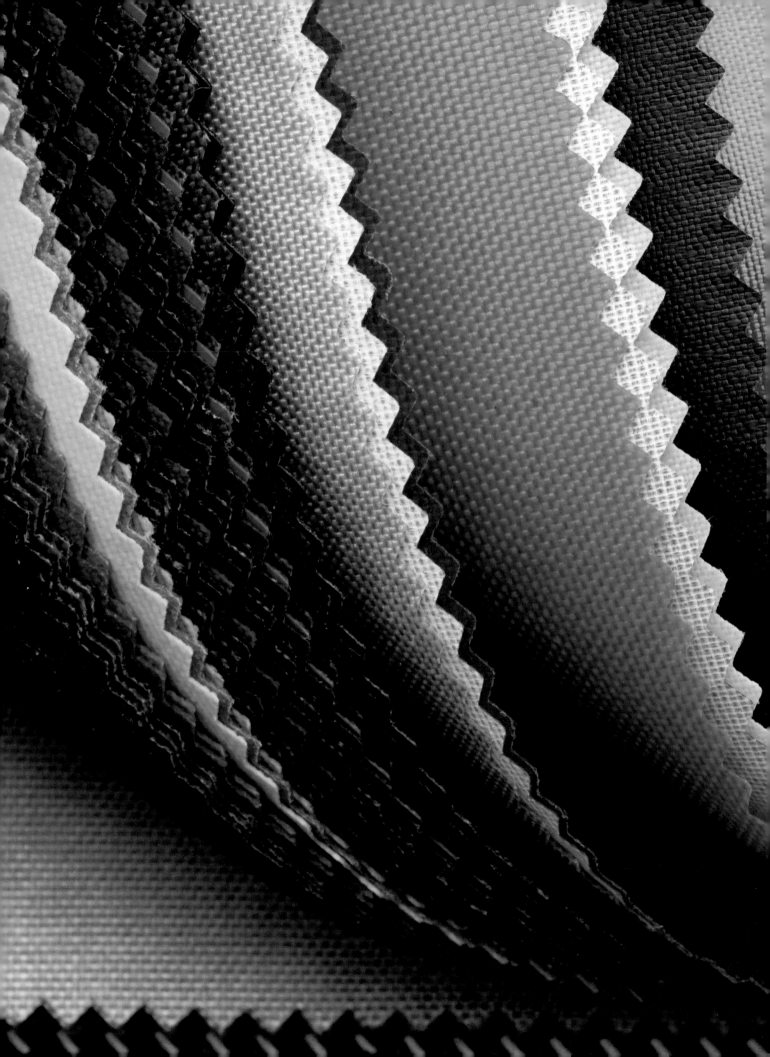

Rob Thompson
Photographs by Martin Thompson

Manufacturing Processes for Textile and Fashion Design Professionals

Thames & Hudson

First published in 2014 in hardcover in the
United States of America by Thames & Hudson Inc.,
500 Fifth Avenue, New York, New York 10110

thamesandhudsonusa.com

Library of Congress Catalog Card Number
2014930114

ISBN 978-0-500-51741-3

Printed and bound in Malaysia by Infinity Press Sdn Bhd

Contents

Part One

Fibre and Yarn Technology

Part Two

Textile Technology

Part Three

Construction Technology

Part Four

Materials

How to Use this Book

In collaboration with some of the most innovative companies and individuals from around the world, this book demonstrates the techniques used to produce the optimum fibre, fabric and finish. The rich content brought together in this original and encyclopaedic guide provides the essential technical knowhow and creative insight to inspire design projects.

The book falls into two sections. The first covers processes, from fibre to finished article, and includes Part 1: Fibre and Yarn Technology, Part 2: Textile Technology and Part 3: Construction Technology. Each process is deconstructed to demonstrate the design opportunities and make sense of technical constraints. Selecting a process has implications for the finished item, as well as the flow of production. Therefore, technologies are explored in isolation as well as cross-referenced, to ensure that the full range of opportunities is understood.

The processes range from ancient crafts to computer-guided technologies. While our approach to certain material

and three-dimensional challenges has remained unchanged for millennia, new technologies are creating unforeseen opportunities. The principles of each – from the everyday to cutting-edge – are explored without bias. This provides an inspiring starting point for projects, as well as containing information to tackle technical and design challenges.

The second section covers materials, the subject of Part 4, which provides a reference against which each process can be examined and individual materials compared. Each material is utilized in a range of applications and manufactured in different formats, finishes and qualities. The base ingredients – whether the

fibre is cellulose-based, protein-based or man-made, for example – are used to determine each material's characteristics and performance in application.

The materials are compared in several ways, including environmental impact. While it is impossible to fully understand a finished item's potential impact, positive or negative, without considering the total life cycle (production, use and disposal), it is feasible to compare the preparation of raw materials. In combination with the process insights demonstrated earlier in the book, this provides the means to refine our approach to established techniques as well as inspire entirely new approaches to the technology of textiles.

Introduction The development of textile manufacturing in relation to clothing, interiors and architecture is explored to provide insight into current and future trends. Emerging textile technologies with the potential to transform fashion, medicine and transport are examined in relation to core materials and processes.

Part Each of the four parts, divided according to the life cycle and format of textiles, is colour-coded for ease of reference: Part 1: Fibre and Yarn Technology (turquoise), Part 2: Textile Technology (coral), Part 3: Construction Technology (yellow) and Part 4: Materials (grey).

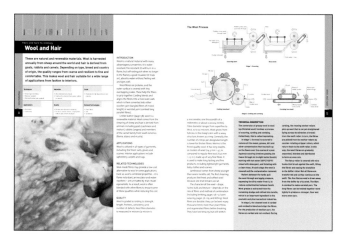

Process Manufacturing processes are explored in detail. Diagrams of the mechanisms involved are given, alongside analytical text that covers typical applications, related technologies, quality, design, materials, costs and sustainability or environmental impact.

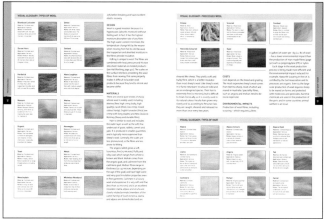

Visual Glossary Examples of the types of colours, details, finish and effects that can be achieved are photographed in close-up detail. Where relevant, comparisons are made between techniques to show the visual differences that can be produced with the featured processes and materials.

Case Study Leading factories, workshops and studios from around the world demonstrate their expertise. Each stage of the production process, from handmade lace to warp-knitted compression garments and woodblock printing to digital techniques, is documented with photographs and descriptive text.

Material An overview of the chemistry and the physical and mechanical properties of each group of materials is provided. The principal material families include plant fibres; fibrous wood, grass and leaves; natural protein fibres; leather and fur; regenerated fibres; synthetic materials; and special purpose materials.

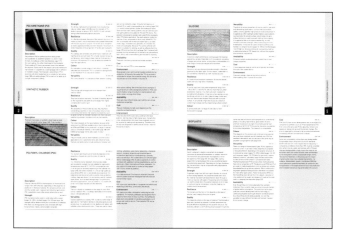

Profiles Each material family is divided into classes – for example, synthetics are thermoplastic, thermosetting, elastomeric or bioplastic – and each member of those subcategories is explored in more detail, including their relative strength, resilience, quality, colourability, versatility, availability, cost and sustainability.

Sources Finally, a range of sources highlights the best books to read to gain further knowledge in specific areas of technology. The details of each of the featured manufacturers are provided, and a glossary explains commonly occurring terminology.

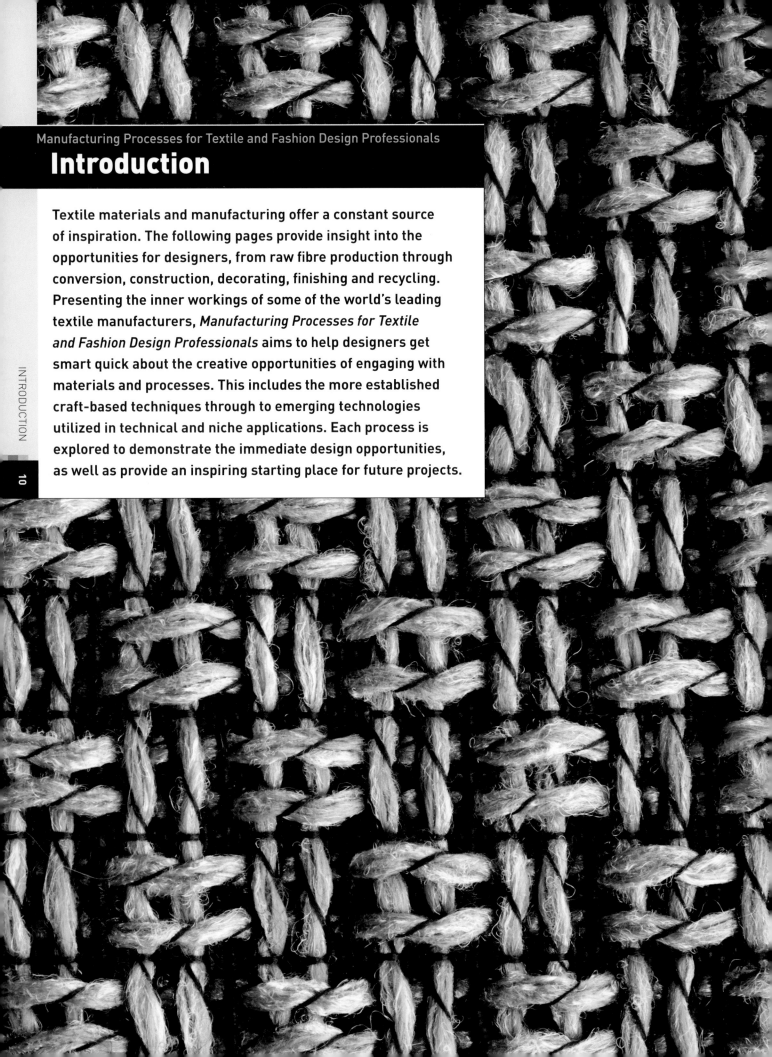

Introduction

Textile materials and manufacturing offer a constant source of inspiration. The following pages provide insight into the opportunities for designers, from raw fibre production through conversion, construction, decorating, finishing and recycling. Presenting the inner workings of some of the world's leading textile manufacturers, *Manufacturing Processes for Textile and Fashion Design Professionals* aims to help designers get smart quick about the creative opportunities of engaging with materials and processes. This includes the more established craft-based techniques through to emerging technologies utilized in technical and niche applications. Each process is explored to demonstrate the immediate design opportunities, as well as provide an inspiring starting place for future projects.

Parts 1–3, focusing on fibre, textile and construction, follow the journey from raw material to finished product. The case studies feature the inner workings of factories producing all types of products, from lightweight quilted fabrics (page 199) to block-printed upholstery (page 259) and fashioned cashmere sweaters (page 133).

Construction techniques continue to be improved and sophisticated computer numerical control (CNC) manufacturing technologies optimize production, from one-off to high-volume. Even so, the beautiful simplicity of an ancient basket (page 120) is the result of the same drivers that guide the development of the most innovative lightweight composites. In other words, while the principles remain the same, the opportunities grow with advances in technology.

For any given application, designers have multiple processes at their disposal. In addition, each technique is applied in a variety of ways. As an example, weft knitting (page 126) is a high-speed manufacturing technique utilized in

the production of fashioned sweaters and circular-knitted undergarments (page 138). While it has evolved as the most practical and economical means of production for these items, its versatility is utilized in diverse applications, ranging from seamless upholstery to super-light running shoes (see image above).

Part 4 focuses on materials, including natural and man-made fibres, films and sheet materials.

Natural materials include plant-based fibres (page 478), fibrous wood, grass and leaves (page 484), natural protein fibres (page 490), and leather and fur (page 498). Semi-synthetics include regenerated fibres, such as viscose (page 508) and acetate (page 509). The synthetics section covers plastics, which have transformed the textile industries since the commercialization of nylon (page 516) by DuPont in 1938. Other well-known types include acrylic (page 516), polyester (page 515) and elastane (page 517). The impact of these fibres is not limited to apparel and interiors; they have transformed many

Knitting The Nike Flyknit technology utilizes high-strength polyester fibre knitted with varying density to provide the right balance of support and ventilation. In addition, the one-piece upper reduces the waste associated with the conventional approach to producing running shoes.

other familiar products, including the humble teabag.

More recently, plastics derived from biobased ingredients – generically called bioplastics (page 519) – have emerged as a sustainable alternative to synthetics. Nylon and polyester may be produced from bio ingredients. The resulting filament is chemically very similar to those derived from crude oil. Polylactic acid (PLA), on the other hand, is wholly bioplastic and as a result is biodegradable at the end of its useful life.

In addition to the more commonly used materials are those reserved for special purposes. These include glass fibre (page 526), basalt fibre (page 527), metals (page 527) and carbon fibre (page 526). Also in this section are the so-called super fibres, which are known mainly under their trade names: aramid (page 524) is referred to as Kevlar, Twaron and Nomex; polyphenylene benzobisoxazole (PBO) (page 525) is known as Zylon; expanded polytetrafluoroethylene (ePTFE) (page 525) is known as Tenara; and liquid crystal polymer (LCP) (page 524) is known as Vectran. Their physical properties set them apart from conventional synthetic fibres, opening up new applications in fashion and architecture alike (see image opposite top).

FIBRE AND YARN TECHNOLOGY

The main types of fibre are cotton (page 28), bast (page 34), leaf (page 40), silk (page 46), wool (page 22) and man-made (page 50). Of course, there are many other types of niche fibre, from paper (see image opposite bottom) to coir.

A fibre's characteristics are determined by its chemical and physical make-up. Plant fibres (page 478) are often referred to as cellulose-based, whether harvested from the seed pod, stem or leaf. They consist of cellulose, hemicellulose, lignin and pectin. The proportion of these ingredients determines the stiffness, strength, handle and colourability of the finished product.

Bamboo is an important cellulose-based fibre and is utilized in several formats. Whereas bast fibre produced from the stem (page 483) is a sustainable material, viscose (page 508) is produced

through a series of mechanical and chemical processes, which have significant environmental impacts. It is important the two are not confused, because the properties, even though desirable, are quite different.

Fibres derived from animals, categorized as protein-based (page 490), are made up of amino acids formed into long polymer chains. There are two main types, keratin (wool, hair and fur) and fibroin (secreted). Keratin is derived from mammals, such as sheep, goats, camels and rabbits; fibroin is obtained from insects.

There are many varieties of keratin fibres, the quality of which derives from the age of the animal, breed, health, diet and climate. Well-known types include Merino (finest and softest of all sheep's wool), cashmere and mohair (goat), Angora (rabbit) and vicuña (one of the rarest fibres of all owing to its near-extinction in the 1960s). These fibres are prized for their softness and insulating properties. In addition, keratin protein fibres are inherently flame-resistant (cellulose-based fibres burn readily unless treated, see page 226).

Many types of insects produce silk. As with fur, the characteristics of each species have evolved to accommodate specific needs related to their environment, predators and so on. Several moth larvae produce a silk cocoon suitable for textiles. The most appropriate of these, in terms of domestication and filament quality, is the mulberry silk worm. The fibre's high affinity for dyeing and lustrous surface results in a brightly coloured yarn, whose strength and beauty are utilized in fine-quality textiles.

Man-made fibres are regenerated or synthetic. Fibres regenerated from naturally occurring cellulose or protein polymers, also referred to as semi-synthetics (page 506), include viscose, cellulose and azlon. Viscose was the first man-made fibre and was initially developed to replace more expensive natural fibres such as cotton and flax. One of the major drawbacks is the negative environmental impact of production.

Lyocell is a more recent semi-synthetic development and commercial production

Above
Super fibres Created by Studio Echelman in 2010, the '1.26' sculpture project premiered at the Biennial of the Americas in Denver, Colorado. Utilizing Honeywell Spectra, an ultralight fibre 15 times stronger than steel by weight, its lightweight structure allows it to temporarily attach directly to the façades of buildings. Because this monumental sculpture is made entirely of soft fibre, it moves and dances with the wind.

Right
Paper fabric The 'Self Fold' series of textiles by Philippa Brock is inspired by her experimentation with paper-folding techniques. These innovative digital, industrial-power-loom-woven textiles include paper in the weft, which gives the material structure and, with the other yarn interactions, causes it to crease into a three-dimensional structure as it comes off the loom.

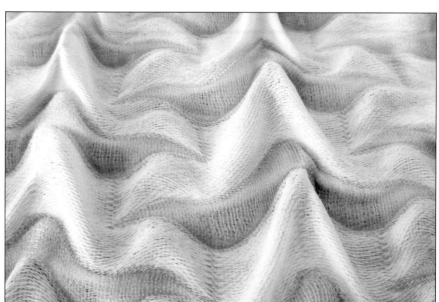

began in the 1990s. Using fewer harmful
chemicals and producing fibre in a nearly
closed-loop process significantly reduces
the environmental impact. It is often
referred to by the trademark name Tencel.

Synthetic plastics (page 510) are
derived from crude oil and utilized in the
production of fibre, film (page 180), sheet
and foam (page 172). They are extremely
versatile and used in all types of textile
application, from commodity to technical.

In addition to being able to tweak
the material properties, the size and
shape of the filament may be adjusted
from microdenier to large-diameter
monofilaments. This is carried out for
various reasons, for example to produce
a smooth flat fibre to reflect more
light and so appear more lustrous and
vibrant. Composite filament yarns are
produced by co-extruding two or more
materials. This is to take advantage of
the properties of two (bi-component
filament), three (tri-component filament)
or more materials.

Developments in Fibre and Yarn
Spinning (page 56) converts fibres into
a coherent yarn, and plied yarn (page 60)
consists of two or more strands twisted
together. Yarns have been developed with
the optimum mix of different ingredients,
tailored to specific applications or
requirements. The benefit for designers
is the diversity of high-quality yarns
available off the shelf.

There are several practical reasons
for mixing fibres, such as producing a
highly absorbent textile that is suitable
for welding (by mixing cotton and
polypropylene for example), or creating
the ability to cross-dye a yarn or fabric
multiple colours (by using two or more
fibres that react to different dyes).

Bridgedale high-performance socks
are knitted from WoolFusion (see pages
142–43). This specially developed two-ply
yarn combines the comfort and insulation
of wool fibre with the moisture-wicking
properties of Coolmax polyester filament
(produced with a profile that promotes
capillary action and so draws moisture
away from the body). This helps to keep
the wearer dry and comfortable even in
extreme conditions.

Using the natural colour of the
fibre eliminates dyeing (page 240) and
the associated process steps. Uniform
colour is built up by blending fibres and
pattern is created by combining different
coloured yarns.

Wool is available in a range of colours,
ranging from off-white to brown and
black. By contrast, plant fibres are typically
light beige to grey in colour. In recent
years, the natural colour variation of
cotton has been rediscovered and organic
varieties are being grown (albeit in very
small quantities) in cream, red, green
and brown.

The brightest and most brilliant
synthetic fibre colour is produced by

mixing the colour with the plastic
prior to spinning. The more inert
fibres, such as polypropylene (PP) and
polyethylene (PE), can be coloured only
in this way. However, it is not always
possible, or practical, because to make
this worthwhile large batches of a single
colour must be produced.

Coatings (page 226) are applied to
yarns (as well as to the finished fabric) to
enhance specific properties. In the case of
man-made fibres, coatings are typically
applied in-line, directly after spinning.
Coatings provide invisible benefits, such
as water repellence and antimicrobial
properties.

Metal coatings are used to enhance
surface properties by improving
reflectivity, wear resistance and corrosion
resistance. For example, pliable nylon
yarn is woven into textiles that are
coated with antimicrobial silver, or highly
conductive copper (see image above). On
the other hand, gold is a precious metal,
non-toxic and will not tarnish. Rather
than making entire objects from this
expensive material it is applied as a thin
and durable coating onto a less expensive
base material.

TEXTILE TECHNOLOGY
Textiles are fabricated from nonwoven
fibres (page 152), unidirectional filaments
(page 234), woven (pages 76–105) or
knitted yarn (see Weft Knitting, page
126, and Warp Knitting, page 144), stitch
bonding (page 196) or laminating (page
188). What sets these materials apart is
the method of construction.

Alternatively, sheets of material, such
as leather (page 158), rubber (page 168)
and film, are applied directly.

Warp and Weft

Woven fabric is constructed by interweaving warp (lengthwise yarn) and weft (widthwise yarn). Looms operate with the warp held under tension so that the weft can be passed through without snagging as the selected warps are raised and lowered. Mechanized looms require that the warp be sized – to increase strength and stiffness – so that it can withstand the applied loads. The size is removed by scouring (page 202), prior to finishing, dyeing and printing.

The weft is not held under the same tension. So, while the warp determines the linear colour arrangement, widthways patterns are not so limited. This is enhanced by floating the weft over several warp yarn, thus bringing its colour to the front; or, by using transparent warp, the weft is visible even when under-lapped.

The simplest fabric construction is a plain weave, which requires two heddle bars (splitting the warp in two). Fancy loom weaving (page 84) is used to produce complex and intricately patterned fabrics. The warps are divided into a larger number of groups (dobby, page 86), or raised and lowered individually (jacquard, page 87) for the ultimate control (see image above).

This arrangement, with the warp and weft running perpendicular (except in the case of multiaxial fabrics, page 91), determines the material characteristics. The physical properties of woven fabrics align with the yarn and are highly controllable, because the yarns run parallel and straight. Therefore, woven fabrics are more suited to formal shirts than body-hugging sweaters, stitched (page 354) as opposed to three-dimensional furnishings, and stable lightweight fabrics for architectural, geotextile or nautical application.

Knitted fabric consists of either a single inter-looped weft, or multiple looped warp yarns (except inlaid yarn, page 197). The loop structure allows freer movement of the yarn and so is more easily distorted, owing to the stretch of knitted fabric. Therefore, knitting is used to make garments that fit more tightly to the body and stretch to accommodate movement. This also means that knitted

Above
Fancy weaving This figured-silk velvet, skilfully woven on a manually operated jacquard loom, combines loop and cut pile on a plain-woven ground. Complex fabrics such as this example from Prelle in Lyon, France, can take weeks, or even months, to weave.

Right
Seamless knitting With each stitch in a warp-knitted garment being formed by a separate yarn, a range of interlocked, inlaid and open structures may be produced. By incorporating various loop patterns, different densities of net are integrated seamlessly into a solid ground.

fabrics tend to lose their shape quicker than woven fabrics.

Weft knitting uses a single end of yarn, which is inter-looped in widthways rows (courses). Each loop is dependent on its neighbours on either side, above and below. With modern V-bed knitting (page 134) almost all types of stitch can be achieved on a single machine. By combining different stitches it is possible to produce flat, fashioned, tubular or three-dimensional fabrics without seams.

Warp knitting, on the other hand, requires an individual yarn for each needle. These track sideways to form the lengthways inter-looping structure (wales). In this way, the stitches in a

warp-knitted fabric are less dependent on their neighbours, producing a more stable textile. It is more complex to design for, and less commonly used than weft knitting, but is ideal for producing interlocked, inlaid and open structures, such as used for hosiery and sportswear (see image above).

Nonwoven, Unidirectional and Spread Tow

Fibres are converted directly into nonwovens by mechanical entanglement or bonding (adhesive or thermal). These are versatile processes, capable of using all types of material. There are many reasons for taking this approach to

Recycled material
With needle punching it is possible to convert almost any type of material, including mixed recycled content, into fabric.

These materials will not be as consistent as those made from virgin material, but are perfectly adequate for padding and insulation, for example.

producing grain-free and high-strength textiles, including cost (such as synthetic leather and packaging), technical (such as tear-resistant sheets and cosmetics) and aesthetic (such as hat blocking, page 420, and novelty yarn).

Nonwovens have fewer process steps. This helps to reduce cost and the material's environmental footprint. Dry-processed nonwovens have the lowest impact of all, because they do not consumer water or chemicals during production. In addition, needle punching (page 157) is used to make nonwoven fabrics from recycled fibres, providing an extended life for fibres that might otherwise be landfill (see image above).

Unidirectional fabrics are constructed with yarns running in parallel to align all of the yarns' strength. This approach is typically used in conjunction with high-strength materials, such as carbon, aramid and glass, to produce textiles used in lightweight composites (see Composite Press Forming, page 446).

Spread tow fabric (STF) (page 234) is made up of unidirectional filaments, reducing the fabric to the thickness of an individual strand. In the pursuit of ultimate strength to weight, composite laminates are designed with spread tow, built up in layers, running along lines of predetermined stress.

A recent development from North Sails, which is leading composite material innovation for high-performance applications, is Thin Ply Technology (TPT). This unique textile is capable of producing structures with ultimate strength to weight and as a result is being adopted for ambitious projects in the automotive, aerospace (see image below) and marine industries.

Sheet, Film, Foam and Multilayered Material

Synthetic sheet material ranges from very thin film measuring just a few microns thick to sturdy sheets, from thin flexible foam to rigid blocks, and multilayered extrusion. The properties of the base polymer have an equally varied range of flexibility: from breathable to hermetic; slip-resistant to non-stick; and insulating to conductive.

Leather and rubber are unique high-performance materials with the potential to be renewable. Leather ranges from tough cowhide to flexible goatskin. Skin from younger and smaller animals tends to be softer and suppler. For example, horse leather is stiff and impractical for many applications, but pony is supple enough for apparel, bags and gloves (see image opposite top).

Rubber and latex are derived from the sap of the rubber tree. They have the ability to stretch several times their length and return to their original shape. While many synthetic replacements have been created – in an attempt to overcome the shortcomings of these naturally varying materials – they remain important materials, such as for gloves and footwear.

Synthetic foam (page 172) is a versatile material, with applications ranging from sports shoes to mattresses and cosmetics to yacht hulls. The material, density,

Thin Ply Technology (TPT) A project to build a solar-powered aircraft to fly around the world has adopted North Thin Ply Technology (TPT) for the structure. The Solar Impulse prototype has a 60 m (197 ft) wingspan; the 80 m (263 ft) wingspan version is currently being built (Boeing 747 wingspan is 64.4 m / 211 ft). TPT is likely the only material capable of producing such a lightweight aircraft, which weighs around 1,500 kg (3,000 lb). Photo printed with the permission of Solar Impulse.

hardness and elasticity are selected
according to the requirements
of the application.

   The ability to tailor a material's
properties through combining multiple
functional layers creates solutions
for both technical and decorative
challenges. Laminating and stitch
bonding permanently combine sheets
of material to produce a composite
fabric. This versatility is utilized in the
production of materials for outdoor gear
(lightweight, breathable and waterproof),
interior fabrics (covers and mattresses)
and tensile structures (UV light-resistant,
waterproof and high-strength).

   Materials developed with the optimum
properties for specific applications are
marketed under trademark names,
such as Gore-Tex. The set-up, testing
and development required to produce
multilayered textiles limits the
development of bespoke laminates for
short production runs.

   This approach is not limited to
synthetic materials. Indeed, cotton is
laminated with a thin film of plastic to
produce waterproof and washable fabrics,
such as used to make lightweight apparel,
linings and bags.

Print and Finish

Decorative and protective finishes are
applied at various stages in production,
from fibre to finished item. They add
value to the base material; the amount
depends on the function. In many cases it
is essential for maintaining the highest-
quality end product.

   Decorative finishes include dye (page
240), print and embroidery (page 298).
Each produces a distinctive visual quality.
Therefore, the majority of printers will use
a range of techniques, the most suitable
being selected according to quality, cost
and volume.

   Screen printing (page 260) is used
to produce high-quality items. It is
surpassed only by block printing for wet
on dry applications. Colour is applied by
pigment, dye or discharge. In addition,
screen printing is used to apply adhesive
in preparation for flocking (page 280),
chemical gel for devoré, foaming inks and
water-based inks (see image right).

Top
Light and durable
leather Pony leather,
similar to goatskin
and deerskin, is
unmatched by
synthetic alternatives
in terms of its balance
of suppleness and
durability. This makes
it an ideal material for
high-quality gloves and
accessories.

Above
Water-based
printing Designer
Marie Pedersen (Silk
& Burg) produces a
range of beautiful
textiles, such as
these Scandinavian
cushion covers, using
water-based ink on
unbleached linen.
Using water-based
inks reduces the
environmental impact
of screen printing.
They do not contain
harmful chemicals,
solvents or plastics and
can be washed away
with water.

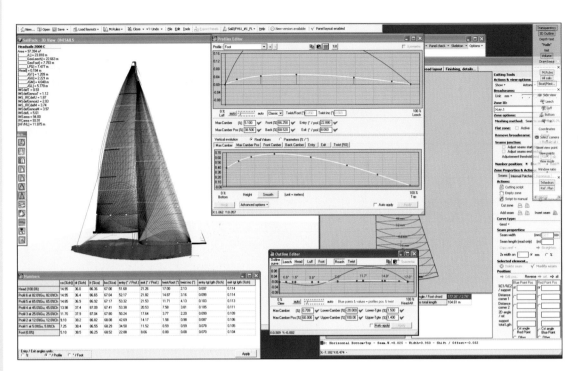

Racing sail design
Sails are designed using computer software. In this case, OneSails creates the optimum 'tailored shape' for the sails of a large ocean racing yacht using a series of two-dimensional inputs, such as the design of the foot of the sail chord and leech (trailing edge), to produce a three-dimensional output (see pages 374–75). The actual flying shape will alter depending on the different loads applied.

Digital printing (page 276) has emerged as a high-quality and cost-effective alternative to screen printing for low- to mid-volume applications, which in turn evolved as a viable alternative to block printing (page 256), the most expensive of all.

Coating is used to enhance specific properties, such as waterproofing, stain resistance or aesthetics. The end use of the item will determine the most appropriate ingredients. For example, domestic interior fabrics require a high resistance to UV light (so they do not fade over time), those used for medical applications must be hygienic, and clothing fabrics have to withstand repeated wearing and washing.

In some cases, the finish is purely functional, such as providing a surface for printing (see Scouring, page 202), raising the nap (page 216), improving shape retention, reducing friction or applying a weldable thermoplastic layer (see Laminating, page 188).

CONSTRUCTION TECHNOLOGY
Factories and workshops are constantly developing new approaches to construction, as well as evolving established processes. Technologies born out of analogous industries give rise to new approaches to construction. These developments, such as applying laser cutting (page 328) and welding (page 376), present designers with an opportunity to break away from archetypical form and function. The following section provides an overview of the most exciting developments relating to cutting, joining and molding.

Cutting Developments
Pattern cutting (page 315) is an art form in itself. Required for any part that has multiple intersecting panels, patterns are created by hand, with computer software, or by a combination of the two. Complex shapes, including people's bodies, may be measured with a 3D scanner (pages 472–73), to provide very accurate data.

Computer-aided design (CAD) and finite element analysis (FEA) software are utilized in the development of three-dimensional items for functional and technical applications. Examples include composites, canopies and sails (see image above).

One-off, bespoke and low volumes may be cut out manually or with a computer-guided laser (page 328) or knife (page 316). Laser cutting is rapid and precise. It is particularly useful in the production of clothing and upholstery, when a high level of flexibility is required. With computerized nesting, the optimum layout of multiple patterns is created and cut out, greatly increasing efficiency.

Die cutting (page 336) is very rapid and used in a broad range of industries, including packaging, leather goods, toys and stationery. The cutting action is instantaneous and multiple layers of material may be cut through in a single stroke, making it very cost-effective.

Joining Developments
Joining processes include mechanical, adhesive and welding techniques. The range of possibilities depends on the materials being combined. Stitching is the most versatile. Used to seam, embroider, hem, overlock and flatlock, machine stitching (page 354) is rapid and precise in capable hands. It is combined with adhesive bonding (page 370) for applications that demand enhanced seam performance.

Adhesive bonding techniques have been developed that are capable of joining materials without the need for a stitched seam. Used alone, seams are less bulky and very strong. Adhesives are engineered with the same performance as the base materials they are being used to join, such as elasticity, waterproofing and tear strength. This unique ability is used in a range of applications from outdoor gear and footwear to upholstery.

Thermoplastic materials, such as polyester, nylon and PE, are suitable for welding. A range of techniques is used – the choice depends on the application and specific material – to produce strong, homogenous bonds, which may be embossed, shaped or three-dimensional. Combined with laminating, virtually any material can be welded.

The processes themselves are very efficient: thermal energy is directed to the seam by convection, conduction or friction. And welding can potentially eliminate the use of additional materials, which makes thermoplastic items much more efficient to recycle.

Molding Developments

Sheets of materials are shaped over a mold to create structural or stiff parts. Molding alters the handle of materials, because curvature introduces rigidity. For example, blocked hats (page 420) are formed from pliable felt and leather shoes (page 430) from supple leather.

In the case of composite laminates, molding is the only practical way to produce the finished shape. High-strength fibres are encapsulated within a resin matrix and shaped over a mold. The combination of polymer and fibre creates a material that is stronger than the sum of its parts. This technique is used to produce a wide range of products, from lightweight racing sails (see 3D Thermal Laminating, page 454) to structural automotive parts.

The most exciting of these developments utilizes a thermoplastic matrix (see page 446). Compared to conventional composite laminates (based on thermosetting resin) they are less expensive to manufacture and more easily recycled and can have higher impact strength and fatigue resistance. They are used for purely technical purposes (such as structural parts), as well as for visual applications (such as sports products).

Filament winding is a unique process. High-performance yarns saturated in high-strength plastic are applied to a rotating mandrel. Owing to the nature of the process it is generally used to make parts that are symmetrical around an axis of rotation. Even so, designers are continually pushing the boundaries of this technology (see image below).

Upholstery is molded to the shape required so no fabrication or cutting is needed; this eliminates assembly and cutting. This reduces waste as well as providing greater control over surface quality. Likewise, dip molding (page 442), which is utilized in the production of gloves and condoms, produces seamless rubber profiles in a single immersion. The shape and size are tailored to fit.

Additive manufacturing (page 466), also referred to as 3D printing, has long been discussed as the next manufacturing revolution. This set of layer-building processes gives designers greater flexibility than before, especially when working with bulk shapes (such as shoe soles and jewelry). The quality is still somewhat limited, but designers are finding ways to use this phenomenal fabrication technique. Shapes are produced that would otherwise be impractical, if not impossible, to create with conventional reductive processes, such as CNC machining (page 449). The more familiar we become with the technology, the greater the range of possibilities that present themselves.

The knowledge packed into this book, provided by technical directors and machine operators alike, provides designers with an unprecedented insight into manufacturing. Textiles are explored throughout their life cycle, demonstrating how smart choices can promote more sustainable and competitive products. Taking this material-led approach means designers work more efficiently with what is available, while challenging conventions, to produce superior products from start to finish.

Filament winding This fabrication technique is generally used to produce high-strength and lightweight composite parts with a certain geometry, such as cylindrical pressure vessels and long thin blades for wind turbines. Seifert and Skinner & Associates (see pages 464–65) utilize this forming process across a broad range of applications and geometries, thus creating opportunities for designers.

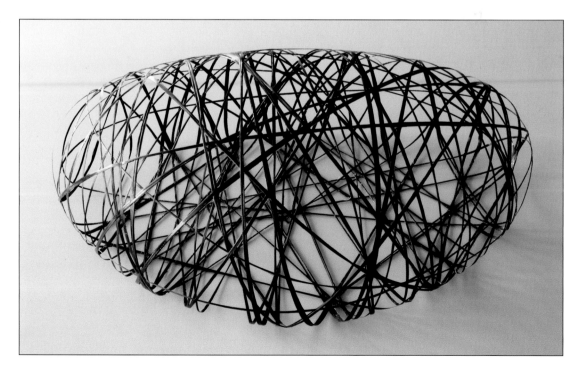

Part

1

Fibre and Yarn Technology

Fibre and Yarn Technology

Wool and Hair

These are natural and renewable materials. Wool is harvested annually from sheep around the world and hair is derived from goats, rabbits and camels. Depending on type, breed and country of origin, the quality ranges from coarse and resilient to fine and comfortable. This makes wool and hair suitable for a wide range of applications from fashion to interiors.

Techniques	Materials	Costs
• Shearing or combing • Woollen (carded) • Worsted (carded and combed)	• Wool (derived from sheep) • Hair (such as derived from goats, rabbits or camels)	• Moderate to high depending on raw material

Applications	Quality	Related Technologies
• Garments • Upholstery	• Durable and resistant to breaking • Variety of natural colours, ranging from white to grey, brown and black	• Plant fibres • Silk • Synthetics

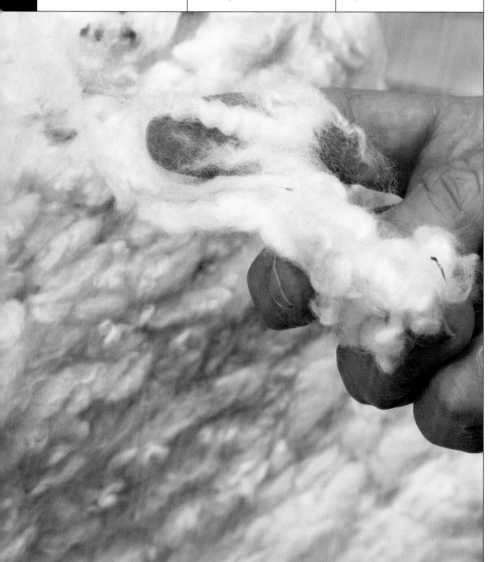

INTRODUCTION

Wool is a natural material with many advantageous properties. It is water-resistant, fire-resistant (it will burn in a flame, but self-extinguish when no longer in the flame), a good insulator (it traps air), absorbs water without feeling wet and dyes well.

Wool fibres are proteins, and the outer surface is covered with tiny overlapping scales. These help the fibres to grip together. Carding blends and aligns the fibres into a more even web, which is then converted into either woollen yarn (tangled fibres of mixed length) or worsted yarn (combed long parallel fibres).

Unlike leather (page 158), wool is a renewable material. Wool comes from the shearing of sheep and hair is derived from animals including goats (cashmere and mohair), rabbits (angora) and members of the camel family from South America (llama, alpaca and vicuña).

APPLICATIONS

Wool is utilized in all types of garments, including the finest suits, gloves and scarves. Interior applications include upholstery, carpets and rugs.

RELATED TECHNOLOGIES

Man-made fibres may provide a low-cost alternative to wool in some applications. Even so, wool's combined properties – it is flame-retardant, an insulator and water-repellent – are unrivalled by man-made equivalents. As a result, wool is often blended with other fibres to impart some of these qualities while reducing the cost.

QUALITY

Wool is graded according to strength, length, fineness, consistency and number of defects. Wool fibre diameter is measured in microns (a micron is

The Wool Process

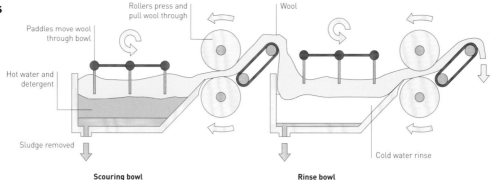

Paddles move wool through bowl

Rollers press and pull wool through

Wool

Hot water and detergent

Sludge removed

Cold water rinse

Scouring bowl

Rinse bowl

Stage 1: Scouring

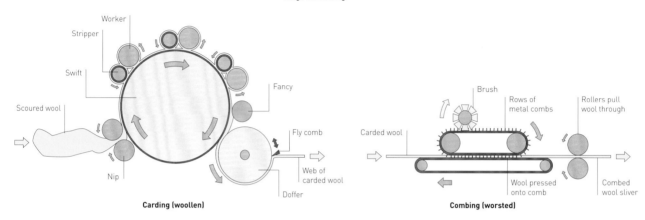

Worker

Stripper

Swift

Scoured wool

Nip

Fancy

Fly comb

Web of carded wool

Doffer

Carding (woollen)

Brush

Rows of metal combs

Rollers pull wool through

Carded wool

Wool pressed onto comb

Combed wool sliver

Combing (worsted)

Stage 2: Carding and combing

a micrometre, one thousandth of a millimetre, or about 0.00004 inches). Fibre diameter ranges from superfine to thick, 10 to 50 microns. Wool grows from follicles in the sheep's skin with a wavy structure, known as crimp. Generally, the number of crimps per designated length is lower for thicker fibres. Merino is the finest-quality wool; it has long staples, or clusters of wool (114.3 mm / 4.5 in. compared to regular fine wool, 63.5 mm / 2.5 in.), made up of very fine fibres. It is used to make long-lasting and fine products, including lightweight garments, baby clothes and scarves.

Lambswool comes from sheep younger than seven months old. The first shearing produces the finest and softest wool because one end remains uncut.

The character of the wool – drape, lustre, bulk and texture – depends on the mix of fibres and method of combination (including knitting, pages 126–51, loom weaving, pages 76–119, and felting). Wool fibres are durable: they can be bent many thousand times more than plant fibres and regenerated fibres before breaking. They have low tenacity, but will stretch

TECHNICAL DESCRIPTION

The conversion of greasy wool to wool top (finished wool) involves a process of scouring, carding and combing. Collectively, this is called topmaking.

In stage 1, the wool is scoured to remove all the sweat, grease, dirt and other contamination that has built up on the fleece over the course of a year. Aqueous scouring involves passing the fleece through six to eight tanks (bowls), starting with hot water (60°C/140°F) mixed with detergent, and finishing with a clean rinse. At each stage the wool is cleaned and the contamination removed.

Rollers between the tanks pull the wool through and apply pressure, squeezing the dirty water from it, to reduce contamination between bowls. Wool grease is extracted from the remaining sludge and refined into lanolin, which is an important ingredient in the cosmetic and pharmaceutical industries.

In stage 2, the cleaned wool is carded and combed to blend and align the fibres. For the production of woollen yarn, the fibres are carded and not combed. During

carding, the rotating worker rollers pick up wool that is not yet straightened (lying across the direction of travel) from the swift roller. In turn, the fibres are picked from the worker rollers by counter-rotating stripper rollers, which return them to the swift roller. In this way, the wool fibres are gradually separated, blended and distributed to form an even mix.

The fancy roller is covered with wire hooks that brush against the swift, lifting the fibres and easing the transition to the doffer roller. Not all fibres are transferred and so they continue on the swift. The fine fibrous web is drawn away from the doffer by a fly comb. The fibre is combed to make worsted yarn. The long fibres can be twisted together more tightly to produce a stronger, finer and more even yarn.

VISUAL GLOSSARY: TYPES OF WOOL

Bluefaced Leicester

Material: Combed wool
Application: Various
Notes: A rare British breed with soft, fine and luxurious fibres around 25 microns in diameter.

Devon

Material: Combed wool
Application: Various
Notes: Off-white, heavy and durable wool. The fibres are around 40 microns thick and up to 200 mm (7.8 in.) long.

Dorset Horn

Material: Combed wool
Application: Various
Notes: Medium-density wool with fibres around 35 microns thick and an irregular crimp.

Gotland
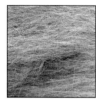
Material: Combed wool
Application: Various
Notes: A soft and lustrous wool with strands 30 to 45 microns thick and 150 mm (6 in.) in length.

Herdwick

Material: Combed wool
Application: Various
Notes: A mountain breed of sheep that yields rugged and durable wool with a mixed fibre quality.

Manx Loaghtan
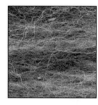
Material: Combed wool
Application: Various
Notes: A rare breed of sheep from the Isle of Man with medium-weight wool and short staples.

Massam

Material: Combed wool
Application: Various
Notes: Massam fleeces are long, with fibres up to 400 mm (15.7 in.), making the wool ideal for handwork.

Merino

Material: Combed wool
Application: Various
Notes: Famous for its soft and luxurious feel. These fibres are just 21 microns in diameter.

Organic

Material: Combed wool
Application: Various
Notes: The final textile is free from allergenic, carcinogenic or toxic chemicals – especially important for babies.

Norwegian

Material: Combed wool
Application: Various
Notes: Versatile and resilient wool with fibres 35 microns thick and up to 120 mm (4.7 in.) in length.

Texel

Material: Combed wool
Application: Various
Notes: A hardy, dense fleece suitable for a range of applications, including carpet.

Welsh

Material: Combed wool
Application: Various
Notes: Hard-wearing wool from the only sheep that produces a pure black fibre.

Wensleydale

Material: Combed wool
Application: Various
Notes: A long, durable and lustrous wool with fibres up to around 50 microns in diameter.

Whiteface Woodland

Material: Combed wool
Application: Various
Notes: A rare breed of sheep from the UK that produces coarse and hard-wearing wool.

25% before breaking and have excellent elastic recovery.

DESIGN

Wool is a good insulator because it is hygroscopic (absorbs moisture) without feeling wet. In fact, it has the highest moisture absorption rate of any fibre. The high water content minimizes the temperature change felt by the wearer when moving from hot to cold because the trapped air and absorbed moisture in the fibres provide insulation.

Felting is unique to wool. The fibres are combined with heat, pressure and friction into sheet materials or molded products (see Hat Blocking, page 420). The scales on the surface interlock, preventing the wool fibres from moving. This same property makes it difficult to launder wool products because they tend to shrink and become stiffer.

MATERIALS

There are several pure breeds of sheep whose wool is widely used, including Merino (fine, high crimp, bulky, high quality), Jacob (thick, low crimp, mixed colour, hardy), English Leicester (thick, low crimp with long staples) and New Zealand Romney (heavy and durable fibre).

Hair is similar to wool and includes the outer layer as well as the soft, fine undercoat of goats, rabbits, camels and yaks. It is produced in smaller quantities and is typically more expensive than sheep's wool. Generally, the scales are less pronounced, so the fibres are less prone to felting.

The angora rabbit grows a soft, luxurious, fine (15 microns), fluffy and silky coat, which ranges from white to brown and black. Mohair comes from the angora goat, and cashmere from the cashmere goat. Mohair fibres range in thickness (25–45 microns, depending on the age of the goat) and have high lustre and very good insulation properties even in fine garments. Cashmere is a luxury wool and expensive. It is very soft and fine (less than 20 microns) and is an excellent insulator. Llama, alpaca and vicuña are closely related animals (members of the camel family) of South America. Llama and alpaca are domesticated and are

VISUAL GLOSSARY: PROCESSED WOOL

Greasy
Material: Wool
Application: Various
Notes: Raw wool contains sweat, grease, dirt and other contamination, which is removed by scouring.

Scoured
Material: Mixed wool
Application: Various
Notes: The wool is clean, but the fibres are tangled and there is still some contamination to be removed.

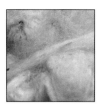

Combed
Material: Mixed wool
Application: Apparel
Notes: Wool is combed to align the fibres fully so they can be spun into smooth and strong worsted yarn.

Naturally Coloured
Material: Mixed wool
Application: Various
Notes: Natural colour ranges from white to grey, brown and black, with many shades in between.

Dyed
Material: Mixed wool
Application: Various
Notes: Wool is dyed a range of colours. The lustre and quality will depend on the breed of sheep.

sheared like sheep. They yield a soft and bulky fibre, which is a better insulator than most sheep's wool. Like sheep's wool, it is flame-retardant. Vicuña are wild and are an endangered species. Their hair is extremely fine (12 microns), but is difficult to treat chemically, so it is mostly used in its natural colour. Vicuña hair grows very slowly and so, according to Peruvian law, they are caught, sheared and released no more than once every two years.

COSTS

Cost depends on the breed and grading. The most expensive sheep's wool comes from Merino sheep, most of which are reared in Australia. Speciality fibres, such as angora and mohair, tend to be more expensive.

ENVIRONMENTAL IMPACTS

Production of wool fibres, including scouring – which requires 4 litres (1 gallon) of water per 1 kg (2.2 lb) of wool – has a lower environmental impact than the production of man-made fibres (page 50) such as polypropylene (PP) or nylon.

Each stage in the wool production process is being made more efficient and the environmental impact reduced. For example, Haworth Scouring in the UK is certified by the Soil Association and its processes are organic. Even so, the large-scale production of wool requires sheep to be reared on farms and protected with medicines and pesticides; harmful chemicals are used in the production of the yarn; and in some countries animal welfare is an issue.

VISUAL GLOSSARY: TYPES OF HAIR

Alpaca
Material: Combed baby alpaca
Application: Various
Notes: Hair from young alpaca can be as fine as 15 microns in diameter and feels very soft.

Alpaca Colours
Material: Combed baby alpaca
Application: Various
Notes: Alpaca comes in a variety of inherent colours from white to grey and brown.

Alpaca Dyed
Material: Combed Alpaca
Application: Various
Notes: Dyeing produces a uniform-coloured fibre. Multiple colours are blended for a natural appearance.

Llama
Material: De-haired and combed llama
Application: Various
Notes: The scales on the surface of llama and alpaca hair are not as pronounced as wool and so feel smoother.

Angora
Material: Angora rabbit
Application: Various
Notes: Angora is very soft, fine and only 10–15 microns thick.

Yak
Material: De-haired, bleached and combed yak
Application: Various
Notes: Fibre from the soft undercoat is 15–20 microns thick and up to 50 mm (2 in.) in length.

Cashmere
Material: Combed cashmere
Application: Various
Notes: It is very fine, typically no more than 18 microns, which is comparable to superfine Merino.

Mohair
Material: Angora goat hair
Application: Various
Notes: The smooth and strong fibres are up to 150 mm (4–6 in.) in length and as fine as 25 microns in diameter.

Camel
Material: Combed and de-haired camel
Application: Various
Notes: The de-haired undercoat is 19–24 microns thick and feels soft and luxurious.

→ Grading and Scouring Wool

Shepherds such as Gareth J. Daniels, shown here, shear sheep each summer (**images 1** and **2**). It is physical work and must be done well to ensure the least stress to the sheep and to maintain the highest quality of fleece. For example, taking two passes with the clippers instead of one will result in shorter fibres, known as second cuts, which are not desirable for spinning high-quality worsted yarn.

Wool grows in clumps called staples. At the British Wool Marketing Board staples are pulled from the fleece to check the quality (**image 3**), in particular the strength, length, diameter of fibre and consistency of crimp. These qualities, along with breed and colour, are used to grade the wool. For example, wool from British and New Zealand sheep is hard-wearing and resilient and so is mostly used for carpets, whereas Australian wool is considered to be of a finer quality and more

suitable for the apparel market. In the UK the British Wool Marketing Board collects, grades and markets wool on behalf of all wool producers.

The fleeces are sorted and graded according to fibre quality. The highest-quality wool comes from the back neck and the lowest quality from around the back legs.

The wool is blended and 'opened up' (**image 4**) to prepare it for scouring (**image 5**). There is a significant amount of lanolin in wool, which has to be removed: 10 kg (22 lb) of fleece produces around 7 kg (15.4 lb) of washed, clean wool (**image 6**).

1

2

3

4

Featured Companies

British Wool Marketing Board
www.britishwool.org.uk
Haworth Scouring Company
www.haworthscouring.co.uk

5

6

1

2

Case Study

→ **Carding and Combing**

The bundles of wool are loaded into the carding machine (**image 1**). The large swift roller rotates in the middle. It is surrounded by slow-moving worker rollers with smaller stripper rollers (**image 2**). The workers rotate relatively slowly, so that the fibres do not return to the same place on the swift roller with each rotation. Thus the wool is blended, any remaining contamination is removed and the fibres are evenly distributed.

The fibres are transferred from the swift to the doffer and removed as a continuous fibrous web (**image 3**), which is lightly twisted into slivers that are stored in drums (**image 4**). The mixed fibres can be converted directly into woollen yarn by twisting (light airy yarn suitable for weaving, knitwear or felt-making). Alternatively, the sliver is gilled (thinned out) and passed through metal combs to remove short fibres and align the long fibres (**images 5**, **6** and **7**). Combed wool is twisted into strong, hard worsted yarn.

3

4

5

6

7

Featured Company

Haworth Scouring Company
www.haworthscouring.co.uk

Fibre and Yarn Technology

Cotton

Cotton is the most commonly used natural fibre. The soft white fibre is relatively easy to extract from the plant, making it economical and efficient to produce. Organic farming methods eliminate the use of genetically modified organisms, as well as the vast quantities of pesticide and fertilizer associated with industrial production.

PREPARATION

28

Techniques	Materials	Costs
• Harvesting by hand or machine • Ginning • Carding and combing	• Lint (staple fibre) • Linters, hulls and seeds	• Moderate

Applications	Quality	Related Technologies	
• Apparel • Towels and bedding • Upholstery	• Soft and supple • Highly absorbent • Stronger when wet	• Man-made fibres • Bast fibres	• Silk • Wool and hair

INTRODUCTION

Cotton fibre is derived from the seed pod of the cotton plant (genus *Gossypium*). It is a staple fibre, measuring between 13 and 38 mm (0.5–1.5 in.) in length, depending on the species. Like all staple fibres it is twisted to make yarn (page 56). Long-staple cotton is considered higher quality and is often identified by the origin or species on the label, such as Egyptian or pima.

Cotton yields a high-quality and versatile fibre that has been utilized in the production of textiles for more than 5,000 years. It is unmatched for its combination of softness, suppleness and strength as a result of its non-structural role in nature.

The Cotton Ginning Process

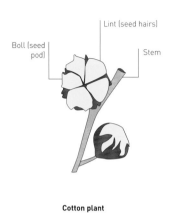

Cotton plant

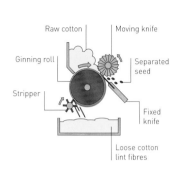

Roller gin

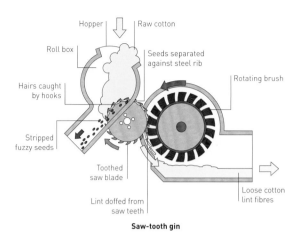

Saw-tooth gin

Ginning

Nowadays it is synonymous with t-shirts, jeans, cosmetics, towels and bedding.

Most cotton is off-white and requires bleaching and scouring (page 202) to improve whiteness before dyeing (page 240). It is virtually pure cellulose, which means that when dyed it will produce a high-quality and consistent colour. However, there are several different species of cotton, and some grow with natural colour variation, including shades of red, green and brown. Utilizing the natural colour reduces the overall environmental impact of fibre production.

APPLICATIONS

As a result of its relative low cost, wide availability and ease of processing, cotton is used in a diverse range of applications including woven (pages 76–95), pile (page 96) and knitted (pages 126–51) textiles.

Many familiar types of fabric used in apparel, interiors and upholstery are traditionally made with cotton. Examples include calico (a low to medium thread-count cotton used for printing, pages 256–79), diaper cloth (woven small repeating patterns), towel (such as terry), drill (a durable twill-weave fabric), denim (warp-dyed twill), muslin (woven from carded cotton that is often embossed, page 220, or printed) and percale (woven from high-quality long-staple yarns that may be carded or combed). The highest-quality cotton is used for fine shirts, suits and bedding.

TECHNICAL DESCRIPTION

Cotton lint grows from the seeds of the cotton plant. Its role in nature is to aid the dispersal of the seeds in the wind. The boll (seed pod) contains several seeds each with several thousand hairs attached. As the hairs grow they split open the boll and burst out.

Cotton is hand-picked or harvested by machine and the lint is separated from the seeds, hulls and other non-fibrous matter (trash) using a gin. The hand-operated roller gin has been used for centuries to process long-staple cotton fibres. Traditionally called a churka gin, it consists of two rollers, or a ginning roller and moving knife, through which the long hairs are drawn. The small gap between the rollers does not allow the seed to pass through, causing it to separate. The full length of cotton lint is preserved as fibre.

While it is practical to process long-staple fibres using a roller gin, 1 kg (roughly 2 lb) of short-staple cotton takes a day to process by hand. It was not until the invention of the mechanized saw-tooth gin in 1793, for which Eli Whitney is credited, that short-staple fibres – now constituting the majority of cotton production – could be processed in a time-effective manner.

In operation, raw cotton lint is loaded in the hopper and from there fed into the roll box. The hairs are caught by multiple spinning toothed blades stacked side by side. The blades are spaced by ribs, against which the seeds are separated from the lint, because the gap is too small to allow them to pass through.

As the blade rotates, a brush sweeps (doffs) the lint from the teeth. The process operates at high speed, with each gin producing several tons per hour, and yields a smooth and uniform fibre.

The seeds, referred to as fuzzy seeds because of their hairy appearance, are collected below. The short fibre that remains on the seed is a valuable by-product – known as linters, it is used in papermaking, batting and nonwovens – as are the seeds themselves.

Cotton is used as a fibre in nonwovens (wipes and cloths, page 152) and as flock fibres (page 280) to produce a suede-like finish on other materials. It is applied as thread in sewing (pages 342–69), lacemaking (pages 106–19) and embroidery (page 298) applications.

By-products of cotton production include linters (fibres less than 3 mm / 0.12 in. long), hulls and seeds. Linters are valuable raw materials used for batting (mattresses and upholstery), man-made fibres (see Filament Spinning, page 50) and paper. The seeds are converted into fertilizer and animal feed.

RELATED TECHNOLOGIES
Even though cotton is not as strong and durable as hemp and flax, colourful and lightweight as silk, or resilient and insulating as wool, it is more cost-effective and widespread in application thanks to its superior balance of properties, its colourability and the relative ease with which it can be processed. As a result, it remains the dominant natural fibre for apparel and interior fabric applications.

Its dominance has only been challenged by man-made fibres, and in particular polyester, over the past few decades. Such is the familiarity of this natural fibre that man-made materials are often produced to mimic its qualities.

QUALITY
Cotton is graded according to length, uniformity, colour (whiteness), fineness (diameter), strength and contamination. Quality varies according to where it was grown, as a result of differences in climate, cultivation and soil. There are also many different breeds commonly in use.

Cotton has very good absorbency and so is used in the production of terry towels and robes. Yarn with a lower twist will have higher absorbency, although reducing the twist also reduces strength.

Cotton fibres do not recover well from creasing; they have low resiliency. Coatings (page 226) are applied to help reduce wrinkling, but they may also affect the natural quality of the material. Compact weaves and knits tend to be more susceptible to creasing than loose and open styles.

Cotton fabrics and garments are prone to shrink over time and with repeated washing. This is reduced during fabric production with controlled stretching (tentering, page 210) and compressive shrinking (also known under the trade name Sanforizing, pages 214–15).

Cotton is typically coloured with vat dyes (page 240), which are available in a wide range of colours and have very good fastness. Mercerizing (page 202) is almost always carried out before dyeing or printing, because it improves the strength and surface lustre.

Being a natural fibre, cotton is prone to yellowing and losing strength when exposed to sunlight for long periods.

DESIGN
Cotton fabrics range from lightweight sheer to comfortable woven cloth and hardy denim. This versatility means cotton is found in a wide range of applications as the primary fibre. Its widespread adoption means large parts of the fabric industry have evolved around the processing of cotton, including finishing, dyeing and printing. It is often referred to as the benchmark against which other fabrics must compete.

The fibre is hydrophilic, which means it will absorb water. This is a desirable quality in warm-weather clothes, because the fabric rapidly draws moisture away from the body.

As cotton absorbs water the fibres swell and become stronger. This property is utilized in uncoated wind and rainproof fabrics, such as Ventile. Made from the highest-quality long-staple cotton using a dense Oxford weave (a symmetrical basket weave where one yarn may cross two yarns), Ventile has been adopted by military, medical and outdoor professionals for weatherproof outerwear.

Cotton does not recover well after stretching. Therefore, parts of garments exposed to tension, such as the knees and elbows, tend to become baggy over time.

MATERIALS
The highest-quality cotton fabrics are produced from long-staple cotton with an average fibre length of more than 28 mm (1.1 in). Extra-long staple (ELS), as

it is often referred to, is a premium fibre grown in relatively small quantities.

Egyptian cotton is used to describe cotton produced in Egypt. It includes ELS (*G. barbadense*), as well as other lower-quality fibres. Therefore the term Egyptian cotton does not guarantee consistently high quality.

Pima cotton (*G. barbadense*), named after the American Indian Pima people and previously known as American-Egyptian, yields very high-quality and long-strand fibre. It is cultivated in the United States, Australia, Israel and Peru, among other countries. Pima is also referred to by the trade name Supima, named after the association that promotes luxury pima cotton. Sea Island cotton, an ancestor of pima, is highly susceptible to boll weevil attack and so is produced in much lower quantities.

Upland cotton (*G. hirsutum*) is the most common in the United States and yields a medium to long staple up to around 32 mm (1.25 in.). Asiatic cotton (*G. herbaceum* and *G. arboreum*), cultivated in India and Central and Eastern Asia, yields a short staple typically no more than 20 mm (0.8 in).

COSTS
Cotton is relatively less expensive than other natural fibres. The cost depends on the grade: Egyptian cotton and fibres cultivated using organic methods are several times more expensive than commodity types owing to the relatively low volumes produced each year.

Commodity cotton is often blended with polyester and other man-made fibres to reduce cost. The Natural Blend trademark indicates a cotton content of 60% or more.

ENVIRONMENTAL IMPACTS
Industrially produced cotton requires vast quantities of pesticides, fertilizer and water. As a result of its being produced in large quantities, several cotton-producing countries rely on cheap labour, and in many cases this has included slave labour and prisoners. Add to that the water and chemicals required to bleach, scour, mercerize and dye cotton fabric, and it is clear why cotton is considered the most

Denim
Material: Cotton
Application: Apparel
Notes: Its balance of lightness, suppleness and durability makes it an ideal workwear material.

Fancy
Material: Cotton and polyester
Application: Apparel
Notes: A mix of novelty cotton yarn with transparent polyester strip yarn is used to produce this fancy fabric.

Plain-Weave
Material: Cotton and polyester
Application: Apparel
Notes: Metallized polyester strip yarn is incorporated as weft to create a linear repeating pattern.

Jersey
Material: Cotton
Application: Apparel
Notes: Knitted t-shirts are produced as a continuous tube, eliminating seams in the body.

Knit
Material: Cotton
Application: Apparel
Notes: The lightness and bulk of cotton provide comfort and insulation in seamless knitted gloves and linings.

Corduroy
Material: Cotton
Application: Apparel
Notes: Its softness and drape are utilized in corduroy and velveteen.

Braided
Material: Cotton
Application: Accessory
Notes: Cotton thread braided into cord is utilized as the weft in woven fabric to make accessories.

Embroidery
Material: Cotton
Application: Accessory
Notes: Colourful and strong cotton thread is utilized in intricate handwork as well as machine embroidery.

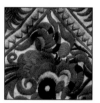

Openwork
Material: Cotton
Application: Apparel
Notes: Its lightness, strength and stability are utilized in decorative clothing, such as this cutwork blouse.

Colour
Material: Cotton
Application: Accessory
Notes: Being almost pure cellulose, cotton can be dyed a range of vivid colours.

Print
Material: Cotton/flax
Application: Interior
Notes: Smooth fabrics are ideal printing material. This high-quality cotton/flax tea towel is printed by Thornback & Peel.

Batik
Material: Cotton
Application: Apparel
Notes: Used for millennia to make fabric, cotton continues to be decorated using traditional techniques.

unsustainable and environmentally damaging natural fibre.

Genetically modified (GM) grades of cotton have been developed to try to reduce the reliance on pesticides. Modifying cotton with the *Bacillus thuringiensis* bacterial gene, referred to as Bt cotton, provides resistance to most pests, but not all. It has been widely adopted and as a result the use of pesticides has been reduced, while biodiversity in cotton fields has increased.

Fair trade and organic cotton are providing a sustainable alternative, albeit in smaller quantities and at a premium price. Each year, organic cotton production increases, and the quality and consistency are improved. Demand is driven largely by the popularity of organic cotton for baby clothes, diapers and underwear.

Naturally coloured cotton eliminates the need for bleaching and dyeing, thus significantly reducing the negative impacts of fabric production. However, one of the challenges of naturally coloured cotton is that it needs to be kept isolated from whiter grades, otherwise there is a chance of contamination.

Cotton fabrics can be recycled. Fabrics that contain cotton blended with other fibres are converted into single-use wipes and industrial products.

→ Organic Cotton Production in Kyrgyzstan

Cotton production was widespread in Central Asia until the collapse of the Soviet Union in the 1990s. In southern Kyrgyzstan, for example, cotton production fell by around two-thirds owing to the impact on the economy. Set up in 2003, the Bio Cotton project aimed to reinvigorate cotton farming with a focus on organic production methods. As a result, thousands of farmers and workers now make their living from cotton grown without the use of mineral fertilizers, synthetic pesticides or genetically modified (GM) crops.

Lint (cotton fibres) starts to develop on the seeds after flowering, which occurs 100 to 150 days after planting. It grows in bolls (seed pods) that burst open once the fibres have matured (**image 1**). The fibre cells die and collapse to form a twisted ribbon-like strand.

Machines are used to harvest most of the cotton produced in the United States, Australia and Israel. The remainder, which accounts for approximately two-thirds of world production, is mainly picked by hand. Hand-picked cotton is superior quality, but is much more time-consuming and so only economically feasible in regions where the wages are low, such as in Central and Southern Asia.

Farmers involved in the Bio Cotton project in Kyrgyzstan hand pick cotton several times each season, taking only the mature bolls each time (**image 2**). The cotton lint is teased from the husks (**image 3**), but it is inevitable that the cotton will include some foreign matter at this stage (**image 4**).

The Bio Cotton farmers gather the cotton and deliver it to a centralized ginning facility (**image 5**). All of the farmers in the surrounding area utilize the same facilities to keep the cost of production down. They cultivate a variety of upland cotton with a fibre length of around 32 mm (1.25 in.).

Ginning is a rapid process, so the cotton is collected in huge stacks prior to processing (**image 6**). When sufficient quantity has built up the ginning process begins. The cotton is transferred to the roller gins via a suction pipe (**image 7**). Each ginning stand has a hopper, into which the cotton is continuously fed.

The long fibres are separated from the seeds and dispensed onto a conveyor (**image 8**). They are collected in a large hopper (**image 9**) and pressed into bales, each weighing 200–250 kg (440–550 lb), ready for export (**image 10**). Before the fibre is converted into nonwoven or spun into yarn, the impurities are removed and the fibres aligned (**image 11**) by carding and combing.

PREPARATION

32

1

2

3

4

5

6

7

8

Featured Organization

**Agricultural Commodity and
Service Cooperative (ACSC)
Bio-Farmer**
www.organicfarming.kg

9

10

11

Fibre and Yarn Technology

Bast Fibres

Long, durable fibres are extracted from the stems of certain tall-growing plants and twisted into yarn. They are renewable and sustainable and have been utilized in the production of high-quality textiles since ancient times. The most commonly used plants are flax, hemp, jute, kenaf, ramie and nettle. Each has its own unique qualities and characteristics.

Techniques		Materials	Costs
• Retting	• Decortication	• Flax, hemp, jute, kenaf, ramie and nettle	• Moderate
• Scutching and hackling	• Carding and combing		

Applications	Quality	Related Technologies
• Apparel	• Strong, but variable	• Man-made fibres
• Upholstery	• Absorbent	• Leaf fibres
• Technical	• Low resilience	• Cotton

INTRODUCTION

This is an important group of fibres that includes flax, hemp, jute, kenaf, ramie and nettle. Bast fibres are obtained from the stems by retting. The long fibres are then mechanically extracted and separated, then carded and combed (page 23). These are manual processes, restricting large-scale production to countries where labour is cheap.

Flax, hemp and ramie are the strongest and stiffest of the plant fibres. They are converted into fine yarn used for lightweight fabrics, while linen is produced from the highest-quality flax. They are also often blended with other fibres, such as cotton, to combine the

The Bast Fibre Production Process

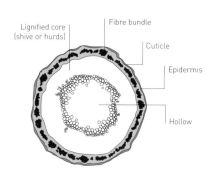

Lignified core (shive or hurds)

Fibre bundle

Cuticle

Epidermis

Hollow

Cross-section of a plant stem

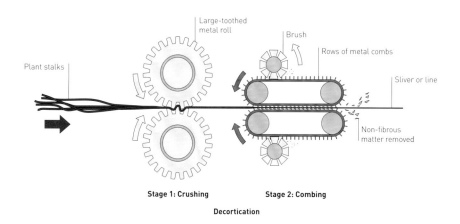

Large-toothed metal roll

Brush

Rows of metal combs

Sliver or line

Plant stalks

Non-fibrous matter removed

Stage 1: Crushing

Stage 2: Combing

Decortication

TECHNICAL DESCRIPTION

Plant stems suitable for bast fibre production are made up of a lignified core – termed hurds or shive – surrounded by a spiralling fibre bundle structure, which is protected from the elements and from microbial attack by a waxy outer layer made up of the epidermis and cuticle.

The fibre bundles are released from the stem by a process known as retting. This is slightly different for each species, but based on the same principle: over a period of days, microbes degrade the hemicellulose and pectin matrix that bonds the fibre bundles together around the lignified core.

Wet retting (also known as biological retting) involves submerging the stems in water for around a week and then leaving them to dry in the field. If not properly controlled, the acid runoff contaminates surrounding waterways. Even though this technique yields the highest-quality fibre, it has been mostly discontinued in favour of the dew-retting process. With this technique, the stems are pulled from the soil (to obtain as much fibre as possible)

and spread out in the field for a period of time that is dependent on the climate and the weather.

Retting in the fields has many drawbacks, including occupying the land for weeks and discoloured fibres. Alternative techniques are therefore being developed, such as using chemicals or enzymes to break down the inter-fibre bonds. However, these have so far proved too expensive to use on most commercial crops.

The thoroughly retted fibre is now ready for mechanical cleaning and separation. Traditionally the processes were termed scutching and hackling, but modern motor-driven machines are known as decorticators. In stage 1, the stems are crushed between a pair of heavy metal rolls to break them open and reveal the fibres. Throughout the process, the stems are aligned to maintain the full length of the fibres. When carried out by hand this part of the process is known as scutching.

In stage 2, the shorter fibres and broken shards of woody core are removed by

combing. The fibres pass through a series of rollers with progressively finer teeth. It is called hackling when done by hand with metal-wire combs. Once cleaned and separated, the sliver of long fibres, also called lines, are suitable for carding and combing.

Shorter fibres may be converted into yarn using similar equipment to that used for cotton (page 29). Known as cottonizing, this is more cost-effective, but produces inferior-quality yarn.

desirable properties of both. Jute yields a hardy and inexpensive fibre traditionally used to produce hessian. It is the most widely consumed bast fibre, second only to cotton. Kenaf is similar, but more expensive.

APPLICATIONS
These are durable fibres that have been employed as rope, twine and heavy-duty fabrics over the years. Hemp, flax, and to a lesser extent nettle and ramie, are converted into textiles suitable for apparel, upholstery, bedding, blankets and covers. Jute and kenaf are used for industrial fabrics, carpet backing, sacks, webbing, upholstery, geotextiles and agro textiles.

Because of the low environmental impact of these materials, and their hardy properties, they are continually being explored in new areas of application. Whereas flax and hemp are providing a cost-effective and high-strength alternative to glass fibre in composite laminates, kenaf and hemp have re-emerged as alternatives to wood pulp in the production of high-quality paper.

None of the rest of the plant is wasted once the fibres have been extracted. The woody core is utilized in particleboard and animal bedding or incinerated to reclaim the embodied energy. Shorter fibres are converted into pulp for paper and board. And the seeds are used in animal feed.

RELATED TECHNOLOGIES
Bast fibres have unique properties among the plant fibres. They are tough, but capable of being converted into the finest linens. The biggest competitors are man-made fibres, which have come to dominate many of the applications where bast fibres previously thrived.

In recent years, bast fibres have seen a marked resurgence in development thanks to their obvious environmental benefits. Yet this has not been sufficient to reverse the declining rates of production worldwide. Hemp faces added challenges, because of its close association with marijuana (a psychoactive drug): hemp cultivation is heavily regulated and even banned in some places, including parts of the United States.

QUALITY
Fineness, lustre, length and lightness of colour are all good measures of quality in bast fibres. They are typically between 50 and 100 microns in diameter and can be as long as the plant itself: hemp fibres reach lengths of up to 4.5 m (15 ft) long, while flax may be as short as 25 mm (1 in.). They are spun into strong staple yarn (page 56).

The relationship between length and diameter of the fibre bundle, referred to as aspect ratio, is important for technical applications, such as biocomposites. Fibres with high aspect ratio provide greater strength for the same weight and volume.

As natural fibres, they have inevitable variations in strength, colour and texture. The properties of the fibre may be enhanced with coatings (page 226), such as fire retardancy or rot proofing.

DESIGN
Bast fibres are used in all types of textile construction, including nonwoven (page 152), woven (pages 76 and 90), fancy (page 84), knitted (pages 126–51) and stitch-bonded unidirectional (page 196), depending on the application. This versatility creates many opportunities for design in all types of application.

Used as composite fibre reinforcement, new yarn formations and fabric constructions have emerged to improve performance. The advantage of unidirectional fibres is they do not pass over, under or around one another, and so travel the shortest distance between two points. 'Twistless' spinning, currently only possible with flax, ensures the highest degree of fibre alignment, impregnation and performance.

The lightness and lustre of these fibres depends on the species. Flax and ramie are produced as bright white fibres by successive bleaching phases. This provides a good base for dyeing. However, the relatively high proportion of non-cellulosic polymers in the fibre bundles means the colour tends to be muted compared to cotton and silk.

MATERIALS
High-quality linen, used in bed sheets and suits for example, is traditionally made from flax. In addition, flax is utilized to make industrial and technical fabrics.

Hemp is also an important industrial fibre, while the plant itself yields many other useful crops including seeds, oil, leaves and a woody core. It has natural antimicrobial properties and has been used for thousands of years to make high-quality fabrics often mistaken for linen.

Jute and kenaf are hard-wearing and preferred for packaging, such as sacking cloth and hessian. Jute was traditionally one of the least expensive fibres, but today is considered expensive compared to man-made equivalents.

Ramie yields a long and light-coloured fibre that can be bleached to a bright white. However, it is relatively expensive, owing to the labour-intensive production cycle, which limits the range of applications. Nettle is similar to ramie and is being re-explored as a renewable fibre for apparel and technical textiles.

COSTS
Bast fibres are relatively inexpensive to cultivate, but time-consuming and labour-intensive to extract from the plants. This makes them more expensive than cotton, and considerably more expensive than commodity synthetic fibres, depending on location.

Mechanical decortication can be used without retting. This is more cost-effective, but produces inferior-quality fibre. Yield per hectare is also an important factor. For example, hemp grows tall and dense, and produces around 3 tons of fibre per hectare, which is around twice as much as flax.

ENVIRONMENTAL IMPACTS
These plants grow with little or no added fertilizer, pesticides or herbicides. They also make up an important part of crop rotation, thanks to their beneficial effects on following crops. It is best to use the fibre untreated if possible, known as 'natural', to avoid the use of bleaching and dyeing chemicals. If not, water-based systems without heavy metals or other toxic ingredients are available.

VISUAL GLOSSARY: BAST FIBRE FABRICS, FINISHES AND USES

Hessian

Material: Jute
Application: Sacking
Notes: Jute is a tough and widely available fibre used in packaging, carpet backing and upholstery.

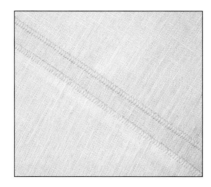

Linen

Material: Flax
Application: Table cloth
Notes: The highest-quality linen fabrics are traditionally made in Ireland and Belgium.

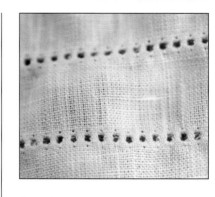

Cutwork

Material: Flax
Application: Table cloth
Notes: Bast fibres have been used for generations to make traditional fabrics, including embroidery and cutwork.

Fancy

Material: Flax
Application: Apparel
Notes: With painstaking preparation, stiff and hardy bast fibres are converted into beautiful fancy weaves.

Terry

Material: Ramie
Application: Bath scrubber
Notes: Natural fibres are absorbent and so used to make towel and terry. Ramie has a light, lustrous finish.

Dye

Material: Flax
Application: Scarf
Notes: Fibres and yarns are often blended, because even though they dye well, the colour will vary.

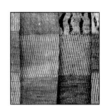

VISUAL GLOSSARY: COMPOSITES EVOLUTION BIOTEX FABRICS

4x4 Plain Weave

Material: Flax
Application: Biocomposite
Notes: Traditionally a coarse fabric made of jute and hemp, this flax is utilized in modern biocomposites.

Twill

Material: Flax
Application: Biocomposite
Notes: Twill weave is one of the most versatile fabrics for composites. It conforms well and has high strength.

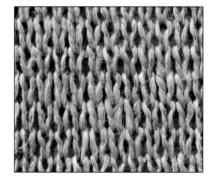

Knit

Material: Flax
Application: Biocomposite
Notes: Knitted fibre reinforcement is used for complex three-dimensional parts that require the fabric to drape very easily.

Biaxial

Material: Flax
Application: Biocomposite
Notes: Stitch-bonded unidirectional fabrics create high strength in specific directions in composite constructions.

Nonwoven

Material: Flax
Application: Various
Notes: Shorter fibres are converted into nonwovens, which are suitable for a range of industrial applications.

Veil

Material: Flax
Application: Biocomposite
Notes: Lightweight nonwoven fabrics provide a uniform surface finish without the typical weave pattern.

BAST FIBRES

37

→ Converting Flax into Fibre

Flax thrives in the temperate climate of western, central and eastern Europe (**image 1**). Planted in the spring, it is ready to be harvested in the summer. The plants are packed very tightly together, forcing the stems to grow tall, straight and thin. This yields higher-quality fibre. Very few chemicals are required to grow the crop, other than a small amount of pesticide.

Flax is harvested and the stalks are laid down in the field to ret (**image 2**). Retting is a natural process, whereby microorganisms dissolve the pectin that binds the fibres to the stem and each other. After four to six weeks the stalks are baled and delivered to the mill for decortication (**image 3**).

The stalks are made up of about one third fibre (**image 4**), which is extracted by decortication. Once separated, the loose fibres are fed into the carding machine (**image 5**) and processed much like wool (page 23).

Carding produces a sliver (**image 6**), which, if required, is further processed by combing and bleaching. Once the desired quality of fibre is achieved the sliver is converted into a roving (loosely twisted) and spun into yarn (**image 7**). Flax is strong (and gets stronger when wet) and durable and has a pleasant feel with natural colour variation.

1

2

3

4

5

6

7

Featured Company

EKOTEX
www.ekotex.com.pl

Fibre and Yarn Technology

Leaf Fibres

A relatively minor group of materials derived from long-leaved plants, they are an important historical fibre in many cultures, used to make apparel, footwear, hats and crafts. Produced by smallholders, the plants have strong fibre bundles that are stripped out of the leaf by hand or by mechanical decortication and converted into yarn.

Techniques	Materials		Costs
• Stripping by hand or machine • Decortication	• Abacá • Piña (pineapple)	• Sisal (or henequen) • Panama	• Moderate to high cost due to availability and labour-intensive production

Applications	Quality	Related Technologies
• Apparel • Interiors • Rope	• Variable • Strong and stiff • Fine and light-coloured	• Bast fibres • Man-made fibres

INTRODUCTION

Similar to bast and cotton, leaf fibres are cellulosic. Their role in nature is structural: they are strong and stiff as a result of having to keep the plant upright. The fibres used to make yarn consist of multiple individual strands of cellulose wrapped together in bundles.

Surrounded by leafy matter – more than 90% water – the fibres are time-consuming and labour intensive to extract. The plants are mostly cultivated by smallholders and processed in a centralized facility; the quality is highly variable and difficult to maintain.

Sisal (and henequen from Yucatán, Mexico) is a succulent plant from the same family (genus *Agave*) as that used to make tequila. It yields a hardy light-brown fibre. Many other members of the *Agave* genus are converted into fibre. Sisal is the most widely used of these and of leaf fibres generally, but still makes up only around a tenth of 1 per cent of natural-fibre production.

The majority of abacá, also known as Manila hemp, comes from the Philippines. The leaves grow several metres long and the fibres run the entire length. The fibres are strong and naturally off-white.

Piña and panama are fine-quality leaf fibres. Whereas panama is used almost exclusively to make hats of the same name, piña is hand-woven into sheer fabrics (page 90) and intricately hand-stitched embroidery (page 298). It comes from the Red Spanish variety of pineapple plant, named during the Spanish rule of the region, and hence why both the fibre and the textile are known as piña (Spanish for pineapple).

APPLICATIONS

Leaf fibres are used to make rope, twine, apparel, footwear and interior textiles. Craft items are another important

The Leaf Fibre Extraction Process

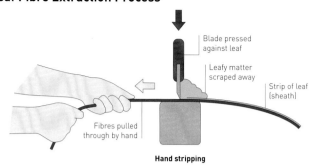

Blade pressed
against leaf

Leafy matter
scraped away

Strip of leaf
(sheath)

Fibres pulled
through by hand

Hand stripping

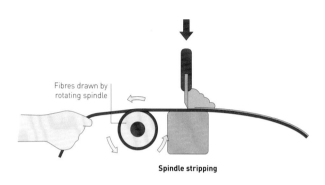

Fibres drawn by
rotating spindle

Spindle stripping

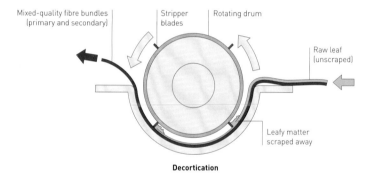

Mixed-quality fibre bundles
(primary and secondary)

Stripper
blades

Rotating drum

Raw leaf
(unscraped)

Leafy matter
scraped away

Decortication

TECHNICAL DESCRIPTION
**Three different methods of extraction
are used: hand stripping, spindle
stripping and decortication.**

**In hand stripping, thin strips
of sheath (tuxies) are drawn under
a sharp knife to remove non-fibrous
material and liberate the fibres. It is
labour-intensive and yields around
20 kg (44 lb) per labourer per day.**

**Spindle stripping is a semi-
automated process. Tuxies are fed in
by hand and pulled past the blade by
a motorized spindle. This increases
production to around 100 kg (220 lb)
per operator per day.**

**Decortication uses a motorized
drum equipped with stripper
blades. The sheaths are fed in and
the fibre bundles are extracted.
Without separating the leaf into
tuxies beforehand, the resulting
fibres include a mix of primary and
secondary types (from the outer and
inner parts of the leaf). So while this
is the most cost-effective method, the
fibres are of relatively lower quality.**

---

end use, including rugs, mats, baskets
and accessories.

Abacá and sisal are the most widely
used. Abacá is converted into sinamay
to make hats (see Hat Blocking, page
420); woven and wicker upholstery; and
hardy floor mats. The fibres are absorbent
and highly resistant to tearing. These
qualities are utilized in high-quality,
high-end papers and nonwoven fabrics
(page 152). Sisal is used in diverse
applications, including the bristles in
fine brushes, rugs, upholstery, rope and
metal-polishing cloths.

Different leaf fibres have surprisingly
contrasting properties, which affect the
areas of application. For example, abacá

was once widely used to make rope for
marine applications as a result of its high
strength and resistance to salt water.
Sisal, however, has very poor resistance to
salt water and so is unsuitable.

Piña and panama are used in hat
making. Piña is used as fine textile in the
Philippines, and is hand-woven to make
shirts, dresses and table linen.

RELATED TECHNOLOGIES

They are labour-intensive and time-
consuming to produce. Traditionally
used for local needs, such as practical
items, ceremonial garments and interior
textiles, leaf fibres have been largely
replaced by man-made fibres (page 50)

as a result of their low cost and apparent
convenience.

QUALITY

These are some of the strongest natural
fibres. Quality ranges from hardy sisal to
fine and luxurious piña. They are durable
and supple.

Sinamay is a type of woven abacá used
in hat blocking. The sheets are pressed
with heat into three-dimensional shapes
in a single operation. It is one of the few
fibres that can be worked in this way
while retaining its strength and integrity.

The stiffness of these fibres is utilized
in diverse end uses, from sheer fabric to
floor mats. This property is the reason

VISUAL GLOSSARY: LEAF FIBRE FABRICS, FINISHES AND USES

Sliver
Material: Abacá
Application: Various
Notes: Once extracted from the leaf, these fibres are processed into yarn in much the same way as bast fibres.

Colour
Material: Abacá and polyester
Application: Apparel
Notes: Abacá is dyed a range of colours using natural ingredients. Here it is woven with metallized polyester.

Sinamay
Material: Abacá
Application: Hats
Notes: Abacá is converted into sinamay, utilized in hat blocking, and shipped all over the world from the Philippines.

Lightweight
Material: Piña
Application: Apparel
Notes: Piña is a fine and stiff fibre used in hand-woven sheer fabrics. It is ivory-white and dyed a range of colours.

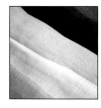

Handwoven
Material: Tikog grass
Application: Banig floor mat
Notes: This is a traditional handwoven mat from the Philippines. The grass leaves are dyed bright colours and woven to form intricate patterns.

Embroidery
Material: Piña
Application: Apparel
Notes: Piña was traditionally used to make barong formalwear (hand-embroidered fabric).

Paper
Material: Abacá
Application: Money
Notes: Its strength, lightness and fineness are utilized in the paper and pulp industry, such as for banknotes.

Nonwoven
Material: Abacá
Application: Teabag
Notes: Abacá remains strong even when wet and so is used in the production of nonwoven teabags.

---

they are being explored as an alternative to asbestos and glass-fibre reinforcement in laminated composites (page 463).

DESIGN
Leaf fibres have remained a predominantly manually worked material. They are selected, extracted and fabricated by hand. Knowledge about how to get the best results from these materials resides with skilled artisans in small communities.

Sisal is mainly used in its natural colour state, because the fibres do not produce uniform or satisfactory colour when dyed (page 240).

Piña is naturally off-white and suitable for dying a broad range of colours. Natural dyes are most commonly used. Piña has good resistance to water, salt and abrasion, but poor resistance to acids and alkalis. It is comparable with the highest-quality linen and considered one of the finest natural fibres.

Whereas abacá is used in a range of dyed colours for hat making, panama tends to be simply bleached. Used to make the famous hat, panama requires a painstaking manual process to ensure adequate softness and suppleness. Dyeing would undermine this.

MATERIALS
Leaf fibres are derived from a handful of tropical plants including pineapple, sisal, abacá and panama. In New Zealand, the long leaves of the phormium are used to make fibres. First exploited by the Maori, the plant is known as 'flax' in New Zealand, but it is nothing like the flax that is used to make linen. In the past it was used in various forms to make rope, twine, fishing nets, baskets and outerwear.

COSTS
Extracting these materials is labour-intensive and time-consuming. They are therefore expensive to procure. Piña and panama are the highest quality and the most expensive.

ENVIRONMENTAL IMPACTS
The juicy leaves of some of these plants are susceptible to pests and diseases and so require the application of insecticide. Fertilizer is also commonly used during growth and after each harvest.

Leaf fibres are renewable and bring diversity to otherwise monocrop plantations, such as sugarcane in the Philippines. They also provide smallholders with a valuable source of revenue.

The Philippine Fiber Industry Development Authority (PhilFIDA) has joined forces with other government organizations, as well as non-government organizations (NGOs) and commercial businesses, to revive leaf-fibre textile production. Such initiatives have a positive impact on the economy, on local communities and on the surrounding environment.

→ Abacá Production in the Philippines

Abacá is closely related to the banana plant. However, the fruit of the abacá is inedible, while banana leaf fibres are blended with other natural fibres to produce yarn for textiles. The plant reaches maturity after 18 months to two years (**image 1**). It may then be harvested every three to four months.

The leaves are cut off at the base of the plant (**image 2**), because the fibre bundles run the full length, dispersed through the outer and middle parts of the leaf. The outside layers of the sheath are removed. In a process known as tuxying, a knife is used to separate the layers, which are then pulled apart along the entire length (**image 3**). The strips of fibrous leaf are known as tuxies and each sheath produces up to four of these (**image 4**).

The fibre bundles are stripped from the leaf using a sharp blade. In this case, a semi-automated process increases output. The operator feeds the tuxies past the knife and around a spindle, which draws them through quicker than pulling by hand (**image 5**).

Once separated, the fibres are hung up to dry in the sun (**image 6**). This can take a matter of hours in the right weather conditions and is essential to ensure a high-quality and rot-resistant fibre. The dried material is baled for shipping (**image 7**).

The Philippines supplies about 90% of the total world abacá requirement. Its fibres are graded according to whether they were stripped by hand or machine. These fibres (**image 8**) were extracted by hand and designated as S2 (streaky two), S3 (streaky three) and G (soft seconds).

1

2

3

4

5

6

7

8

Featured Organization

Philippine Fiber Industry Development Authority
fida.da.gov.ph

→ Extracting Piña for Fancy Fabrics

Piña is obtained from the succulent leaves of the pineapple plant. The leaves are pulled from the base (**image 1**). Hand scraping produces the highest-quality fibre. First, the sharp spiny outer edges are removed (**image 2**). The soft part of the leaf is scraped away with a sharp blade, or as in this case, a shard of pottery (**image 3**).

The fibre bundles are revealed beneath and expertly peeled from the inside surface of the sheath (**image 4**). They are washed to remove water-soluble material and hung out to dry (**images 5**, **6** and **7**).

The dry fibre, which is naturally ivory-white in colour, is typically around 600 mm (23.6 in) long. The bundle is combed to separate the individual strands and prepared for hand weaving (**image 8**).

Piña is traditionally used to make intricately decorated fabrics: this weaver is producing delicate brocade (**image 9**).

1

2

3

4

5

Featured Organization

**Philippine Fiber Industry
Development Authority**
fida.da.gov.ph

Fibre and Yarn Technology
Silk

Silk has been used in the production of textiles since ancient times. The majority is harvested from the cocoon of cultivated mulberry silkworms. Each cocoon yields a continuous filament up to about 1,000 m (1,094 yd) long. Silk is a luxurious material, highly prized for its strength, lustre and affinity to brightly coloured dye.

Techniques	Materials	Costs
• Sericulture • Reeling	• Filament yarn • Staple yarn • Farmed or wild silkworms	• Filament yarn is high-cost. Staple fibres are less expensive

Applications	Quality	Related Technologies
• Apparel • Interiors • Velvet	• High tensile strength • Natural lustre • Low abrasion resistance	• Man-made fibres • Plant fibres • Wool and hair

INTRODUCTION

Silk is a protein fibre: like wool, it is made up of amino acids. Silk has unique properties. It is a very fine filament, measuring approximately 10 microns in diameter with very high strength to weight, making it one of the strongest natural fibres. It reflects light to give a bright and lustrous appearance. It has good moisture absorption, meaning it has very good affinity to dyeing (page 240).

Silk is harvested from the cocoon of the mulberry tree moth caterpillar (*Bombyx mori*). In most production, the silkworm caterpillar is sacrificed so that a continuous length of filament can be extracted. Alternatively, the caterpillar is

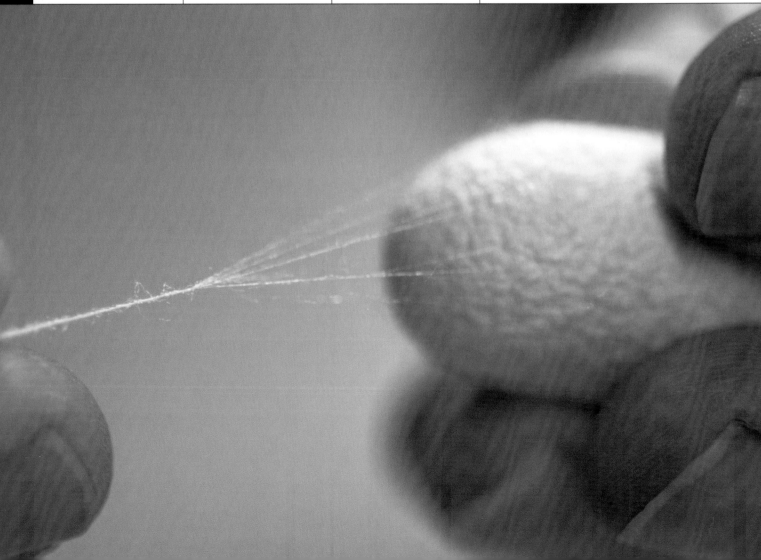

The Silk Production Process

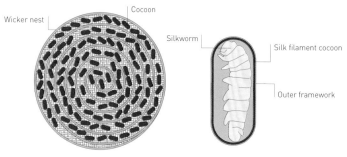

Stage 1: Nest

Stage 2: Silkworm cocoon

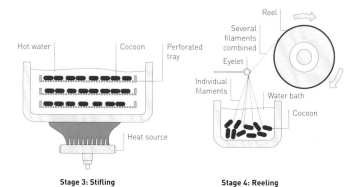

Stage 3: Stifling

Stage 4: Reeling

TECHNICAL DESCRIPTION
In stage 1, it takes three to four days for the silkworm to produce a continuous filament around 1,000 m (1,094 yd) long to completely encapsulate it. They make their cocoons in nests supplied by the farmer, which are often constructed by basket weaving (page 120). In stage 2, the cocoon is completed. It consists of an outer structural framework of rough short fibres and an inner silk filament cocoon.

After eight to nine days the caterpillars are half way through their metamorphosis into moths and the cocoons are ready to be stifled. In stage 3, they are steamed or cooked to kill the developing moth before it is ready to hatch. In stage 4, individual filaments are carefully brushed from the surface of a few cocoons. They are soaked in water to dissolve the sericin. Several filaments are gathered together through an eyelet and taken up on a spool. Several filaments are combined into a single yarn because they would be too fragile to handle on their own. The consistency and weight of yarn determines its final quality.

allowed to develop and hatch. The moth's breaking out of the cocoon means that the filament will no longer be continuous, but may still be used by spinning into staple yarn (page 56), although this material has a lower value.

Most silk is produced in China, and has been since ancient times. The highest-quality silk comes from Italy and France.

APPLICATIONS
Silk has been used for thousands of years in high-value items. Apparel applications include ceremonial gowns, nightwear, jackets, suits, blouses, shirts, lingerie, hosiery, ties and other neckwear. Indoors it is used to make drapery, upholstery, bedding, rugs and wall coverings. It is used in embroidery (page 298) to make bright and shimmering patterns, velvet for both apparel and interiors (see Fancy Weaving, page 84), and lacemaking (pages 106–19).

RELATED TECHNOLOGIES
Silk is a luxurious material and quite expensive. Many attempts have been made to mimic its qualities with man-made materials. The most successful are fibres extruded (see Filament Spinning, page 50) with a triangular profile, similar to the natural structure of silk, such as viscose and nylon.

Viscose has a softness comparable with silk and is utilized as a cost-effective alternative. Even though it is derived from biological and potentially renewable sources, such as wood pulp and bamboo, it cannot be considered a natural material, because large quantities of chemicals are used in production. The dominant method is by hydrolysis alkalization with multiple bleaching phases. Chemicals are used to extract the cellulose, which is then extruded into fibre. Mechanical means have lower environmental impact, but with current technology the fibre quality is inferior.

A great deal of research has been done into synthesized mass-production alternatives, such as using the protein from spider's silk to make synthetic fibre, as well as mimicking how a silkworm manufactures filament. There could be many advantages to producing synthetic silk, such as being able to rapidly produce large quantities without killing silkworms; to very accurately control the strength and density; and to be able to recycle the material at the end of its life.

QUALITY
Silk has very high strength for its weight: it is one of the strongest natural filaments. It is one of the few fibres that become weaker when wet, and has poor elastic recovery. This means that when stretched it does not return to its original state but remains permanently deformed. It does not have very good resistance to abrasion. However, as a luxury material this is not usually a problem owing to the way it is used.

Silk can be dyed a range of bright colours and its natural lustre means they

Natural Colour
Material: Silk
Application: Various
Notes: Silk comes in a range of natural colours including shades of brown, yellow, gold and white.

Dyed Colour
Material: Silk
Application: Various
Notes: Silk has good absorption and high affinity for dyes. It has a natural lustre that intensifies the colour.

Tussah Silk
Material: Silk
Application: Various
Notes: Tussah silk comes from wild giant silk moths and is naturally light yellow to brown in colour.

Gummed Silk
Material: Silk
Application: Various
Notes: Silk left with the natural sericin coating is used in the production of paper, felt and other craft materials.

Degummed Silk Cocoons
Material: Silk
Application: Various
Notes: Cocoons used for hand spinning are degummed in hot water to remove the natural sericin coating.

Degummed Silk
Material: Silk
Application: Various
Notes: The sericin is removed in the process of converting silk into yarn to make fabrics.

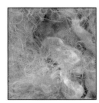

Noil
Material: Silk
Application: Various
Notes: Rough waste taken from the outside of the cocoons, or produced during combing or spinning.

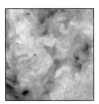

appear vivid and saturated. Whereas silk from the mulberry moth larva has a triangular prism-like cross-section that reflects light in all directions, Tussah silk is flat like a two-sided strip.

Traditionally, silk was measured in terms of momme (mm), a measurement system developed in Japan. Momme is defined as the weight in pounds of a piece of fabric 91.4 m (100 yd) long and 1.14 m (45 in.) wide. Thus, 15-momme silk is notionally taken from a piece of silk that would weigh 16.8 kg (15 lb) if it were 91.4 x 1.14 m. Heavier fabric would have a higher momme weight.

DESIGN
The fine and lustrous properties of silk are best exploited in satin weaves, whose long floating yarns catch the light and show brilliant colour. As a sheer fabric it is used to produce fine gauzes as well as to form the backing layer for devoré (decorative effect created by selectively dissolving areas of fabric, see page 263).

Silk is also used to make high-quality pile. For example, the finest handmade velvet (see Fancy Weaving, page 84) made from silk has up to 118 warp ends per cm (300 per in.), because the fine yarns can be packed very tightly together. By contrast, synthetic equivalents may have only 31 warps per cm (80 per in.), and while they may look saturated with colour, they will feel quite different.

MATERIALS
Cultivated silk is harvested from the mulberry silkworm. There are several commercial species, but these are the most common, because they produce the whitest and finest silk filament. They are killed with heat before reaching the moth stage and the silk filament is unravelled from the cocoon. It is a natural material and so there is some variability.

Staples obtained from broken cocoons or left over are lower quality and less expensive. They are often combined with other fibres during spinning (page 56) to impart some of the softness and strength of silk.

Wild silk is considered lower quality than cultivated types. It comes from China, India and Thailand and is harvested from several species of silkworm that feed on different types of tree, such as oak and plum, as well as mulberry. The silk is harvested from the cocoons after the caterpillar has metamorphosed into a moth and broken out of the cocoon. The length of staple depends on the species

of silkworm and how it breaks out of the cocoon. Wild silk is used in its natural colour, such as shades of brown, yellow and green, or bleached and dyed.

COSTS
Silk is fine and woven into dense fabric: a single garment may contain the silk of as many as 50,000 cocoons. Much of the production is labour-intensive, making this an expensive material.

Staple fibres are much less labour-intensive to extract than a single filament and so are less expensive. They are spun like cotton or wool.

ENVIRONMENTAL IMPACTS
Silk is renewable and biodegradable at the end of its useful life. Silk production (sericulture) requires large quantities of mulberry trees to feed the caterpillars.

To make a continuous filament of silk the caterpillar must be stifled with heat before it can develop into a moth and break out of the cocoon. Using farmed or wild staple silk means the caterpillar has been allowed to grow into a moth (although some staple comes from the waste of cultivated filament). When used without dyeing, these yarns have very low environmental impact.

It is a labour-intensive process, so care should be taken to ensure that silk is sourced from factories that have high standards for employees.

→ Sericulture in the Philippines

Sericulture involves breeding moths, hatching the caterpillars from eggs, feeding them leaves, and once they produce their cocoon, stifling the caterpillars and unravelling the filament and combining into yarn.

Every step of the process is very carefully controlled to ensure the highest-quality yarn. Mulberry moths (*Bombyx mori*) lay their eggs on specially prepared paper. They are kept in a temperature- and humidity-controlled room. The caterpillars of the mulberry moth hatch and are fed young fresh mulberry leaves (**image 1**). They eat constantly and grow to around 10,000 times their original weight within about one month (**image 2**).

When the silkworms have eaten enough food they begin to make their cocoon. They do this by excreting silk from two glands in their head, which mixes and hardens with exposure to air to form a fine silk filament (**image 3**) coated with adhesive gum (sericin).

The adhesive makes up about one third of the cocoon by weight.

Each cocoon is a slightly different shape and size. After three or four days the silkworms have finished their cocoons and the trays are stacked to allow the filament to harden (**image 4**).

With the silkworm inside, the cocoons are carefully removed from the frames (**image 5**). They are collected (**image 6**) and heated to kill the developing moths. They are soaked in water to dissolve the sericin and unravelled by hand (**image 7**). It is up to the operators unravelling the cocoons to combine filaments into a uniform plied yarn.

The plied silk yarn is wound into skeins, which may then be further processed by dyeing (**image 8**) in preparation for weaving into a wide range of products.

1

2

3

SILK

49

4

5

6

Featured Organization

**Philippine Fiber Industry
Development Authority**
fida.da.gov.ph

7

8

Fibre and Yarn Technology
Filament Spinning

Consisting of long-chain polymers, man-made fibres are synthesized from regenerated natural materials or petroleum by-products. The polymers are converted into usable fibres by spinning, whereby a filament is extruded and hardened as it cools or by exposure to chemicals or solvents. They are extremely versatile and have replaced natural fibres in many applications.

FORMATION

50

Techniques	Materials	Costs
• Melt spinning • Dry spinning • Wet spinning (chemical or solvent)	• Regenerated • Synthetic • Bioplastic	• Low to high depending on base material and complexity of processes

Applications	Quality	Related Technologies
• Apparel • Interiors • Technical	• Uniform and high quality • Properties tailored with additives and finishes	• Plant fibres • Wool and hair • Silk

INTRODUCTION
Since viscose was first manufactured in the 1890s, developed as a low-cost alternative to silk, there has been significant development in the quality and controllability of man-made fibres. More recent innovations include fibres produced in an almost closed-loop process to minimize environmental impacts, and some of the strongest materials ever created – 'super fibres' – which outperform steel several times over for the same weight.

In the process of spinning, liquid polymer is squeezed through tiny holes in a spinneret and drawn into continuous filaments. Inspired by nature, this process recreates how a caterpillar spins the silk for its cocoon (see Silk, page 46).

The exact technique depends on the base material. Thermoplastics, such as polypropylene (PP) and polyester, which are formed by melting and extrusion, re-solidify as they pass through cool air on leaving the spinneret. Other types of plastic go through some form of chemical change, or reaction, in their conversion from dope (polymer and solvent mix) to filament. Such spinning techniques require chemicals or solvents to be used, which allow the polymer to be processed as a liquid.

There are a few specialized materials that cannot be spun by these conventional methods. For example, fluoropolymer will not melt and is practically inert, so is spun in a temporary carrier and sintered (coalesced into a solid mass without passing through the liquid phase) to produce high-performance filament.

APPLICATIONS
Man-made fibres are extremely versatile as a result of being able to tweak the properties or engineer entirely new ones,

The Filament Spinning Process

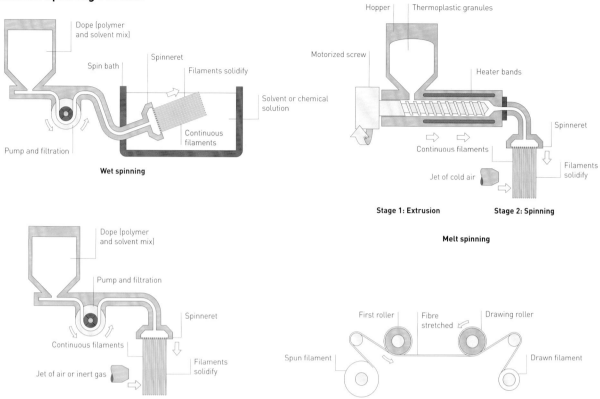

Wet spinning

- Dope (polymer and solvent mix)
- Spin bath
- Spinneret
- Filaments solidify
- Solvent or chemical solution
- Continuous filaments
- Pump and filtration

Melt spinning

- Hopper
- Thermoplastic granules
- Motorized screw
- Heater bands
- Spinneret
- Continuous filaments
- Jet of cold air
- Filaments solidify
- Stage 1: Extrusion
- Stage 2: Spinning

Dry solvent spinning

- Dope (polymer and solvent mix)
- Pump and filtration
- Spinneret
- Continuous filaments
- Jet of air or inert gas
- Filaments solidify

Drawing

- First roller
- Fibre stretched
- Drawing roller
- Spun filament
- Drawn filament

TECHNICAL DESCRIPTION

Melt spinning is used to process thermoplastics that can be melted and extruded. This includes polypropylene (PP), polyethylene (PE), nylon and polyester. The plastic is compounded with all the necessary additives and colourants before spinning. In stage 1, plastic granules are fed into the heated barrel from the hopper. The material is melted and consolidated as it is moved towards the spinneret by the rotating screw.

In stage 2, the liquid polymer is squeezed through the many tiny holes in the spinneret. As the plastic cools, aided by a jet of cool air, it re-solidifies to form multiple continuous filaments. The plastic is relatively weak until it is drawn out to several times its extruded length. At a specific temperature, which depends on the plastic, the yarn is stretched without breaking. Up to a certain point, this greatly increases strength and resistance by improving the orientations of the polymer structure.

Materials not suitable for melt extrusion are converted into fibre by solvent spinning. Fibres processed in this way include regenerated plastics, acrylic, modacrylic, elastane, aramid, liquid crystal polymer (LCP) and polyphenylene benzobisoxazole (PBO).

In the process of dry spinning, the polymer is suspended in a suitable solvent. This viscous mix is referred to as dope and includes all of the necessary additives and ingredients. It is pumped to the spinneret, where it is forced through a series of tiny holes. The solvents evaporate as the dope passes through warm air, or inert gas, and the plastic hardens. The solvents are valuable and so are reclaimed in most cases. This helps to reduce, but does not eliminate, the volatile organic compounds (VOCs) that are off-gassed during production.

Like melt-spun filaments, the strength of certain (but not all) plastics is greatly improved by drawing, which may be carried out during or after spinning.

Wet spinning is used to form plastics that are hardened through a chemical or solvent reaction after passing through the spinneret. The polymer is prepared as a dope as before. It is forced through the spinneret into a chemical or solvent bath, where the filaments solidify. The advantage is that in liquid form, harmful chemicals (such as acids and salts, but not solvents) are more easily controlled and may even be reclaimed or recycled.

Acrylic, lyocell and para-aramid (lower-strength meta-aramid is produced by dry spinning) are examples of fibres produced using this technique. The chemicals used (and thus potential environmental impact) depend on the material. For example, whereas viscose production creates potentially harmful by-products including sulphur, metal salts (copper and zinc) and ammonia, lyocell is processed in a weak solvent (N-methylmorpholine-N-oxide) that is recovered and reused to minimize potential environmental impact.

and so dominate fashion, interior and technical textiles. Applications range from single-use garments to space suits.

Synthetics such as PP, polyethylene (PE), nylon and polyester are used to enhance the performance of woven outerwear, knitted sportswear, pile carpets and even paper-based materials. Purely synthetic fabrics are used to produce lightweight material for tents, parachutes, balloons and medical products.

So called super fibres, known mainly by their trademark names such as Dyneema (ultra-high-molecular-weight polyethylene/UHMWPE) and Kevlar (aramid), are creating whole new areas of application thanks to their superior strength to weight. Examples include lightweight ropes used in rigging and parachute cord, racing sails, bulletproof vests, armoured vehicles, and Formula One racing cars.

Regenerated fibres preceded synthetics. Owing to the limited base ingredients – naturally occurring cellulose or protein – they are not as versatile and so have been largely replaced by synthetics in such applications as apparel and interior fabrics. Recent innovations however, such as producing lyocell in an almost closed-loop process, mean these fibres are still finding new applications.

RELATED TECHNOLOGIES
Man-made fibre is the ultimate competitor. Developed to overcome the limitations of natural fibres, they have since been engineered for entirely new areas of applications where no other type of material would be suitable.

Fabrics and construction techniques have evolved around naturally occurring materials over many hundreds, or even thousands, of years. As a result, fibres such as cotton, flax, silk and wool continue to outperform synthetics and regenerated fibres in many familiar applications. For example, the moisture absorption properties coupled with the highly insulating nature of wool fibres makes them ideal for applications close to the skin, while the natural lightness and strength of silk combined with lustrous surface quality produce luxury fabrics that are both technically high-performing

and beautiful. Indeed, the beauty and desirability of these materials have fuelled trade routes and built cities.

Synthetic leather and fur provide an alternative to using animal by-products. They can be manufactured as nonwoven (page 152), knit (pages 126–51) or weave (pages 76–105), and recent developments in fibre properties have led to materials that can be hard to distinguish from the real thing. Of course, on closer inspection the difference is noticeable, but synthetic equivalents have many benefits other than not being derived from animals, such as lower cost, resistance to insects and bacteria, abrasion resistance, and the ability to tailor the fibre properties to the application.

QUALITY
The size of filament yarn, cord and rope is measured as weight per length. The three standards are denier, tex and decitex (dtex). Denier refers to g/9,000 m; tex refers to g/1,000 m; and dtex refers to g/10,000 m filament yarn. These systems are used to indicate the weight of single filaments in a construction, as well as the complete yarn. Anything smaller than 1 denier (1.1 dtex) is categorized as microfibre.

The highest-quality colour is achieved by dyeing the base polymer before it is spun into fibre in a process known as solution dyeing, dope dyeing or melt dyeing. In this way, the colour is integral to the fibre structure and therefore has better saturation and colourfastness. The more inert fibres, such as PP and PE, can be coloured only in this way. However, it is not always possible, or practical, because to make this worthwhile large batches of a single colour must be produced. Several types have good affinity for dyeing (page 240) once in fibre form, in particular nylon, polyester and acrylic.

High-performance materials are often available only in one or two colours, either because they cannot be dyed, or because they are produced in such low quantities that dyeing is not considered economically worthwhile.

DESIGN
The performance of man-made fibres is affected at several stages in production, including material selection, additives, spinning technique, drawing and finishing. All of these variables may be taken into account. Many of the most popular functional properties, such as tenacity, resilience and elasticity, are packaged together under trademark names. These become synonymous with specific applications, such as Tyvek (produced from UHMWPE, it is lightweight and highly resistant to tearing), Tactel (a quick-drying nylon) and Coolmax (polyester filament with a profile that helps to wick moisture away).

The cross-section of fibres may be adjusted during spinning. The outside shape may be round, triangular or square to affect the lustre or durability. For example, trilobal nylon, produced with a triangular cross-section, produces a brighter finish on flocked (page 280) and woven fabrics. Fibres may be produced hollow, or with voids, to improve insulation or aid moisture wicking.

Alternatively, multiple different materials may be combined in the same filament to take advantage of their different properties. Known as bicomponent, multicomponent or multifilament fibres, they are produced by mingling two or more materials in the spinneret just prior to forming. Thus, the elasticity of elastane may be combined with the high tenacity of nylon.

Once spun into continuous filament there are several modifications that can be applied. Drawing is a critical step: fibres are stretched several times their original length to align the crystalline structure. This improves the strength and resistance of certain materials.

Thermoplastics that may be formed with heat and pressure are knit-de-knitted (the yarn is knitted, heat-set to permanently take on the loops and kinks, and then unravelled), crimped (the yarn is fed into compression boxes causing it to buckle) and false twisted (yarn heat-set with a high twist rate) to set a shape in the fibre.

Known as texturing, these techniques are used to change the surface profile, add bulk, and improve stretch and recovery. They increase the cover of plied yarn for the same linear density and this is used to achieve a similar handle to natural staple yarns (page 56). For example, smooth and straight filaments are often not suitable for apparel, because the finish is not desirable. Alternatively, a filament core is wrapped with a textured or staple yarn (known as core-spun). This technique is often used in the production of high-strength thread for machine stitching (page 354) apparel. Such fibre modifications perform equally well in technical applications, such as controlling oil spills or preventing subsidence.

MATERIALS

Filaments are used to make nonwovens, twisted into plied yarn, or cut into shorter lengths and twisted into staple yarn.

Some of the most commonly used synthetics include PP, PE, nylon, polyester, acrylic and elastane. The properties range from light and strong to tough and resilient. High-performance fibres include UHMWPE, aramid, liquid crystal polymer (LCP), polyphenylene benzobisoxazole (PBO) and polytetrafluoroethylene (PTFE). They are lighter than conventional synthetic fibres and used for applications that demand higher strength or temperature resistance, for example.

Regenerated fibres include viscose, lyocell, acetate and azlon. They are not as strong as synthetics, although modifications have improved strength to weight. The importance of these fibres is that they are derived from natural and potentially renewable materials. The extent of environmental impact depends on the manufacturer and on legislation in the region where they operate.

COSTS

The cost of man-made fibres ranges considerably, from low-cost commodity types to very high-cost performance fibres. The price of specific materials fluctuates according to the price of oil and global demand. As a result, currently inexpensive materials, such as PP and PE, are becoming gradually more expensive to produce.

VISUAL GLOSSARY: MAN-MADE FIBRES AND FINISHES

Bulk Yarn
Material: Nylon
Application: Apparel
Notes: Filaments are texturized by false twist, knit-de-knit or stuffer box to add bulk and cover for the same density.

High Bulk Yarn
Material: PP
Application: Filtration
Notes: Fine-denier staple yarn (3 denier), produced by open-end spinning, wrapped onto a filter cartridge.

Crimped
Material: PP
Application: Various
Notes: Crimped PP fibres are strong and highly resilient. They are used as geotextiles and to control oil spills.

Colour
Material: PP, nylon and wool
Application: Floor covering
Notes: Synthetics are produced in a range of vibrant inherent colours, or dyed after spinning.

Mixed
Material: PP, nylon and wool
Application: Floor covering
Notes: A combination of synthetic and natural fibres produces a durable pile surface for commercial carpet.

Nonwoven
Material: PP
Application: Floor covering
Notes: Thermoplastics are combined into nonwovens directly from extrusion. Heat bonds the filaments.

ENVIRONMENTAL IMPACTS

The type of fibre determines whether the spinning technique uses solvents or melting to form the polymer into filaments.

Solvent spinning processes are either wet (into solution) or dry (into air). Solvent spinning is used to form viscose as well as acrylic and elastane.

Lyocell fibre is the result of a recent innovation in spinning. Dilute solvents are used in place of chemicals, and are reclaimed and reused in an almost closed-loop process, dramatically reducing the environmental impacts. At the end of its useful life, lyocell may be composted or recycled.

Melt spinning does not generate volatile organic compounds (VOCs) like solvent spinning and is more energy-efficient. This means it is generally less polluting and more cost-effective. Fibres produced in this way include PP, PE, nylon, polyester and polylactic acid (PLA).

The majority of plastics are derived from crude oil. Plastics produced from biological ingredients, such as maize or potato, are referred to as bioplastics or biobased. Conventional plastics such as nylon and polyester may be produced from partially biobased ingredients. PLA, on the other hand, is wholly biobased.

Biodegradable plastics form another class of products, which may be biobased or derived from crude oil (by incorporating additives that accelerate degradation, but the way the material breaks down is quite different). Not all bioplastics are biodegradable.

Replacing the use of oil with biobased materials means that the raw materials are renewable, but not necessarily sustainable. Some of the challenges associated with crops used to make bioplastics include genetic modification (GM) and land and water use.

→ Melt Spinning Polypropylene

The process of melt spinning starts with granules of thermoplastic (**image 1**). The material is either supplied to the factory with all the necessary ingredients already included, or compounded on site to incorporate the required additives and colourants (**image 2**).

The plastic is melted by extrusion and fed to the spinneret, where it is squeezed through the hundreds of tiny holes (**image 3**). Each of the strands is pulled upwards and solidifies as it passes through the air (**image 4**).

Melt spinning is just the start. Once the fibre is formed it is drawn over rotating wheels (**image 5**). The wheels rotate at different speeds, the last spinning quicker than the first. This causes the filament to be stretched to several times its original length.

At this point the plastic is still warm and easily manipulated. It is pressed into a stuffer box as it cools (**image 6**). Compacted into a chamber, the filaments are folded as they cool and so take on a crimped texture (**image 7**).

Once the fibre has cooled and been finished with any necessary coatings, it is bound into bales and stacked ready for distribution (**image 8**).

1

2

3

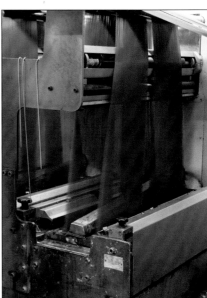
4

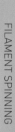

5

6

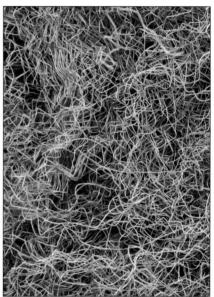

7

8

Featured Company

Drake Extrusion
www.drakeuk.com

Fibre and Yarn Technology

Staple Yarn Spinning

Short fibres, known as staples, are spun into continuous lengths of yarn. A single type of fibre is used to produce pure yarn, or multiple materials are blended to take advantage of the different properties of each. Continuous filaments, such as silk and synthetics, can be cut into short lengths and processed in the same way to get the benefits of a staple yarn.

FORMATION

56

Techniques	Materials	Costs
• Carded (woollen) • Combed (worsted)	• All types of fibre including natural and man-made	• Combed is typically higher cost than carded • Material selection affects cost

Applications	Quality	Related Technologies
• Fabrics and garments • Upholstery and carpets • Technical textiles	• Carded is soft and bulky • Combed is smooth and strong	• Man-made fibres • Nonwoven • Silk

INTRODUCTION

There are three main categories of staple yarn: carded, semi-combed and combed. Carded yarn consists of short-staple fibres spun together, and so is softer, bulkier and not as hard-wearing. For example, muslin is a loosely woven fabric made from carded cotton yarn. Passing the fibres through an additional combing process prior to spinning, to remove the short staples and better align the fibres, produces combed yarn (page 23). It is used to produce strong and smooth fabrics for clothing, or thread for lacemaking (pages 106–19), for example.

Wool, and in some cases synthetic, yarn is referred to as woollen (carded), semi-worsted (semi-combed) and worsted (combed), even though the carding, combing and spinning processes are essentially the same. Flax yarn (see Bast Fibres, page 34) made up of short staples is known as tow or hackled, and long-staple flax yarn is known as line or well hackled.

Staple yarn production is essentially the same for all types of fibre, including wool, cotton, flax and synthetics. However, the preparation processes vary according to the nature of the materials. For example, the production of wool yarn includes scouring, carding (and combing) into slivers, followed by spinning (see case study, page 59); whereas in the production of cotton yarn, the bales are opened and cleaned before the fibres are carded into slivers, which are drawn to blend and align the fibres, followed by combing and spinning.

APPLICATIONS

Staple yarns are used for weaving (pages 76–105), knitting (pages 126–51), embroidery (page 298), crocheting, cording, braiding, rope making (see Plying and Twisting, page 60), carpets, and as

The Staple Yarn Spinning Process

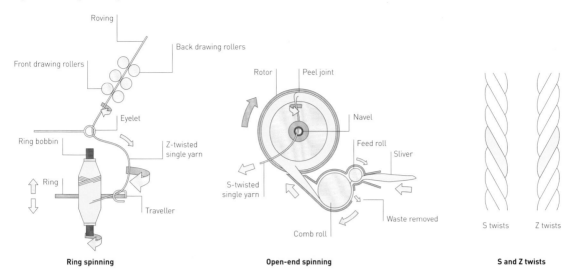

Ring spinning

Open-end spinning

S and Z twists

threads for sewing. Combed yarns tend to be smoother and stronger and utilized for higher-quality fabrics and garments.

RELATED TECHNOLOGIES

Man-made fibres (see Filament Spinning, page 50) are produced as a continuous filament (monofilament), or fibres twisted or grouped together (multifilament). Converting them into yarn is therefore more straightforward than staple fibres. They are used for applications that demand a controlled and precise yarn, such as lingerie and smooth fabrics. Synthetic filaments are cost-effective and have all of the benefits associated with plastics, such as colourability and elasticity. They are textured to impart some of the handle of staple yarns, increase comfort and add stretch and recovery, for example.

Nonwoven fabrics (page 152) are produced from webs of fibres held together by mechanical, adhesive or thermal bonds. They eliminate the need for yarn formation and so are typically less expensive to make than woven or knitted fabric. However, owing to the nature of construction they tend to have less drape and feel stiffer. But they can be molded (see Hat Blocking, page 420).

QUALITY

The characteristics of spun yarn are determined by a combination of material properties and spinning technique (count, ply and construction). Materials

TECHNICAL DESCRIPTION

Short fibres are twisted into staple yarn by spinning. In ring spinning, the roving (loosely twisted yarn) is twisted and wound onto a bobbin in a single operation. The yarn is then fed through a series of rollers, rotating at different speeds. Known as drawing, this causes the fibres to mix, align, elongate and further twist together. The yarn passes through an eyelet, a traveller and is taken up on the spinning ring bobbin (the package). The traveller spins around a ring, which moves up and down to evenly distribute the yarn. One complete twist is applied with each rotation of the bobbin. Centrifugal force causes the yarn to balloon outwards as it spins.

Clockwise rotation of the bobbin produces Z twists (right) and anticlockwise rotation produces S twists (left).

The other primary spinning process is open-end spinning (also known as break or rotor spinning). In operation, a sliver is fed directly into the rapidly turning rotor, where centrifugal force deposits the fibres into a groove. The spinning action causes the fibres to twist together as they are drawn out through the navel. Since it does not require a roving and the twist is not inserted by the rotating package, the process can operate up to six times faster than ring spinning and the length of yarn is unlimited. The yarn is bulkier and tends to be weaker than ring-spun yarn.

are selected to provide specific qualities, such as wool for providing insulation (warmth), nylon for its durability, cashmere for softness, or acrylic because it is low-cost.

The number of twists (or turns) per inch (TPI) or per centimetre (TPCM) will affect the strength of the yarn. Tightly twisted yarns are stronger and finer, but will quickly lose their strength if they become over-twisted, because of the shearing effect between the fibres. Bulkier yarn, such as woollen, will have fewer TPI than worsted, because it contains more air between the fibres.

Spun yarns are either single, ply, cord or novelty (used to create fancy effects). Single yarns are the most common. They consist of single staples twisted together into a continuous yarn, and so can be separated if the yarn is untwisted. Ply yarns, cord and rope are produced by twisting, or folding, two or more singles together (page 60). They are called two-ply when two singles are combined, three-ply when three singles are combined and so on. They are stronger than singles and used for high-quality fabrics, sewing thread and rope.

Woollen Yarn (Carded)
Material: Wool
Application: Apparel
Notes: Wool that has been carded and spun forms a soft and bulky yarn with low strength.

Worsted Yarn (Combed)
Material: Wool
Application: Apparel
Notes: Aligning the fibres by combing produces a stronger yarn: there is greater friction between the fibres.

Plant Fibre Yarn
Material: Flax
Application: Biocomposite laminate
Notes: Flax is strong (and gets stronger when wet) and durable and has a pleasant feel.

Synthetic Staple Yarn
Material: PP
Application: Socks
Notes: Synthetic filaments are cut into short lengths to make them suitable for spinning.

Intimate Blend
Material: 90% Merino wool and 10% nylon
Application: Socks
Notes: Fibres are blended to combine the properties and characteristics of each.

Mélange Yarn
Material: Wool
Application: Apparel
Notes: The fibres are dyed different colours before spinning to make a speckled colour pattern.

FORMATION

58

DESIGN

Fibres are blended to combine the qualities of different materials, such as wool and polyester to provide warmth, durability and abrasion resistance. Combining the fibres produces a superior yarn. Fibres can be different lengths, type, diameter and colour. Blends are created to improve performance (such as strength or comfort), to make production more efficient (such as spinning or weaving), to enhance handle or appearance, and to reduce cost. They are referred to as an intimate blend when the individual characteristics of each are lost to the combination. Weaving or knitting pure yarns of different generic types creates a mixture in the finished fabric.

Individual fibres, or the yarn itself, can be dyed with colour (page 240). Dyeing the fibres means they can be blended to create a wide range of effects, such as flecks of colour (tweed), speckled colour (mélange), multiple shades of the same hue, or even multiple colours (variegated or random-dyed). Dyeing the yarn after spinning produces different coloured effects. For example, two-ply yarns with different colours can be combined to create a marbled effect, or the yarn can be selectively dyed to produce stripes and patterns in the finished fabric, such as ikat (see Resist Dyeing, page 248).

MATERIALS

All types of materials, including natural and man-made fibres, are converted into yarn by spinning. Natural fibres, except silk, are short and so have to be spun together to form yarn. Man-made fibres and silk, which are produced as a continuous filament, are cut into short lengths to make them suitable for spinning into staple fibres. This is to increase comfort, insulation, absorbency and stretch, while maintaining consistency. Alternatively, they are twisted directly into singles or ply yarns from filament.

COSTS

The processes are rapid and labour costs are low for mechanized production. Cycle time depends on the fibre and the spinning process. Open-end spinning is typically highly automated and very rapid (150,000 rpm). Ring spinning, used to produce smaller packages of high-quality yarns, is slower (25,000 rpm) and involves more labour, so is more expensive.

Combed yarns are typically more expensive, owing to the additional processing that is required. Fibre material affects price: expensive fibres are often blended with less expensive materials to reduce cost.

ENVIRONMENTAL IMPACTS

The production and preparation of the materials determines the environmental impact, because spinning itself requires very little energy and produces virtually no waste. With all natural materials, certification is key to determining the overall impact of the production processes. For example, organic cotton is grown without chemical fertilizers, pesticides or fungicides. By contrast, regular cotton is the most heavily sprayed field crop and accounts for more than 10% of global agrochemical consumption, which has a significant impact on the surrounding environment.

The source is critical to determine the quality of working conditions. For example, more labour-intensive and lower-skilled processes, such as ring spinning, are likely to be carried out in countries with lower labour costs.

→ Carding and Ring Spinning Worsted Ply Yarn

The blended wool (pages 242–43), which in this case is brown, is passed through a five-stage carding process. In between each stage the wool is realigned against the direction of travel to ensure the colour is fully mixed (**images 1** and **2**). Gradually, the wool is reduced to a very fine fibrous web, which is separated by leather belts (**image 3**). The size is determined by the application. During roving, the fibres are rolled into loosely twisted yarns and taken up on spools (**images 4** and **5**). The outer surface of wool fibres is covered with tiny overlapping scales, which help them grip together as they are twisted into yarn.

Next, two yarns are combined to produce a ply yarn (see Plying and Twisting, page 60). They are drawn through an eyelet and are twisted together by the rotation of the bobbin (**images 6** and **7**). The finished two-ply yarn is taken up on ring bobbins (**image 8**). After spinning, it is transferred onto a suitable package ready for weaving.

1

2

3

4

5

6

7

8

Featured Company

Mallalieus of Delph
www.mallalieus.com

Fibre and Yarn Technology

Plying and Twisting

These yarn formation techniques, also called folding and doubling, are used to combine singles into ply yarn, cord and rope. Combining two or more singles produces a stronger, more balanced yarn, used for knitting, weaving and braiding. Materials may be combined to take advantage of their complementary properties, such as colourability, tenacity and insulation.

Formats	Materials	Costs
• Ply yarn • Cord and rope • Novelty yarn	• All types of yarn including staple and filament • Natural and man-made materials	• Low to moderate depending on raw materials and complexity

Applications	Quality	Related Technologies
• Yarns for weaving, knitting and braiding • Cord and rope • Thread for embroidery and lacemaking	• High, depending on materials and processes • From strong and smooth to soft and bulky	• Braiding • Filament spinning • Staple yarn spinning

INTRODUCTION

Single yarns are twisted together to improve balance and increase strength. Typically, yarn production is known as plying, and cord and rope making is known as twisting. Each forming process, such as weaving (pages 76–105), knitting (pages 126–51) and braiding (page 68), will be set up to accommodate specific types of yarn or cord.

Twisted yarns are described as ply, cord, rope or novelty. Ply yarns, also known as folded or combined yarns, are two or more single yarns twisted together in opposing directions. They are called two-ply when two singles are combined, three-ply when three are combined and so on. They are

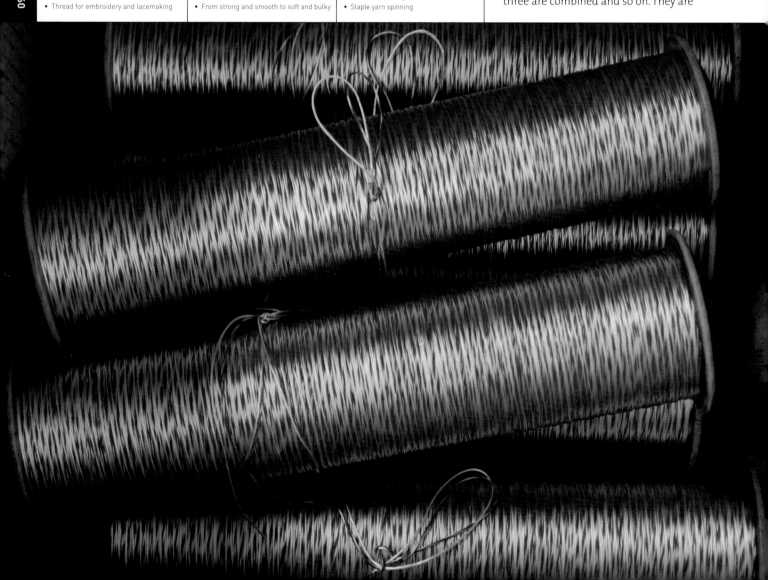

The Plying and Twisting Process

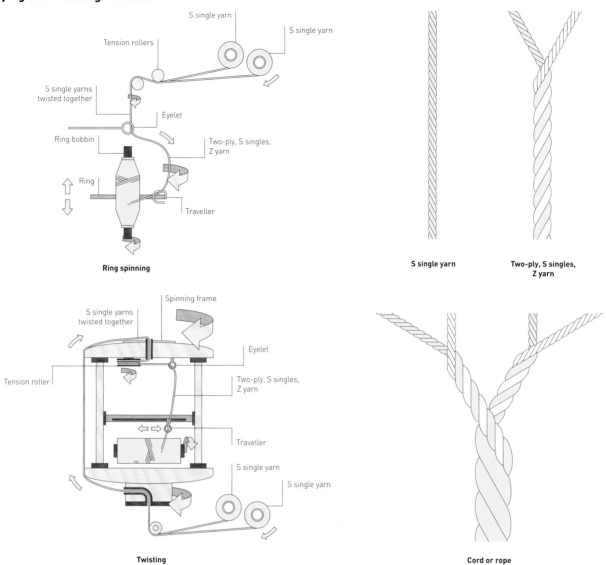

Ring spinning

S single yarn

S single yarn

Tension rollers

S single yarns twisted together

Eyelet

Ring bobbin

Two-ply, S singles, Z yarn

Ring

Traveller

Twisting

Spinning frame

S single yarns twisted together

Eyelet

Tension roller

Two-ply, S singles, Z yarn

Traveller

S single yarn

S single yarn

S single yarn

Two-ply, S singles, Z yarn

Cord or rope

TECHNICAL DESCRIPTION

Yarns are combined into ply yarns by ring spinning or twisting. In both cases, the packages of singles, which are typically wound onto tubes, are mounted on a frame so the yarn can come away freely.

In ring spinning, two or more yarns pass between tensioning rollers, through an eyelet and the traveller, and are taken up on the ring bobbin. The rotation of the bobbin draws the yarns together, causing them to twist with each rotation. Centrifugal force causes them to balloon outwards. The spinning traveller, which is located on a ring, is pulled around by the yarn, thus guiding it onto the bobbin as it is lifted up and down, which ensures even distribution of the yarn. Clockwise rotation applies Z twists (right) and anticlockwise rotation applies S twists (left). For a balanced yarn, two or more yarns are combined in opposing directions. For example, two S singles are combined with Z twists to make two-ply, S singles, Z yarn.

The number of twists per inch (TPI) is determined by the ratio between the speed of the tension rollers and that of the spinning bobbin. Ring spinning produces very high-quality ply yarn, but is limited by the speed of the rotating bobbin.

Twisting is a more recent development and can operate at higher speeds. With this process, the same principle is used to produce very fine-ply yarn or thick rope. The size that can be made depends on the scale of the equipment. In operation, the yarns are fed in at the bottom and around the outside of a spinning frame. Each rotation of the frame applies a single twist. The plied yarn is drawn through the centre of the top and guided onto the bobbin. The end result will look much the same as, if not identical to, ring-spun ply yarn.

The yarn is wound onto a package and either taken directly to manufacture, or rewound onto a suitable bobbin or cone.

stronger than singles and used for high-quality fabrics, thread, cord and rope.

Twisting is used to form the core of braided cord and rope (see Braiding, page 68), or the entire construction, known as three-strand.

Novelty yarns consist of a core surrounded with a decorative yarn, such as bouclé, or combine unmatched yarns for a unique visual effect, such as corkscrew. They are also sometimes referred to as novelty or complex yarns. They are used to add texture and visual intrigue to knitted and woven fabrics.

APPLICATIONS

Plying is a fundamental step in the production of yarn for many knitting, weaving and braiding applications. Many knitters, weavers and braiders will ply and twist their own combination of materials, to ensure a high-quality and consistent yarn that is suited to their production techniques. This also allows greater flexibility.

Twisted rope, such as three-strand, has been surpassed by braided rope, which is mechanically superior. Even so, twisted rope is relatively inexpensive and sometimes used for the visual quality.

RELATED TECHNOLOGIES

In garment construction, replacing plying and twisting would mean changing from a knit or weave to a nonwoven (page 152) or laminated construction (page 188), because it is a fundamental part of the manufacturing.

Certain processes can accommodate yarns in parallel formation. For example, an eight-plait rope may consist of four doubles combined in parallel formation, or eight singles. Similarly, high-performance laminated composites may be constructed using untwisted bundles of continuous filaments (tow), tapes of parallel filaments (spread tow, page 234), or stitch-bonded (page 196) unidirectional fabric.

Modern rope making utilizes braiding and weaving more than twisting. These are generally more expensive than twisting, but they are flexible and produce rope with superior mechanical and visual properties.

Alternative means of producing fibre-based yarns include braiding and weaving. Strip yarn is produced by cutting plastic film, woven fabric, or other sheet material into narrow lengths. Plastic strip is used in functional applications (such as circular woven packaging, page 83), as well as decorative novelty yarn (such as metallized polyester).

QUALITY

To create a balanced yarn, two singles with S twists are combined in Z direction with a balanced number of twists (or turns) per inch (TPI; also known as twists per centimetre), or vice versa. This produces the smoothest and most lustrous yarn.

Plying singles in the same direction as they were spun (known as active twist), such as S singles into S two-ply yarn, will cause the fabric to skew. This effect is used intentionally to create three-dimensional effects in fancy fabrics.

TPI will affect the strength of the yarn. Tighter-twisted single yarns are stronger. However, plied yarn may be used with very few TPI, or even parallel, depending on the mechanical and visual requirements of the application. Typically, yarns with more TPI are stronger, more elastic and more resistant to abrasion than yarns with fewer TPI. Yarns with fewer TPI feel softer and have better absorbency.

Staple fibres are short and so produce yarns that are more bulky, absorbent and elastic than filament yarns. The appearance may also be preferable in some applications. For example, polypropylene (PP) staple filaments are twisted together to mimic the look and feel of natural fibre, such as hemp.

Material selection plays a significant role in the final look, feel and technical performance of the yarn.

DESIGN

Plied yarn, cord and rope combine various colours, materials, qualities and effects. In addition, each single yarn may be made up of a combination of materials with contrasting properties.

Fibre selection and the combination of yarns will determine the range of mechanical and visual qualities that

can be achieved. It is not uncommon to create a unique yarn for an application. For example, Bridgedale knits high-performance socks with WoolFusion, a specially developed two-ply yarn that combines wool with polyester to provide comfort, durability, insulation and moisture-wicking qualities in a complex knitted structure (see case study on page 142). Alternatively, warp yarn in loom weaving has to be strong enough to withstand the stress and abrasion applied by the process, whereas light and delicate yarns can be used as the weft, or filling yarn in woven fabric. These considerations have a direct impact on the choice of materials and twisting process.

Colour is also determined by the choice of fibre. For example, wool and nylon can be acid dyed together, whereas PP has to be extruded as a solid colour. Therefore PP is limited to standard colours, or minimum order quantities of roughly 1 tonne will be likely to apply.

Yarn size is measured in the same way as singles: filament yarn, cord and rope is measured in denier, tex or decitex (dtex); and staple yarn as cotton count, woollen count, worsted count and linen count. Twisted yarns range from a few microns to thick rope. The limit is determined by the manufacturer's equipment and capability.

MATERIALS

Plying and twisting can combine all types of filament and staple yarn. Both man-made and natural materials are suitable. Final selection is determined by the requirements of the construction processes and application.

Fine metal wire is plied into yarn just like natural and synthetic fibres. Heavier-gauge wire is twisted into rope (or cable) and used for marine and architectural applications.

COSTS

Labour costs are relatively low for automated processes: one operator is able to look after multiple machines. Choice of material will have the greatest impact on cost. For example, technical fibre, such as polyphenylene benzobisoxazole (PBO), known under

Balanced Two-Ply Yarn
Material: 2 x bamboo viscose
Application: Socks
Notes: Z singles have been combined with S twists to create a balanced yarn that is smooth and strong.

Complex Two-Ply Yarn
Material: Polyester and nylon
Application: Sports socks
Notes: Coolmax (polyester filament with a profile that helps to wick moisture away) is combined with nylon to make knitted sports socks that are both hard-wearing and comfortable.

Corkscrew Yarn
Material: Wool
Application: Upholstery
Notes: Combining two yarns that are different sizes creates a two-ply yarn with a unique visual effect.

Bouclé Yarn
Material: Wool
Application: Upholstery
Notes: Two or more yarns are loosely bound together to create a bouncy random construction.

Metallic Yarn
Material: Polyester monofilament wrapped in a metallized polyester film
Application: Apparel
Notes: Metallic yarn is formed by wrapping metallized plastic foil around a filament core.

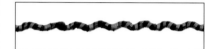

Metallic Corkscrew Yarn
Material: Silk with gold cover
Application: Interior
Notes: The silk is twisted into a corkscrew and wrapped in gold leaf for decorative effect.

PU Coated
Material: Cotton, polyester and PU
Application: Apparel
Notes: Used for knitting, weaving and braiding, the PU coating creates a yarn that feels like leather or plastic.

Cord
Material: Cotton
Application: Interior
Notes: Two colours of cotton yarn are twisted to make decorative cord for interior application.

Multicoloured Cord
Material: Various
Application: Interior
Notes: Multiple different yarns, left over from various applications, are twisted into three-strand cord.

Twisted Fabric
Material: Knitted fabric
Application: Interior
Notes: Waste fabric is cut into strips, knotted to make long lengths, and twisted into rope.

Rope
Material: Polyester
Application: Marine
Notes: A general-purpose rope comprising three twisted bundles of twisted polyester filaments.

Rope from Staple Yarn
Material: PP
Application: Marine
Notes: Synthetic filaments are dyed, cut into short lengths and twisted to resemble traditional hemp rope.

the trade name Zylon, is many times more expensive than polyester, a commodity material. Since the material cost is normally the greater part, the per-metre cost increases with size. The per-weight cost decreases with larger fibres, because the smaller the fibre the longer it takes to process a given mass.

ENVIRONMENTAL IMPACTS
Plying and twisting processes create very little waste. The energy used in production is offset against the improvements in fibre performance, making products more durable and longer lasting.

It is important to consider the source of materials made with processes that require a high level of labour, or unskilled labour, such as ring spinning. For economic reasons, large quantities from such processes are often carried out in countries where labour costs are low.

→ Ring Spinning Ply Yarn

Ring spinning is used to make singles or combine singles into ply yarn. When two or more yarns of different colours are combined the twist is clearly visible (**image 1**). In this case, Marlow Ropes is combining two parallel singles of green polyester with two parallel singles of PBO. This is the natural colour of this technical fibre, also known as Zylon (**image 2**). The four Z-twisted singles are fed into the ring-spinning machine in pairs (**image 3**).

They are drawn in by rollers, after which the rotation of the bobbin causes the fibres to twist together (**image 4**). The four singles are combined into a four-ply with S twists to create a balanced yarn.

The yarn passes through a traveller, which spins around the rotating bobbin (**image 5**). The traveller is pulled around the ring by the yarn being taken up on the bobbin. Therefore, the weight of the traveller affects the tension of the yarn. The various colours indicate the different weights (**image 6**). Selection depends on the type of material being plied. The finished four-ply yarn is removed (**image 7**), ready for transferring onto a package suitable for converting into braided rope.

FORMATION

64

1

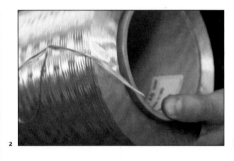

2

3

4

5

6

7

Featured Company

Marlow Ropes
www.marlowropes.com

→ Twisting Ply Yarn

Bridgedale combines singles into balanced and complex yarns to make high-performance socks. It folds its own yarns to ensure the highest quality and consistency, which is fundamental for knitting such intricate products.

This is forming S-twist two-ply yarn, with 85 twists per metre. The purple yarn is Coolmax (polyester) and the white yarn is nylon (**image 1**). Combined, they make a strong, moisture-wicking and breathable yarn (see image, Complex Two-Ply Yarn, page 63).

The singles are fed into the twisting machine (**image 2**) and wound onto a cone (**image 3**). A transversing mechanism rotates at the same speed as the cone and distributes the plied yarn uniformly along its length (**image 4**). When set in motion, the frame rotates at high speed, causing the two fibres to twist together. The finished yarn is ready to be knitted (**image 5**).

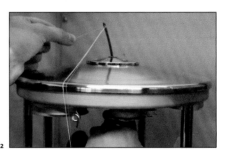

PLYING AND TWISTING

65

Featured Company

Bridgedale Outdoor Ltd
www.bridgedale.com

→ # Twisting Rope

Large-diameter rope can be made up of many groups of yarns combined together in a single process. This involves several twisting phases, whereby the filaments and yarns are progressively combined into larger sections until the final number of strands is wound into rope. Or, they may be combined in stages, depending on the exact requirements.

Packages of ultra-high-molecular-weight polyethylene (UHMWPE), also known under the trade names Dyneema and Spectra, are loaded onto a frame (**image 1**). Twisted UHMWPE is used in the core of rope, which is then covered with a braided jacket, or as the entire construction, such as three-strand rope. UHMWPE fibre is often the first choice for tensile applications, such as in marine, defence and industrial products.

The untwisted bundles of filament yarn (tow) are drawn into the twisting machine through an eye (**image 2**). Twist is applied with each rotation of the frame and the cord is taken up on a package (**image 3**).

The cord is made up of many filaments combined together (**image 4**). A machine of this scale is used to make cord measuring 22,000–110,000 dtex, whereas ring-tube spinning is typically used to make yarn up to 22,000 dtex.

Next, seven packages of the twisted UHMWPE cord are loaded onto a second, larger, twisting machine (**image 5**). It is possible to combine many more, but in this case only seven are required. The fibres are drawn in and twisted together (**images 6** and **7**). Here, the twisted UHMWPE is converted into twelve-plait braided rope (**image 8**), but it could equally be used as a twisted rope.

1

2

3

Featured Company

Marlow Ropes
www.marlowropes.com

Fibre and Yarn Technology

Braiding

Braids are constructed of three or more yarns interlaced diagonally and lengthways. They are interlaced in opposing directions to create a balanced rope that will not twist under tension. Any flexible yarn can be used. The raw materials range from high-performance synthetic fibres to natural materials, such as cotton and strips of leather.

Techniques	Materials	Costs
• Flat	• Filament and staple yarn	• Low to moderate depending on the yarn
• Cable	• Strips of material	• High for mandrel braiding
• Tubular	• Natural and man-made	

Applications	Quality	Related Technologies
• Rope, cable and cord	• High strength to weight	• Plying and twisting
• Military and industrial	• Static to dynamic	• Filament winding
• Marine and sports	• Low twist	• Weaving and knitting

INTRODUCTION

There are two main types: flat braids, in the form of narrow strips of fabric; and tubular braids, which may be hollow, solid or shaped over a mandrel (reusable metal rod). Flat braiding makes up one of the four principal stitches in handmade lacemaking (page 112), along with half, cloth and tally. In handmade lacemaking it is also referred to as plaiting.

Tubular braiding is used to make technical products, such as ropes and cables. Material selection and the physical characteristics of the braid are determined by the requirements of the application. For example, marine rope for sailing and racing boats has to be

The Tubular Braiding Process

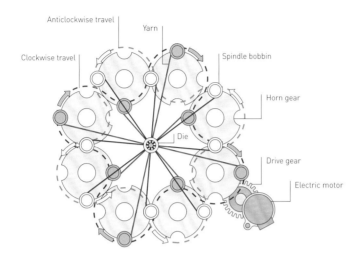

Anticlockwise travel

Yarn

Clockwise travel

Spindle bobbin

Horn gear

Die

Drive gear

Electric motor

TECHNICAL DESCRIPTION
Tubular braiding interlaces groups
of yarn by passing each one over and
under the adjacent groups. Several
patterns of interlacing are used, such
as plain (single bobbins) and twill
(bobbins run in pairs). Patterns affect
the interlacing structure and thus the
mechanical properties.

The yarns are loaded onto spindle
bobbins, which stand upright. They are
located in the horn gears, which are
driven by an electric motor. Each horn
gear rotates counter to the adjacent
gear. The bobbins are transferred
between the gears as they rotate.
Tubular braids are typically made with
an even number of yarns. Half rotate
clockwise and interlace with the other
half, moving in the opposite direction.

The yarns are gathered through a
metal die (though not always) to form
the braided cord. Braid angle (pitch) is
a measurement of the relative angle
of the groups of yarn compared to
the longitudinal centre line. It ranges
from 10° to 85° and is governed by
the relationship between the speed
of rotating gears and that of the
take-up roll. Lower angles (higher
take-up speed) result in a soft rope
with high axle stiffness, and higher
angles have greater hoop strength
(circumferential).

lightweight and capable of withstanding
heavy loads without the risk of failure.

Technical braids are engineered to be
reliable under relatively heavy loads and
stressful conditions. The same technology
can be used to produce decorative parts.

APPLICATIONS
Braiding is used to make all manner
of ropes and cables. Unlike twisted rope
(page 66), braids will not untwist when
put under tension. This quality makes
them desirable for all sport-related
and technical applications, including
rigging on sailing boats and oil rigs,
parachute cord, climbing rope, mooring
line, fast-rope for abseiling, and reinforced
data cable.

It is used to make decorative and
technical cord and cable for interior
applications including trim, pull cords
and curtain tiebacks.

Braids in various materials are used
to make and finish accessories, such as
cords, jewelry, straps and handles.

Pre-impregnated fibres (high
performance fibres impregnated with
resin) are braided over a mandrel to form

precise and lightweight composite parts
for high-performance applications, such
as sailing boat masts, booms and kickers;
wind turbine blades; and bicycle frames
and car chassis.

RELATED TECHNOLOGIES
In the case of rope and cord, braiding
competes directly with twisted rope. In
comparison, braided rope has superior
flexibility and mechanical performance.
However, the visual quality of twisted
rope may be preferable – braiding
looks modern and technical – such as
in the case of restoration projects and
traditional-looking applications.

Braiding and twisting can be used
alongside one another, such as in the
production of rope with a plied core
and plaited jacket (see page 71).

Braiding and weaving are combined
in garment construction. Flat braids are
used as trim; joined together by machine
stitching (page 354). Rope braids are
used for waistbands and hood cords, for
example. Similar items are produced by
weaving (pages 76–105), weft knitting
(page 126) and warp knitting (page 144)

narrow widths of fabric. In addition,
braiding may be used to combine strips of
material to make narrow fabrics or cord.

Braiding and filament winding
(page 460) are used for similar high-
performance tubular composite
constructions. The advantage of braiding
is that the yarns are interlaced, as
opposed to simply wound. Therefore,
braiding is used to produce self-
supporting structures, which can
be molded with vacuum and pressure
to form complex profiles.

QUALITY
Braids are flexible. The longitudinally
interlaced structure means they can be

formed around corners with minimal disruption to the weave pattern. For example, tightly formed knots rely on the ability of braided rope to maintain integrity while bent under heavy load.

Braids are balanced, because the yarns are interlaced diagonally as well as lengthwise. Therefore, they cannot untwist like plied yarn. This is critical for high-load safety applications, such as climbing ropes.

Material selection will have a significant impact on elasticity, abrasion resistance, colourfastness and surface friction (grip). Elasticity, or the lack of it, is an important consideration for many applications. For example, dynamic ropes used by climbers have engineered elasticity, so that if they slip the rope gives just enough to cushion their fall. By contrast, static ropes used to haul heavy loads should have as little stretch as possible, because failure under tension would cause the rope to whip and this can be extremely dangerous.

DESIGN
Braided yarn, cord and rope combine various colours, materials, qualities and effects. In addition, each single yarn may be plied from multiple materials with contrasting properties.

Colour is determined by the choice of fibre. For example, plastic filaments such as nylon, polyester and polypropylene (PP) are available in a range of standard colours including bright and fluorescent shades. Higher-performance fibres, such as aramid and polyphenylene benzobisoxazole (PBO), are typically available only in their natural colour.

Tubular braiding, used to make rope, may be solid, hollow, or shaped over a mandrel. Solid braids are the most straightforward and are the result of combining three or more yarns or strips without any additional material. Alternatively, braiding may be used to form a jacket over an existing core, such as twisted filaments (sometimes referred to as kernmantle rope), steel wire or rubber filaments.

Braided cores are coated with polyurethane (PU) or silicone to improve resistance to abrasion, liquids and

ultraviolet (UV), and thus eliminate the need for a braided jacket (which adds weight).

Hollow braids are formed over a removable core. This technique is used to make the jackets (also known as casing or armour) for hosepipe and electrical cables (after braiding, the delicate fibre-optic core is pulled through with the removable mouse line, a length of cord used to pull a rope core through its cover).

Non-cylindrical braids are formed over a mandrel. This technique is used to make high-performance composite parts. Thanks to the interlacing yarns, braiding produces parts with exceptional torsional strength and impact resistance. Like filament winding, the angle can be adjusted depending on the mechanical requirements. It is a specialist technique that is used for automotive and performance sports products. As a result, it is a high-cost and currently limited to low-volume production.

MATERIALS
Material selection is determined by the requirements of the application. The required strength, stretch and fatigue resistance will influence the choice of fibre, as will cost, weight and lifespan.

Rope making utilizes man-made fibres (see Filament Spinning, page 50) ranging from commodity to high-performance. Nylon has good strength to weight, high elongation (30–40%), good resistance to UV and good resistance to creep. However, it is hydroscopic, which means it absorbs water. When wet, nylon loses 10–20% of its strength.

Polyester has lower elongation than nylon (15–20%). It is considered a good all-round material, low-cost, available in a wide range of colours and commonly used to make plaited jackets.

PP is low-cost and has lower tensile strength. It is lightweight (it floats on water), has moderate resistance to creep and good resistance to UV. Ultra-high-molecular-weight polyethylene (UHMWPE) comes from the same polymer family, but has superior properties: it has abrasion resistance, low elongation (3–5%) and exceptional strength to weight. It is commonly used for tensile

applications and is often referred by the trade names Dyneema and Spectra.

Aramid fibre is mainly known under the trade names Kevlar and Twaron. It has exceptional strength to weight and very low creep and elongation (2–4%). However, it has poor UV resistance, making it vulnerable when used outdoors. It is available in gold and black (Technora).

Liquid-crystal polymer (LCP) is also known under the trade name Vectran. It has high strength and very low creep (virtually immeasurable), with good resistance to fatigue, flexing, abrasion and high temperature (up to 330°C/625°F).

PBO, known under the trade name Zylon, has the highest strength and modulus of any polymer fibre (higher than aramid). It has zero creep and very low elongation (2–4%), but has poor resistance to abrasion and UV so is typically used in the core.

Polytetrafluoroethylene (PTFE) is the most expensive filament yarn. It is commonly known under the trade name Teflon, and is a high-performance filament with very low surface energy. It is blended with other filaments to produce a rope with good slip and grip. This quality is desirable for marine applications, where the rope needs to slip around a winch drum, as well as gripping and easing smoothly.

Other than polymer filaments, metal wire is braided or encapsulated as a core. Steel wire and copper wire are used for high-strength applications in the construction, transportation and power distribution industries.

Natural fibres such as hemp and sisal are suitable for braiding, although they are more commonly converted into twisted rope and used for decorative applications.

COSTS
For conventional flat and tubular braiding the costs are governed by the material and dimensions. For example, PTFE is many times more expensive than PP. The price of synthetics constantly fluctuates depending on supply and demand.

Braiding over a mandrel is much more expensive. However, it may be cost-effective compared to other composite

12-Plait Rope
Material: LCP
Application: Film and theatre
Notes: This is a high-strength rope that is quick and easy to splice (join the ends of rope by interweaving individual strands), because it does not have a jacket.

12-Strand Core with 16-Plait Jacket
Material: UHMWPE core and polyester jacket
Application: Marine
Notes: Polyester comes in a range of colours and has good resistance to abrasion, making it an ideal cover material.

8-Plait Core with 16-Plait Jacket
Material: UHMWPE core with UHMWPE/PP jacket
Application: Marine
Notes: Plying together different materials of different colours produces a more random appearance.

8-Plait Core with 16-Plait Jacket
Material: PP core with polyester jacket
Application: Marine
Notes: The PP core is removed for splicing to maintain diameter. The higher-strength polyester remains.

12-Plait Core with 24-Plait Jacket
Material: UHMWPE core with PBO jacket
Application: Marine
Notes: PBO has the highest strength to weight of any polymer filament. It is used for demanding applications.

8-Plait Core, 16-Plait Inner Jacket and 24-Plait Outer Jacket
Material: UHMWPE core with polyester inner and outer jacket
Application: Marine
Notes: The staple fibre polyester jacket produces a rope with a soft hand, or feel. This provides improved grip.

8-Plait Core with 16-Plait Jacket
Material: UHMWPE core with UHMWPE/PP jacket
Application: Marine
Notes: Combining PP and UHMWPE produces a very light and high-strength rope that will float on water.

Parallel Core with 24-Plait Inner Jacket and 24-Plait Outer Jacket
Material: PU-fused aramid core with aramid inner and outer jacket
Application: Marine
Notes: The PU-fused aramid core creates a very stiff rope that is used to transmit turning force along its length.

Twisted Parallel Cores with 16-Plait Cover
Material: Nylon
Application: Rescue
Notes: The twisted parallel cores produce a rope with very low stretch, known as static rope.

Latex Filament Core with 16-Plait Cover
Material: Latex filament core with polyester jacket
Application: Marine
Notes: The latex core is held under tension during braiding. Once released it relaxes to the correct length. Thus, in application the core can stretch several times its length without damaging the jacket.

12-Plait Core with 24-Plait Jacket
Material: UHMWPE core with polyester/PP jacket
Application: Marine
Notes: The white UHMWPE core is coated with PU, turning it black. The coating improves strength and stops stranding.

8-Plait Core, 16-Plait Inner Jacket and 24-Plait Outer Jacket
Material: UHMWPE core with polyester inner and outer jacket
Application: Marine
Notes: One core may be taken out through the side to produce a forked end to the rope, and the ends spliced separately.

laminating techniques, because producing seamless tubular profiles reduces labour-intensive assembly later on in the production process.

ENVIRONMENTAL IMPACTS
Braiding combines yarns into a construction that has improved strength to weight. This reduces the amount of material required. The process itself is efficient and generates very little waste.

Braids made in a single plastic material are readily recycled and converted into new products. For example, dynamic ropes are made with nylon, which can be recycled by melting and re-pelletizing for filament extrusion. Items that contain a mix of materials pose a challenge at the end of their useful life, because they cannot be directly recycled. Typically, the material is chopped into staple fibres and remanufactured into nonwovens (page 152).

→ # 12-Strand Single-Braided Dyneema

In preparation, packages of UHMWPE, commonly known as Dyneema, are twisted and wound onto bobbins. These are loaded into the horn gears. In this case, 12 bobbins are loaded onto a carrier with six horn gears (**image 1**). The tops of the bobbins are colour-coded.

The twisted yarns are gathered by a die, which sits above the bobbins in the centre (**image 2**). As the bobbins rotate around the gears the yarns are passed over and under one another to form a tight braid (**image 3**).

The pitch (**image 4**) is controlled by the rotation of the gears driving the bobbins relative to the speed of the take-up roll. Tension is maintained by springs holding the bobbins. If the take-up speed is increased relative to the rotation of the bobbins then the pitch decreases (lower angle between the yarns).

The braided rope is drawn from the take-up roll under tension to maintain accurate pitch (**image 5**) and gathered in a container ready to be coiled (**image 6**).

1

2

3

4

5

6

FORMATION

72

Featured Company

Marlow Ropes
www.marlowropes.com

→ Braiding a Colourful Polyester Jacket

Jackets are braided over a core to provide additional resistance to abrasion and UV. They are also used to add colour, which is useful for identification. Polyester is durable and available in a range of bright and saturated colours, making it an ideal jacket material.

Grey, black and fluorescent yellow polyester bobbins are loaded onto the gears (**image 1**). The black yarn is sandwiched between two yellow yarns to create a striped pattern. The jacket is a 24-plait construction, so requires 24 separate bobbins (**image 2**).

The jacket is braided directly onto a 12-strand UHMWPE core, fed in through the bottom of the machine. They are gathered in the steel die (**images 3** and **4**). The polyester yarns are wound onto the bobbins in pairs (parallel formation) and braided in pairs (twill weave).

The pitch is measured to check the rope is braided at the correct speed (**image 5**). The core and cover need to match very neatly to leave adequate room for splicing (**image 6**).

1

2

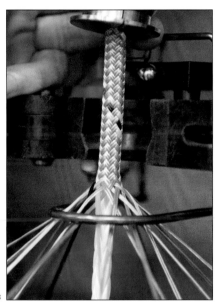

3

4

5

6

Featured Company

Marlow Ropes
www.marlowropes.com

Part

2

Textile Technology

Textile Technology

Loom Weaving

Weaving is the process of passing two or more sets of yarns, strands or strips of material over and under one another. The intertwined structure can range from soft and pliable to rigid, depending on the raw materials and finishing processes. Basic weaves include biaxial and triaxial configurations, used to make textiles for apparel and for interior and technical applications.

Techniques		Materials	Costs
• Plain (flat and tubular) • Twill	• Satin • Triaxial • Multiaxial	• All types of staple yarn and filament yarn • Warp must be strong	• Biaxial weaves are low, depending on the yarn • Multiaxial weaves are moderate to high

Applications	Quality	Related Technologies	
• Apparel • Interior textiles • Technical textiles	• Firmer than knitted fabric • Strongest in the direction of the warp	• Fancy weaving • Handloom weaving • Weft knitting	• Sheet materials (leather, non-woven, laminate)

INTRODUCTION

Loom-woven fabrics are produced from warp yarns (running lengthwise) and weft yarns, also called filling yarns (running widthwise), interlaced together. The three principal weaves are plain, twill and satin (see Handloom Weaving, page 90). The structure of virtually every woven fabric is derived from these three basic weaves. Stretch is introduced with elastomeric yarn, and rigid fabrics are produced by incorporating metal wire or other stiff material as warp or weft or both.

The majority of fabrics are biaxial (the warp and weft are at 90°). Multiaxial fabrics, such as triaxial (three directions) or quadriaxial (four directions), developed

The Basic Weaving Process

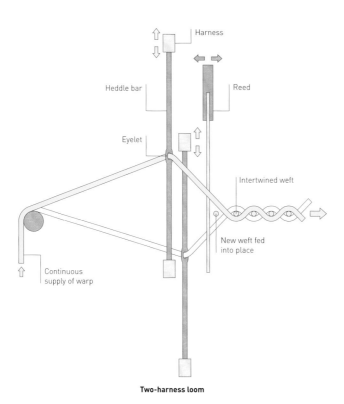

Two-harness loom

Harness

Heddle bar

Reed

Eyelet

Intertwined weft

New weft fed
into place

Continuous
supply of warp

for technical applications including high-strength composite structures (page 446) and sailcloth, are stable in more directions, corresponding to the additional yarns.

Three-dimensional fabrics are made flat and then folded, shrunk or otherwise distorted after weaving. Alternatively, it is possible to weave a tube. For example, pillowcases and agricultural sacks are both woven as a seamless tube.

Fancy weaves include pile, piqué and jacquard (page 84). While basic weaves are naturally patterned as a result of the construction, fancy weaves tend to have more elaborate inherent patterns or texture.

Woven fabrics are finished using a wide range of processes including dyeing (page 240), laminating (page 188) and printing (pages 256–79).

APPLICATIONS

Plain-weave fabrics include lightweight sheer fabrics (such as chiffon and ninon), open-space fabrics (such as cheesecloth), basket weave (chequerboard pattern), batiste (lightweight balanced weave),

printed cloth (medium-weight fabrics such as chintz and calico), muslin, percale (bedding fabrics, printed or unprinted), ripstop (filament-yarn weave with slightly larger warp and weft yarns at intervals to stop ripping), hopsacking (coarse open weave) and gingham (balanced medium-weight fabric that uses different-coloured warps and wefts to create a checked pattern).

Twill weaving is typically used for more durable fabrics, such as denim (warp-dyed cotton), flannel, herringbone (twill reversed at regular intervals), chino (hard-wearing with a slight sheen) and houndstooth (even-sided twill with warp and weft in contrasting colours), because dirt is less visible against the diagonal pattern and they tend to recover from wrinkles better than plain-weave fabrics.

Satin-weave fabrics are used in garments, lingerie, bedding and upholstery for their smooth feel and lustrous appearance.

The versatile properties of woven fabrics are utilized in high-performance composite laminates, such as carbon-

TECHNICAL DESCRIPTION

In preparation, the warp yarns are sized (coated with solution) to increase stiffness and strength, and wound onto a warp beam (also called loom beam or weaver's beam), which is loaded onto the back of the loom. Each warp yarn is carefully threaded through the loom mechanism, held under tension.

Loom weaving consists of five movements repeated: warp let-off (releasing the tension on the warp beam), shedding (raising and lowering the heddles), picking (feeding the weft through the warp yarns), beating up (pushing the weft to its final position), and fabric take-up on a fabric beam.

The first, warp let-off, releases the precise amount of yarn for the next shed. Each warp yarn is passed through a heddle (length of wire or cord), attached to a harness. They are operated in groups by computer numerical control (CNC), or by a pedal.

The harnesses are moved up and down on a crank, according to the weave pattern, which determines whether the warp or the weft is visible from the topside. A plain weave needs two harnesses, a twill weave three or more and a satin weave five or more (see Handloom Weaving, page 90).

With each pick, one or more weft yarns are fed into the shed (space between the reed and the warp yarns). On shuttle looms it is fed by a shuttle, which passes the yarn from side to side; on shuttleless looms, the wefts are fed from side to side by a rapier (metal strip with yarn gripper at one end), a jet of compressed air or a water jet.

The reed comprises a series of blunt blades that sit between each warp yarn. They pack each weft tightly into the cloth fell (open edge of the fabric). The weft is held in place by the reed as the lower heddle bar moves up and the upper bar moves down, locking the weft between the warps. The process is repeated as the fabric is taken up on a fabric beam. The rate of take-up, along with machine speed, determines the fabric's density (thread count).

Balanced Plain Weave

Material: Cotton
Application: Men's shirt
Notes: A balanced 1/1 weave, whereby equal weight warp and weft yarns interlace with each neighbouring yarn.

Plain Weave with Open Structure

Material: Jute
Application: Shopper
Notes: Known as hessian (or burlap in the USA), jute bast fibres are woven into a tough textile with an open structure to reduce weight and increase flexibility.

Plain Weave with Multicoloured Weft

Material: Wool and wool/polyester
Application: Upholstery
Notes: Alternate weft yarns consist of one black and one plain yarn, plied together, creating a mottled appearance.

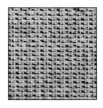

Plain Weave with Unbalanced Yarn Weft

Material: Wool
Application: Upholstery
Notes: Two different types of yarn are twisted together to make an unbalanced yarn. If used in the weft this creates an uneven texture.

Plain Weave with Bouclé Yarn

Material: Wool/nylon
Application: Upholstery
Notes: The surface is covered with entangled loops because the weft is made up of two bouclé yarns plied together.

Plain-Woven Metal Wire

Material: Copper
Application: Conductive textile
Notes: Modified looms are capable of weaving metal wire in the same way as natural or man-made yarns.

Basket Weave

Material: Nylon
Application: Garments
Notes: This is a 2/2 basket weave, produced by plain weaving pairs of warp and weft yarns.

Basket Weave with Different-sized Yarns

Material: Wool
Application: Upholstery
Notes: Different-sized warp and weft yarns are combined to produce a textured surface.

Ripstop

Material: Nylon
Application: Outdoor clothing
Notes: Slightly larger warp and weft yarns are placed at intervals to increase tear resistance.

fibre-reinforced plastic (CFRP) and glass-fibre-reinforced plastic (GFRP). The strength of composites is determined by the direction of the weave. The ideal pattern is selected according to the application, such as whether a given three-dimensional shape requires a large amount of drape. Multiaxial weaves provide high strength in more than two directions.

The weight of fabrics has evolved to suit specific applications. For example, lightweight fabrics (less than 34 g/m² or 1 oz/yd²) include netting and sheer; 'top' weight (around 100 g/m² or 3 oz/yd²) is used in sportswear, blouses and shirts; 'bottom' weight (around 200g/m² or 6 oz/yd²) is used to make trousers and skirts; heavy weight (around 340 g/m² or 10 oz/yd²) is used for workwear; and very heavy weight (over 475 g/m² or 14 oz/yd²) is used in upholstery. Higher weight does not always equal higher quality and vice versa.

RELATED TECHNOLOGIES

Weaving is a versatile process used to produce the majority of fabrics for all types of applications. One drawback is that woven fabrics have to be cut and joined to make three-dimensional shapes (apart from seamless tubular weaving). In contrast, knitting (see Weft Knitting, page 126, and Warp Knitting, page 144) is used to produce flat fabrics as well as complete three-dimensional shapes such as sweaters, gloves and socks. Another advantage of knits is that the loop structure has very good stretch and recovery. But this same quality means they are more prone to stretching and sagging.

Sheet materials, such as leather (page 158), plastic (see Plastic Film Extrusion, page 180) and nonwovens (page 152), provide an alternative to woven fabrics in certain applications. For example, high performance plastic films are isotropic, waterproof, lightweight and durable. Woven fabrics are laminated (page 188) with film to utilize these properties for applications ranging from lightweight outerwear to inflatables.

Fancy weaving differs from basic weaving because a texture or complex pattern is formed as part of the structure of the cloth. A wide range of qualities and effects can be achieved, including fabrics with an irregular structure (momie), extra warp or weft yarns, floating yarns, complex patterns with up to 30 warp arrangements (dobby) and designs that require individual warp control (jacquard).

QUALITY

The quality of fabric ranges from loosely woven cotton sacks to the finest silk lingerie. Mechanized loom weaving is typically carried out on computer-guided looms, which produce consistent and high-quality materials.

Plain weave is the simplest and is strong and hard-wearing. There is no difference between the face and the back, unless the fabric has been printed or treated in some way. Twill and satin fabrics have a front side (face) that is different from the back side (reverse). For example, in satin weaving the floating yarns will be on the front side. Finishes, such as calendering and

Foil front   Back

Foiled
Material: Wool and cotton
Application: Apparel
Notes: A simple weave with unequal yarns produces a relief surface profile, which is emphasized by calendering with metallic foil.

Twill Weave
Material: Flax
Application: Composite laminating
Notes: A 2/1 twill weave, whereby each warp and weft floats over two yarns and interlaces with the third.

Triaxial Weave
Material: Carbon fibre
Application: Composite laminating
Notes: Triaxial weaves have greater dimensional stability, helping to produce lightweight and high-strength composite structures.

Woven Tape
Material: Nylon
Application: Safety harness
Notes: Narrow fabrics are woven in strips. The edges are finished by interlooping the weft with each pick in a process that combines weaving with knitting.

embossing, will be more pronounced on the front side. For this reason garments and upholstery are usually constructed so the front side faces outwards.

Fabrics may also have a different appearance when viewed from the 'top' or the 'bottom'. The direction of the pile affects this most of all, for example in brushed, napped (page 216) or emerized fabrics. This should be taken into account during pattern cutting (page 314).

Thread count, fabric count or fabric density is a measure of yarns per square centimetre or square inch. Balanced fabrics have the same number of warp yarns and weft yarns per cm²/in.². Unbalanced fabrics have proportionately more of one yarn than the other. This is either because different sizes of yarn are used for the warp and the weft, or the warp is more or less spread out than the weft. Higher thread count usually indicates higher-quality fabric, which is stronger and more resistant to abrasion and has a better hand.

The warp yarn must be able to withstand the stress of being pulled tight throughout the process. Warp yarns are usually filament yarns (page 50), or tightly twisted staple yarns (page 56), which can withstand the stress and abrasion. If a weaker yarn is required, such as a novelty staple yarn, then it is preferable to use it as the weft.

The outside edge is known as selvedge. There are many styles of finish, depending on the method of construction, used to create a strong edge that will prevent the fabric from fraying. In shuttle loom weaving the weft is continuous and loops back and forth to produce 'shuttle selvedge'. A 'fringe selvedge' is made up of lengths of weft with cut ends. The exposed weft ends are fed back into the fabric construction to produce a 'tucked-in selvedge', or left exposed, such as in the case of a 'leno selvedge'.

The selvedge will have different performance from the rest of the cloth: it will be stiffer and may be thicker. Therefore it is typically removed during pattern cutting. In some cases it may be used to reduce the need for hemming because the edge is durable and will not fray easily.

Owing to the complexity of the processes there are many issues that can occur during weaving, resulting in defects in the fabric. Examples include bowed or skewed yarns, looping, sticking, hook marks, broken threads, slippage and surface abrasion. Fabrics with high numbers of floating threads, such as satin, are more prone to snagging because the yarns are exposed.

DESIGN

Within each piece of woven fabric there is a choice of yarn, pattern, density (thread count) and crimp (the displacement of the warp by the weft, or vice versa, determined by the warp tension, affects fabric thickness).

Fabric produced from the same yarns can have a different appearance, for example by changing the order of the warp yarns in the heddles or the rhythm of the harnesses' movement up and down. In addition, two or more yarns can be used in the same fabric to produce linear repeating patterns.

Looms with the same number of harnesses are used to produce different fabrics. For example, whereas passing pairs of neighbouring warp yarns through heddles on the same harness produces a weft rib weave (pairs of warp pass over each weft), raising alternate warp yarns for two sheds at a time produces a warp rib weave (each warp passes over two weft yarns). Alternatively, ribs may be formed in plain-weave fabrics by incorporating larger-diameter yarn in either the warp or the weft.

Many textiles are multifunctional, and there are copious opportunities associated with construction and application. Each type of weave has a different appearance, drape and robustness. Combined with different materials, an endless number of structural properties can be achieved. This makes woven structures suitable for self-supporting as well as reinforced applications. Different types of yarn can be combined to produce fabrics with in-built mechanical performance, such as ripstop, which has slightly larger warp and weft yarns at intervals (5–8 mm/ 0.2–.03 in.) to stop the fabric tearing. It is used to make ultra-lightweight sailcloth and hot air balloons, for example.

MATERIALS
All types of materials can be woven. Conventional looms use filament and staple yarns including natural

materials, man-made fibres, metals and combinations of all these. It is also possible to weave with strips of such materials as paper or plastic, although this will probably require modifying a loom or weaving by hand.

COSTS

Basic-weave fabrics are the least expensive. Material selection and thread count have a significant impact on cost.

Mills refer to minimum production runs as metreage (or yardage). Basic fabrics require large minimum orders of 1,000 m (1,100 yd) or more, whereas fancy fabrics could have a minimum run of 200 m (220 yd) or so.

Labour costs are moderate to high because a high level of skill is required. Modern looms operate at speeds of 200 to 1,000 picks (weft insertions) per minute. Faster cycle times typically reduce cost, but this depends on volume (this is less significant for lower volumes).

ENVIRONMENTAL IMPACTS

Sustainability is determined by yarn selection, preparation, weaving process and finishing techniques.

In some cases, the yarns may need to be treated before weaving, such as adding size (coating to increase stiffness and smoothness) or lubricants to warp yarns. This must be removed and recycled after weaving. Companies such as Rohner Textil have demonstrated that, with consideration for the complete life cycle, products manufactured using complex industrial processes can have a minimal, or even neutral, impact on people and the environment.

Shuttle loom processes are noisy and workers have to wear ear protection to work near the machines. Shuttleless looms are much less noisy and are generally more efficient overall. The waste produced during weaving cannot be directly recycled (it must be used for an alternative application or burned as fuel). Reducing weaving faults by improving the

quality control, or accepting lesser-quality fabrics, reduces waste.

Woven fabrics that contain a mix of materials pose a challenge at the end of their useful life. For example, a natural yarn combined with a plastic yarn means that the fabric is only partly biodegradable and partly recyclable. Separating the two materials so that they can be properly processed is not straightforward. In some cases it is possible to recycle parts of garments or products by removing panels and using them to make new products. Using this technique, it is possible to make an item of equivalent or even higher value (upcycle). Even so, it is not possible to make the identical product because there is wastage in the cutting process.

→ Weaving a Checked Pattern

The warp and weft yarns are subjected to different conditions during weaving and so require different preparation. The warp yarns have to withstand the tension applied during weaving and so are treated with size (solution that is removed after weaving) to improve strength and smoothness.

In a process known as winding, the worsted yarn is transferred onto cones (**image 1**), which can hold more than the packages the yarn was originally spun onto. The cones of warp yarn are loaded onto a creel frame (**image 2**). Each end of yarn is fed through a reed (**image 3**), according to the placement of colour in the weave pattern, and wound onto the warp beam. This process is known as creeling (or warping) and the aim is to produce a sheet of long parallel yarns (often longer than the cones and so multiple lengths are joined together). This determines the basis of the pattern to be woven (**image 4**).

The warp beam is loaded onto the loom (**image 5**). In a process known as dressing the loom, each thread is passed through an eyelet in the heddle bar. The heddle bars move up and down on a harness to intertwine the weft yarn, which runs perpendicular. The weft yarn is passed through the shed (space between the reed and the warp yarns) by the rapier (**image 6**). The newly inserted weft yarn is beaten up to its final position by the reed (**image 7**) to produce a tight and uniform weave (**image 8**).

With these techniques a wide range of colours and patterns can be created, as long as they run in line with the warp (stripes or bands), or diagonal (twill) or perpendicular (check, stripes or bands) to it.

1

2

3

4

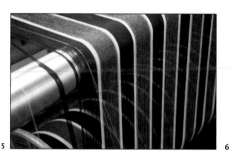

5

6

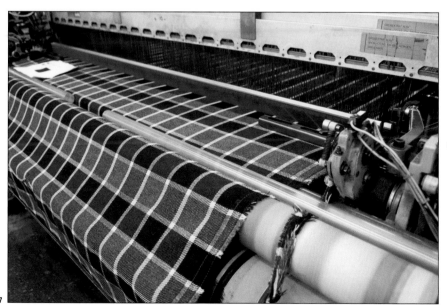

7

8

Featured Company

Mallalieus of Delph
www.mallalieus.com

→ Weaving Narrow Widths

Latex yarn (grey) is interwoven with the polyester (brown) on a narrow loom to create an elastic strip with controlled stretch. Narrow fabrics are either woven in strips – on narrow looms or with independent shuttles on a wide loom – or they are woven full width and trimmed to the correct dimensions. The latter approach requires that the width is determined prior to weaving, so strong selvedges may be formed at the correct dimensions.

Latex is supplied in strip form, which is split into individual filament yarns in preparation for weaving (**images 1** and **2**). Handling it as strip avoids it buzzing up into knots and becoming unworkable.

Combined with a non-stretch polyester yarn, the latex is fed through a set of heddle bars, which move up and down on harnesses in order to make the basket-weave pattern (**image 3**). Within each cycle the same weft is passed back and forth by a rigid needle (**image 4**) to form a strong unbroken selvedge along both edges (**image 5**).

The latex and polyester construction is woven under tension and pulled tight by rollers (**image 6**). As the tension is released the latex relaxes and so pulls the weft together to form a textured surface (**image 7**).

The fabrics are taken up on rolls (**image 8**). As well as conventional loom weaving, narrow fabrics are produced by jacquard weaving (page 84), warp knitting (page 144) and stitch bonding (page 196).

1

2

WEAVE

82

3

4

5

6

7

8

Featured Company

H. Seal & Company
www.hseal.co.uk

→ Circular Loom Weaving

Seamless tubes, such as those used for making packaging, bags, hoses and pillowcases, are produced on circular looms.

The warp is supplied straight from cones on a creel stand because the yarn must be distributed right around the circumference of the tube (**image 1**). In this case, 2mm (0.08 in.) wide strips of polyethylene (PE) are woven into tough and tear-resistant industrial packaging. The number of strip yarns depends on the diameter of the tube, the density of the weave and the width of the strips.

The clear PE warp is fed under tension. A series of shuttles moves around the circumference of the tube at high speed (1,000 picks per minute or 140 m/460 ft per hour). Shuttles are driven mechanically or electromagnetically. As they rotate, the warp is moved above and below by groups of heddles, interlacing the weft 1/1 to create a plain weave (**image 2**). The blue weft, visible inside the machine and between the warps, is carried around the circumference on bobbins in line with the shuttles, providing them with

a constant supply of yarn. After each weft insertion, which is a continuous cycle, wires slot between the warp yarns and beat up the yarn to its final position.

Solid colour patterns are formed with a clear warp and coloured weft, whereby the colour comes through the warp to give the appearance of colour throughout (**image 3**).

1

2

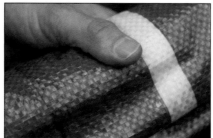

3

Textile Technology

Fancy Loom Weaving

Figured, relief, lampas, damask, double-faced and leno net fabrics are produced using specially adapted looms, such as the dobby, jacquard and multi-warp. The type of fabric and complexity of pattern depends on the loom. As well as conventional yarns, these highly decorative fabrics often make use of novelty yarns for added decorative effect, such as metallic, chenille and textured.

Techniques		Materials	Costs
• Cam	• Jacquard	• All types of staple yarn and filament yarn	• Moderate to high depending on the complexity and the yarn
• Dobby	• Specialized	• Warp must be strong	

Applications		Quality	Related Technologies	
• Apparel	• Novelty yarn	• Strongest in warp direction	• Embroidery	• Loom weaving
• Interior textiles	• Technical textiles	• Firmer than knitted fabric	• Handloom weaving	• Weft knitting

INTRODUCTION

Fancy weaving techniques divide the warps into a larger number of groups than conventional loom weaving (page 76). Each warp yarn passes through an eyelet in a heddle bar (length of wire), attached to a harness, and raised and lowered (shed) in preparation for each pick (weft insertion). The motion of the harnesses on the simplest mechanized looms is provided by cranks (suitable for plain weaves). More complex weaves are possible with cam-operated harnesses (up to 14 warp arrangements) or dobby looms (up to 30 warp arrangements, making possible virtually any design). Jacquard looms employ independent heddle bar

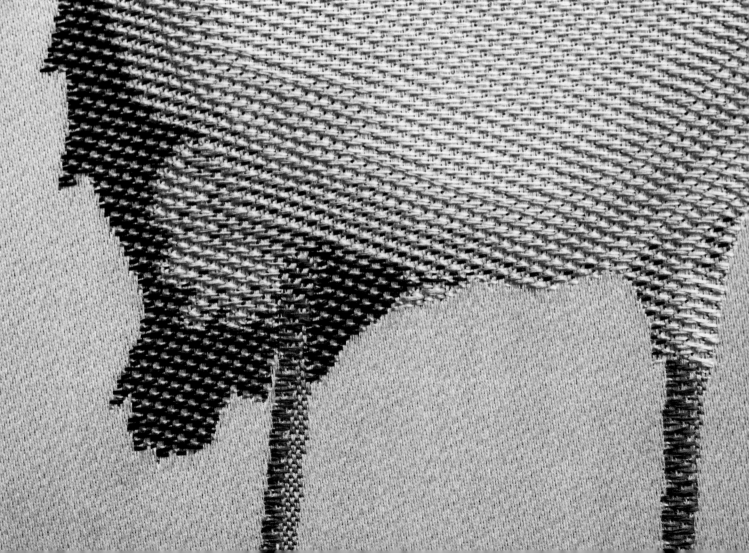

The Jacquard Weaving Process

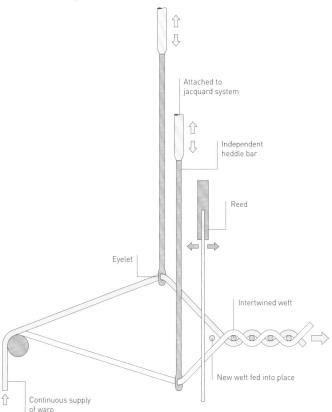

Attached to
jacquard system

⇑
⇓

Independent
heddle bar

⇑
⇓

Reed

Eyelet

← →

Intertwined weft

New weft fed into place

⇑ Continuous supply
of warp

TECHNICAL DESCRIPTION

The principal difference between jacquard and conventional loom weaving is that the heddles are controlled independently, as opposed to being mounted in groups on harnesses. Therefore, each heddle can be operated individually, or in groups, depending on the pattern required.

The position of each warp is controlled by CNC or manually inserted jacquard card (see Handloom Weaving, page 90). In each picking cycle, a series of warps is lifted and the weft is fed in between. Different colours of weft are used in combination with the constant supply of warp to produce a wide range of patterns, from simple check to highly figured designs. The highest proportion of yarn (whether warp or weft) on the surface determines the visible colour and thus creates the pattern.

control, making them the most versatile of all. Modern jacquard looms are CNC (computer numerical control) and capable of producing any type of weave.

Double-faced textiles are made with multiple warps, whereby two fabrics are woven and interlaced with the additional warp. Using fancy yarns and varying the weaving tension produces three-dimensional patterns. These techniques are all mass-producible. More complex techniques, such as brocade, are made by hand weaving (page 90).

APPLICATIONS

Fancy loom weaves are used mainly for apparel and interior textiles. Although essentially the same techniques are used for both, there are differences. For example, interior applications require yarns with very good colourfastness, because they will be long-lasting. By contrast, textiles used to make apparel must have very good water repellency. These two qualities cannot always be achieved in the same yarn with the same dyeing process (page 240).

Patterned and figured textiles are used in a wide range of applications, from silk ties to linen cloth. Luxury textiles, such as lampas, are used as wall hangings and for upholstery.

Mesh and netting are used as upholstery, sheer curtains, packaging and geotextiles. Leno weave is slit into thin widths to make chenille yarn.

RELATED TECHNOLOGIES

Like conventional loom weaving, fancy weaving techniques are versatile. However, they are limited to flat textiles. If they are to be made into three-dimensional items then panels are cut and joined together by stitching (page 354) or bonding (page 370). Knitting (page 126) overcomes this by producing seamless three-dimensional parts straight from the loom. This approach has been adopted in the automotive and furniture industries for covering foam upholstery, as well as knitted mesh upholstery.

Printing (pages 256–79) is a cost-effective alternative for reproducing patterns and graphics on fabric. It is

flexible and versatile, but the colour will be visible only on the printed side, leaving the other side dull. Warp printing (page 268) soaks dye right through the yarn, producing a mirror image front and back, but will be slightly fuzzier than conventional printing, because of being unpicked and rewoven.

QUALITY

Fancy fabrics combine elements of plain, twill and satin weaving, and so will share many of the same qualities. They are set apart by distinctive patterning, texture and structure. Density ranges from open mesh to heavy, double-faced cloth.

The yarn is dyed before weaving, so patterns are inherent in the structure of the fabric. Complex and intricate patterns are infinitely reproducible.

Floating glossy filament yarns, such as silk and polyester, will produce a bright and colourful surface that cannot be achieved in any other way. However, particularly long floats are prone to snagging, which limits some very decorative textiles to wall coverings.

Dobby Pattern
Material: Nylon
Application: Apparel
Notes: Small repeating patterns, fewer than 30 wrap arrangements, are made on a dobby (or jacquard).

Jacquard Pattern
Material: Silk
Application: Apparel
Notes: Complex figured designs are produced on a loom with a jacquard attachment. The floats produce a bright, lustrous colour.

Figured Pattern with Metal Coated Yarn
Material: Polyester
Application: Apparel
Notes: Novelty yarns, such as metal-coated polyester, are jacquard-woven to create dramatic patterned effects.

Relief
Material: Silk
Application: Interior
Notes: Long floats produced with the weft yarn rise up over the warp to create a relief pattern.

Front   Back

Honeycomb
Material: Wool
Application: Apparel
Notes: Floating yarns in a structured pattern produce a relief surface profile also known as honeycomb.

Lampas
Material: Silk
Application: Interior
Notes: The gold-coloured weft yarns are tied into a ground made up of extra warps, removing them from view.

Textured Yarns
Material: Nylon
Application: Exfoliating cloth
Notes: The weft is textured by knitting, heat-setting and de-knitting. The loops give a rippled surface to the cloth.

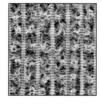

Scalloped Elastic Trim
Material: Elastane and nylon
Application: Lingerie
Notes: The elastic trim has a scalloped edge, woven to mimic fancy embroidery running along the edge.

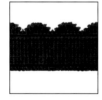

Chenille Yarn
Material: Silk
Application: Interior
Notes: Leno weave is cut into strips to make fancy yarn. As a woven yarn it adds texture and intrigue to fabrics.

Woven fabrics are dimensionally stable. This is especially important for technical meshes and absorbent materials.

DESIGN
Patterns and textures are created with floating yarns. They may be the same colour as the ground, in which case the pattern is made by changes in surface texture and light reflectance. For example, damask is made with one type of warp and one type of weft and the pattern is created with satin floats. Therefore, it will be reversible, because the back side is the opposite (negative) of the face.

Multicoloured figured designs are made with several different wefts. For maximum colour freedom in mechanical loom weaving, additional warp beams are used. The additional warps are used to construct a separate ground, which the long floats can be woven into. For example, lampas is a highly decorative textile made with up to four separate warp beams. Traditionally made from silk and precious metals (gold and silver), it is used to reproduce motifs and figured designs for wall coverings and upholstery. Designing these weaves can be a very lengthy process. This affects the lead-time, making it difficult to create bespoke solutions for fashion ranges. However, this is not usually such a problem for interior applications.

Known as a leno weave, mesh and netting is made with the addition of a 'doup' to a plain or dobby loom. Warps are worked in pairs and crossed between each pick (weft insertion) to lock the weft in place. This produces a stable fabric with little movement between warp and weft.

MATERIALS
Fancy weaves are more expensive than plain types and so are typically made from high-quality yarn, such as silk, flax, cotton and man-made filaments. Novelty plied yarns (page 63) include corkscrew, slub and bouclé. In addition, yarns may be produced from fabrics or plastic film cut into strips. Yarn is coated with a thin film of metal to produce a bright and reflective surface for highly decorative applications.

COSTS
Fancy weaves are moderate- to high-cost, depending on the process, complexity and yarns. Modern looms operate at speeds of 200 to 1,000 picks per minute. Faster cycle times typically reduce cost, but this depends on volume (less significant for lower volumes). Complex designs take longer to weave.

ENVIRONMENTAL IMPACTS
Environmental impacts are the same as for conventional loom weaving (page 76) and are determined by yarn selection, preparation, weaving process and finishing techniques.

→ Jacquard Weaving with Novelty Yarn

In the past, designs were produced by hand. Each warp and weft overlap was calculated and punched into a card. The holes in the card, which formed a continuous loop, determined whether a warp was raised or lowered. Prelle is an expert in silk weaving by hand (pages 94–95), as well as in modern computer-guided looms that take design directly from CAD data.

Using novelty yarn in the weft, such as this chenille (**image 1**), creates a textured surface on the woven fabric. Combined with jacquard, intricate patterns can be made, to create a cost-effective, visually intriguing and tactile finish (**image 2**).

As in plain weaving, the silk warp is supplied on a beam from the back of the loom (**image 3**). In jacquard weaving, each warp is fed through an eyelet in an independent heddle, which is controlled from above by a series of computer-guided electromagnets (solenoids).

The weft is fed in front of the rapier (**image 4**), which picks it up and transverses the width of fabric between the reed and warp yarns. In this case a double rapier is used, whereby the two long strips meet in the middle, passing the yarn between them (**image 5**). The reed presses the loose weft yarn into the cloth fell and its final position (**image 6**). The finished fabric is inspected to check for weaving defects as it is taken up on a beam (**image 7**).

1

2

3

4

5

6

7

Featured Company

Prelle
www.prelle.fr

→ Jacquard Weaving Figured Silk Textile

Damask is a reversible fabric made from silk, cotton, wool or flax. In this case, Prelle is weaving fine silk damask. Two colours are used: blue warp (**image 1**) and gold weft (**image 2**). The weft floats over the warp to produce a gold pattern on the face (**images 3** and **4**). Therefore, the back side is the reverse of the face: blue motifs on a gold ground (**image 5**). The rest of the fabric is plain weave.

Introducing an additional-colour weft, as in this figured satin (**image 6**), produces a single-sided textile. The heddle bars raise up selected black warp yarns, and the two colours of weft (cyan and gold) are alternated back and forth by the rapier. The raised yarns will interlock on top of the coloured weft, holding it in place and creating the next line of the figured design (**image 7**).

The warp determines the underlying colour and the weft provides the colours for the pattern (**image 8**). The warp is constant, whereas the colour of the weft (and even the type of fibre) can be changed throughout.

1

2

3

4

5

6

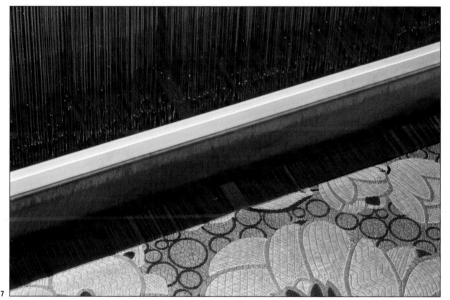

7

8

Featured Company

Prelle
www.prelle.fr

Textile Technology

Handloom Weaving

Handloom weaving is practised as a cottage industry, as well as making up a fundamental part of the factory system. There are many different techniques and numerous fabrics can be made, such as for apparel, upholstery and interiors. For particularly complex and fancy weaves, such as figured silk velvet, it may be the only practical means of production.

Techniques	Materials	Costs
• Traditional • Jacquard	• All types of staple and filament yarn, as well as novelty types	• Moderate to very high • Cycle time is slow for heavily figured designs

Applications	Quality	Related Technologies	
• Accessories • Apparel • Interiors	• Depends on the skill of the weaver • Natural variation	• Embroidery • Fancy loom weaving • Hand weaving lace	• Loom weaving

INTRODUCTION

Although mechanized looms (page 76) have mostly replaced handlooms, there is a still a great deal that can be achieved only with manual techniques. Hand weaving is used to make all types of fabric, from plain weave to intricately figured brocade.

There are two main categories of loom: traditional types that use harnesses to control groups of warp yarn, and those with a jacquard attachment to allow the weaver to control each warp independently. It is possible to weave virtually any pattern on even the simplest loom by tying groups of warps together and raising them by hand. The jacquard system allows for complex patterns to be repeated more quickly and accurately.

Manual processes are more versatile than mechanized looms. They allow designers and weavers to experiment with patterns and yarns before committing to an idea, while mechanized looms allow less room for error.

APPLICATIONS

Hand weaving is used to make prototypes and one-offs, as well as finished items.

Fabrics that are not practical to make on mechanized looms, such as figured velvet, brocade and tapestry, are still made by hand. Also, open and loose weaves with soft yarns, which may be too delicate to produce by machine, can be carefully woven by hand.

Handlooms continue to play a role in the Prelle factory in Lyon. They still use the original hand-weaving mill, built in 1880. There, weavers were able to recreate the gold and silver brocade in Louis XIV's bedchamber in Versailles. This particular example took 29 years of research and weaving at the rate of 30 mm (1.2 in.) per day. Another example is the relief peacock feather brocade of the curtains

Hand Weaving Techniques

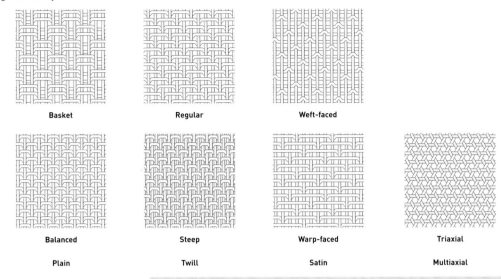

Basket   Regular   Weft-faced

Balanced   Steep   Warp-faced   Triaxial

Plain   Twill   Satin   Multiaxial

and wall hangings in Marie Antoinette's bedchamber in Versailles.

In such countries as Indonesia, Malaysia, India and China, hand weaving is maintained as a cottage industry. It is used to make clothing, interior textiles, upholstery, accessories and rugs. The value of these items ranges from the everyday to the ceremonial.

RELATED TECHNOLOGIES

Handloom weaving makes up a fundamental part in design development for mechanized-loom weaving. Reproducing fabrics on mechanized looms is much faster and highly repeatable. However, they are much less versatile. Therefore, in many cases handlooms have been adapted with power-assisted harnesses to combine the best of both.

Hand weaving is capable of producing highly figured designs, which may be similar to fancy loom weaving (page 84) or embroidery (page 298). Fancy loom weaving is used to produce patterned, textured and pile fabrics for apparel and interiors. It is much faster than hand weaving, but not capable of the full range of techniques used in hand weaving.

Leaving spaces between groups of warp or weft yarns, or both, creates open structures similar to lace (pages 106–19). Hand weaving is much quicker to work than handmade lace, although very complex weaves can take just as long.

Embroidery is an embellishment technique. Modern machines are very

TECHNICAL DESCRIPTION

Woven fabrics consist of warp yarns (running lengthwise in the direction of the loom) and weft yarns, also called filling yarns (running widthwise). Hand weaving is used to create a huge variety of weaves. Even so, the majority of fabrics are based on plain, twill or satin, or a combination of these.

Plain weave is the simplest of all. Each warp goes over and under the adjacent wefts, and vice versa, known as a 1/1 weave. Therefore, plain fabrics do not have a face or a back, unless they have been finished in some way. Balanced plain-weave fabrics are the most common type: ranging from ultra-light sheer (less than 34 g/m² or 1 oz/yd²) to heavyweight cloth (around 340 g/m² or 10 oz/yd²). Unbalanced fabrics have more yarns in one direction than the other.

Weaving groups of two or more warp with the same number of weft produces a basket weave. There are fewer interlacings, making these fabrics more pliable.

In twill weave, each warp or weft floats across two or more weft or warp yarns. With each pick (inserted weft), the interlacing is moved on one yarn, creating a distinctive diagonal pattern and a face and a back. The diagram illustrates 2/1 warp-faced twill. The angle, from reclining to steep, is determined by the density of the weave and balance

of the fabric, whereby regular twill is balanced. Even-sided twills, or 2/2, such as houndstooth, can be reversed, but the diagonal pattern will be the opposite way around on the back.

Twill weaves are pliable and durable, although a twill weave with the same count as a plain weave will not be as strong, because it has fewer overlaps between the yarns.

Satin (and sateen) weaves have at least four floats for every overlap, or 4/1. They are either warp- or weft-faced. The yarns that come to the surface are called 'raisers' and those underneath are 'sinkers'. Few overlaps means the yarns can be packed very tightly, producing a lustrous surface. It will snag easily owing to the long floats. Satin is typically made with filament yarn, such as silk, whereas sateen uses spun yarns (cotton or flax).

Triaxial weave is made with two warps and one weft. The yarns are interlaced at 60° to create a fabric with greater dimensional stability. It is light and strong and typically uses only half the number of yarns as a plain weave. This gives it a lower thread count and makes it quicker to produce.

Plain Coloured Weave
Material: Silk and cotton
Application: Apparel
Notes: Plain weaves are the simplest to make. The front and back of balanced weaves look the same.

Satin Floats
Material: Silk and cotton
Application: Apparel
Notes: Floats bring colour to the front. The weft is continuous and so is visible throughout the width of fabric.

Figured Fabric
Material: Silk and cotton
Application: Interior
Notes: Complex patterns are formed on handlooms by controlling each warp independently.

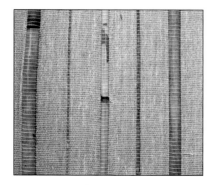

Bamboo
Material: Flax and bamboo
Application: Interior
Notes: Lengths of bamboo are split and incorporated as the weft. The linen is unbleached and naturally grey.

Grass
Material: Dried grass stems and cotton
Application: Interior
Notes: Grass stems dye to produce irregular colour. Here, the lengths are held together with cotton warp.

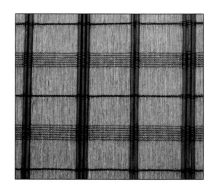

Wood
Material: Linen, cotton and wood
Application: Interior
Notes: Lengths of wood and linen weft are interwoven with a cotton warp to produce a gridded pattern.

fast and versatile, working with a range of threads and ribbons to reproduce designs. Threads are stitched over the top of existing fabrics, in all directions. By contrast, hand weaving incorporates patterns and textures into the structure of the fabric itself.

QUALITY
The quality of hand weaving is largely dependent on the skill of the weaver. Of course, being made by hand means these fabrics are more likely to have imperfections than mechanized loom-woven fabrics. But it is these qualities that set handmade fabrics apart. Even so, very high-quality and expensive

handmade fabrics made by expert craftspeople, such as wall coverings, will be visually perfect.

The materials will affect the quality of fabric. While certain yarns are easier to weave, others add bulk or texture to the surface. A balance must be achieved between feasibility and desirability.

DESIGN
The type of loom determines the size of fabric that can be woven. For example, looms with hand shuttles are only practical for making fabrics as wide as the weaver can stretch his or her arms. Flying shuttles make it possible to weave wider fabrics, and more quickly.

With every pick the fabric design and texture can change. Simple yarns, including staple (page 56), man-made (see Filament Spinning, page 50), metallized and metal, are commonly used. Introducing different dyeing techniques, such as mottled patterns produced by ikat (page 250), and novelty yarns, allows designers to create effects with colour, texture and shape.

The warp is fixed throughout, but the weft may be changed with each pick. The choice of yarn will greatly affect the appearance of the finished fabric. For example, using the same weave pattern, Elaine Ng Yan Ling experimented with coloured and

Diamond Pattern
Material: Nylon and viscose
Application: Headwear
Notes: A traditional Chinese diamond pattern from Guizhou requires the warp be divided into four groups. All six of the samples on this page are the same weave pattern, but with different yarns.

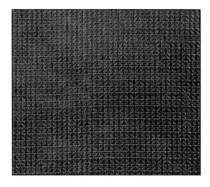

Metallized Yarn
Material: Nylon, metallized yarn and copper yarn
Application: Headwear
Notes: The viscose weft has been replaced with bright metallized yarn plus one copper yarn every fourth pick.

Copper Yarn
Material: Nylon, metallized yarn and copper yarn
Application: Headwear
Notes: The copper weft (left and right), enamelled with colour, shimmers in the light making the pattern less visible.

Alternating Yarns
Material: Nylon, metallized yarn and copper yarn
Application: Headwear
Notes: By alternating the type and colour of weft with each pick the pattern becomes very subtle.

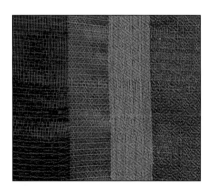

Bold Stripes
Material: Nylon, metallized yarn and copper yarn
Application: Headwear
Notes: Grouping the weft together highlights the different appearance of each yarn with the same weave.

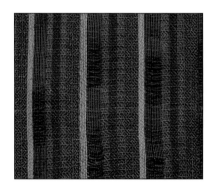

Narrow Stripes
Material: Nylon, metallized yarn and copper yarn
Application: Headwear
Notes: Thinner stripes, relative to the size of the weave pattern, reduce the visibility of the repeat.

metallic weft to produce a series of beautiful fabrics with dramatically contrasting appearance (see main image on previous page, and above).

Hand weaving creates the opportunity to explore three-dimensional fabrics with clever yarn combinations – for example by combining different sizes of yarn, bulky novelty yarn, yarns with conflicting (unbalanced) twist, or yarns that react to washing or heating at different rates (shrink or expand). The texture and shape of the textile can be tailored according to the difference between the yarns and relative tension in the weave.

MATERIALS
The widest range of materials can be woven by hand. While the warp should be strong enough to maintain tension in the fabric, the weft can be any material. Braid (page 68), strips, straw and even bark may be interwoven in the weft.

COSTS
Hand weaving takes longer than weaving with a machine. Therefore equivalent materials will tend to be more expensive. The source will also affect cost: fabrics from economically developing countries will likely be much less expensive than equivalent materials made in economically more developed countries.

Hand weaving is used to produce some of the most valuable fabrics available.

ENVIRONMENTAL IMPACTS
These hand techniques produce very little waste and there is virtually no machine energy required. As the items are made with a great deal of care and can take many months or even years to complete, the pieces are cherished and long-lasting.

→ Weaving Figured Velvet

Jacquard weaving was first demonstrated in the 19th century and has since revolutionized the way figured and fancy fabrics are made. Rather than tying groups of warps together and controlling them by hand with harnesses, lengths of punched card pass through the jacquard head and control whether individual warps are raised or not. This makes very complex patterns highly repeatable and quicker to make. Exactly the same technique is used to make modern figured fabrics. The only difference is that modern textiles are made on electric-powered computer-guided looms.

Complex figured velvet continues to be made by hand at Prelle, assisted by the jacquard process. With this technique, loop and cut pile are combined with a plain, twill or satin woven ground (**image 1**).

This design for an interior application consists of 7,900 warp ends (**image 2**).

The silk warp that will make the woven ground is supplied on two beams (46.7 dtex). The bobbins hold the silk yarns that make up the pile (140 dtex). A single silk weft is used on a shuttle (35.7 dtex).

The yarns pass through eyelets in the heddles (lengths of wire or cord) (**image 3**), which are controlled by the jacquard head above. With each pick, holes in punched cards control whether the heddle bar is raised (**image 4**). The stack of sheets of punched card makes one repeat in the pattern, which is 1.5 m (4.9 ft) and takes approximately one week to weave. In the end, this piece of fabric will be 20 m (66 ft) long and take six months to complete, which includes dressing the loom.

Inserting a wire in the shed between the warps forms a loop (**image 5**). This is either left as a loop, or cut to expose the ends (**image 6**). The finished velvet is carefully brushed to raise the nap and taken up on a cloth beam (**image 7**).

1

2

3

4

5

6

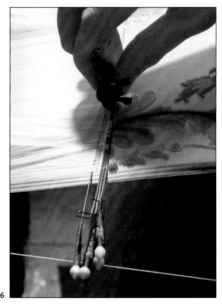

7

WEAVE

94

Featured Company

Prelle
www.prelle.fr

→ Weaving 17-Colour Brocade

Brocade, also called broché, is distinguished by complex figured patterns made with discontinuous weft yarns (**image 1**). In this example, the floating wefts are woven onto a plain background.

The warp is fed in from a single beam (**image 2**). This loom, which dates back to 1820, is equipped with a jacquard head. Each warp is passed through an independent, weighted heddle bar, controlled by punched cards (**image 3**). With each row of punched holes, a select group of warps is raised and coloured weft is passed through the shed with a shuttle (**image 4**). As each colour section is finished it is tied off. The pattern is slowly built up in this way from the back side (**image 5**). It takes around one day to weave 35 cm (14 in.) of this design. Less colourful and complex designs will be quicker to produce. The wefts are purely decorative, making a satin pattern on the face of the fabric (**image 6**).

1

2

3

4

5

6

Featured Company

Prelle
www.prelle.fr

Textile Technology

Pile Weaving

Pile created from additional warp or weft in woven fabric is high quality and used to make luxurious velvet, velveteen, corduroy and carpet. The pile is integral to the structure of the weave and can be loop, cut, coloured and patterned. The principal techniques are weft pile, slack tension, double-cloth and over-wire.

Techniques		Materials	Costs
• Weft pile: cut floats • Warp pile: double-cloth, slack tension, over-wire	• Axminster	• Silk, cotton, wool, polypropylene (PP) and nylon pile yarn	• High to very high depending on the complexity and materials

Applications	Quality	Related Technologies
• Apparel • Interiors • Carpet	• Cut or loop • Density depends on technique and yarn count	• Flocking • Napping • Tufting

INTRODUCTION

Weft pile includes fabrics such as corduroy and velveteen. The pile is created by cutting floats during or after weaving. Warp pile includes velvet, produced from double-cloth; terry, made with slack-tension warp; and Wilton carpet, which is woven over wire. Axminster is slightly different from warp- and weft-pile construction: tufts are created with additional yarns inserted into the ground fabric during weaving. It is solely used to weave carpet.

These techniques are all mass-producible to varying degrees. More complex combinations, such as densely woven figured velvet, continue to be made by hand (page 90).

In combination with jacquard or computer-guided looms, complex and colourful patterns are reproduced. In addition, using the over-wire technique a mix of cut and loop pile may be produced in a single piece and combined with different colours.

APPLICATIONS

Carpet is the largest area of application and includes Axminster, Wilton and Brussels. Velvet is used to decorate interiors too, but is limited to applications that will not cause abrasion to the delicate surface.

Woven pile is used to produce apparel fabrics, such as velveteen and corduroy. It is also possible to make fringe and fake fur, although these two are more commonly produced by other techniques.

Terrycloth is used to make towels, robes and upholstery. Low-twist yarn produces the most absorbent material. Being made from staple yarn (page 56) it is prone to shedding fibres. Therefore, it is not typically used for glass cleaning and similar applications where stray fibres could cause problems.

The Pile Weaving Process

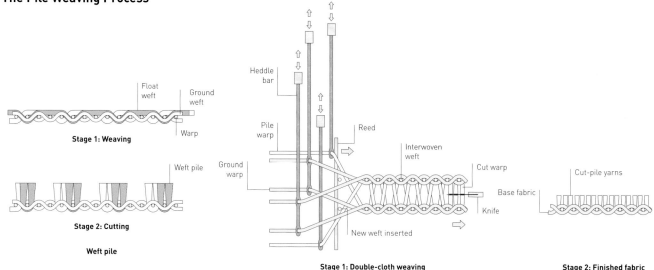

Stage 1: Weaving

Stage 2: Cutting

Weft pile

Stage 1: Double-cloth weaving

Stage 2: Finished fabric

Double-cloth

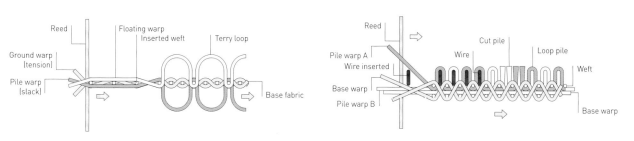

Slack tension

Over-wire

TECHNICAL DESCRIPTION

Weft-pile fabrics are constructed with long floats. In stage 1, the location and length of float determines the extent of the cut pile. In stage 2, the floats are cut and the pile is brushed to open up the yarn ends, or sheared to a uniform length.

Corduroy has distinctive wales (ridges) that run the length of the fabric (warp direction), and velveteen has uniformly distributed binding points. Velveteen is quite hard-wearing, and has a similar appearance to velvet, but the pile lies flatter and it is generally made from cotton.

Warp pile is constructed as double-cloth, slack tension or over-wire. Double-cloth warp pile is used to make dense uniform-cut pile, including velvet and fur-like fabrics. In this process, two even fabrics are woven, joined together by an interlacing pile yarn (warp). As the fabric is constructed a razor

cuts the pile, tracking back and forth with each pick. Thus the length of pile will be half the distance between the two layers of base fabric.

Slack-tension weaving is used to produce terry (as well as seersucker). The loops can be on one side or both sides, depending on the warp arrangement. The base fabric is woven with two sets of warps. With each pick the reed beats the weft a set distance and does not compress it to the edge of the fabric until a predetermined number of insertions has been made (three in this case). Once all the wefts are in place, interwoven with the warps, they are compressed together. The pile warps are slack and so form into loops. The base warps are under tension and so form into a tightly woven ground. The weaving sequence and slackness of pile warp

determines the length, density and pattern of loops. The over-wire method is used in hand weaving (page 90) as well as mechanized looms (page 76). At least two additional warps are required: the total depends on the number of different colours. As the base fabric is woven from the ground warp and wefts, heddle bars raise the selected pile warps above the others. As this occurs a length of wire is inserted, so that when the heddle bars are lowered, the wire is encapsulated in the weave to form a loop of yarn. Thus the depth of wire determines the length of pile. A weft to lock the loop in place follows and the process is repeated. The wire is left in for several rows to ensure the fabric is stable. Then it is withdrawn automatically and if equipped with a knife at the far end it will slice through the loops to form cut pile.

Pile   Back side

Corduroy
Material: Cotton
Application: Apparel
Notes: Corduroy has distinctive linear wales as a result
of cutting the weft floats to make pile.

Pile   Back side

Printed Fringe
Material: Polyester
Application: Apparel
Notes: The weft is woven with long floats, which are cut
to produce a fringed surface to the fabric.

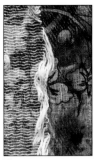 

Pile   Back side

Printed Base with Plain Fringe
Material: Wool and flax
Application: Apparel
Notes: Acid dye penetrates the wool base, but does not
affect the flax floats. After printing they are cut.

Velveteen
Material: Cotton
Application: Apparel
Notes: Produced in the same way as corduroy, except
the floats are not arranged in a linear pattern.

Terry
Material: Cotton
Application: Towel
Notes: Loops are formed by an extra warp, which is
slacked between picks and held in place by the weft.

Double-cloth weaving, used to make velvet, is also utilized in technical textiles. Instead of cutting the interlacing yarns to make pile, the double-layered fabric is kept intact. The three-dimensional structure improves strength to weight in, for example, composite laminate constructions.

RELATED TECHNOLOGIES

Pile fabrics are used in a wide range of applications, with diverse requirements. To meet the different needs, they are therefore constructed and finished using various techniques.

A soft nap is formed on the surface of woven and knitted fabrics by raising the fibres with brushes or teasels (see Napping, page 216). It is more cost-effective than woven pile and as a finishing process it is applied to a wide range of fabrics, for example in the production of velour and fake fur. A surprisingly life-like finish is produced by napping knitted fabric made up of a blend of yarns.

Terry is also commonly produced with knitting techniques. It is used to make cushioning, insulating and absorbing fabrics for apparel and interiors.

Flocking (page 280) is a low-cost technique used to create pile surfaces on a wide range of materials. The surface is sensitive to abrasion, because the fibres are held to the base material only by their ends. Therefore, it is mainly used for decorative purposes.

In the carpet industry, tufting (page 286) is by far the most economical method for producing pile. Its cost-effectiveness led to the adoption of wall-to-wall carpets in northern Europe and North America during the mid- to late 20th century. Axminster and Wilton

weaving techniques produce higher-density and higher-quality pile, while jacquard and computer-guided looms open up a range of design opportunities not so practical with tufting.

Smooth carpet, produced without pile, may be fabricated using several other techniques, such as nonwoven (page 152), plain weaving (page 76), fancy weaving (page 84), hand weaving (page 90) and knitting (pages 126–51).

QUALITY

Pile is both functional and decorative. Soft and absorbent materials are made with low-twist loop pile, whereas deep cut pile is more comfortable and insulating. Silk produces a decorative pile with rich colour and a soft hand unlike any other material. However, care must be taken to ensure the pile is not flattened unintentionally, because this would affect its lustrous surface appearance.

The type of yarn and the weaving technique will determine the density and height of pile that can be achieved. For example, the finest velvet made from silk has up to 118 warp ends per centimetre (300 per inch). By contrast, synthetic pile may have only 31 warps per centimetre (80 per inch), which makes it feel quite different; but it is much faster to weave and so less expensive.

Loop is more durable than cut pile, although terry is prone to snagging. Longevity is also affected by the choice of material. For example, high-traffic

Wilton
Material: Wool pile with jute and cotton ground
Application: Carpet
Notes: The cross-section illustrates how the continuous warp pile interweaves the sturdy jute weft.

Wilton Cut Pile
Material: Wool
Application: Carpet
Notes: A dense cut pile made up of 4.7 rows per 10 mm (12 per 1 in.). Colours are combined using jacquard.

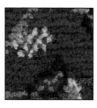

Wilton Loop Pile
Material: Wool
Application: Carpet
Notes: Loop pile, also known as Brussels, forms a compact and hard-wearing surface.

Wilton Cut and Loop Pile
Material: Wool
Application: Carpet
Notes: Cut pile is woven over wires with a blade. The number of warps over the wire depends on the design.

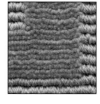

Wilton Cut and Loop with Colour Pattern
Material: Wool
Application: Carpet
Notes: Jacquard patterns are combined with a mixture of cut and loop pile exploiting the full variation.

Wilton Back Side
Material: Jute and cotton/polyester blend
Application: Carpet
Notes: The back side of Wilton shows sturdy jute interwoven with cotton/polyester base warp.

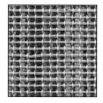

Axminster
Material: Wool/nylon blend with jute ground
Application: Carpet
Notes: Each tuft is anchored to a weft and held in place by the sturdy woven ground.

Axminster Pile
Material: Wool/nylon blend
Application: Carpet
Notes: Axminster is only ever cut pile. It is typically 100% wool or blended with synthetic for added durability.

Axminster Jacquard
Material: Wool/nylon blend
Application: Carpet
Notes: Complex coloured patterns are reproduced using jacquard or with computer controlled looms.

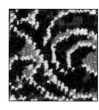

Axminster Pattern
Material: Wool/nylon blend
Application: Carpet
Notes: Patterns may be linear or curved. Jacquard is used to imitate the appearance of twill in the pile.

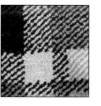

Axminster Back Side
Material: Jute and cotton
Application: Carpet
Notes: Like Wilton, Axminster is made up of sturdy jute ground, which will be coated with latex to finish the carpet.

PILE WEAVING

applications require a hardy and resilient pile that will retain its appearance even after heavy wear. By contrast, decorative applications are concerned more with colour and finish.

DESIGN
Design options depend largely on process selection. There are many variations possible, including the type of pile, density, height, colour and pattern.

Weft-yarn pile is formed from floats and so is always cut. Therefore there is a tradeoff between length of pile and density: longer pile means that more of the base fabric will be exposed below. This is not the case for warp-pile techniques.

Weft-yarn pile is woven in a uniform colour. Surface effects are created by varying the shape of pile, or by printing (see Digital Printing, page 276) the surface before or after cutting the floats.

Of the warp-yarn pile techniques, the wire method is perhaps the most versatile. Up to seven colours may be combined and with jacquard or computer-guided looms a range of patterns may be produced, including a mix of cut and loop pile.

Woven terry produced by slackening one or more warps appears much the same as plain-woven fabric, except that a pile may be formed on one or both sides. Colour is integrated into the pile yarn and so runs lengthways, or it may be mixed between the warp, weft and ground. Changing the colour of the warp or weft creates stripes. The height of the loops is controlled by the amount of extra yarn let out by the pile warp beam. Localizing the extra warps creates relief patterns.

The widest range of colour options is possible with Axminster. Up to 12 colours can be used. This presents a challenge

when setting up: arranging the many thousand ends by hand can take days. With modern CNC (computer numerical control) looms the changeover time has been dramatically reduced, thus making this process more versatile. This makes it possible to produce one-off and bespoke carpets, for example to fill a hotel lobby.

Finishing techniques are used to enhance the pile. For example, steaming wool and cotton after cutting produces a much softer and denser-feeling pile. Alternatively, flattening the pile will affect the look and feel. Crushed velvet takes advantage of this, produced by twisting the fabric when wet.

All types of cut pile may be printed, depending on the material. Applying a vacuum improves dye penetration. Loop pile may also be printed, but the results are likely to be less satisfactory owing to the uneven surface.

→ Double-Cloth Weaving Silk Velvet

Traditionally, the highest-quality velvet was pure silk. Many variations exist today, including weaving silk pile onto synthetic ground, or synthetic throughout. Therefore, loom manufacturers have stopped producing looms capable of making full-density silk velvet. Prelle found that the only way to maintain production was to recondition the jacquard looms originally built at the beginning of the 20th century. These looms have enabled Prelle to continue to manufacture full-density silk velvet to the highest standards.

Warp-pile velvet is made with three warp beams (**image 1**). Two warps are used to construct the ground (38 dtex) and one to make the pile (77 dtex). The warps are fed into the loom under tension (**image 2**). It is a traditional loom and produces velvet 65 cm (25.6 in.) wide with a dense pile 1.7 mm (0.07 in.) high.

The velvet is woven as a double-cloth, joined by the pile warp (**image 3**). With each pick, a weft is inserted into the top and bottom sheds. Then, the two sides are cut apart by slicing the interwoven pile with a sharp blade, which passes back and forth in a wooden case next to the operator (**images 4** and **5**). Cutting the warp in this way produces an even pile that does not require tip shearing.

The finished velvet is taken up on two rolls (**image 6**). The top and bottom rolls will have slightly different visual characteristics, because the pile is at a slight angle. Putting them side by side will highlight this.

Mechanical pile is produced in solid colours (**image 7**). Of course, warp and weft can be a different colour, and this affects the visual colour of the finished pile. Or the finished velvet may be printed (see Digital Printing, page 276). The only way to produce inherently figured dense silk velvet is by hand weaving (page 90).

1

2

MATERIALS

Woven pile is fabricated from filament and staple yarns, including natural and man-made types. Choice of material is governed by the application. For example, fabrics used in high-wear applications, such as commercial carpet and upholstery, demand the most resilient and long-lasting (and cost-effective) materials, which include wool, nylon and polypropylene (PP).

Silk is used for its luxurious colour and lustre. The appearance is affected by the direction of the pile, which can be advantageous or otherwise. It must therefore be handled carefully to avoid flattening the fibres.

COSTS

Pile weaving is typically more expensive than fancy types, although there is considerable crossover, depending on the complexity of the design and rate of production. The price also depends on the raw material cost. For example, silk will be around four times more expensive than synthetic pile, but the difference in quality is apparent.

The cost of pile carpet depends on the density of rows per cm/in. and the length of pile. Wilton is typically used to produce the highest-quality carpet and five rows per cm (12 rows per in.) is considered dense. Axminster is capable of producing carpets with a higher density of rows,

but the yarn is cut to length and does not run throughout the construction. Material selection is largely governed by the application. For example, 100% wool is used only for expensive domestic situations, otherwise it is blended with a man-made fibre to reduce cost.

ENVIRONMENTAL IMPACTS

Woven pile is much heavier than conventional fabrics, because more material is used for the same area. Otherwise, the environmental impacts are the same as for other weaving techniques and determined by yarn selection, preparation, weaving process and finishing techniques.

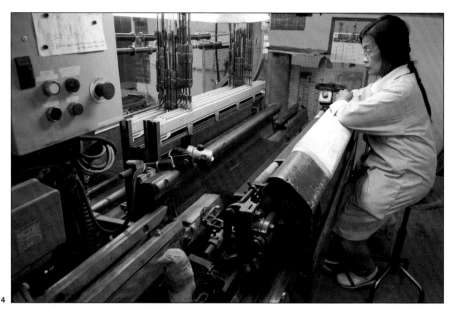

Featured Company

Prelle
www.prelle.fr

→ Over-Wire Weaving Wilton Carpet

Wilton is considered the finest carpet construction. The pile may be loop or cut and a wide range of designs is achievable with jacquard weaving (the same as fancy weaving, page 84).

This technique requires several sets of warp, which may be supplied on beams, or directly from bobbins held on a creel (**image 1**). Wilton carpets may be up to 4.5 m (15 ft) wide and require many thousand yarns.

Patterns are determined by the colour of warp at the surface. Each end is fed through a jacquard-controlled heddle (**images 2** and **3**), which raises or lowers the yarns accordingly. In this way virtually any pattern can be reproduced.

The blue-and-white repeating pattern is woven with wool-loop pile. With each pick a metal strip (wire) is inserted through the warps (**image 4**). Those above the metal will become loops. A length of weft is fed into the shed and beaten into place by the reed, securing the loop and metal strip (**image 5**).

Once several more rows have been formed the metal strip is removed (**image 6**). On mechanical looms this is a rapid process. If cut pile is required then metal strips with a small blade on the far end are used, so that as they are withdrawn they cut through the yarns. The woven carpet is loosely folded in preparation for finishing (**image 7**).

1

2

3

4

5

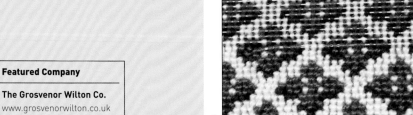

6

7

Featured Company

The Grosvenor Wilton Co.
www.grosvenorwilton.co.uk

→ Wilton Design, Colour and Finishing

Over-wire weaving design uses jacquard principles with the addition of choosing cut or loop pile. The design is finalized, or chosen from an existing version (**image 1**), and the appropriate colours are selected from swatches of dyed yarn (**image 2**). Package dyeing in skein form produces the highest-quality colour and ensures penetration right through the fibres so that when they are cut the colour is consistent.

Multiple-colour patterns require more complex set-up of the creel (**image 3**). Once set up though, the process is relatively speedy. For example, this repeating pattern in loop pile, made up of seven colours, runs at around 2 m (6.6 ft) per hour (**image 4**).

Once weaving is complete the carpet is steamed, sheared smooth and carefully checked for errors. Minor faults are mended by hand, mimicking the weaving process with needle and yarn (**images 5**, **6** and **7**).

Finally, the carpet is brushed and steamed to burst open the pile. A latex coating is applied to the back side and it is heated on a drum to fully cure (**image 8**). This ensures a hardy material that can be handled, installed and maintained without causing damage to the structure.

1

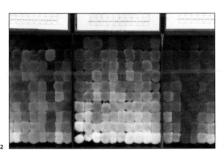

2

3

4

5

6

7

8

Featured Company

The Grosvenor Wilton Co.
www.grosvenorwilton.co.uk

TECHNICAL DESCRIPTION

Axminster is solely used to weave dense pile carpet. In operation, the gripper picks the selected yarn tuft from a jacquard-controlled carrier, whereby it is cut to length. The gripper rotates downwards and places the tuft against the breast comb. Two wefts are carried across the lowered warps by a rigid needle. The weft yarns are beaten into place by the reed to secure the tuft. The gripper rises up and releases the tuft, forming it into a U-shape, in preparation for picking a new one.

Colour is determined by the placement in a carrier, from which the gripper picks the yarn. Therefore, with computer-controlled holders, bespoke and one-off designs are possible. This opens up a wide range of opportunities for design.

The Axminster Process

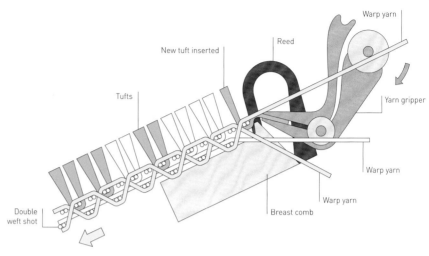

Axminster

Case Study

→ Axminster Carpet Production

Carpets are mocked up using what is known as a kibby (**image 1**) to allow final checking of the design prior to full production. Tufts are punched through small holes in a metal plate to simulate its woven appearance.

The design is transferred onto jacquard cards, which are loaded onto the loom (**image 2**). Each hole in the card represents a tuft, which comprises two ends of the same yarn. This technique is used to produce one-offs as well as high volumes of carpet. Modern CNC looms have removed the need for cards altogether and so reduce set-up time.

Bobbins of pile yarn are prepared and loaded onto the creel (**image 3**). A single Axminster loom may require up to 9,000 bobbins, depending on the width. Each of the yarn ends is fed into the carrier (**image 4**). The carrier holds the yarn end in readiness for the gripper to take the required length when the computer, or jacquard, selects it (**image 5**).

Grippers rotate up to the carrier and the correct colour is aligned. Once it has the yarn securely fastened a specific length is drawn out and cut (**image 6**). The gripper transfers the length of yarn in-line with the warp (**image 7**) and the double weft is passed behind using a rigid needle (**image 8**). The weft is beaten into place by the reed, securing the tuft.

The weaving action of the base fabric is constant. The colour is formed by the tufts of yarn inserted from the creel sitting behind the loom (**image 9**).

A complex set of finishing operations is required to produce the highest-quality Axminster. First of all, the pile is brushed and the tips are sheared to produce a smooth finish. It is then checked for faults and any minor corrections are made by hand. Afterwards, the carpet is brushed again and then steamed (**image 10**) to

open up the pile. The back side is coated with latex to secure the tufts and increase rigidity. Finally, the pile is tip sheared to a level finish.

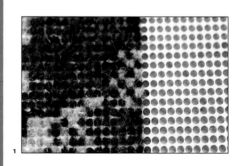

1

2

3

4

5

6

7

8

9

10

Featured Company

Calderdale Carpets
www.calderdalecarpets.com

Textile Technology
Machine Weaving Lace

Lace and net are openwork textiles created either by hand or by machine. Modern lace is manufactured on modified weaving looms or knitting machines. Complex patterns and figures are reproduced in the structure of the textile itself, unlike embroidery, which is applied to a base material as an additional layer of embellishment.

Techniques	Materials	Costs
• Weaving (bobbinet, Nottingham and Leavers)	• Natural fibres, such as cotton and flax • Man-made fibres, such as polyester and nylon	• Machine-woven lace is high cost • Material selection also affects cost

Applications	Quality	Related Technologies
• Bedspreads, tablecloths and placemats • Net, curtain and sheer • Lingerie and hosiery	• High quality • Flat to relief patterns • Dense to open structure	• Embroidery • Knitting • Hand weaving lace • Schiffli

INTRODUCTION

Lace is a highly valued material used to make both decorative and technical textiles. Lace was traditionally made by hand (page 112). The principal techniques, including bobbin, needle, tatting and crochet, are still practised by many enthusiasts, but not commonly used for commercial applications, because they are time-consuming processes and so expensive to make in sufficient quantity.

Modern machine-made lace is produced by weaving or knitting (see Weft Knitting, page 126, and Warp Knitting, page 144). Machine-woven lace – also known as bobbinet tulle, Nottingham lace or Leavers lace,

The Machine Weaving Process

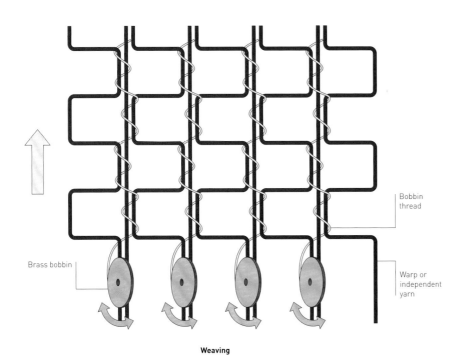

Brass bobbin

Bobbin thread

Warp or independent yarn

Weaving

TECHNICAL DESCRIPTION
Net is formed on a loom equipped with bobbins. They wrap around the warps, tying them together as the weave progresses. With the jacquard independent heddle system, it is possible to make decorative lace with figured patterns in the net. This system creates lace of the closest quality to handmade achievable with a mechanized loom. There are essentially two techniques: warp-based and independent yarns. They are more or less the same, except that the warp system can be used to make more intricate nets and patterns and is therefore better suited to floral and non-geometric designs. Independent-yarn looms may have more than 100 beams, whereas warp looms will be equipped with two warps and only around 50 beams.

Both techniques use the same system. The yarns are moved back and forth by independent heddles, which are operated with the jacquard system. As the weave progresses upwards a bobbin is passed around the group of yarns, binding them together. The weave is beaten up with a reed and the cycle is repeated. The heddles can move several rows with each stitch and so a wide range of patterns and effects can be reproduced. As the yarns move vertically upwards, the pattern can only continue horizontally or downwards. Therefore, more than one yarn is required to complete a circular design, for example.

Like conventional loom weaving, the warp and yarn are supplied to the loom on beams. The bobbins are prepared with up to 120 m (394 ft) of thread. When they run out, they are replaced and the process can continue.

after John Leaver, who invented the machine – is expensive and made only by a few specialist manufacturers. It has superior quality and handle, and, as with handmade lace, very complex patterns can be reproduced. Using the jacquard technique, many hundred warps, or independent yarns (which can be different sizes and colours), are individually moved back and forth, and bound together with a fine bobbin thread as the weave progresses vertically.

APPLICATIONS
Decorative applications include interior textiles, such as bedspreads, curtains, tablecloths and placemats. Machine weaving is used to make garments for special occasions, such as wedding dresses and ceremonial gowns. Fine lingerie and hosiery is made entirely from lace, or may have a lace trim.

RELATED TECHNOLOGIES
Embroidery (page 298) is a technique used to embellish fabrics and sheet materials. It is carried out by hand or machine. It is possible to embroider and

cut out simultaneously, such as with eyelet embroidery, to create a similar appearance to lace, but the structure of the material will be quite different and the mechanical properties weakened because of cutting. Alternatively, a sacrificial fabric (for example, a net constructed from water-soluble yarn) is embroidered on, and dissolved away afterwards to leave behind an interlaced and openwork structure. However, there are still some environmentally harmful chemicals used with certain techniques and it is important to take this into consideration. This type of fabric is also referred to as Schiffli lace, after the machine it is made on.

Machine-knitted lace is typically less expensive and so more widely used. An openwork structure is formed with computer-controlled weft- or warp-knitting machines. In weft knitting, loops are transferred to create an open structure. Warp-knitted net and lace is created with one or more yarns per needle, interlocked with each stitch. Tricot is a general term used for warp-knitted textile, named after the knitting

→ Machine Weaving White Cotton Leavers Lace

Lacemaking on the loom starts with preparing the bobbins (**image 1**): 120 m (394 ft) of nylon thread is wound on the brass bobbin (**image 2**). The nylon filament has been pre-stretched to reduce the diameter and increase tensile strength.

The bobbins are mounted onto carriers (**image 3**) and then onto the loom (**image 4**). They are packed tightly and a single loom may hold more than one thousand, depending on the length of beam. A bobbin is required for each yarn.

In this case, independent yarns are combined into lace. Each one is wound onto a beam, which is mounted underneath the loom (**image 5**). The yarns are fed into the loom and passed through the jacquard-operated heddles (**image 6**). The lace is made up of white cotton yarn, with a medium-weight net ground and heavier outline (gimp). The yellow threads are polyester, which are used to join sections of trim together as they are woven. This can be removed later on.

At Cluny, based in Derbyshire, England, the card jacquard system is used, which was in operation before CNC machines were developed. As the weave progresses, the cards are circulated. Where there is a hole punched in the card, the pin drops in and so the heddle bar is retracted. The pattern and number of holes define the pattern that will be reproduced (**image 7**).

As the yarns are moved back and forth, they are bound together with each stroke of the loom, forming the pattern (**image 8**). Strips are made utilizing the full width of the loom. They are held together with polyester thread, which is unravelled to reveal the finished trim once the full length has been completed (**image 9**).

1

2

3

4

5

6

7

8

9

Featured Company

Cluny Lace Co.
www.clunylace.com

Independent Yarns
Material: Cotton and nylon
Application: Apparel
Notes: Independent-yarn lace weaving is best suited to geometric repeating patterns.

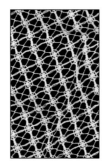

Warps
Material: Cotton and nylon
Application: Apparel
Notes: Warp-based weaving is suitable for making complex patterns and figures.

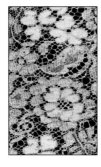

Edging
Material: Cotton and nylon
Application: Interior
Notes: The edges may be finished with loops (pearls or picos), twists (pictured) or a straight line. Sacrificial yarn is woven along the edge and removed.

Dyeing
Material: Cotton and nylon
Application: Apparel
Notes: Cotton lace is typically woven in white or off-white and piece-dyed.

machine. Net is constructed by knitting half the number of threads on alternate needles. Known as raschel, seamless tubular structures are warp knitted with an openwork pattern. And by adding jacquard (independent guide bars), complex and intricate patterns are feasible. These techniques are used to make one-piece sportswear and undergarments with a net structure: eliminating seams means they are more comfortable and lightweight.

Lace is imitated with a range of finishing techniques such as printing (pages 256–79), embossing (page 220), chemical etching and laser cutting (page 328). Colour and relief patterns are reproduced on knitted, woven and nonwoven materials. Printing does not affect the mechanical properties of materials, whereas etching or cutting into the surface will reduce the strength of the material.

QUALITY
Machine-woven lace is made up of continuous lengths of warp yarns or independent yarns. The openwork structure is held in place by looping or binding the yarns together at intervals,

making this a robust construction. Thanks to the complex nature of lacemaking, and the range of interdependent factors in production, the finished quality is largely dependent on the skill of the machine operator, known as a twist hand.

DESIGN
Only a few designers remain with the skills required to create new patterns for machine-woven lace. Today, most woven lace is produced from design archives dating back 100 years or more. Tens of thousands of patterns already exist. The equipment determines the patterns that can be created. For example, warp weaving is used to create more intricate nets and patterns, such as floral designs, than weaving with independent yarns.

Like loom weaving (page 76), there are many variables that the designer can choose from, such as the type of yarn, density of yarn, edge finish and finishing processes (dyeing and coating). Both staple and filament yarns are used, but for certain applications only filament yarns may be suitable. For example, nylon filament yarn is used on the bobbin in woven lace, because it needs to be very fine, uniform and strong.

Woven lace may consist of hundreds of interlacing yarns; the total number depends on the density of weave and width of beam. Each of the yarns, or warps, may be interchanged to create a coloured net ground, or outline (gimp). Lacemaking may be used with other techniques, such as pile weaving (page 96), to make fabrics combining dramatically different visual and physical properties.

MATERIALS
Decorative woven lace is made from cotton, but in practice all types of yarn are suitable, including flax, polyester, viscose and metallic.

COSTS
Machine-woven lace is high-cost. The exact price depends on the intricacy of the design and how long it takes to produce. Weaving looms produce roughly 2 m (6.6 ft) of fabric per hour.

ENVIRONMENTAL IMPACTS
Machine-made lace may generate a little more waste than conventional weaving, because the set-up often requires fine-tuning and production runs will typically be shorter than for plain fabrics.

→ Warp-Woven Leavers Lace with Gold Outline

All of Cluny's lace patterns were originally designed by hand (**image 1**). The sketch is translated by a draughtsman into the technical weave pattern (**image 2**). In turn, this is translated into the jacquard system for punched cards (**image 3**). The twist hand, who operates the looms, uses these calculations to ensure the weave is being made correctly as the machine progresses.

This loom uses two warps and 50 independent yarns. The heavier yarns (gimps) are made up of gold metallized yarn wrapped around a nylon filament core (**image 4**). The floral design is woven 'full beam' (**image 5**). In other words, the full width of the loom is utilized (4 m / 13 ft), rather than making a series of trims joined together to make a continuous sheet.

The gold yarn emphasizes the design by tracing its outline (**image 6**). White cotton yarns make up the interlocking structure and filled-in areas. There are several levels of density in this pattern, from very open to tightly packed yarns.

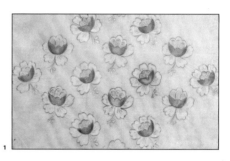

1

2

3

4

5

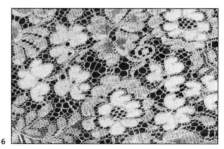

6

Featured Company

Cluny Lace Co.
www.clunylace.com

Textile Technology

Hand Weaving Lace

Lace is a highly valued material that was traditionally made by hand. The principal techniques are needle, bobbin, tatting and crochet. There are many styles, often named after the towns and cities where they were practised. Enthusiasts and hobbyists, who still practice the art of handmade lacemaking, keep the traditions and skills of these antique crafts alive.

Techniques		Materials		Costs
• Bobbin	• Tatting	• Cotton	• Silk	• High to very high
• Needle (and cutwork)	• Crochet	• Wool	• Novelty yarns	
		• Flax		

Applications	Quality	Related Technologies	
• Interior textiles	• Very fine quality	• Braiding	• Knitting
• Ceremonial garments	• Flat to relief patterns	• Embroidery	• Machine weaving lace
• Lingerie	• Dense to open structure		

INTRODUCTION

Nowadays most lace is machine-made (page 106). It is used to make highly decorative garments, lingerie, accessories and interior textiles. The main techniques are needle, bobbin, tatting and crochet.

Needle lace is made of buttonhole stitches worked with a needle and thread. The design is attached to a firm backing and threads are couched (laid down on fabric and fixed in place by stitching with another thread) along the lines of the design. The lace is then built up of rows of buttonhole stitches, each row linked to the previous row and attached to the couched threads. Making holes in fabric and filling them in with needle lace is

known as cutwork. An example of this is Reticella from Italy, although this name is also used for needle lace worked on a geometric grid of couched threads.

Bobbin lace is made with multiple threads, each wound on a separate bobbin, and is worked on a pattern attached to a firm pillow. The stitches, which involve two pairs of bobbins (i.e. four threads), are held in place as they are made with pins pushed through the pattern into the pillow.

Tatting is worked with thread wound on a shuttle. The designs are built up of knotted rings and chains connected together with picots (loops of thread) as the pattern is worked. The tatting knot is a double knot, made up of two half hitches, one the mirror image of the other, carried on another thread, which forms the internal core.

Crocheted lace is made by hand with a crochet hook. The basic structure is a simple chain of interlocking loops but by adding different combinations of crochet stitches to the basic chain, a wide variety of lace patterns can be produced.

Several traditional laces are named after the town or area where they were, or still are, made. Examples include Alençon (needle lace with a mesh ground), Bedfordshire (very decorative bobbin lace without a net ground), Branscombe (a tape lace), Buckinghamshire (bobbin lace with a net ground), Chantilly (very fine bobbin lace usually worked with black silk), Honiton (one of the most delicate English bobbin laces), Venetian Gros Point (needle lace often said to look like carved ivory), and Valenciennes (delicate but hard-wearing bobbin lace).

Lace evolved in many locations in different countries, so there are different names for the same stitch and pattern; conversely, one name may be used to describe several different things.

APPLICATIONS
Handmade lace is seldom used commercially owing to the high cost and long cycle time. Queen Victoria's wedding dress was made from Honiton lace and took many months to complete. Honiton lace was particularly well suited, because the floral design (or similar) can be made in pieces, and joined together with a net or plaited ground to form larger pieces. Today, handmade lace decorates handkerchiefs, bags, accessories, jewelry, shawls, bedspreads and tablecloths.

RELATED TECHNOLOGIES
Handmade lace has a unique quality. The tools and techniques for machine-made lace have continually evolved in an attempt to produce materials of similar quality and refinement. In some cases, it can take an expert to tell the difference between handmade lace and certain machine-made equivalents, namely Leavers and raschel (see Warp Knitting, page 144).

QUALITY
Because of the complexity of lacemaking and the range of interdependent factors, the finished quality depends on the skill of the lacemaker. Handmade lace tends to have very slight undulations in the structure, caused by the small variation in tension and twist that is inevitable with hand techniques. This gives the fabric its beautiful and distinctive quality. Patterns may be dense or sparse, depending on the design, tension and yarn.

Bobbin and needle lace may be designed with a relief profile. Even so, they make relatively flat textiles compared to knotted and crocheted lace.

DESIGN
Handmade lace is limited to very few applications, because nowadays it is carried out by a handful of enthusiasts and cottage industries. Design freedom is defined by the skill and experience of the lacemaker. Each technique creates a particular aesthetic, such as radial, open or heavy-lined, and several may be combined to create an infinite number of variations. Most handmade laces are based on antique designs, but a few experts continue to develop new patterns and aesthetics.

Decorations, such as beads and sequins, are added to the thread before and during the process. A range of different effects can be achieved, depending on how the decorations are introduced. Lace is converted into three-dimensional parts by adding starch, or an adhesive, such as epoxy or polyurethane (PU). Alternatively, wire is included in the outline (gimp), or the entire yarn structure, and can be bent and manipulated into three-dimensional shapes.

Colour may be introduced with the yarn, or yarns, or the finished item may be dyed a solid colour. Bobbin lace, which has many ends, provides the greatest potential for introducing pattern through colour. However, using different coloured yarns will show up any inconsistency in the number of twists, because the colour will move in the wrong direction. With needle lace, introducing small areas of colour is easier but there is less potential for combining several different colours.

In the past, geographic location determined the choice of technique, because the lacemakers would have a preference or certain skillset. Even so, some areas used both needle and bobbin, such as Brussels lace, which may be either or even a combination of the two.

MATERIALS
Traditionally, flax, wool and silk were used. Nowadays, cotton is the primary material for handmade lace. Depending on the application and desired effect, hemp, synthetic, metallic and other novelty yarns are also used. Almost any material can be experimented with, such as metal wire and twisted paper.

COSTS
Manual processes vary according to the technique and skill of the maker, but it can take days, or even weeks, to make a single piece. It requires a high degree of skill and so tends to be very expensive.

Using thicker yarn reduces the amount of time needed to make any lace item of a given size, because fewer stitches will be needed. The time taken to make each stitch will be the same whatever the size of the yarn.

ENVIRONMENTAL IMPACTS
These hand techniques produce very little waste and there is virtually no machine energy required. As the items are made with great care and can take many hours, pieces are cherished and long-lasting.

The Handmade Lace Process

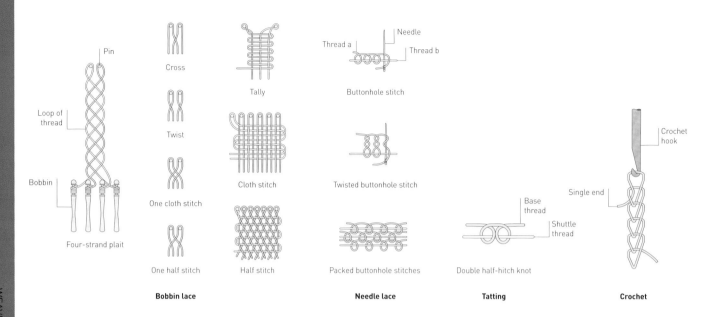

Pin
Loop of thread
Bobbin
Four-strand plait

Cross
Twist
One cloth stitch
One half stitch

Tally
Cloth stitch
Half stitch

Bobbin lace

Thread a — Needle — Thread b
Buttonhole stitch
Twisted buttonhole stitch
Packed buttonhole stitches

Needle lace

Base thread
Shuttle thread
Double half-hitch knot

Tatting

Crochet hook
Single end

Crochet

WEAVE
114

TECHNICAL DESCRIPTION

BOBBIN LACE

Bobbin lace is worked with multiple threads, each wound onto a separate bobbin, on a pattern attached to a firm pillow. The threads, which are always worked in pairs, are fixed at the start of the pattern so can move freely, creating the possibility for producing a wide range of patterns and densities. More pairs can be added, or some removed, as the work progresses. All the stitches involve two pairs (four threads) and once the stitches have been made they are held in position by pins pushed through the pinholes in the pattern into the pillow. When the lace is finished it is released from the pillow by removing the pins.

The basic moves in bobbin lace are: cross (left over right between pairs) and twist (right over left within a pair). Stitches are made up of combinations of these moves. Working 'cross-twist' gives half stitch, 'cross-twist-cross' gives cloth stitch, and working several half stitches with the same two pairs gives a plait. Weaving one thread of a pair under and over its partner and the

threads of another pair gives a small solid shape known as a tally.

A typical piece of bobbin lace will have pattern motifs, sometimes outlined with a thicker thread (gimp), worked in cloth stitch or half stitch where stitches are close together. Depending on the type of lace, these motifs will be joined together with plaits or a net ground with spaces between stitches or groups of stitches.

NEEDLE LACE

Needle lace is made using only a needle and thread. The basic unit is the buttonhole stitch in either its simple or its twisted form. Stitches can be packed closely together to give solid areas or spread out to give a net.

The pattern is fastened to a firm backing fabric and foundation threads are couched (fixed down to fabric by stitching with another thread) along all the lines of the design with a thread that goes through the pattern and the underlying fabric. The pattern is then worked with rows of buttonhole stitches, each linked to the

previous row and to foundation threads at both ends. Stitches can be worked in both directions or in one direction only, in which case the thread is attached at the end of the row, then taken straight back and attached ready to start the next row. The pattern motifs can be joined with either a net ground or bars of thread covered with buttonhole stitches. The edges of the motifs can also be embellished with padding threads covered with buttonhole stitches. Once the lace is finished it is released from the pattern by cutting the thread that couched down the foundation threads.

Needle lace may be combined with other lacemaking techniques. Or it may be used to fill in cutout areas of fabric, in a technique known as cutwork. It is also used in tape laces, where machine-made tapes tacked along the lines of the design replace the couched foundation threads.

TATTING

Tatting is worked with thread wound on a shuttle. The designs are built up of knotted rings and chains connected together with picots (small loops of thread) as the pattern is worked. The tatting knot is a double knot, made up of two half hitches, one the mirror image of the other, carried on another thread, which forms the internal core. Each knot is made by moving the shuttle thread around another thread but is then 'capsized' so that the shuttle thread forms the inner core.

Simple patterns of rings can be worked with a single shuttle wound with thread, while more complicated ones require a shuttle and a ball of thread or two shuttles.

CROCHET

Crochet lace is made up of interlocking loops in a continuous length of yarn. A crochet hook is used to form each loop. A simple crochet chain is started with a loop made by tying a slip knot around the hook, wrapping the yarn around the hook, and pulling a loop of yarn through the first loop, to form a new loop. The process is repeated with a loop of yarn being pulled through the new loop, and so on. Once the chain is the required length, the crochet is built up by taking the hook through an existing loop before the yarn is pulled through to make a new loop. Different stitches are formed depending on whether the new loop is pulled through in one step (slip stitch) or several (double crochet or treble crochet). Stitches can be grouped and varied to produce openwork patterns and three-dimensional items.

The density of the loops is determined by the tension applied by the maker, but also depends on the size of the hook and yarn used. Thicker yarn requires a larger diameter hook. Crochet lace is typically made with yarn, although any suitably flexible material may be used, such as twisted paper or split leaves. Crochet lace is often made up of several motifs, which are later joined together to make the finished piece.

VISUAL GLOSSARY: HANDMADE LACE

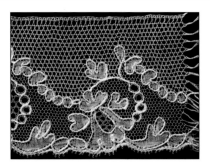

Buckinghamshire (Bucks Point) Lace
Material: Cotton
Application: Fabric trim
Notes: The hexagonal net ground in bobbin lace is produced using a half stitch with two extra twists. The more densely woven areas are made up of cloth stitch (or linen stitch). The loops along the edge are called picots (or pearls).

Wire Lace
Material: Copper wire
Application: Jewelry
Notes: Almost any material can be used if the lacemaker can handle it.

Tape Lace
Material: Cotton and glass beads
Application: Decoration for a bag or other fashion accessory
Notes: Tape lace is a type of needle lace. Bobbin lace with the same appearance is called braid lace.

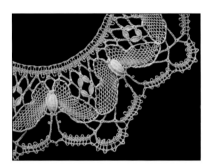

Bedfordshire Lace
Material: Cotton
Application: Coaster
Notes: Bedfordshire bobbin lace is not based on a net ground. Instead, the butterflies and other elements are linked together with plaits and tallies. Half stitch, outlined with a thicker gimp thread, is used for the butterfly wings and cloth stitch with some added twists for the two edges. The butterfly bodies are tallies.

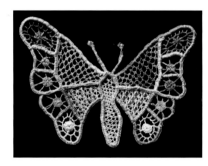

Needle Lace
Material: Cotton and glass beads
Application: Decoration
Notes: Needle lace is made up of a variety of buttonhole stitches. The appearance may be very similar to bobbin lace.

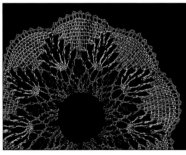

Torchon Bobbin Lace
Material: Cotton
Application: Coaster
Notes: Bobbin lace uses many ends, providing opportunity for producing dramatic colour effects. However, this increases the complexity of the work, because the twists in each stitch must be exact to keep each coloured yarn in the correct position.

Tatting
Material: Cotton
Application: Decoration
Notes: Tatting is made up of a series of half-hitch knots. Different shapes are created by connecting rings and chains of knots with picots (small loops of thread between knots) as the pattern is worked.

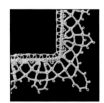

Plaited Lace
Material: Cotton
Application: Frame
Notes: Handmade lace trim is used to decorate picture frames and doilies, for example.

→ Making Buckinghamshire Bobbin Lace

Jean Leader is an experienced lacemaker who has been practising for more than 30 years. She uses traditional designs, as well as creating new ones. Here, she demonstrates making Buckinghamshire (Bucks) Point lace. It is a delicate lace worked in fine thread, with an outline in thicker thread (**image 1**).

Lacemaking starts with a pattern (**image 2**). The dots mark where pins will be used to support stitches and the line indicates the heavier gimp thread that will outline parts of the lace design.

The design is transferred onto card and covered with blue plastic film. This creates contrast with the white thread, making it easier to see. A row of support pins is inserted into the pillow, and the thread joining a pair of bobbins is looped around one of them (**image 3**). The support pins are set one position back from the start and will be removed as soon as pins are placed to anchor the stitches worked on the starting line (**image 4**). The first line of stitches and pins is staggered, so the join may be more easily concealed.

As the lacemaking progresses, the cloth, used to cover the card pattern, is moved backwards towards Jean (**image 5**).

The thicker gimp thread is held in place between the pairs of threads, without contributing to the stitches (**image 6**). The lace is finished by tying the ends into the first row of stitches (**image 7**).

The lace is made up of a net, which is created using half stitches with two extra twists. This produces a hexagonal mesh (**image 8**). The small dense squares are tallies; the solid area, in the corner, is cloth stitch.

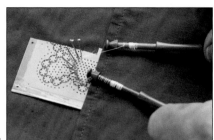

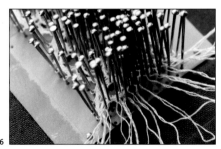

Featured Maker

Jean Leader
www.jeanleader.net

→ Needle Lace

The basic technique for all needle laces is the same. The structure is made up of buttonhole stitches, which the lacemaker can work spread out or close together depending on the effect required. A continuous length of thread is used. Variegated colour is achieved with multicoloured thread (**image 1**).

As with bobbin lace, the pattern is placed underneath blue film, contrasting the white thread (**image 2**). The pattern covered with film is tacked onto a fabric backing. Then the foundation thread is couched onto the pattern, around the outline (**image 3**). The first row of buttonhole stitches is made onto the foundation thread, and the second row is made over the first (**image 4**). In this way, the net ground is built up and is self-supporting (**image 5**). The outline is padded with a bundle of threads, which is attached to the foundation thread with buttonhole stitches (**images 6** and **7**). The couching thread is then cut away and the piece is removed from the pattern (**image 8**).

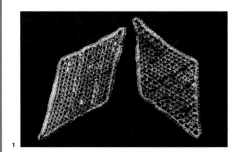

1

2

3

4

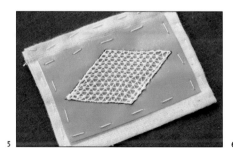

5

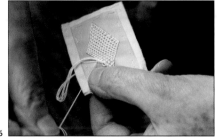

6

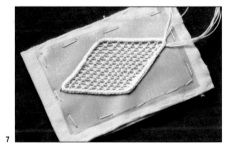

7

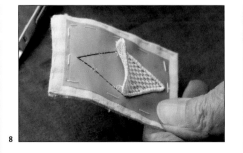

8

Featured Maker

Jean Leader
www.jeanleader.net

→ Tatting a Net Bag

Tatting is thought to have evolved from knotting, a favourite pastime in the 18th century, which used a large shuttle wound with thread to make knots of various sizes along the thread. Tatted lace, such as this net bag, is made up of sturdy double half-hitch knots (**image 1**). The pattern for this bag is started in the centre, and spirals outwards. There is no way to produce designs like this by machine.

A shuttle is used to make the knots (**image 2**). It is passed around the base yarn (or ball yarn) in a double motion to create the loops of the knot (**image 3**). Once formed, the knot is 'capsized' from the shuttle thread onto the base thread, and pulled tight (**image 4**).

After a series of double knots are formed, in this case five, a short length of thread is left between the knots to form a picot (**image 5**). This will be used to join the lengths of knotted thread together to form lace.

Four lengths of five knots are constructed with picots (small loops) in between. It is pulled into a ring, which forms the first petal in the centre of the bag (**image 6**). The next ring is started and after five double knots have been formed, it is joined to a picot on the first ring using a hook (**images 7** and **8**). Six petals are knotted and joined (**image 9**) before the next row in the bag is started. This row and subsequent rows are made up of rings and chains, and are joined to picots on the previous row as they are made (**image 10**).

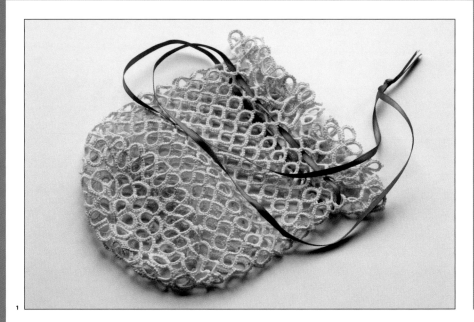
1

2

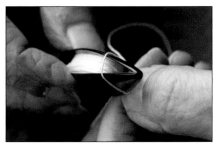
3

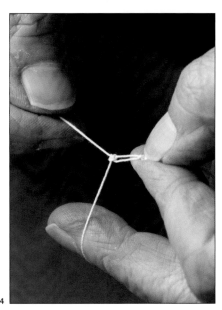
4

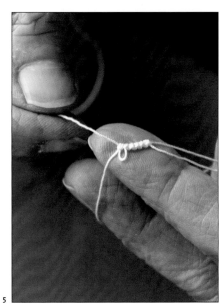
5

6

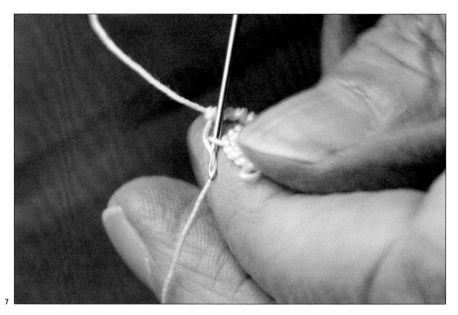

7

8

9

10

Featured Maker

Jean Leader
www.jeanleader.net

Textile Technology

Basket Weaving

The techniques used in basket weaving depend on local materials and traditions. All types of natural, pliable, fibrous materials are used, including willow, straw and leaf. Compared to loom weaving and knitting, the raw materials are relatively inflexible, although the finished woven textile may still be quite soft and pliable.

Techniques		Materials		Costs	
• Plain	• Fancy	• Natural fibres	• Wood	• Labour costs are low to high	
• Twill	• Openwork	• Bark	• Plastic	• Material cost depends on source and processes	

Applications		Quality		Related Technologies	
• Interiors		• Variable, handmade		• Filament winding	• Warp knitting
• Hats		• Lightweight and durable		• Injection molding	• Hand weaving
• Containers		• Semi-rigid to rigid		• Nonwoven	lace

INTRODUCTION

Basket weaving covers a wide range of traditional techniques, rituals, materials and applications. The process of interweaving pliable materials to make lightweight baskets, furniture, nets and hats has been practiced for millennia. Some practices have changed very little, such as weaving locally sourced willow to make wicker baskets for storage and transportation.

Natural materials have unique and varied properties. This means that no two items are identical. Even so, very skilled weavers are capable of producing precise structures and intricate patterns in a highly repeatable manner, making it difficult to tell two apart.

Basket weaving is not limited to natural materials. Virtually any type of material can be woven, as long as it is pliable and can be handled by the weaver. For example, high volumes of identical plastic containers are woven with strips of polypropylene (PP).

APPLICATIONS

Baskets are used for ceremonial and decorative applications as well as being practical items used, for example, for storage, fishing, cooking, packaging, shade, upholstery and bags.

Basket weaving is utilized in the production of many types of hat. In some cases hats are made directly from strands of material, such as cone hats worn by Chinese farmers. Alternatively, hats are molded from pre-woven structures in a process known as blocking (page 420). This technique is used to make the classic fedora, among others.

Basket-woven plastics are mainly used for low-cost storage containers (domestic and industrial applications), as well as bags and accessories.

The Basket Weaving Process

Plain checked

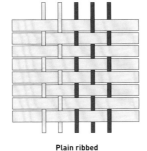

Plain ribbed

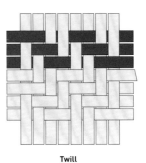

Twill

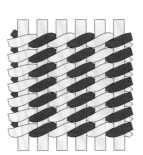

Twined

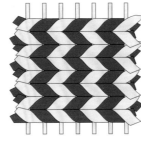

Twined reversed

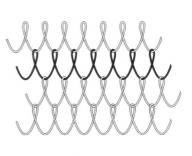

Coiled

RELATED TECHNOLOGIES

Basket weaving is used to produce many different items, from sheet material to finished products. And it remains the primary means of production for many materials and products, because more recent developments have not yielded a suitable or practical alternative.

Basket weaving is used to make all sorts of containers, such as for storage and transportation. Similar items are produced in plastic by injection molding, thermoforming and rotation molding. However, these techniques are only really suitable for high-volume production, because the metal molds are expensive and the process is complex to set up.

Lace, nets and bags – made by a weaver (see Hand Weaving Lace, page 112), by warp knitting (page 144) or by injection molding – are used for similar applications to basket weaving. Structures made with these processes are very lightweight and can be very strong, depending on the materials. However, they do not have the structural rigidity of baskets.

TECHNICAL DESCRIPTION

Basket weaving is a series of steps repeated over and over. A plain weave is the most straightforward and is made of palm, birch bark, leaf or reed, for example. The warp (vertical strands) and weft (horizontal strands) are the same width and are placed at right angles, alternately passing over and under one another, to produce a checked pattern. Using rigid warp ribs and flexible weft strands produces the familiar willow basket texture.

Sheets of plain-woven basket may be produced on specially adapted mechanized looms. This technique is used to produce upholstery materials, for example, much more cost effectively than can be achieved by hand.

Twill weaving is also at right angles. Each warp is passed over at least two wefts and under one weft. Passing each warp over and under the same number of wefts will result in a reversible pattern (the same on both sides). These techniques produce a diagonal pattern, known as twilled or herringbone.

Twining uses rigid warp strands. The wefts are grouped together in twos or threes and wound around each other as they are alternately woven over and under the warps. Alternatively, the twist in the wefts is reversed in each row, producing a more densely packed weave.

Coiled baskets are made by looping string or rope together. Traditionally, flexible materials such as cotton, willow, raffia and straw are used.

QUALITY

Applications range from disposable items – such as baskets made to package or carry fruit – to items used many times over, and quality ranges accordingly.

Baskets are lightweight and durable: every fibre contributes to their strength. They are known as hard textiles, because the fibres are typically semi-rigid to stiff, although they can also be very pliable. For example, panama hats are famed for being soft enough to fold up without losing their shape.

The visual quality is largely dependent on the materials, which in turn determine the techniques used. Many consider basket weaving an art form. The use of natural materials results in variation in colour and appearance. Even so, experienced weavers are capable of producing consistent materials to very high standards. Such weavers usually select their raw materials while they are still growing, because the quality of the material is critical for the end result.

Natural materials are perishable and so have a finite life. They dry out, making them brittle and prone to cracking along crease lines. Unlike woven and knitted textiles, it is not certain how long basket weaving has been practised, because any trace of ancient baskets biodegraded long ago. Even so, basket-woven items that are well looked after can last many years.

DESIGN

In every part of the world, the available materials and performance requirements of basketry have resulted in a very wide range of styles, shapes and finishes. To some, such as the American Indians, traditional weaving is seen as an expression of culture, with techniques ranging from temporary basketry made on site with immediately available materials, to intricate ceremonial works of art.

The choice of material impacts on the basket weaving, because each material has particular properties as well as methods of handling. Raw materials are used as they are found, in their 'green' state, or they are specially prepared. It can take days or even years to make materials ready after they have been prepared.

Leaves come in a variety of forms, which are woven using different techniques. Large whole leaves are folded and stitched (it is also possible to mold some types into three-dimensional shapes); fronds are woven while attached to the central rib; or strips of leaf are separated, prepared and woven. Wood may be used in a variety of forms too, such as whole rods, strips, veneer, or the bark treated and used separately. Fibrous materials are woven as individual strips, or bundled together in groups.

With hand processes, any type of material may be integrated, if not by itself then by combining with materials that can be woven, or as a non-structural element.

Weave density ranges from solid to open. The patterns and openness of the structure will depend on the weaving technique used. It is also possible to weave over an existing object, such as a container, encapsulating it within the intertwined fibres. For example, Chianti wine bottles are woven with plants taken from the surrounding marshlands. Filament winding (page 460) uses a similar principle to encapsulate a mandrel in between high-performance fibre-reinforced composite.

Basket weaving makes flat as well as seamless three-dimensional items. Virtually any size and shape can be made by hand, although the weave tends to be symmetrical. All of the weaves used in handloom weaving (page 90) – plain, twill, satin, double-faced and fancy – occur in basketry. Patterns and colours are inherent to the structure, or they are applied after weaving by spray coating. The fineness of the raw materials will affect the intricacy of woven patterns. Stencilling and hand painting allows for more complex figured patterns, unlike woven types that have to follow the geometric structure.

In traditional basketry, decorations such as feathers, beads and tassels are affixed during weaving. The weave pattern combined with the shape of the material – round, square or flat – determine surface texture. Edges are finished with a border, which determines the shape of the basket. Several

techniques are used, such as hemming, twining, knotting and braiding.

MATERIALS

Any suitably pliable material may be used, including natural, man-made and metallic types. The most commonly used materials are natural fibrous types, including grass, leaves, bark, rattan, willow, reed and bamboo. The typical materials depend on location. For example, willow is common in Europe, bamboo in China and birch in subarctic countries, such as Finland.

COSTS

Basket weaving is a hand process. Therefore, high-volume products tend to be made in countries where manual labour is less expensive, such as India, Mexico, Taiwan and Pakistan. Even so, the techniques are carried out throughout the world and weavers exist in most countries including the United States and in Europe. As a result labour costs range from low to high.

Cycle time is moderate to long depending on the size and complexity of the design. Using longer or larger materials reduces cycle time, because structures can be built up more quickly and it avoids running out of material and starting new elements.

ENVIRONMENTAL IMPACTS

Plant materials are harvested from renewable sources and minimal transportation is required. There is no energy-consuming machinery required; weaving is carried out by hand. Natural materials are biodegradable and do not transmit any harmful elements into the environment at the end of a basket-woven object's useful life.

The source of material is important, because basket weaving may take place a long distance from where the materials were harvested. For example, raffia is woven all over the world but comes from palms native to Africa. Using locally available materials reduces or eliminates transportation. With consideration, basket weaving may have no negative impacts on people or the environment.

Plain
Material: Birch veneer
Application: Packaging
Notes: This is the simplest weave: overlapping alternate warp and weft creates a checked pattern.

Rigid
Material: Palm
Application: Satchel
Notes: Self-supporting structures are created with stiff materials, or by layering materials to make them rigid.

Soft
Material: Birch bark
Application: Storage
Notes: The bark from certain species of tree is suitably strong to be made into baskets and packaging.

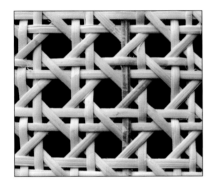

Multiaxial Openwork
Material: Split rattan core
Application: Upholstery
Notes: Pre-woven sheets of cane are produced on a loom and mimic handmade techniques.

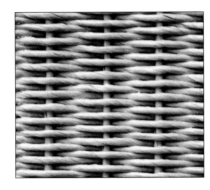

Loom Woven
Material: Paper and metal wire
Application: Upholstery
Notes: Flat sheets of loom-made hard textiles combine twisted paper warp and paper-covered wire weft.

Twill
Material: Rattan
Application: Bread basket
Notes: Split cane, dyed shades of brown to create a natural pattern, is twill woven into a semi-rigid basket.

Stained
Material: Rattan
Application: Bag
Notes: Rattan is stained like wood. The contrast emphasizes the weave pattern.

Painted
Material: Rattan
Application: Storage basket
Notes: After weaving and assembling with hinges, legs, catches and so on, the entire basket is spray painted.

Dyed
Material: Grass
Application: Slippers
Notes: Grass is dyed in much the same way as natural fibres to produce a range of bright colours.

→ Weaving a Willow Basket

Even though there is a global market for basket-weaving materials, it is of course best to use local and renewable material. Willow rods are harvested and graded by length (**image 1**). Brown willows are rods dried and used with their bark intact. Buff willow has been boiled for several hours and the bark removed. White willow is peeled without boiling. The rods are soaked in water for several days in preparation for weaving and then stored in a cool, damp environment for a day or two to relax.

Basket weavers often start by using their feet to hold the first rods in place. This English shopping basket has an oval base, which is constructed of three parallel sets of thick bottom sticks interwoven with flexible rods (**image 2**). The upright 'stakes' are inserted alongside the bottom sticks and bent upwards using a hook-shaped knife to make a crease line (**image 3**). A ring is placed around the ends to hold them upright and at the correct diameter.

More willow weavers (rods) are woven tightly around the stakes in a process known as 'upsetting' (**image 4**). The sides of the basket are then woven using a mixture of buff and white willow for decoration (**images 5** and **6**). The top is finished with a border of stakes, which are folded down and back-woven to secure the ends. This style is known as a simple common border (**image 7**). Throughout the process the weaver sprays the willow with water to keep it moist and flexible.

A roped handle is formed and woven into the border. It is a robust construction made of a thick core wrapped with thinner flexible rods (**image 8**). The joints are tied in using rods that have been twisted to loosen the fibres and make them more pliable (**image 9**).

All the colours are natural, deriving from different varieties of willow and the various preparation techniques (**image 10**).

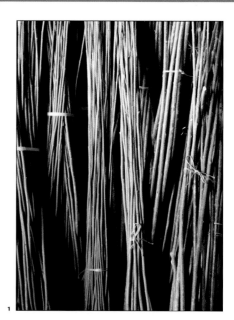

1

2

3

4

5

6

7

8

9

10

Featured Company

English Willow Baskets
www.robandjuliekingbasketmakers.co.uk

Textile Technology

Weft Knitting

Knitting is a versatile process based on interlooping continuous lengths of yarn. Machine processes are based on either weft or warp yarn formation and are capable of producing complete garments in a single process. A range of loop combinations is used to make precisely engineered fabrics from natural and synthetic fibres.

Formats	Materials	Costs
• Plain (jersey) • Three-dimensional • Openwork (net and lace)	• Staple, filament and novelty yarn • Natural and synthetic • Inlaid yarn	• Moderate to high depending on the complexity and yarn

Applications	Quality	Related Technologies	
• Apparel • Interiors • Technical	• High stretch • Continuous yarn • Fashioned (shaped)	• Circular weft knitting • Fancy loom weaving • Loom weaving	• Machine weaving lace • Warp knitting

INTRODUCTION

Knitting is used to make flat, tubular and three-dimensional fabrics. Unlike loom weaving (page 76) – which is based on interlacing warp and weft – weft knitting produces fabrics from continuous lengths of yarn looped together. Therefore, each loop is dependent on its neighbours: above, below and either side.

The loop structure allows yarn to move freely, even under tension. This produces fabric with high stretch, up to several hundred per cent its original dimensions. Additional yarns that do not make up part of the loop structure may be inlaid to reduce stretch, improve elasticity and recovery, or affect handle, density or visual appearance.

In weft knitting, also known as fill knitting, the loop structure is formed from a continuous length of yarn, which is passed from side to side to form courses (direction of weft). By contrast, warp knitting (page 144) uses one or more yarns per needle, and the loop structure is constructed along the length of fabric forming wales (direction of warp).

Weft knitting is versatile and so is widely used to make apparel, interior textiles and technical textiles. Single yarns or multiple yarns are interlooped with a range of techniques to produce simple (jersey) or complex fabrics. The simplest weft-knitted fabrics for garments include plain jersey, stockinet and lisle. Complex, or fancy, fabrics are produced with a combination of different yarns and stitches. They may be intricately patterned (jacquard), ribbed, pile (knit terrycloth and fake fur) and double-sided with an interlocked structure.

Fabrics are shaped during knitting by producing adjacent courses or wales of different lengths. Garments are known as fully fashioned when courses are shortened or lengthened by transferring loops to create panels, or complete garments, with a body-shaped profile. Three-dimensional fabrics are produced with a combination of loop transfer within courses and different lengths of wales. The number of stitches made by a needle determines the length of each vertical wale. Therefore, holding loops on some needles, while others continue to knit, produces wales of varying length. This technique is used to manufacture seamless upholstery and medical products, for example.

APPLICATIONS

Weft knitting is used to make plain and patterned fabrics for apparel (such as t-shirts, sweaters, sportswear, hats, gloves and fleece); lace (apparel, net, sheer); and technical textiles, which includes nets, upholstery and geotextiles. Socks are made on circular weft-knitting machines (page 126).

RELATED TECHNOLOGIES

Loom weaving is used to produce textiles for many of the same applications, but they rarely compete directly, because the qualities of knitted and woven materials are different. Weaving is preferred for applications that require a large area of flat fabric with good cover, such as upholstery and interiors; low stretch, such as suits and jackets; sheer (less yarn is required for the same coverage); and combining novelty yarns that are not suitable for knitting. Knitting outperforms weaving in the production of fashioned, elastic, bulky, tubular and three-dimensional fabrics.

Like weaving, knitting is used to make openwork textiles for lace or more technical applications. Knitted lace is more economical than weaving and there is a wide range of design opportunities linked to the different loop structures. For example, lightweight sportswear with localized stretch, support and breathability is produced in a single operation from multiple yarns. In addition to creating openwork structures in the plain-knitted ground, the two may be combined in a single layer to create a honeycomb layer on top of the ground, known as semi-breakthrough.

Knitting has been developed to mimic many fancy weaving techniques, such as the reproduction of intricate patterns similar to jacquard, intarsia (areas of colour, which is similar to tapestry in effect), pile (cut and loop), and double-faced fabrics.

QUALITY

Knitted fabrics have more stretch than woven fabrics, because the loop structure allows freer movement of the yarn and is more easily distorted than woven warp and weft. Therefore, knitting is used to make garments that fit more tightly to the body and stretch to accommodate movement. This also means that knitted fabrics tend to lose their shape more quickly than woven fabrics.

Single-knitted fabrics have a 'technical face' and 'technical back'. The technical face is the outside surface as knitted, but may not be the outside surface in application. Doubled-sided fabrics may have mirrored, or totally different, patterns on each side, depending on the stitch design.

Owing to the interlooping structure, knitted fabrics are prone to snagging. Broken yarn causes ladders (or runs) to occur, which affects the visual and technical performance of the fabric. Unravelling can be an advantage during production: edges are temporarily held together by open loops, which are removed just before joining parts together to expose the 'live' loops that are seamlessly joined into another knitted structure. This technique is used to join cuffs to sleeves and link sleeves into the body of sweaters, for example (page 386).

The edges of knitted fabrics are finished with a range of techniques. The first row is cast on and held by the following row of loops. And the last row is cast off. To stop it unravelling it is finished by linking into another knitted structure; by binding off to secure the loops; or by knitting with a fusible yarn, which is heated and melted to form a sealed edge.

Secured edges are known as welts. There are many different types of loop used, which all have different qualities, but they all stop the fabric from unravelling. The edges do not have to

be parallel and welted edges can follow complex winding edge profiles, such as a cardigan neckline. If the edges are cut they are finished by overlocking, hemming, or with a French seam (see Hand Stitching, page 342, and Machine Stitching, page 354). The stitch pattern must allow sufficient extensibility to avoid compromising the stretch of knitted fabric. Popular stitches include chain stitch, three-thread overlock and five-thread flatlock (flat seams). Exposed edges within the fabric itself, such as buttonholes, are locked in by the loop structure, or cut out and sealed with a buttonhole stitch (page 344).

The distance between the needles, called pitch, will affect the thickness of yarn that can be knitted. The length of the machine, divided by the total number of needles, determines the gauge. The fineness of fabric that can be knitted is proportional to the machine gauge. Higher-gauge machines have more needles for the same distance, which means they tend to be slower, but capable of producing finer and more densely knitted textile.

The complexity of knitting means that fabrics are prone to faults caused during manufacture, such as dropped stitches, skew and shrinking.

DESIGN

Computer-guided processes are very flexible. Each job is transferred to the machine from the design department, via computer-aided manufacture (CAM) software, and knitting begins. These machines do not need to be set up again for a new operation, unless the yarn needs to be changed, or if fly (tiny dust-like fibres) from natural fibres, such as wool, needs to be cleaned away. This makes them suitable for prototypes, one-offs and batch production, as well as for mass production of identical parts. It should be noted that each equipment manufacturer uses bespoke software. This means that designs may not be transferred directly between the different manufacturers' equipment.

Weft knitting is based on four primary loop structures: plain, rib, interlock and purl (see technical description on

The Weft Knitting Process

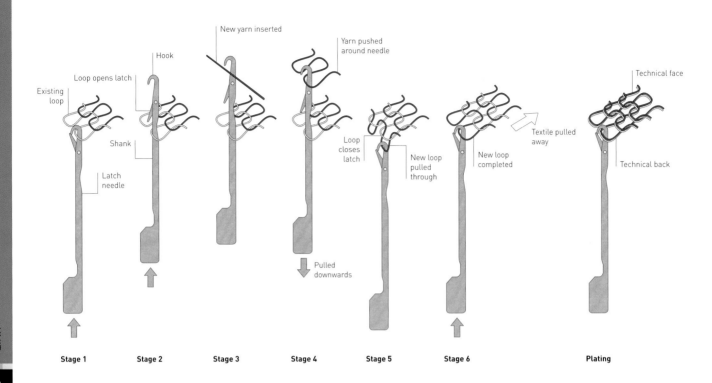

Existing loop · Loop opens latch · Hook · New yarn inserted · Yarn pushed around needle · Technical face · Shank · Latch needle · Loop closes latch · New loop pulled through · New loop completed · Textile pulled away · Technical back · Pulled downwards

Stage 1 · Stage 2 · Stage 3 · Stage 4 · Stage 5 · Stage 6 · Plating

TECHNICAL DESCRIPTION

Machine knitting builds a textile row by row. Each row is referred to as a course. In stage 1, each loop is located on a needle. In stage 2, the needle is pushed upwards to its clearing height. This motion causes the existing loop to open the latch and run down the shank until it clears the latch. In stage 3, yarn is fed in on the latch side of the needle. In stage 4, the yarn is pushed either side of the needle by a loop-forming sinker. Therefore, in stage 5, as the needle lowers back down, it catches the yarn. A new loop is formed as the needle draws the yarn down through the existing loop. As this happens, the existing loop closes the latch and rises over the top of the needle, known as cast-off or knock-over, to form a new loop in the course, held in place by the loop that is now formed on the needle.

To assist with loop formation the textile is continually under tension, pulled away from the needles. In stage 6, a new course of loops is successfully formed and the sequence can start again. The loops form wales, which run the length of the fabric.

Each loop is dependent on its neighbours above, below and either side.

Plating involves knitting two yarns side by side in the same loop structure. The two yarns typically have different properties. For example, two different-coloured yarns in a plated structure will produce a different-coloured technical face and technical back. Alternatively, the outer-facing yarn may be more abrasion resistant and the inner yarn more comfortable. With circular yarn, it is difficult to get perfect alignment, whereby only one is visible on each side, because they tend to slide over one another.

There are four main types of stitch: plain, rib, purl and interlock. Plain forms the basis of single jersey and is used to make the majority of knitted fabrics. The technical face is smooth and made up of a v-pattern running up the wales. The back has a series of interlocking semicircles as visible courses. A broken yarn will cause a run, or ladder, whereby the fabric structure will start to unravel. Single jersey fabrics are produced on all types of knitting machines.

Ribs require two sets of needles, because each rib is formed by a series of face-loop wales and back-loop wales, which are formed simultaneously. The ribs stand out on either side, because the face-loop wales move over and in front of the back-loop wales. This technique is used to make high-stretch items, such as cuffs and waistbands, because the rib structure produces a dense fabric with good stretch and recovery. Stretching the fabric widthwise reveals the back-loop wales in between.

Purl fabric has courses made up of alternating face loops and back loops. This produces an effect that is perpendicular to ribs. In other words, the purled ribs run across the width rather than along the length of fabric, and the shape is made up of intersecting semicircles rather than v-shaped ribs. Like ribbed fabric, it is reversible. The loops tend to draw together, reducing the length of the fabric, and so create additional lengthwise stretch. Pulling the fabric lengthwise reveals the back-loop course.

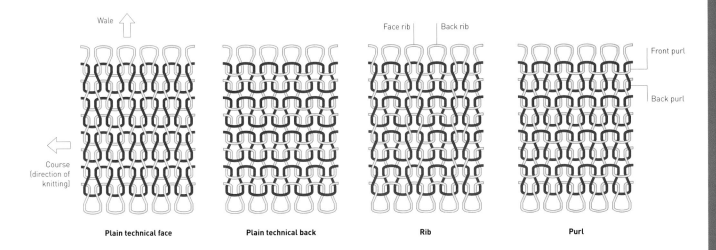

Wale

Course (direction of knitting)

Face rib | Back rib

Front purl

Back purl

Plain technical face   **Plain technical back**   **Rib**   **Purl**

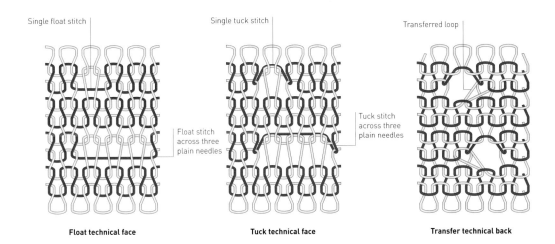

Single float stitch

Float stitch across three plain needles

Single tuck stitch

Tuck stitch across three plain needles

Transferred loop

Float technical face   **Tuck technical face**   **Transfer technical back**

Interlocking two or more yarns from either side forms double-knit fabrics. In operation, the two sides are knitted separately and the yarns are transferred from side to side within each course to create an interlocking structure. This technique is employed to form double-faced fabrics.

In combination with all the different types of stitches, loops may be held on needles for one or more courses. The yarn is floated over one or more loops, or tucked into subsequent courses, to form pile, reproduce graphics, create an open structure (lace or mesh) or affect surface texture.

A transfer stitch is formed by picking up and moving loops to adjacent needles in the same row, either left or right. As knitting continues, an eyelet is created by the absence of the loop. When the row width is maintained, loops must be doubled, thus creating a bulking of the knit. The number of loops that are transferred determines the location. So, in other words, transferring one or four will result in an overlap next to the eyelet, or four loops away, respectively.

'Increasing' is when an opening is formed and the row is lengthened, and 'decreasing' is when the row is shortened. This technique forms the basis of producing 'fully fashioned' garments (shaped to the body).

Held loops are retained on the needle for at least one extra knitting cycle (the number of cycles it can be held is determined by the tension). After a set number of knitting cycles the held loop is cast off and intermeshed into a higher course.

page 128). In addition, loops may be held to produce a float or tuck stitch. Combining these in different ways produces the technical and visual properties of knitted fabrics. Therefore, the same idea may be reproduced in many different ways, resulting in different properties. For example, reproducing a graphic on the face of a fabric can be done by varying the loop structure (to change the surface texture), by introducing a different-coloured yarn (jacquard), or with openwork.

A major advantage of knitting is that flat, fashioned and three-dimensional fabrics may be produced on the same machine with a range of finishes, colours and patterns.

Tubular fabrics are made in one of two ways. Either they are knitted on a flat bed with a front and back fabric joined at the edges to make a continuous loop – gloves and one-piece sweaters are made in this way – or they are knitted on a circular weft-knitting machine (page 138). With this process, continuous courses of loops are formed in a spiral. In addition to making tubular articles, this technique is used to produce high volumes of plain-knitted fabrics, such as jersey. Multiple courses are knitted simultaneously and the tube is slit to make flat fabric.

The type of knitting machine determines the range of visual and technical qualities that can be achieved. The most common machines are bar frame, V-bed and circular. Bar-frame machines have a single straight bed of needles and are therefore unable to knit ribs or interlock stitches. They produce high-quality plain fabrics (single jersey). V-beds have two sets of needles that intersect at right angles. Computer-guided machines are very flexible and capable of producing all types of stitch and mimicking virtually any type of weaving, including the most complex jacquard (page 85). Fabrics formed with two sets of needles are known as double jersey (or double-knits). They can be made up of any combination of stitches, including plain, rib, interlock, purl, tuck and float. Circular-knitting machines are limited to making continuous lengths of tubular fabric to a set dimension (ranging

from socks to full-size garments, or large tubes split into sheets).

Seamless knitted three-dimensional items, such as sweaters, may be produced in parts and linked together with chain stitch (page 386). This technique is used to join sleeves onto the body of sweaters, for example. Or garments may be produced in a single knitting operation, by knitting the front and back and interlocking the edges. This technique is used to make sweaters, sportswear, gloves and hats. This reduces post-production labour time, although some stitching and finishing may still be required.

Colour patterns and graphics are achieved in weft-knitted fabrics by printing, dyeing or using different-coloured yarns. With multiple yarns, all manner of designs can be reproduced, down to changing the colour of individual loops. Flexibility depends on the capabilities of the machine. Changing the colour of the yarn being fed in will create horizontal stripes. Areas of solid colour are created with the intarsia method. This involves multiple yarn feeds, which are brought in and out as the colour is required; this is therefore limited to machines with a sufficient number of feeds. The yarns track back and forth, making unbroken courses, and each colour area overlaps slightly to create a strong fabric.

Complex colour patterns and graphics are produced with individual stitch selection, known as jacquard. Computer-guided straight V-bed and circular machines containing two sets of needles can reproduce all types of stitch. Combined with several yarns, intricate and complex colour patterns are formed in the structure of the knit. Fabrics made up of more than two colours or types of yarn are constructed with a range of stitch types. When the colour is not needed on the face, or the back, it is interlocked or held to take it out of view. Adding colour and increasing complexity reduces production rates.

It is possible to change the appearance of knitted fabrics without affecting the loop structure. Techniques include laying-in yarn, pile, openwork and plating (see technical description on page 128).

Yarns are inlaid to improve dimensional stability. Those that are not suitable for knitting, such as loosely twisted, novelty or bulky yarns, may be incorporated for visual and tactile enhancement. This is possible because inlaid yarns do not contribute to the loop structure. Laid-in elastic yarns improve stretch and recovery and so contribute to better-fitting garments and hosiery.

Pile is formed either during knitting, or by napping during finishing (page 216). In knitting, excess yarn is fed into float stitches to form terry loops. The loop yarn may be part of the ground structure, or an inlaid yarn. Knitted pile includes fleece and towel (terry loops) – and plush fabrics.

Like woven fabric, they are finished with a range of processes including tentering (page 210), calendering (page 220) and napping.

MATERIALS

Natural and man-made staple yarns (page 56) and filament yarns (page 50) are all suitable. Ideally, yarns will be smooth, fine and strong. Filament yarns are the most consistent and have lower shrinkage, so are the most straightforward to knit. They produce a consistent, stiff and resilient fabric. Texturing filament yarns improves bulk, stretch and recovery without compromising the efficiency of knitting.

Staple yarns should be combed or worsted to improve knitting performance. They produce soft fabrics with high drape. Specially adapted machines are capable of knitting smooth novelty yarns and even metal wire.

COSTS

Yarn is continuous and has to be high quality to ensure consistent and reliable fabrics. This increases cost compared to weaving.

Modern computer-guided knitting machines have very high production rates. Complex patterns and double-knits take the longest to knit. And knitting designs with colour adds cost, because machine speeds are reduced.

Circular knitting is the fastest way to make plain fabric (many courses are knitted in tandem and the tube is split to make flat fabric).

Plain-Knit
Material: Aramid
Application: Heatproof glove
Notes: Knitting with blended or plied yarns produces mottled colour effects.

Repeat Pattern
Material: Cashmere
Application: Apparel
Notes: A two-colour repeating pattern made by floating each colour behind the other in alternate wales.

Herringbone
Material: Cashmere
Application: Apparel
Notes: Knitting is used to make traditionally woven colour patterns, such as dogtooth and herringbone.

Two-Colour Pattern
Material: Cashmere
Application: Apparel
Notes: The placement of colour and pattern are limited by stitch and yarn and the imagination of the designer.

Jacquard Pattern
Material: Cashmere
Application: Apparel
Notes: With individual stitch selection, many colours can be combined into complex and intricate patterns.

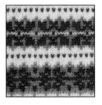

Purl
Material: Cashmere
Application: Apparel
Notes: Purl is made up of alternating face and back loops, creating ridges across the width of the fabric.

Cabling
Material: Cashmere
Application: Apparel
Notes: Relief patterns, such as cable stitch, are made up of cords of face loop wales, not unlike ribs.

Relief Pattern
Material: Cashmere
Application: Apparel
Notes: Relief patterns run the full length and width of the fabric. These types of stitch consume more yarn and so produce heavier fabric.

Relief Twill
Material: Cashmere
Application: Apparel
Notes: Virtually any pattern is possible in weft knitting, including mimicking traditional basketry.

Relief Colour
Material: Cashmere
Application: Apparel
Notes: Different colours and stitches are combined to produce dramatic relief patterns.

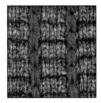

Looped Pile
Material: Nylon and wool
Application: Apparel
Notes: Knitted loops, formed with excess yarn fed into float stitches, provide cushioning and insulation.

Lace
Material: Cashmere
Application: Apparel
Notes: Intricate openwork patterns are created with a combination of stitches, held loops and transferred loops.

Knitting is utilized to produce finished or semi-finished articles. This reduces or eliminates post-production processes, such as pattern cutting and machine stitching.

ENVIRONMENTAL IMPACTS
Knitting is a relatively efficient process and generates little waste. The total environmental impacts are determined by yarn selection and processing, knitting process and finishing.

The source of yarn is significant, because natural yarns should be from sustainable sources and processed accordingly. The amount of recycled content in synthetic yarns varies according to the polymer and supplier.

Compared to conventional pattern cutting and sewing, three-dimensional knitted upholstery, garments and shoe uppers can be made with fewer parts and significantly less waste. Using such fabrics can reduce or eliminate other materials, such as polyurethane (PU) foam padding.

Knitted fabrics that contain a mix of materials pose a challenge end-of-life. For example, a natural yarn combined with plastic yarn means that the fabric is only partly biodegradable and partly recyclable. Extracting the two materials so that they can be properly processed is not straightforward. In some cases it is possible to recycle parts of garments or products by removing panels and using them to make new products. Even

though it is possible to make a product of equivalent or higher value, it is not possible to make the same product, because there is wastage in the cutting process and knitted fabrics are prone to unravelling.

→ Knitting Plain Jersey on a Straight Bar Frame

In preparation, the waistband and cuffs are knitted separately on a ribbing machine (**image 1**). They are loaded by hand onto a set of transfer needles, known as bar filling (**image 2**). Each loop (or pair of loops in the case of rib) is taken up on a separate needle. The expendable white yarn, which was used to maintain the 'live' loops on the edge of the ribbed waist, is unravelled (**image 3**). The ribbed parts are loaded in groups and unloaded onto a second set of needles on the bar frame one by one (**image 4**).

When ready to commence, the machine automatically aligns and transfers the loops onto the needles that will perform the knitting operation (**image 5**). This two-stage transfer process allows the operator to prepare each bar frame while it is still occupied with the previous knitting cycle. This type of machine uses bearded needles. The first row of loops is formed with a sequence of movements: the needles rise up to clear the existing loop; a new yarn is laid on the bearded side of the needles (behind when looking at the machine); a sinker between each needle pushes the yarn forwards, folding it onto the shank of the needle (**image 6**); and the needles are drawn downwards catching the yarn and forming a new loop through the existing loop (**image 7**). With each cycle a feeder lays the red cashmere yarn against the needle and in front of the sinkers (**image 8**).

Fully fashioned garments are shaped by transferring loops inwards to reduce the length of the course, or outwards to increase the length of the course, known as decreasing and increasing respectively. Alternatively, wales are added to the edge with additional loops to increase the width of a course.

Transferring is done by a set of specially designed needles, which pick up the loops and move them to adjacent needles while knitting is paused (**image 9**). Eyelets left by increasing are clearly visible in the knitted structure (**image 10**). With each new course, drawing the fabric away from the needles aids the formation of new loops. A seamless join is formed between the ribbed waist and the plain jersey body (**image 11**).

Typically, multiple identical parts are knitted simultaneously on bar frames (**image 12**). This reduces cycle time. However, this type of equipment, which is controlled by punched cards and takes considerable time to set up, is only really suitable for high-volume production. With eight frames, two sets of four parts (front, back and sleeves to make two complete sweaters) are produced every 15 minutes. A certain amount of pattern and openwork can be achieved, depending on the capabilities of the machine.

1

Featured Company

Johnstons of Elgin
www.johnstonscashmere.com

2

3

4

5

6

7

8

9

10

11

12

TECHNICAL DESCRIPTION

Two sets of needles can knit the same loop structure as a single set. Unlike bar-frame machines, with modern V-beds almost all types of stitch can be achieved on a single machine.

The beds of needles are set at a right angle with the hooks facing outwards. In stage 1, an interlock stitch is formed, whereby the same yarn passes between both needle beds creating an interlooped double-sided fabric. Alternatively, a plain, rib or purl stitch may be produced.

In stage 2, the needles raise up to clear the loops formed in the hook. The loops open the latch and are drawn down the shank as the new yarn is fed in. As the needles are drawn back down, the loop on the shank is cast-off, forming a new loop. With each course the fabric is drawn away from the needles to encourage loop formation.

The placement of each new feed is determined by the pattern being knitted. It can be different in every row, depending on the colour, pattern and stitches that are used.

The V-Bed Knitting Process

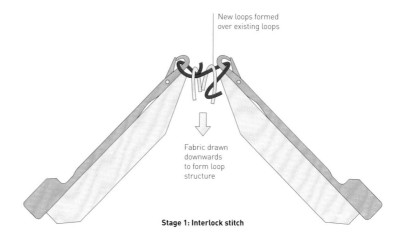

New loops formed over existing loops

Fabric drawn downwards to form loop structure

Stage 1: Interlock stitch

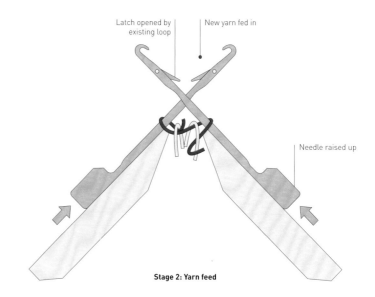

Latch opened by existing loop

New yarn fed in

Needle raised up

Stage 2: Yarn feed

→ Fancy Knitting on a V-Bed

Jacquard patterns, welts, ribs, purl stitches and seamless double-sided parts, such as these gloves (**image 1**), are knitted with a computer-guided V-bed. All types of stitch can be formed, saving the time it takes to make lots of separate parts and join them together.

In addition, one-piece garments with complex colour and pattern combinations – such as cabling, which is a traditional pattern produced by transferring wale loops to make raised ribs on a back loop ground (**image 2**) – are constructed in a single process. A complete sweater takes around half an hour, depending on the number of courses and complexity of design.

The process starts on the computer, working with the equipment manufacturer's software (**image 3**). Once ready, data is sent directly to the V-bed and knitting commences. This makes it practical for very short production runs, unlike bar frame.

V-bed consists of two sets of needles, set at right angles (**image 4**). Unlike bar frames, they have a knitting head. The full complexities of the stitches that can be achieved are visualized by the inner cam mechanisms of the computer-guided carrier (**image 5**). The tracks between the cams guide the needles up and down in a metachronal wave (sequential ripple). Depending on the machine, the cam carrier may hold two, three or more heads. Each of these includes a set of cams capable of stitching or transferring. Thus, multiple courses may be stitched simultaneously, while additional heads transfer loops ready for the next cycle.

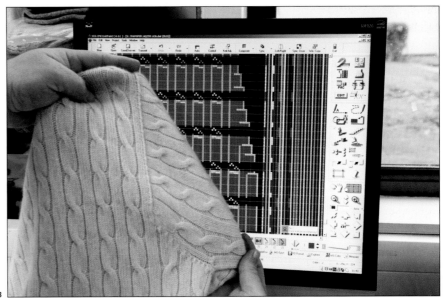

Unlike mechanical jacquard selection systems, a computer-guided cam carrier does not need to travel to each end of the machine to pick up instructions for the next course. Using a technique known as variable stroke, narrow widths like glove fingers are knitted much more quickly.

The head controls which yarn is fed to the needles from multiple feeds (**image 6**), at what point, which needles are knitting and which are holding onto previous loops, and loop transfers.

Featured Company

Johnstons of Elgin
www.johnstonscashmere.com

→ Technical Knitting Three-Dimensional Fabrics

Shaped and three-dimensional fabrics for technical applications, such as upholstery and geotextiles, are weft knitted in a single operation on computer-guided V-bed machines. The same techniques may be used to make one-piece garments and shoe uppers.

Teknit, a specialist knitting company acquired by Camira Fabrics, developed the first three-dimensional upholstery covers for General Motors in 1989. Later, it developed solutions for knitted covers without foam padding, such as for chair backs (**image 1**).

Modern computer-guided machines are quiet and efficient and have a small footprint, so even relatively small mills can output significant numbers of parts (**image 2**). It takes roughly 20 minutes to knit a complete cover, depending on the complexity.

CAM data is sent directly to the knitting machine. The cam carrier controls all of the knitting operations (**image 3**). It travels from side to side and, depending on the number of heads within it, several courses may be stitched in a single pass. In this case, there are three heads: two are knitting and the third is transferring loops between needles. As the carrier passes over the needles, they are raised upwards (unless selected to hold the stitch, in which case they stay down) and the new yarn is fed in front of the hook (**image 4**). A sinker occupies the space between each needle. One of the breakthroughs that enabled three-dimensional parts to be knitted in a single piece was changing the way tension is applied to the knitted fabric. Instead of drawing underneath with a take-down roller, pressure is applied by the sinkers, which push each loop from above. Therefore, the shape of the fabric does not affect the tension that can be applied. Meanwhile, brushes stop the latches flicking up and closing the hooks when knitting at high speed (**image 5**).

Polyester yarn is popular for this type of upholstery (800–1,000 dtex), because it has good dimensional stability even under tension. The ends are fed into the machine via tension control (**image 6**). Several ends are fed to each feed and combined in parallel formation. This produces the correct gauge and quality.

As the knitting cycle progresses, the outermost wales knit the least number of courses. The part is designed to fit around a foam seat without cutting and stitching (**images 7** and **8**). Staggering the different number of wales versus courses to make a three-dimensional part reduces the visible stitch pattern to produce a smooth profile.

1

2

3

4

5

6

7

8

Featured Company

Camira Fabrics
www.camirafabrics.com

Technical front   Technical back

Double-Sided
Material: Polyester
Application: Upholstery
Notes: Two colours are used to produce a double-knit with mirrored patterns on either side.

Technical front   Technical back

Graphics
Material: Polyester
Application: Upholstery
Notes: A logo is made by reversing the placement of the structural backing yarn (red) with the bulky surface yarn.

Technical front   Technical back

Repeat Pattern
Material: Polyester
Application: Upholstery
Notes: Two colours are used to produce a double-knit with different patterns on either side.

Technical front   Technical back

Relief Pattern
Material: Polyester
Application: Upholstery
Notes: Bulky polyester yarn opens up in the tuck stitches due to the increased length of the loop. Thus, contrasting surface texture is created with a single yarn.

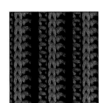 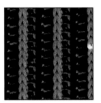

Technical front   Technical back

Ribs
Material: Polyester
Application: Upholstery
Notes: The ribs are widely spaced and produced in alternating colours to create a more pronounced relief profile.

Technical front   Technical back

Mesh
Material: Polyester
Application: Upholstery
Notes: A structural filament yarn is knitted onto the back side for added dimensional stability.

WEFT KNITTING

137

Case Study

→ Knitting Three-Colour Double-Faced Fabric

An almost unlimited range of colour and relief patterns can be reproduced on V-bed machines. The two sets of needles work together to knit double-faced fabrics, and the different colours of yarn are on either the technical face, the technical back, or floated or tucked in between, depending on the stitch.

This sample, produced by Camira Fabrics to demonstrate its technical knitting capabilities, consists of three colours. The technical face and back show yellow and dark-grey vertical stripes (**images 1** and **2**), which are made up of a combination of alternating plain and float stitches (**images 3** and **4**). The border does not have any visible yellow on the face, because it was knitted on the back side (**image 5**). Increasing the length of the float reduces colour density, and vice versa.

1

2

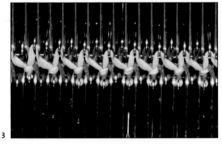

3

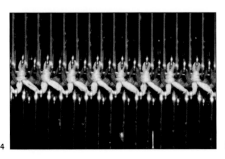

4

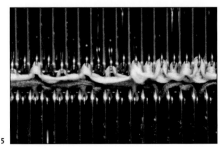

5

Textile Technology
Circular Knitting

With this technique seamless tubes of fabric are knitted with a spiral of courses running around the circumference. It is used to make long lengths of fabric that are slit into open widths during finishing, or to make garments and hosiery in a single operation. Producing items in a single piece eliminates cutting and sewing.

Formats	Materials	Costs
• Continuous lengths (tubular or flat) • Finished one-piece garments (hosiery)	• Staple, filament and novelty yarn • Natural and synthetic • Inlaid yarn	• Low unit cost • Very rapid production

Applications	Quality	Related Technologies
• Apparel	• High stretch • Continuous yarn • Smooth or pile	• Loom weaving • Warp knitting • Weft knitting

INTRODUCTION

Circular knitting is a mass-production process used to make long lengths of fabric. One or two beds of needles are used, creating a range of design opportunities similar to weft knitting (page 126). Large-diameter machines are used to knit continuous tubes, which are slit into sheets; medium-diameter machines are used to make seamless garments, such as vests and t-shirts without side seams; and small-diameter machines make hosiery, such as socks and tights.

Unlike flat-bed processes, which produce tube profiles by joining opposite sides together, circular knitting constructs

The Circular Knitting Process

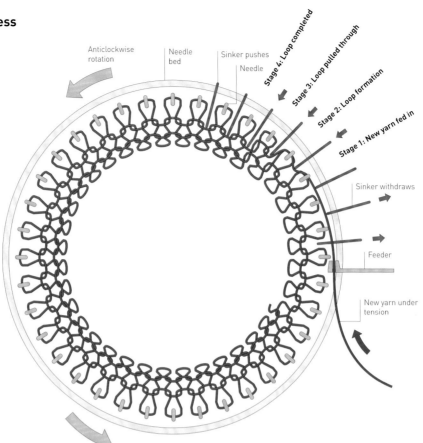

Anticlockwise rotation

Needle bed

Sinker pushes

Needle

Stage 4: Loop completed

Stage 3: Loop pulled through

Stage 2: Loop formation

Stage 1: New yarn fed in

Sinker withdraws

Feeder

New yarn under tension

tubes with a spiral of courses running around the circumference. Fancy fabrics are made with a combination of plain, tuck and float stitches. Machines are equipped with multiple feeds, and different colours can be introduced at any point along the circumference so as to make intarsia or jacquard.

On modern computer-guided machines the diameter and gauge can be adjusted with very little downtime, making the process more flexible.

APPLICATIONS

Circular knitting is utilized in the production of long lengths of single and double jersey used to make apparel such as vests, t-shirts, sweaters, underwear and dresses. Yarns are laid-in and formed into pile, such as in the manufacture of towel and fake fur.

Tubular apparel, such as socks, tights (pantyhose), t-shirt bodies, vests, hats, medical products and food packaging are constructed in a single process.

TECHNICAL DESCRIPTION

The weft-knitting process is the same whether it is on a flat-bed (page 128) or circular-knitting machine. In stage 1, the needle has an existing loop and rises up. As it does so, the yarn is fed in front of it, above the open latch (this diagram shows circular knitting from above). The needles face outwards. In stage 2, a loop is formed around the shank of the needle by sinkers moving towards the centre. In stage 3, the needle catches the loop and draws it through the existing loop to form a new loop. The latch is closed by the existing loop. And in stage 4, the new loop is completed and the needle rises up to receive the next feed. In this way, multiple yarns may be introduced in close proximity. This builds several courses in tandem, accelerating the process. Sock-knitting machines may have six or so, whereas jersey knitters may have more than 50. The yarns may be plain knitted, floated, tucked, plated or removed, depending on the desired effect.

Each new loop is formed as the needle bed rotates anticlockwise at high speed. The sinker aids loop formation and the sock is drawn downwards by a combination of gravity and suction. The yarn is fed in under very accurately controlled tension. This allows for higher knitting speeds and ensures consistent and high-quality fabric production.

The fabric is constructed in spiral formation and is drawn out through the centre of the needle bed. Therefore, there are no edges, except at the start and end. The first row is cast on and so secure. The final row must be finished to stop it unravelling. In sock making, the toe forms the last row and so is linked together to secure the loops (see Linking, page 386).

RELATED TECHNOLOGIES

The quality of knitted fabric is very different from that of woven fabric (pages 76–95). Even so, the two types of material are applicable for many of the same applications. Thus, process selection is determined by the qualities desired in the finished article.

Knitting is ideal for parts that require flexibility, high stretch and recovery, or a porous and airy structure (with the same quantity of yarn, knitting produces less cover than weaving). Knitting is capable of producing openwork structures similar to machine-made lace (page 106).

The choice of knitting technique depends on the quantity, cost and desired performance of the finished part. Circular knitting is unrivalled in the production of seamless tubular profiles and large quantities of consistent fabric. By contrast, flat beds are much more flexible and capable of producing flat, fully fashioned and three-dimensional parts.

QUALITY

Circular knitting produces high-quality and consistent fabric. Quality is determined by yarn, gauge and stitch. Each type of fibre blended into the yarn provides specific properties, such as wool for warmth and comfort, and nylon for wear and abrasion resistance.

Like flat-bed knitting, the fabrics have high stretch and recovery. The loop structure allows relatively free movement of the yarn and is easily distorted. Therefore, knitted garments adapt well to body shape and movement.

Floats and pile are used to add specific properties, such as colour, comfort and insulation. However, they are prone to snagging and care must be taken to ensure their length does not hinder the article's quality and performance.

One-piece items are typically knitted from the top down, such as socks, which are knitted from leg to toe. There is no seam and the last row of loops is finished by linking or stitching (see Linking, page 386), producing a robust loop structure unlikely to unravel.

DESIGN

Circular knitting is used to produce fabrics with the same combination of stitches, colour and pattern as flat bed. Those equipped with single needles are similar to bar frame. And if they have two sets of needles they are capable of the same fancy fabrics constructed on the V-bed, such as interlock, pile and inlaid yarn.

Circular-knitting machines are limited to making continuous lengths of tubular fabric to a set dimension. On some machines it is possible to change the gauge and diameter but this may be a time-consuming operation and so increases cost (depending on the machine). With modern machines and multiple feeds, a wide range of visual and technical properties can be achieved, such as graphics, plating and pile.

During knitting, the diameter of the knitted tube is controlled and changed (within reason) by adjusting the gauge, tension, stitch, elasticity of the yarn, loop length (controlled by needle height) or a combination of these techniques. The amount it can be decreased, relative to the diameter of the knitting machine, depends on the size of the machine (larger tubes have more wales to affect) and capabilities of the equipment. Knitted thermoplastic yarn, such as nylon, may be heat set with a tailored profile to better fit the body. In a process known as boarding, stockings and tights (pantyhose) are stretched over a metal leg form (board) of a specific size and heat set by steam treatment (see Boarding, page 416).

To make three-dimensional shapes, such as sock heels and toes, courses are added across designated wales. This technique is similar to flat-bed knitting seamless three-dimensional upholstery and shoe uppers (page 430).

As in flat bed, colour patterns and graphics are achieved by printing (pages 256–79), dyeing (page 240) or using different-coloured yarn. Complex colour patterns and graphics are produced with individual stitch selection, known as jacquard. Adding colours or increasing complexity reduces production output.

Yarn selection will affect the colour range that can be achieved. For example, wool and nylon are dyed together, whereas polypropylene (PP) has to be extruded as solid colour and then blended as a staple fibre (page 56). This reduces flexibility, because coloured filament yarns will have minimum order requirements, unless they are standard colours already produced in high volume.

In addition to printing, knitted fabrics are finished with a range of processes, including calendering (page 220) and napping (page 216).

MATERIALS

Natural and synthetic staple and filament (page 50) yarns are all suitable. The choice of yarn will be determined by the application, performance requirements, cost and volume. Silk is fine, bright, strong and expensive, whereas man-made yarn is less variable and low-cost and can be engineered to perform a specific function, such as moisture wicking.

COSTS

Long lengths are produced at high speed and so have low unit cost, depending on the yarn. The consistency of the yarn is critical to maintain high-quality knitting. This increases cost.

When producing finished articles in a single process, the costs of pattern cutting and stitching (pages 342–69) are reduced or eliminated.

ENVIRONMENTAL IMPACTS

Knitting is a relatively efficient process and generates little waste. Garments produced in a single process, without cutting and sewing, reduce assembly procedures and waste to almost zero. Like flat-bed weft knitting, the total environmental impacts are determined by yarn selection, processing and finishing.

Technical face

Plain Knit, Fillet Rib and Slack Rib
Material: Wool, polyester, PP and elastane
Application: Technical sock
Notes: Different yarns and stitches are combined to produce a durable and comfortable sock with adequate stretch and support.

Technical face

Stretched Rib
Material: Wool, polyester, PP and elastane
Application: Technical sock
Notes: Stretching the fabric widthwise reveals the knitted ribs – once the tension is released the elastane pulls the fabric back to its original position.

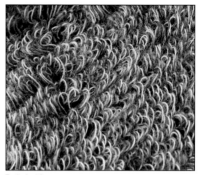

Technical face

Terry Loop
Material: Wool, polyester, PP and elastane
Application: Technical sock
Notes: Terry loops are used to provide padding in strategic locations (to reduce bulk elsewhere) around the sock.

Technical back

Inlaid Yarn
Material: Wool, polyester, PP and elastane
Application: Technical sock
Notes: Elastic yarn is inlaid into the knitted rib structure – running in parallel lines and not contributing to the knitted loop structure – to improve stretch and recovery properties for comfort and fit.

Intarsia and Jacquard
Material: Wool, polyester, nylon and elastane
Application: Technical sock
Notes: With individual stitch control, multiple colours may be incorporated seamlessly (on the face) to reproduce intricate graphics. The pattern follows the shape of the stitches.

Technical face

Technical back

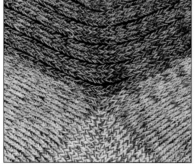

Technical face

Three-Dimensional Knitting
Material: Wool, polyester, PP and elastane
Application: Technical sock
Notes: The heel is made up of additional courses covering only half the wales, forming a pouch.

Openwork
Material: Polyester, nylon, PP and elastane
Application: Technical sock
Notes: A fine openwork structure is formed using lightweight yarns – consisting of textured filament yarns twisted to make bulky plied yarn – and a combination of float and tuck stitches.

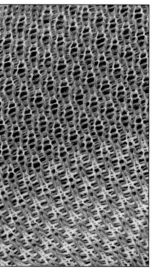
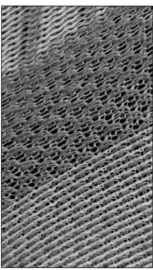

Technical face

Technical back

→ Circular Knitting Technical Socks

Tubular one-piece socks are knitted with a three-dimensional heel and toe. The Bridgedale Trekker is made up of nylon, wool and elastic yarn (**image 1**). The exact mix of materials is determined by the desired performance and location in the sock. For example, wool provides comfort and warmth; nylon and polypropylene (PP) are used to provide abrasion resistance and tensile strength; and elastic yarn ensures the sock fits tightly. They are either used as filaments, blended into staple yarn or plied (page 60) to take advantage of their distinct properties.

The needle bed runs around a brass cylinder (**image 2**). As the sock is knitted it runs down into the cylinder and away from the needles, drawn partly by gravity and partly by suction. The machine is closed during knitting and the head rotates at high speed (**image 3**). The machines are computer guided and the data is transferred directly from computer-aided manufacture (CAM) software (**image 4**).

Knitting progresses from the top of the sock towards the toe. It is knitted in a spiral, with each course building on the preceding one. A combination of stitches is used, such as purl for added bulk (**image 5**), floats and looped pile for comfort and insulation (**image 6**).

The shaped profile of the heel and toe are formed by temporarily halting the circular knitting action and knitting additional courses back and forth in a hemisphere. Courses of varying length are built up to create a pouch (**image 7**). Afterwards, circular knitting recommences to form the tubular foot profile.

The toe is made in the same way as the heel. Once knitted, the last course of loops is transferred onto a toe-closing frame (**image 8**). Held on a hemispherical needle bed, the live loops in the toe are joined together by chain stitching (**images 9** and **10**). The process is known as linking, and completes the circuit to form a seamless and durable sock.

Colours, patterns and graphics are incorporated into the structure using different feeds (**image 11**). Combined with individual stitch control, this provides a range of colour and pattern opportunities.

KNIT

142

1

2

3

4

5

6

7

8

9

10

11

Featured Company

Bridgedale Outdoor Ltd
www.bridgedale.com

Sheet	Layers	Block	Composite	Shaped	Relief	Pattern

Textile Technology

Warp Knitting

In warp knitting, one or more yarns are fed to each needle and the stitches are made simultaneously, making this a very rapid process. With this technology it is possible to make plain, tubular, spacer and openwork textiles for a variety of applications. Coloured and patterned fabrics are produced using raschel jacquard knitting with independently controlled guide bars.

Formats		Materials	Costs
• Plain	• Openwork (net and lace)	• Staple, filament and novelty yarn	• Low to moderate unit cost
• Tubular		• Natural and synthetic	• Very rapid cycle time
• Spacer			

Applications		Quality	Related Technologies
• Apparel	• Interior	• Intermeshed structure less prone to ravel	• Fancy loom weaving
• Footwear	• Technical	• Continuous yarn	• Machine weaving lace
		• Dimensional stability	• Weft knitting

KNIT

144

INTRODUCTION

Machine warp knitting is a more recent development than weft knitting (page 126). The looped structure is formed between multiple yarns running lengthways (wales), as opposed to a single yarn across the width of the fabric (courses), as is the case in weft knitting. Each needle is fed with one or more yarns. So a long-sleeved t-shirt may consist of up to 5,000 ends, depending on the machine gauge. With each stitch, the yarn is tracked from side to side by the guide bar and so intermeshes with neighbouring yarns to produce a cohesive knitted structure.

With each stitch being formed by a separate yarn, a range of interlocked, inserted yarn and open structures may be produced. And it can be made in a range of densities, from heavy cloth to fine lace. By combining different loop patterns, net may be incorporated into a solid ground, such as used to make upholstery and seamless sportswear.

APPLICATIONS

Single needle-bed machines are used to make plain and patterned fabrics that are used flat, or cut and sewn into garments, underwear and upholstery. Lace and nets are used to make garments, underwear, curtains, fishing nets, mosquito nets and packaging. Warp knitting is particularly well suited to produce sequential lengths of fabric with fringed ends, such as scarves.

Industrial applications include reinforcing composite laminates, conveyor belts, safety clothing, tarpaulin, advertising banners and awnings.

Double needle-bed machines are used to make all that single beds are capable of, plus seamless and lightweight sportswear, swimwear, underwear, lingerie and gloves. Jacquard knitting is

The Warp Knitting Process

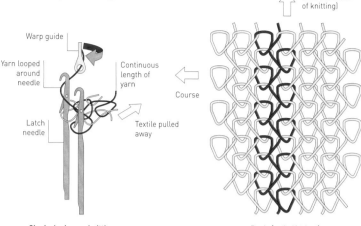

Single-bed warp knitting

Warp guide

Yarn looped around needle

Continuous length of yarn

Latch needle

Textile pulled away

Wale (direction of knitting)

Course

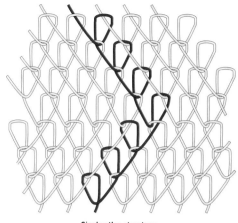

Basic (or half tricot) technical back

Single atlas structure technical back

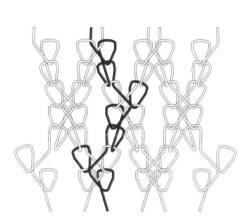

Diamond net technical back

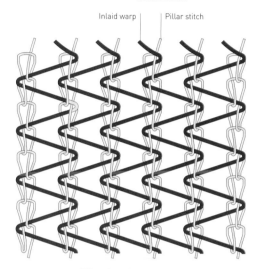

Inlaid warp

Pillar stitch

Pillar stitch with yarn insertion technical back

TECHNICAL DESCRIPTION

Warp knitting forms courses and wales, just like weft knitting. However, the structure of the loops and the range of possibilities for loop formation are different.

Each needle is fed a separate yarn via a warp guide. The guide passes the yarn around the needle with each cycle. The warp yarn is passed from one course to the next, thus creating a zigzagging intermeshed structure. Each loop in the same course is formed simultaneously.

Single-bed machines, known as tricot, are used to make flat sheets of textile. The basic warp knit, which is also known as a half tricot, is produced with a single guide bar. Lapping the stitches in the same direction for a specific number of courses, and back again, produces an atlas structure. The sideways movement produces a zigzag

pattern down the length of the fabric. Due to the tension on the yarn, only one overlap in either direction is usually permitted.

Overlapping the yarns at intervals forms an openwork structure. With conventional knitting machines, the guide bars move in tandem, so each stitch in a course on the same bar will be the same. This means that only balanced nets are possible, unless knitting on raschel jacquard.

A pillar stitch is made up of chains of stitches that form unconnected wales. They must be connected by an inserted yarn, which is overlapped by a second guide bar (for a detailed description of warp knitting with inlaid weft that does not pass around the loops, see Stitch Bonding, page 196). Other types of two-bar fabric include lock-knit, double atlas (the two guide bars

mirror each other's movements to create a balanced fabric) and pile structures (long floats are formed on the technical back, tied into a knitted structure). Tricot machines may have up to four guide bars.

Raschel machines have one or two needle beds. Machines equipped with two needle beds may have up to three guide bars for each bed (six in total). They are set up in a similar way to V-bed weft-knitting machines (page 134) with the needles back to back. With this set-up, yarns may be overlapped between both sets of needles at any point. This is how seamless tubular structures, such as gloves and t-shirts, are produced. With jacquard, each yarn guide is controlled independently. This is how complex patterns are reproduced.

used to make nets and lace employed in all manner of applications, from sportswear and lingerie to shoe uppers, medical products and upholstery.

RELATED TECHNOLOGIES

Warp knitting is not as straightforward to design for as weft knitting, and so remains less widely used. Processes used to join panels into finished garments, such as linking (page 386), are only compatible with the weft-knitted loop structure, although warp-knitted fabrics may be joined by lock stitching (page 355).

Both weft and warp knitting are capable of producing seamless garments in a single operation. A significant advantage of warp knitting is the design opportunities of openwork structures. For example, lightweight garments with localized stretch, support and breathability are produced in a single operation from multiple yarns. In addition to creating openwork structures in a solid ground, the two may be combined in a single sheet to create a honeycomb layer on top of the ground, known as semi-breakthrough.

Warp knitting is faster than weaving and capable of producing larger widths, making it more economical. With regard to lace, there is a wider range of design opportunities, owing to the different loop structures. Machines fitted with independently controlled guide bars (jacquard) are capable of producing very complex and intricate patterns similar to Leavers lace (page 108).

Warp knitting is used to make double-faced and spacer textiles, similar to fancy weaving (page 84). Both technologies are used in composite constructions such as laminating and thermoplastic molding (page 446).

QUALITY

Thanks to the different number of needle beds and guide bar arrangements, a wide variety of fabric qualities can be achieved, ranging from stable to high stretch; dense to open; smooth to lofty; and single- to double-faced.

The amount of stretch depends on the stitch and the yarn. Warp-knitted fabrics have varying amounts of lengthwise stretch and little or no crosswise stretch (weft-knit fabrics stretch more across their width). And certain constructions, such as stitch-bonded nonwovens (page 196) and inlaid weft, may have virtually zero stretch in either direction.

Warp-knitted fabrics are less likely to ravel and run than weft-knitted types, because the loop structure is formed from multiple yarns intermeshed in a zigzag formation.

Lightweight warp knits are common in sleepwear, because they are soft and have good drape. The simplest fabrics are characterized by vertical ribs (wales) on the face and horizontal ribs (courses) on the back.

DESIGN

Basic fabrics, also known as tricot, are knitted with a single set of yarns and single guide bar. Because a stitch is formed on each needle simultaneously, the machine can run at very high speed – up to 2,000 courses per minute. Tricot machines are capable of producing fancy fabrics by floating yarns over up to five wales. For example, satin textiles are soft with high lustre; pile is made by breaking the floats during finishing (see napping, page 216); and nets are constructed by knitting pillars without overlap for one or more courses.

Knitting with two sets of yarn requires two or more guide bars, which provides greater design freedom. As in weft knitting, the two needle beds are set up in a V configuration. Known as raschel, this technique is capable of producing double-faced, tubular and openwork structures. And by adding jacquard (independent guide bars), complex and intricate patterns may be introduced into the structure. Termed raschel jacquard, this is the most versatile form of knitting, but because of the complexity of movement it is very challenging to design for.

Warp knitting is the most versatile process for making nets. They can range from semi-breakthrough to wide open, depending on the structure. Like other warp-knitted structures, they are resistant to slipping, ravelling and runs. Therefore, they are used to produce many types of net, including lightweight packaging,

industrial linings and geotextiles.

Spacer textile is made up of two face fabrics connected by spacer yarn, which is perpendicular to the outer face fabrics. The thickness, density and resistance to compression are determined by stitch and yarn selection.

Yarns are inserted to provide visual or performance-enhancing properties (see also stitch bonding with inlaid weft, page 201). Weft-insertion fabrics have yarn laid in across the wales, and warp-insertion knits have yarns laid in-line with the vertical chains. There are many benefits of using these techniques, such as introducing yarns that otherwise would be impractical to knit (too delicate, too coarse or too fine) and reducing or increasing stretch and recovery, handle, surface or weight. Combined with jacquard, different yarns and configurations are used to provide colour, pattern, graphics, stretch, bulk and so on.

With the yarn insertion technique, it is possible to make uni-, bi- and multiaxial textiles, such as for high-performance composite laminates. One or more sets of yarns may be inserted into the secure knitted structure.

Compared to weft knitting, there is less stress applied to the yarn during production. Therefore, warp knitting is used to construct fabrics from relatively less-flexible yarns, such as glass, aramid and carbon.

The surface of warp-knitted fabrics is finished with conventional techniques, such as calendering (page 220). Pile is made using inserted yarns, or with floating stitches, which are tied into the ground structure to form secure loops. Cut pile may be made by cutting looped pile, or by splitting double-faced fabric to make two cut-pile fabrics. These techniques are used to make velvet, velour and fake fur.

Colour is applied in one of three ways: knitting coloured yarn, piece dyeing (page 240) the knitted item, or printing (pages 256–79). Setting up warp-knitting machines is a lengthy process, because one or more ends are threaded for each needle. Therefore, it is more efficient to knit plain-coloured yarn and dye the finished garment. By mixing two types

Technical face   Technical back

Half Tricot
Material: Polyester
Application: Apparel lining
Notes: Half tricot is a sheer fabric produced with a single guide bar making repeated lapping motions.

Technical face   Technical back

Tricot
Material: Polyester
Application: Sports shoe
Notes: Full tricot has a smooth face with good cover, making it an ideal fabric for printing.

Atlas
Material: Polyester
Application: Laminating
Notes: The lapping movement back and forth produces a zigzag pattern down the fabric.

Technical face   Technical back

Two-Colour Raschel Jacquard
Material: Polyester, nylon and elastane
Application: Sportswear
Notes: Two types of yarn are jacquard knitted and plated to either bring the colour to the front, or hide it inside.

Two-Colour Net
Material: Polyester, nylon and elastane
Application: Hosiery
Notes: The two different types of yarn react to different dye types, so can be coloured after knitting.

Printed Net
Material: Polyester and elastane
Application: Hosiery
Notes: Complex and intricate patterns are reproduced by transfer printing.

Novelty Yarn
Material: Wool, nylon and elastane
Application: Hosiery
Notes: Novelty yarns are incorporated in the knitted structure to create a unique look and feel.

Diamond Net
Material: PE
Application: Fruit packaging
Notes: Plastic strip is knitted together to form strong and lightweight packaging for fruit and vegetables.

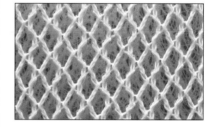

Spacer Textile
Material: PP
Application: Upholstery
Notes: A double-faced diamond net structure is held together by spacer yarns to produce a cushioning fabric.

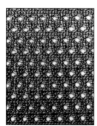 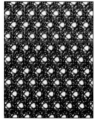

Technical face   Technical back

Honeycomb Mesh
Material: Titanium-coated polyester
Application: Apparel lining
Notes: Nanometal-coated meshes are used for technical applications, such as heat retention lining.

Pillar Stitch Mesh
Material: Polyester
Application: Technical apparel
Notes: Horizontal underlapping reduces stretch particularly in openwork structures.

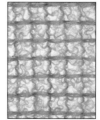 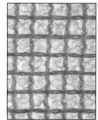

Technical face   Technical back

Pillar Stitch with Textured Inlay
Material: Ramie and nylon
Application: Bath cloth
Notes: Pillar stitches are combined with inlaid warp to form a grid. The textured yarn creates a wavy structure.

Case Study

→ Warping

In preparation for warp knitting, filament yarn is wound onto cones. From a creel, 256 ends are gathered together with a fine comb (**image 1**). The yarn is coated with a layer of synthetic oil (natural yarns are coated with paraffin wax) as it passes over a roll (**image 2**). The coating reduces friction between the yarns and so improves knitting efficiency.

It takes roughly an hour to fill a beam (**image 3**). The northern Italian company Cifra uses machines with 3 m (9.8 ft) beds. Six beams are required to cover the length of the bed (**image 4**). The double-bed machines use up to three beams on either side, so 36 in total.

Each end is carefully threaded through the machine and into the guide bar (**image 5**). It is a delicate process and can take several operators half a day or more to replace the yarns on a machine (**image 6**). Therefore, Cifra warp knits all of its garments in plain white and dyes the finished items if colour is required. Using two different types of yarn, such as polyester and nylon, means that two different colours can be achieved with dyeing (page 240).

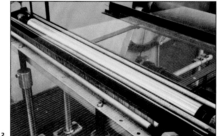

of yarn that are affected by different types of dye, multicoloured patterns can be achieved.

MATERIALS

Man-made filament yarns (page 50) are the most commonly utilized, owing to the high consistency required, although staple and novelty yarns are also used. Frequently used filament yarns include silk, nylon, polyester, polypropylene (PP), polyethylene (PE), viscose and elastane.

Technical yarns are warp knitted for high-performance applications, such as composite-laminated structures. These include carbon, aramid and glass.

COSTS

Warp knitting is a very rapid process, able to produce hundreds of courses per minute. It is possible to produce wide lengths of fabric or several narrow strips. In this way, a single machine may be capable of producing several garments at once, further reducing cycle time.

The high consistency of yarn required increases costs slightly.

Setting up the machine takes longer than weft knitting if the yarns need to be changed, because each needle is fed with a separate end. Knitting a single colour and piece dyeing finished garments saves time and cost by reducing machine downtime.

ENVIRONMENTAL IMPACTS

Like weft knitting, this is an efficient process and generates very little waste. Producing garments in a single operation removes pattern cutting and sewing, further reducing or even eliminating waste completely. Warp-knitted spacer fabrics are used in place of polymer foams, such as polyurethane (PU) in cushioning applications.

The source of yarn is important; natural yarns should be from sustainable sources and processed with consideration for people and the environment. The amount of recycled content in synthetic yarns varies according to the type of plastic and the supplier.

4

5

6

Featured Company

Cifra S.p.A.
www.cifra-spa.net

→ Seamless Warp Knitting

Seamless garments are produced on raschel jacquard machines. The fabric is made up of pillar stitches, underlapped and overlapped with inserted yarn. With this technique, openwork structures are seamlessly combined with plain fabric to make fitted garments (**image 1**).

Plain white yarn, in this case polyester, is fed from the beams into the guide bars (**image 2**). These are fine-gauge machines with 24 needles per 25.4 mm (1 in.). The more needles per inch, the lower the diameter of yarn that can be knitted and the higher the density of fabric that can be produced.

Each guide bar moves through three actions with each stitch. When the needle has cleared the previous loop the guide bars move forwards (**image 3**). Each guide bar is independently computer-guided. They move sideways and pass the yarn around a designated needle, either in front or behind (**image 4**). The needle is pulled downwards, catching the yarns placed in front and forming a stitch (**image 5**). The second bed of needles rises up and the sequence is repeated in reverse.

The process is very rapid: up to 600 courses are completed every minute and the knitted item is drawn downwards under tension (**image 6**). The finished garment emerges from the machine and is ready to be dyed (**image 7**).

Garments are checked to make sure the knitting process is running without errors (**image 8**). A long-sleeved t-shirt like this comprises around 1,322,000 stitches and a single one out of place will show up on the finished product. It is made oversize to allow for shrinkage during dyeing and finishing. The outline is knitted into the garment and the waste is removed (**image 9**).

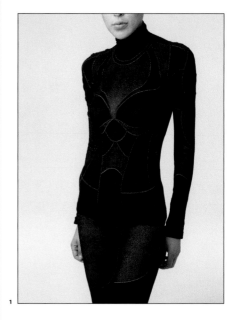

1

2

3

4

5

6

Featured Company

Cifra S.p.A.
www.cifra-spa.net

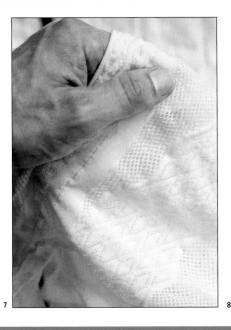

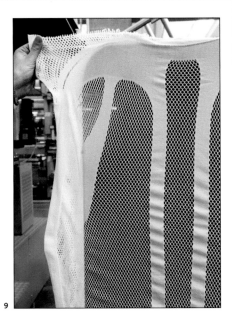

7

8

9

→ **Warp Knitting with Novelty Yarn**

Yarns are inserted for decorative and functional reasons. In this case, a gold-metallized yarn is being inserted into jacquard-knitted hosiery (**image 1**). The yarn would be too fragile to knit as part of the loop structure and so is inserted into the loops as an additional warp.

The fancy yarn is introduced from the front on both sides of the machine (**image 2**). It is knitted into the fabric and the finished item is drawn away under tension (**image 3**). The areas that will be open are visible as long slots. This is how they are knitted: a needle is missed out for a given number of courses and so a window is formed. Once the fabric has relaxed the openwork structure is revealed.

On this narrow-bed knitting machine eight pairs of stockings are knitted simultaneously (**image 4**).

1

2

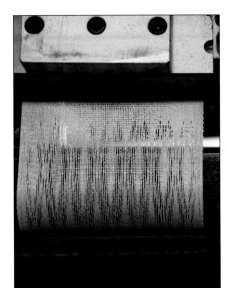

3

4

Featured Company

Cifra S.p.A.
www.cifra-spa.net

Textile Technology

Nonwoven

An important and fast-growing part of the textile industry, this family of materials includes all fibre-based nonwoven and non-knitted textiles. The random-oriented fibres are mechanically interlocked together, or they are bonded together using heat or adhesive. The range of applications is as diverse as the choice of materials.

Formats		Materials	Costs
• Mechanical (needle punching and hydro-entanglement)	• Bonded (thermal or adhesive)	• All types of fibre, depending on the process	• Low cost (with a few exceptions) • Rapid cycle time

Applications		Quality		Related Technologies	
• Apparel • Cushioning and insulation	• Geotextiles • Technical textiles	• Random-oriented fibres • Good tear strength	• Lightweight • Poor drape	• Plastic film extrusion • Leather	• Foam • Laminating

INTRODUCTION

Nonwoven textiles are constructed from synthetic, natural, metallic or a mixture of fibres. There are many techniques used in their construction, but there are two fundamental approaches: fibres are either combined mechanically, such as entangled; or they are bonded with heat or adhesive. Nonwovens eliminate the need for yarn formation and so are typically very cost-effective. Either staple fibres are used, or filaments are taken directly from the extrusion process (see Filament Spinning, page 50).

Mechanical entangling is used to combine all types of staple fibre. Developed around the 1840s, needle punching evolved as a high-speed and cost-effective alternative to wool felt. It is used to combine a single type of fibre, or a mixture to take advantage of the properties of each. Staple length ranges from 20 to 150 mm (0.8–6 in.), depending on the application and selected material. For example, short-fibre nonwovens are used for low-stress applications, such as insulation, whereas long-fibre types can be used for high-load applications, such as geotextiles and automotive.

Bonding includes thermal and adhesive techniques. Thermoplastic fibres soften when heated (or straight from the extrusion spinneret) and are bonded together under pressure to form high-strength and lightweight nonwovens down to less than 2 g/m² (0.8 oz/ft²). Alternatively, natural and thermoplastic fibres are mixed with adhesive resins or powders to form composite nonwovens.

Nonwovens are finished with a range of processes, including calendering (page 220), printing (pages 256–79), flocking (page 280), dyeing (page 240) and die cutting (page 336). Additionally, bulky nonwovens are steam treated to produce a stiff, dense and more durable material.

The Fibre Web Process

Carding

Dry laying fibre web

Air laying

Spun bonding

Direct laying fibre web

Needle punching

Mechanical interlocking

Resin infusion

Web bonding

Thermal bonding

TECHNICAL DESCRIPTION

Nonwovens are fabricated in a three-stage process: forming a fibre web; entangling or bonding the fibres together; and finishing. At all stages, the processes may be used in isolation, or combined to produce a material with unique properties.

Dry fibre web techniques include carding and air laying. Alternatively, fibre webs are formed in a wet process similar to papermaking. All these processes use staple fibres. Carding is similar to the process used to make staple yarns (page 56). Blended fibres are fed between the nip rolls and onto the swift roll. Teeth covering the surface of the rolls align the fibres into a fine web, which is peeled from the doffer by the oscillating fly comb. This aligns the fibres in the direction of travel. Therefore, to increase strength and pliability in the finished material, layers are stacked perpendicular to one another before they are permanently fixed in place by bonding or entanglement.

Air laying is used to produce fibre webs from shorter staples. The fibres are blown onto a screen to produce a fine, lofty web with random orientation. At this stage, the webs are self-supporting, but only loosely held together by friction.

The fibres must be bonded together or entangled to form a usable textile. Mechanical techniques interlock the fibres throughout the material. Needle punching involves piercing the fibre web with thousands of barbed needles, drawing fibres from one side to the other. Passing the web through a second needle bed, working in the opposite direction, produces a fully interlocked fibre structure. The amount of interweaving depends on the length of fibre.

A similar entanglement can be achieved using water jets instead of needles. The high-speed water pierces the web, causing the fibres to curl and become entangled.

Bonding is either thermal or adhesive, and these techniques may be used to supplement mechanically entangling fibre webs.

Adhesive is applied by impregnation, printing, spraying or coating (page 226).

Printing allows for selected areas of the fibre web to be treated, such as for making a more pliable fabric.

Thermal bonding is a commonly used technique. The fibre web is passed between metal rolls. With heat and high pressure the fibres are permanently joined together. The finish may be smooth or patterned, depending on the surface profile of the calender roll. Alternatives include steam treatment and ultrasonic welding.

Spun bonding is a direct laying process, which means no further entanglement or bonding is required. Filaments are extruded through a spinneret, cooled, stretched and dispersed onto a collection surface. The slightly molten fibres bond together as they are laid down. The fibre web may be further processed and strengthened with thermal bonding or needle punching. A variation of this technique, known as melt blown, involves cutting the filaments into staples and air laying them onto the collection surface. The fibre web is not as strong as those made from filaments.

APPLICATIONS

Nonwoven fabrics are important for a wide range of industries. Applications in clothing include suede- and leather-like fabrics, cushioning, padding, liners and insulation. They are used to provide padding for upholstery and mattresses, and as an insulating layer in bedspreads and pillowcases.

Medical and personal hygiene applications include single-use gowns, dressings, bandages, padding, nappies (diapers) and wipes.

They are also used for technically very demanding applications, such as fabricating the fibre reinforcement for composite laminating (page 188), sound absorption (automotive and interiors), liquid absorption (such as cleaning up oil spills), aramid fibre ballistic protection, drainage (geotextiles) and filters.

Conductive fibres, such as carbon fibre, are added to make nonwovens for anti-static applications, such as for resistive heating, electromagnetic interference (EMI) shielding and carpet underlay.

RELATED TECHNOLOGIES

Nonwovens have replaced many established materials in everyday applications. For example, spun-bonded ultra-high-molecular weight polyethylene (UHMWPE) has replaced paper, such as for making maps, as well as being converted into disposable workwear (industrial and medical) and nappies. Lofty nonwovens, such as latex-bonded coir, provide an alternative to plastic foams (page 172) for padding in upholstery (page 404).

Blown films (page 180) can be very thin and micro porous. As a result, they may be used to provide the same properties to laminated fabrics (page 188), or hygiene products.

Synthetic leather is easier to maintain, more consistent and less expensive than natural leather (page 158). Nonwoven synthetic leather and suede are made by needle punching. Other techniques used include coating paper with a foam polyvinyl chloride (PVC) or laminating.

QUALITY

Generally, nonwovens have high tear strength and no grain, will not unravel or run and do not drape well. Depending on the materials and processes used, the properties range from light and strong (sheer) to light and insulating (skiwear), or very dense (sound absorption panels).

Mechanically entangled fabrics have a soft feel and dull surface finish. Thermal treatment increases stiffness by compacting the fibres.

Needle punching puts many tiny holes in the fibre structure, which may be visible in some materials. Needle density averages 170 per cm² (1,090/in.²). One of the most extreme cases is synthetic leather, which may have up to 15,000 punches per cm² (96,000/in.²). The fibres in these types of materials are very fine and so are the needles.

Thermal-bonded fabrics can be matt or gloss, depending on the texture of the calender roll. Ultrasonic-welded fabrics are covered with a fine pattern, left by the tools impressing into the surface. The fibres are strongly bonded and the material maintains a soft feel, because the fibres are looser between the weld points.

Thermoplastics may be thermally bonded straight from the extrusion process, because they will join together while still slightly molten. Fine filaments produce a thin paper-like material, which can be improved with additional heat and pressure. Thick filaments will not lie as flat and so create a looped pile, which is controlled by the speed of the process. Coiled PVC carpet is made in this way.

Adhesive-bonded nonwovens may be compact or lofty, depending on the process and combination of materials. As with composite laminating, a wide range of fibres may be used and properties achieved. Due to the additional cost and complexity, adhesive-bonding processes tend only to be used when there is no other way, such as making high-loft natural-fibre padding for bedding, pillows and upholstery.

Synthetics process more consistently than natural fibres.

DESIGN

Material selection plays a critical role in determining the overall feel and appearance. It also affects the design opportunities. For example, synthetic nonwovens with suitable length staples or filaments – including UHMWPE, nylon and polyester – can be molded into deep profiles. This technique is used to make hats (page 420) and embossed fabrics (page 220).

Two or more types of fibre are combined to utilize the properties of both. The same rule applies to colour: different-coloured fibres will result in a blend, whereas two layers of different-coloured material will create a visible pattern. When combining two or more different-coloured layers, colour from the bottom layer can be pulled though, or exposed, to reveal a pattern on the face.

Flexibility is determined by the capabilities of the supplier. Some suppliers will be willing to run short production runs, to trial materials, and minimum orders of 200 to 300 tonnes or so. Needle punching is very flexible and allows for all sorts of fibre combinations, even in small batches, although a certain amount of material will be required to get set up, and rid the equipment of traces of previously used fibres.

Total thickness depends on the process and fibre. Process selection is partly determined by the required weight of fabric. Typically, thermal bonding is used for nonwovens up to 150 g/m² (57 oz./ft²); needle punching up to 2,000 g/m² (759 oz./ft²); and adhesive bonding for heavier fabrics. Needle punching can produce interlocked sheets in excess of 150 mm (6 in.) thick, built up from many layers.

MATERIALS

Needle punching is used to combine almost any type of fibre, as long as it is suitable length (over 20 mm/0.8 in.). Even so, short fibres can be mixed in with longer fibres for carding, or can be air laid. Natural fibres, crimped synthetics and even metal wire (such as for filters) are needle punched. Fibres can be mixed without too much consideration, as long as the needles are compatible (larger-diameter fibres require larger needles).

Thermal bonding is suitable only for thermoplastics fibres. The most common for nonwovens include PP, PE, polyester and nylon. Natural fibres may be bonded if they are combined with enough thermoplastic, but these materials will not be as strong. If two thermoplastics are combined, they must be compatible.

Adhesive bonding utilizes latex, vinyl acetate and other water-based resins (although they may also be applied as a powder, print or coating). The type of fibre and the requirements of the process help determine the correct bonding adhesive for the application.

COSTS

These processes convert fibres directly into fabric, eliminating the need for yarn formation. This reduces labour and so means that nonwovens are generally lower-cost than equivalent woven or knitted fabrics. Spun bonding unit price is lowest, because it has the fewest processing steps and cycle time is very rapid; the process operates at speeds of up to 1,000 m/min. (3,280 ft/min.).

Material selection will affect cost, because, for example, high-quality virgin materials are more expensive than recycled mixed fibres.

ENVIRONMENTAL IMPACTS

Dry-processed nonwovens have the lowest environmental impact, because they do not consume any water or chemicals during production and renewable natural fibres may be used in place of man-made fibres. In addition, needle punching is used to make nonwoven fabrics from recycled fibres, providing an extended life for used fibres that might otherwise end up in landfill. Mechanically formed nonwovens, such as needle punched and hydro-entangled, can be re-opened and converted back into new material, although with each cycle fibre length is reduced.

Processes that consume water and chemicals, such as wet laid and spun bonded, are developing to reduce the environmental impacts. Anything added to the process increases the likelihood of negative impacts on the environment. Water consumption is reduced wherever possible and all materials are recycled.

Significantly, nonwovens are used to make disposable items. In some cases this is critical, such as single-use medical

garments. For some applications it may be possible to reduce the impact of single-use items through careful material selection. For example, the same fibre produced by two different companies can have dramatically different environmental impacts, including organic cotton versus pesticide-sprayed cotton; synthetic fibres with a high percentage of recycled content (used PET bottles); or

lyocell fibres from certified forests and manufactured in a virtual closed-loop process (Lenzing Tencel). Alternatively, biodegradable thermoplastic or natural fibres may be suitable. Finally, specifying a single material without additives will increase recycling efficiency.

VISUAL GLOSSARY: NONWOVEN FABRICS

Needle Punched

Material: Acrylic
Application: Toys
Notes: Acrylic is available in a wide range of vivid colours.

Loose Needle Punched

Material: Mixed
Application: Upholstery
Notes: The needle-punch holes are visible on loosely bundled nonwoven constructions.

Recycled Fibres
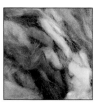
Material: Mixed
Application: Upholstery
Notes: Mixed fibres taken from discarded clothing are blended and needle punched.

Recycled Nonwoven
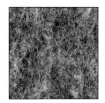
Material: Mixed
Application: Upholstery
Notes: The colour of recycled nonwovens depends on the source and may vary.

Bonded Natural Fibres
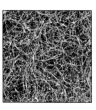
Material: Coconut coir and latex
Application: Upholstery
Notes: A sustainable upholstery material comprising coconut fibres bonded together with flexible latex.

Luffa

Material: Luffa (loofah)
Application: Beauty
Notes: Ripe, fibrous seedpods from the luffa vine are peeled and pressed into mats to be used as exfoliators.

Synthetic Leather
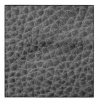
Material: Elastane microfibres
Application: Upholstery
Notes: Nonwoven synthetic leather is made up of micro denier fibres needle punched together and embossed.

Spun Bonded

Material: UHMWPE
Application: Paper replacement
Notes: Strong and light fibres are spun into a random pattern and bonded with heat and pressure.

Thermal Bonded Patterned

Material: UHMWPE
Application: Packaging
Notes: Bonding with an embossed roll produces a strong material with a soft feel.

Thermal Bonded Looped
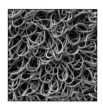
Material: PVC
Application: Carpet
Notes: Also known as silk spurting, this durable flooring is constructed from continuous extruded filaments.

→ Forming a Fibre Web

The first step is to open up bales of blended fibres (**image 1**). Bales weigh 200 to 300 kg (440–660 lb) depending on the fibre type. The crimped white polypropylene (PP) staple fibres are 60 mm (2.4 in.) long and weigh 4.4 dtex.

The fibres are opened up and transferred to a Fernought machine (**image 2**). The teeth-covered rolls draw the clumps of fibre in, separating them into a more even blend (**images 3** and **4**). Next, the fibres are fed into the carding process (**image 5**). The rotating rolls, which are covered with much finer teeth than the Fernought, pick up each fibre that lies across the direction of travel, repositioning it in the web (**image 6**). The process also removes any contamination that has found its way into the mix.

The uniform and aligned web of fibres is teased from the doffer by an oscillating fly comb (**images 7** and **8**). The web, which is self-supporting but still quite fragile, weighs 10.9 g/m² (4.14 oz/ft²). The fibres are aligned in the same direction, but the crimp holds them together. Before needle punching, they are cross-laid to make a stronger nonwoven structure (see opposite page).

OTHER SHEET

156

1

2

3

4

5

6

7

8

Featured Company

Fybagrate
www.fybagrate.co.uk

→ Needle Punching Synthetic Felt

Punching barbed needles through the fibre web to pick up fibres and pass them from one side to the other forms an entangled structure. To demonstrate the principle a machine operator manually inserts a needle through two layers of fibre web (**image 1**). Mixing two colours shows how the fibres of the lower (pink) fibre web are passed through the upper (white) layer.

In operation, the PP fibre web is fed in continuously from the carding process. The layers are cross-laid: each is stacked perpendicular to the next (**images 2** and **3**).

Two beds of needles are used (**image 4**): one punches from above and the other from below. The needles rapidly pierce through the web (**image 5**), entangling the fibres from one side to the other. The finished material consists of 22 layers, thus creating a 240 g/m² (91 oz./ft²) nonwoven (**image 6**).

Needle punching is an efficient process, but inevitably the edges need to be trimmed. Around 5% is cut off and baled, ready to be re-opened and converted into new material (**image 7**). Recycling the offcuts virtually eliminates waste.

Synthetics are available in a range of colours. Acrylic is particularly well suited to bright colours (see title image, page 152). After needle punching the nonwoven is steam treated. The example – pink acrylic (**image 8**) – demonstrates how the steam treatment causes the fibres to draw together to half the thickness of the original sheet. This produces a stiff and dense material suitable for die cutting, among other processes.

1

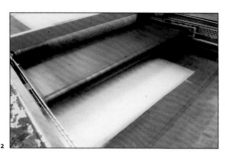

2

3

4

Featured Company

Fybagrate
www.fybagrate.co.uk

5

6

7

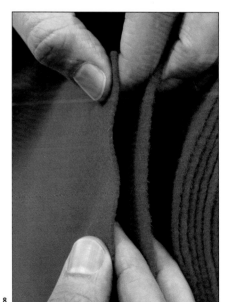

8

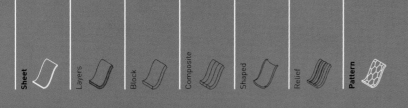

Textile Technology

Leather Tanning

Leather is used as a high-value material in the fashion, furniture and automotive industries. Tanning is the process of curing raw hides – typically sourced from cows, sheep, goats or pigs as a by-product of meat production – into durable leather. It is produced in a wide range of finishes, including dyed, embossed, suede, coated, patent and metallized.

Techniques	Materials	Costs
• Mineral (chrome, aluminium or zirconium) • Vegetable • Synthetic (aldehyde)	• Cow, pig, sheep and goat • Exotic: deer, snake, stingray and alligator	• Low to high, depending on quality, source and animal • Vegetable tanning is the most expensive

Applications		Quality	Related Technologies	
• Apparel • Footwear • Upholstery	• Automotive • Leather goods / accessories	• Each skin is unique • From stiff and heavy to thin and soft	• Bonded leather • Coating	• Nonwoven • Laminating

INTRODUCTION

Leather is a high-performance material and because it is natural, each product made with it will have its own unique qualities. It comes from the skins and hides of mammals, reptiles, fish and birds. Pelt is an animal skin that has the hair, wool or fur fibres still attached.

Tanning is essential to preserve the skin. It is a curing process: the tanning agent forms links between the skin collagen (protein), making the leather non-perishable, durable and resistant to water. Two principal methods of tanning are mineral (chrome) and vegetable. Chrome tanning is the most widely used; it is faster (it takes around 20 hours) and produces more consistent and high-quality leather.

Vegetable tanning uses tannin obtained from biomass such as tree bark (this is where the name originates from). Traditional methods take about one year and produce firm and hard-wearing leather, although it will not be as long-lasting as chrome-tanned leather. The pelts are suspended in tanning pits and moved through a cycle of progressively stronger concentrations of tannin solution, which is known as 'liquor'.

APPLICATIONS

Leather is used in all forms of clothing, protective clothing (aprons and motorcycle gear), footwear (shoes and boots), gloves, bags, accessories (wallets and belts) and upholstery (furniture and automotive interiors). Pelts are used for insulating purposes (collars, scarves, jackets and linings), for comfort (covers and throws) and for luxury items (winter coats and ceremonial outfits).

RELATED TECHNOLOGIES

Artificial leather, or synthetic leather, is a low-cost and inferior-quality alternative

The Chrome Tanning Process

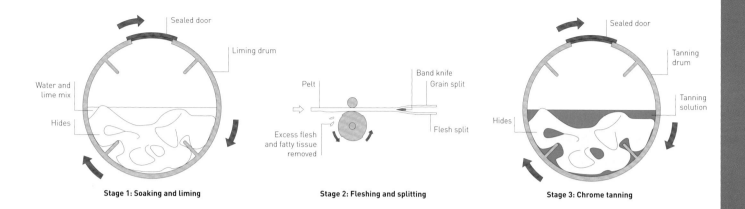

Stage 1: Soaking and liming

Stage 2: Fleshing and splitting

Stage 3: Chrome tanning

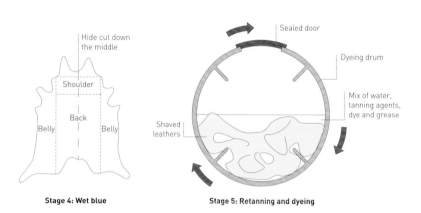

Stage 4: Wet blue

Stage 5: Retanning and dyeing

TECHNICAL DESCRIPTION

The first stage in the process is soaking and liming. The raw hides are loaded into the liming drum and soaked in water. Liming takes place in the same drum with the addition of lime and sodium sulphide. This raises the pH value to remove hair. At this stage, substances that cannot be turned into leather, such as natural oils and protein, are removed. After 24 to 36 hours the soaked, limed and de-haired hide, called 'pelt', is removed from the drum.

In stage 2, the pelt is split using a band knife into 'grain split' and 'flesh split'. The grain split will be turned into 'top grain' and 'full grain' (which is the highest quality and is used, for example, in jackets, shoes, bags and upholstery), and the flesh split is usually processed into 'split leather' (suede used in bags and upholstery). Lower-quality hides that are split and buffed to

a smooth finish on the top side are known as 'corrected grain'. They are usually heavily coloured or embossed and are the least expensive.

Stage 3 consists of five processes, which are carried out in sequence in a rotating drum: deliming, bating, pickling, tanning and basification. In stage 4, the water is pressed out on a sammying machine and the hides are cut down the middle, which makes them easier to handle.

In stage 5 the shaved leathers are loaded into the dyeing drum and go through a process of neutralization, retannage, dyeing and greasing. Additional tanning agents are added to influence softness, feel, tearing value, elasticity and other characteristics. Dye is added to obtain the desired colour and the pigments are mixed according to each batch of leather.

Skin grease, which was removed during liming, is added to improve the colour, lustre and feel of the leather. Hydrophobic greases improve the leather's resistance to water absorption. The breathability of the leather is maintained because the grease wraps around the individual fibres.

After dyeing, the underside of the 'grain split' is made smoother and more even by a 'setting-out' cylinder with blunt blades. The leather is dried to 'fix' the fibres (they become bonded together), causing it to become a little stiffer. It is subsequently passed over a series of heavily vibrating metal pins, in a process known as 'staking', to loosen up the fibres and produce a more pliable material ready for finishing.

for applications such as bags, clothing, upholstery and consumer electronics. It typically consists of woven fabric (cotton or polyester) film-coated (page 226) with polyurethane (PU) or polyvinyl chloride (PVC). The surface is embossed (page 220) to give the appearance of leather grain.

Synthetics are more stable and controllable; they can be molded, such as for laptop cases and covers; they are waterproof and easy to clean. In some ways they are ethical too, because they have not been derived from animals and are not by-products of meat production and farming.

Bonded leather, or reconstituted leather, is produced from offcuts that are ground up, mixed with PU and molded. It is less expensive and is used in place of leather to make parts of shoes, stationery and bookbinding, for example.

QUALITY
Leather is a natural material: each hide is unique, it has a familiar odour and the surface will wear with use. It is durable, elastic and hard-wearing and has high resistance to tearing, puncture and expansion. There are many factors that affect its quality and character, including the type of animal, breed, age and where the leather was taken from (for example, in bull hides, the belly region is thicker and fuller and has fewer blood vessels on the surface). Quality of wool, hair and fur varies considerably, from the short bristly hair on an adult cowhide to soft luxurious lambswool.

The visual quality of hides is affected by the lifestyle of the animal. For example, scars can come from fighting, barbed-wire fences, parasite damage or insect bites. Vein marks and wrinkles are typically seen as undesirable in the finished product. Skin formation and associated markings are referred to as 'grain' and this varies from animal to animal. Careful material grading is critical to ensure a high-quality end product.

DESIGN
The appearance is controlled by the choice of tanning, retannage, dyeing and finishing processes. Dyed, uncoated leather is known as aniline finish. This means it retains the hide's natural finish, grain and blemishes; it is the most natural looking. Typically, only the highest-quality leather is produced with an aniline finish. Semi-aniline leather has a thin protective topcoat, which makes the surface more durable and resistant to wear and staining. Pull-up leather is oiled or waxed, so that over time it will age and appear more 'lived-in'. Pigmented leather is coated with plastic, making it more resistant to scuffing and fading. It is used for demanding applications. Nubuck leather, also known as grain-suede leather, is buffed on the topside to produce a roughened, more hard-wearing surface (see also sueding and napping, page 216).

Ironing increases the smoothness and gloss of the surface of the leather. Increasing the temperature and pressure increases gloss level. Patent leather is created by coating (page 226) a thin layer of plastic onto the surface with a glossy finish.

Hides are typically split into layers, so that they are thinner and more pliable. The top layer, which is the outside surface, has densely packed fibres, making it the most durable and water-resistant. The highest-quality leather, known as 'full grain', has all the appearance of the original skin with grain, hair follicles and blemishes. 'Top grain' refers to leather from the top layer that has been shaved and buffed to produce a uniform finish free from natural imperfections.

The inner layers are known as 'split' leather. They are not as smooth as top grain and tend to wear less well, because the fibres have been cut through. They are usually embossed (page 220) or buffed to create a more durable finish. Suede is produced from split leather, which is napped (abraded with emery sandpaper) to pull up the fibres.

Pelts are usually tanned to retain their natural colour. Over the years, selective breeding on fur farms has led to the colours of fur we are accustomed to today. Less expensive furs, such as rabbit, are sometimes dyed, sprayed or printed to resemble more expensive types.

The size of hide is determined by the type and age of animal. For example, cow hides range from 4 m² to 5 m²

(43–54 ft²) and sheepskins can range from 0.75 m² to 1 m² (8–10 ft²).

MATERIALS
Leather comes mainly from mammals, including cows, sheep, pigs, goats, deer and horses. Skins from other species are used, but are less common and more expensive. They include fish, such as shark, stingray, perch, salmon and sea bass; birds, such as ostrich; and reptiles, such as crocodile, alligator, snake and lizard.

Choice of leather is determined by cost and application. For example, peccary (a native American mammal resembling a pig) leather is very soft, strong and supple, making it an ideal glovemaking material. Cow leather is tough, durable and stiff, making it a good choice for footwear. The distinctive appearance of ostrich and alligator makes them desirable in high-end fashion items.

COSTS
Costs range from low to high, depending on the quality of leather, the tanning processes and the finishing techniques. Pig is the least expensive, though it will depend on the breed. Cow, sheep, goat and deer are moderately expensive, depending on the quality, and exotic leathers are the most expensive.

ENVIRONMENTAL IMPACTS
Tanning uses large quantities of water. Heinen Leather in Wegberg, central Germany, uses 90 litres/m² (24 gallons/10 ft²) and around 0.5 kg (1.1 lb) of chemicals for every 1 kg (2.2 lb) of leather. However, most tanneries consume more than 350 litres/m² (92.5 gallons/10 ft²). Heinen Leather purifies all of the water after use. Harmful substances are filtered out, collected and converted. Most of the water can then be reused, thereby reducing overall water consumption.

Immediately after slaughter, the skin will start to break down because of bacteria. It is preserved by adding salt (the traditional method used to remove the water) or by refrigeration. The best option is to use locally sourced refrigerated hides because salting can cause pollution, for example by increasing chloride levels in rivers.

Suede
Material: Pig leather
Application: Upholstery
Notes: The surface of this dyed leather has been emerized to produce a soft nap.

Embossed
Material: Cow leather
Application: Upholstery
Notes: With high pressure and temperature, patterns are embossed permanently into the surface.

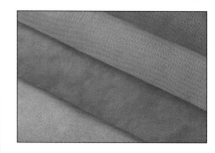

Dyed Colour
Material: Cow leather
Application: Footwear
Notes: All chrome-tanned leather is dyed, either to retain a natural appearance or to match a colour reference.

Metallic
Material: Cow leather
Application: Footwear
Notes: Metal foil is permanently applied to the surface with high heat and pressure. A wide range of colours and finishes can be achieved.

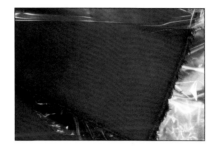

Gloss
Material: Cow leather
Application: Footwear
Notes: Ironing increases the smoothness and gloss of the surface of the leather. Increasing the temperature and pressure increases gloss level.

Patent
Material: Calfskin
Application: Upholstery
Notes: A thin layer of plastic is spray coated onto the surface to produce a glossy hard-wearing leather. The grain is completely concealed.

Hair
Material: Pony leather
Application: Upholstery
Notes: Depending on the animal, the coat ranges from soft dense fur to short coarse hair.

Wool
Material: Sheepskin
Application: Apparel
Notes: Wool is soft and highly insulating. It is left on the skin and used to line gloves, shoes and jackets.

Fur
Material: Rabbit skin
Application: Apparel
Notes: Animals that have thick winter coats yield the softest fur.

The environmental impacts of leather tanning depend on the tannery. With its Terracare brand, Heinen Leather continually reduces the amount of water (72% less now than in 2003), chemicals and energy (45% of energy is derived from biogas produced from limed offcuts) it uses, promoting social responsibility. In doing so, it is a key player in the reduction of environmental impacts of leather production. As a result, it is estimated that their carbon dioxide equivalent emissions are below the limit set by the European Union and less than half the amount expected from a tannery operating in a developing country.

The majority of leather is made from the skins of animals reared for meat, milk or wool. The source of skin is nonetheless critical, to ensure that the animals were treated humanely. And several species of animal that have been used for leather in the past are endangered and illegal to use for leather in many countries.

→ Preparing Raw Hides for Tanning

Heinen Leather is based in Germany and uses only locally sourced (central European) refrigerated cowhide. It buys bull hide weighing 30 to 50 kg (66–110 lb), or up to 60 kg (132 lb) for its heaviest leather, which arrives graded and sorted (**image 1**).

The hides are soaked and limed in batches (**image 2**) and the pelt is mechanically defleshed and trimmed. Unwanted parts of the hide, such as the kneecaps and the root of the tail, are cut off (**image 3**). There is no waste from the hide. All unwanted parts are put to other uses, such as the production of glue or gelatin.

The pelt is cut through its thickness into grain split and flesh split (**images 4** and **5**) and is ready for chrome tanning.

1

2

3

4

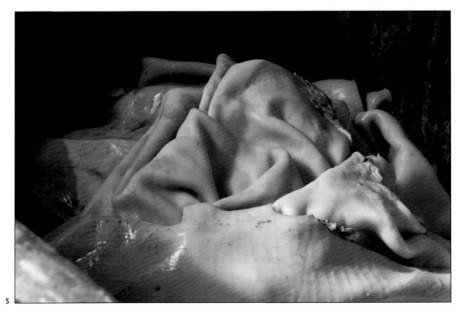

5

Featured Company

Heinen Leather
www.heinen-leather.de

Case Study

→ Chrome Tanning

Heinen Leather uses carbon dioxide to delime the pelts. Hair remnants and proteins are then removed by using enzymes in a process known as bating. This makes the fibres more pliable. During pickling, salt and acid are added to reduce the pH and make the leather ready for the adding of the tanning agent. At low pH the chromium molecules are very small and so penetrate deep into the leather fibres. As the pH rises the molecules grow and form cross-links with the collagen. This act of curing is the tanning process.

Many different types of tanning agent can be used. Trivalent chromium (chromium III), which is used by Heinen Leather (**image 1**), is a natural, non-toxic element thought to have the least environmental impact of all tanning agents (90% of global tanning employs trivalent chromium). Hexavalent chromium (chromium VI) is banned in many countries, including Germany, because it is known to be carcinogenic.

The tanned hides are known as 'wet blue' because the chromium dyes them a light shade of blue (**image 2**). Hide flaws, such as scratches and scars, are easily identified on the wet-blue leathers (**image 3**). The graders, known as 'wet-blue classifiers', carefully sort the leather into the different classes of quality (**image 4**).

The hides are shaved to the correct thickness, which depends on the application (**images 5** and **6**). Soft leather for bags, for example, requires a different thickness from leather for walking boots. The layer that is not required is shaved from the back side of the wet blue. The shavings are either recycled as bonded leather products or the protein is extracted and used for industrial processes.

1

2

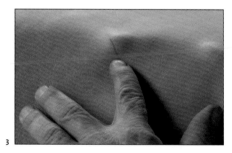

3

4

5

6

Featured Company

Heinen Leather
www.heinen-leather.de

→ Retanning, Dyeing and Finishing Leather

The leather is given specific qualities relevant to the customer's requirements. Each batch is processed separately (**image 1**). A wide range of chemicals is used for dyeing and finishing (**image 2**). Heinen Leather uses only water-based products.

The leather is flattened and stretched out on the grain side by a 'setting-out' cylinder with blunt metal blades. Then, it is placed grain-side down onto a heated metal plate (40°C/104°F) (**image 3**). With the addition of vacuum pressure, the leather is dried (but not completely) and the fibres are 'fixed' in place.

The leathers are hung to complete the drying process slowly (**image 4**). This allows the chemicals added to the leather to form strong bonds and produce a stable and high-quality material. At Heinen Leather the heat used in this process is a by-product of generating electricity at their own two-block power station. After drying the skin fibres have bonded together, making the leather harder and stiffer. Therefore, it is soaked with water (**image 5**) to lubricate the fibres, and staked with heavily vibrating metal pins, which softens the leather by loosening the fibres and improves the surface finish (**image 6**). The softness is affected by the amount of staking and can therefore be tailored to the application. The leather is ironed at around 90°C (194°F) and stretched slightly to improve surface finish.

Soft and stretchy leathers are pulled and stretched on a metal frame, known as a toggle dryer (**image 7**), to make them larger and smoother. They are dried in the stretched position (**image 8**).

Fringes and other unwanted parts are removed (**image 9**). Once trimmed it is known as 'crust' leather and is ready to use. Aniline-finish leather is dyed and the surface is uncoated. Finishing processes include buffing on the top side to produce a roughened, more hard-wearing surface (nubuck); dry-splitting to create a precise thickness accurate to 0.1 mm (0.004 in.); and softening by tumbling in the milling drum to produce a velvety feel (the best-known example is nappa). Leather that has been tumbled in the milling drum to raise the grain (more pronounced finish) is known as milled-grain leather.

Subsequent finishing processes change only the surface and visual qualities of the material, such as spraying or roll coating to add colour, metallic effect, wax or lacquer; and embossing (such as grain pattern) or printing.

1

2

3

4

5

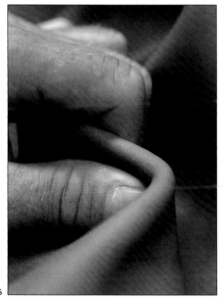

6

7

8

9

Featured Company

Heinen Leather
www.heinen-leather.de

→ Tanning Sheepskin

In the production of sheepskin, as with other types of pelt, the tanning process is adjusted slightly so as to maintain the highest-quality fleece. The techniques have been practised for thousands of years and continue to involve a great deal of handwork.

Skyeskyns is located in the heart of the Scottish Highlands crofting community. Using age-old methods of leather tanning, it continues to produce hand-combed fleeces to the highest standards. The white blackface is the most common local sheep.

Sheepskin is a by-product of meat production and the best fleeces come from lambs around 10 months old that have not yet been sheared (ideally around seven months old, born in the spring and sold before their wool coarsens as the temperature drops). The skins are salted to preserve them and can be kept like this for up to 12 months or so.

The process begins by washing the skins in warm water and detergent. The solution has been specially formulated to remove as much of the dirt, marker dyes and grease as possible. After sufficient soaking, they are fleshed to remove fat and loose bits from the inside surface of the skin (**image 1**). Carried out by hand, this will take longer than mechanical fleshing, but yields a high-quality pelt.

The prepared skins (**image 2**) are washed again and then pickled in a solution of sulphuric acid, formic acid (helps suppleness) and salt (**image 3**). The salt helps to open up the skin in preparation for tanning as well as moderating the action of the acid (the pH is around 1.8 and without salt the skins would eventually break down).

The main difference between tanning wool and tanning leather is that with wool, the chromium is mixed with around 25% aluminium sulphite (alum). This helps to preserve the natural colour of the skin and wool. Sheep leather is not stiff and hard-wearing like cowhide, and so a suppleness agent is added to enhance the flexibility of the material. The pH is gradually raised from 2.2 to 3.8 by the addition of bicarbonate of soda and the pelts are given a hot wash.

The tanned skin is brushed with an emulsion of marine oil (fish oil), lanolin and water, after which the hides are toggled onto drying racks (**image 4**). This is carried out with great care to ensure the skin dries flat. They are left under tension for up to four days.

Prior to finishing, the dry, stiff pelts are tumbled in a warm drum for a couple of days to loosen up the fibres and give a more supple material. The back side is then polished on a large granite curling stone (**image 5**). This produces a neat, white suede finish.

The wool side is brushed with a helical wire comb (**image 6**) to align the fibres and remove any loose wool. It is sprayed with conditioner for a silky lustre (**image 7**) and the skin side is ironed on a heated rotary drum.

The edge is neatened with a sharp knife (**image 8**) and the fleeces are ready to be converted into gloves, shoes and other items of clothing (**image 9**).

1

2

3

4

5

6

7

8

9

Featured Company

Skyeskyns
www.skyeskyns.co.uk

Textile Technology

Natural Rubber and Latex

Latex and natural rubber are elastomeric: they return to their original shape after stretching. Latex is tapped from rubber trees and used for dip molding products such as gloves and condoms, or is further processed into rubber sheet or fibre, which are used to make garments and footwear. Over the years they have been replaced by synthetic equivalents in many cases.

Techniques	Materials	Costs
• Tapping • Coagulation • Pressing and vulcanization	• Latex • Rubber	• Low

Applications		Quality	Related Technologies
• Gloves • Garments • Footwear	• Upholstery • Condoms • Coatings	• High strength • Poor resistance to oil, oxidization and ultraviolet light	• Dip molding • Laminating • Plastic film extrusion

INTRODUCTION

Elastomers are characterized by their ability to stretch several times their length and return to their original shape. Natural rubber is the oldest of these and relatively low-cost by comparison. It is produced from sap (latex) tapped from the Pará rubber tree, as opposed to synthetic elastomers, which are derived from petroleum. Over the years, many synthetic replacements, including neoprene and polyurethane (PU), have been developed to overcome some of the limitations of natural rubber and latex. Even so, they remain important materials for the production of waterproof gloves, footwear and garments. They are also

The Rubber and Latex Production Process

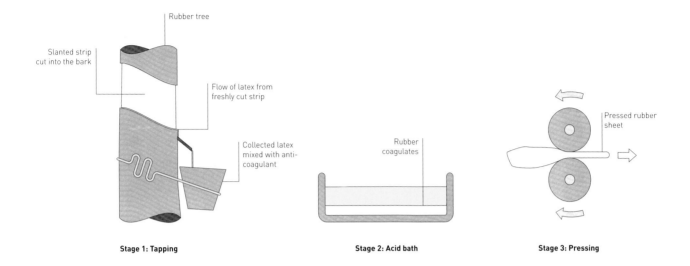

Stage 1: Tapping

Stage 2: Acid bath

Stage 3: Pressing

used for padding and protection in upholstery (page 404).

APPLICATIONS
The majority of latex and natural rubber is used to make tyres, hoses and pipes. The rest is used for a range of industrial and domestic applications. Uses in fashion- and textile-related industries include waterproof coatings (see Fluid Coating, page 226), gloves and swimming hats (see Dip Molding, page 442), upholstery and footwear. Natural rubber fibres are used in garments, although they have been superseded by synthetic equivalents for applications such as elasticated waistbands, swimsuits and sportswear.

RELATED TECHNOLOGIES
Latex and natural rubber are low-cost, durable and excellent general-purpose elastomers. Depending on the format – liquid, molded, sheet or fibre – latex and natural rubber compete with various synthetic alternatives. Latex is used to make dip-molded products, such as gloves, neck seals and condoms. Synthetic alternatives for these types

of application include neoprene and polyvinyl chloride (PVC).

Several types of elastomer are formed as sheet and film materials (see Plastic Film Extrusion, page 180). Those that compete directly with natural rubber include neoprene, PVC and thermoplastic polyurethane (TPU). Like natural rubber, sheet plastics are slit into thin strips to make yarn. Alternatively, elastomers, such as elastane, are produced directly as a filament yarn (page 50). The drawback of natural rubber yarn is its low resistance; it is more vulnerable to degradation and failure.

QUALITY
Rubber is vulcanized by exposing the raw material to sulphur at high temperature. The sulphur forms cross-links with strong bonds between the polymer chains. This ensures that heating and cooling do not destroy the material. By contrast, thermoplastics, such as elastane, lose their strength when heated above their softening point.

Rubber and latex are very durable and capable of being stretched several

TECHNICAL DESCRIPTION
Tapping rubber trees is a highly skilled process. In stage 1, the worker cuts a 30° angle slot about 2 mm (0.08 in.) into the bark around one side of the tree. This wounds the bark and causes it to excrete latex. If the cut is too deep it could damage the tree and if it is too shallow the flow of latex will be reduced. The latex is collected in a cup and mixed with an ammonia solution to stop coagulation. At this point, the latex is either collected and transported for processing into concentrated latex, or further processed to make rubber.

In order to produce rubber sheets, in stage 2 the latex is sieved to remove contamination and mixed with acetic or formic acid. Over several hours the rubber coagulates to form a soft rubber paste. In stage 3, after drying for 18 hours or so, the soft rubber is passed through a series of rollers to remove excess moisture and prepare it for vulcanization.

Rubber Filament
Material: Latex
Application: Apparel
Notes: Natural latex is produced as continuous filaments. It is woven into narrow fabrics and used in ropes.

Dipped Crinkle Finish
Material: Latex coating
Application: Gloves
Notes: Knitted gloves are dipped in latex, which soaks into the fibres, and given a crinkle finish to improve grip.

Core-Spun Yarn
Material: Latex
Application: Apparel
Notes: The latex yarn is wrapped with synthetic filament to produce a high-strength composite yarn.

times their original length without causing permanent deformation; they have excellent mechanical and elastic properties. They have low resistance to oil, oxidization and ultraviolet light. This affects how they are used in garments and footwear, because they are prone to degrading over time.

Carbon black and other additives and fillers are added to rubber to significantly increase durability and tensile strength.

DESIGN
Latex is liquid and so can be molded into three-dimensional shapes or applied as a coating. It is dried and coagulated to form rubber, which is used to mold three-dimensional parts, or as a sheet material. The different formats available make these versatile materials suitable for a range of flat, curved and three-dimensional designs.

As a coating on woven fabrics (page 76), latex is used to add colour and create a waterproof protective layer. It flows around the fibres and so creates a strong mechanical bond. Adding this type of coating will make the fabric stiffer. It is possible to screen print (page 260) latex onto selective areas of fabrics.

Dip molding is used to form latex into three-dimensional sheet parts, which may have an undercut shape (for example, wrap almost entirely around a mold, such

as a hand-shaped glove mold), because the material can be stretched off the mold without damaging it.

Non-vulcanized rubber can be processed into sheets, or compression molded to make three-dimensional products. Unlike dip molding, compression molding is capable of producing bulk parts with varying wall thickness, such as the soles of boots and shoes. Once vulcanized, rubber cannot be molded again, although it can be trimmed and profiled, such as by die cutting (page 336).

Latex is naturally milky white and rubber is yellowish after vulcanization. They are mixed with pigments to produce a range of bright colours. Rubber that has been mixed with carbon appears black.

MATERIALS
Latex and natural rubber are usually vulcanized to produce a more stable and durable material.

COSTS
The raw materials are relatively low-cost. Ultimately, unit price will be determined by the shaping and finishing processes used.

ENVIRONMENTAL IMPACTS
Natural rubber can be fair trade and certified by the Forest Stewardship Council (FSC). For instance, GreenTips's

elastic bands and wellington boots from Sri Lanka are FSC certified.

Rubber is a thermoset material and it is therefore not possible to recycle it directly. Unlike thermoplastics, which can be melted and reprocessed using relatively little additional energy, latex and rubber are shredded. The recycled material, known as 'crumb', is used as a filler material for shock-absorbing floors (such as for playgrounds), insulation, asphalt and aggregate. Rubber products, such as tyres, make a significant contribution to landfill.

Developments include blending rubber with a thermoplastic to make an elastomer with a mix of the properties of both materials, but so that it can be reprocessed by melting and molding. The possibility of de-vulcanizing cured rubber after shredding has also been explored, but not yet optimized and commercialized.

Some people are allergic to the proteins in latex and in the most extreme cases the reaction can be life-threatening.

→ Tapping Latex and Vulcanizing Rubber in Malaysia

A new incision is made at the bottom of the cut area of the bark (**image 1**). This is repeated three to four hours later. The fresh cut will produce latex for several more hours (**image 2**). Properly managed in this way, rubber trees will produce latex for around 25 years. Production is carried out in the forest, among the plantation, either by small producers or in large plantation factories.

The latex is mixed with acid to produce soft, gelatinous rubber. After drying for several hours excess moisture is pressed out by treading on the soft rubber (**image 3**). The roughly formed sheet is then passed through a series of textured rollers (**images 4** and **5**). At this point the rubber will deform easily: it can be formed by extrusion or compression molding, for example, or it can be vulcanized in sheet form.

Vulcanizing is carried out in a smokehouse (**image 6**). After the rubber has been exposed to sulphur for several days at high temperature, cross-links are formed in the polymer, causing it to become a more stable, durable, resilient and long-lasting elastomer. This is known as 'ribbed smoked rubber' (**images 7** and **8**).

Other common types include graded and standardized 'vulcanized rubber', 'pale crepe' for making shoe soles, 'hevea crumb' pressed rubber particles, and 'skim rubber', which is lower quality.

1

2

3

4

5

6

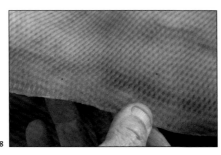

7

8

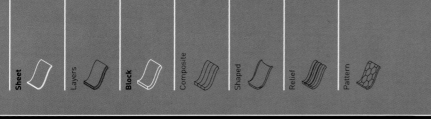

Sheet | Layers | Block | Composite | Shaped | Relief | Pattern

OTHER SHEET

172

Textile Technology
Foam

Produced as slabstock and extruded or molded shapes, synthetic foams have a cellular structure that is created with blowing agents. They may be open- or closed-cell, depending on the production process, and range from soft and elastic to dense and rigid. Foam is extremely versatile and used in many familiar products to provide insulation, cushioning and protection.

Techniques	Materials	Costs
• Forming (slabstock) • Slitting and peeling	• Polyurethane (PU), polypropylene (PP) and polyethylene (PE)	• Low cost (with a few exceptions) • Rapid cycle time

Applications	Quality	Related Technologies
• Footwear and apparel • Upholstery and carpet • Construction and transportation	• Open- or closed-cell, dense to airy, soft to hard, and rigid to elastic	• Weaving • Knitting • Foam molding

INTRODUCTION

Foam is extremely versatile and used in many familiar products to provide insulation, cushioning and protection. Sheet materials are manufactured from large blocks, called slabstock, or peeled from round logs. Shaped products are extruded or molded (page 410). Slabstock is the least expensive format to produce, because the pouring process is quick and efficient.

Foam for textile applications is typically polyurethane (PU), which is a thermosetting material. This means that permanent cross-links are formed between the polymer chains during polymerization, resulting in a resilient and

The Slabstock Production Process

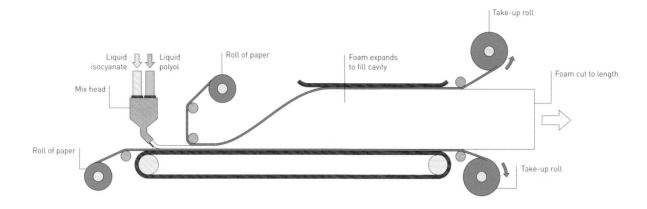

durable material that cannot be melted and reprocessed like thermoplastics. Other types of plastic produced as foam include thermoplastic polyurethane (TPU), expanded polypropylene (EPP) and expanded polyethylene (EPE).

APPLICATIONS

Textile applications of compressible foams include shoe soles, carpet, flooring, mats, upholstery and mattresses. Depending on the density and elasticity of the material, it provides support and comfort. It is applied as an insulating layer in laminated (page 188) and stitch-bonded (page 196) fabrics, such as quilts and blankets. When incorporated in protective or heavy clothing, foam absorbs energy and so reduces impact on the wearer.

Rigid foams are used as core material in sandwiched composite constructions to increase strength to weight. The added thickness spreads the layers of high-strength composite and so increases bending stiffness without adding much weight.

RELATED TECHNOLOGIES

Spacer fabrics, produced by double-cloth weaving (page 97) and knitting (see Weft Knitting, page 126, and Warp Knitting, page 144), are constructed from two layers of fabric separated by interlacing yarns and measure 2 to 12 mm (0.08–0.5 in.) deep. They have similar characteristics

to two-dimensional textiles. The advantage of these materials is that they allow air to circulate freely and so do not encourage the build-up of heat or moisture. They are therefore very useful for applications close to the body, such as trainer uppers, bra cups and backpack padding, as well as technical applications including upholstery, mattresses, medical products and geotextiles.

QUALITY

The type of foam depends on the requirements of the application. Open-cell foams (also known as reticulated) allow air and liquid to pass freely around the structure, because the cells are interlinked. They are used for absorption, such as sponge, vibration dampening and impact protection. Smaller cells result in higher levels of absorption. Closed-cell foams are made up of separate pockets of gas surrounded by plastic. They do not absorb gases or liquids and so are used to provide buoyancy, insulation and high strength under load.

Foams are potentially highly flammable as a result of containing a lot of air passageways. Flame retardants are added for applications where this could be problematic.

Foams are available in a wide range of colours, although they are typically produced in natural colour, which for PU is off-white. Over time PU tends to become yellow, owing to an oxidization

TECHNICAL DESCRIPTION
Sheets of foam are produced from molded blocks, called slabstock. The mix head is supplied with liquid isocyanate and polyol: the reaction of the chemicals produces polyurethane (PU). If additives are required, they are incorporated in the raw ingredients.

Blowing agents are mixed with the raw ingredients. Chemical types are mixed into the raw ingredients and are triggered by reaction with isocyanate or thermal decomposition. They react to give off gas, such as nitrogen (N_2) or carbon dioxide (CO_2), and thus cause the material to foam. Or CO_2 or N_2 may be applied directly to the liquid ingredients. Dense foams are typically formed with chemical blowing agents, and physical blowing agents are used to produce lower-density foam.

Once brought together in the mix head the chemical reaction begins. The liquid is poured onto paper, which is pulled along by a conveyor below. Paper covers all four sides and controls the rate of expansion, thus determining the density of foam.

This is a continuous process, able to produce huge quantities of foam. It is often run intermittently, to reduce storage problems further on. The slab is cut to length and cured for 24 hours before it is converted into sheet.

Closed-Cell Foam
Material: PE
Application: Various
Notes: Closed-cell foams have individual pockets of gas and so do not allow air or liquid to pass through.

Open-Cell Foam
Material: PU
Application: Various
Notes: Open-cell foams have a series of interconnected pores, giving a softer and more absorbent material.

Microcellular Foam
Material: PE
Application: Various
Notes: The size of the cells possible ranges from less than 1 mm to more than 10 mm (0.04–0.4 in.) in diameter.

reaction in the presence of sunlight, but this does not significantly reduce mechanical properties.

DESIGN

Lengths of foam are slit into sheets from 3 mm (0.1 in.) or so. It is known as slitting when cut from a block and peeling when cut from a roll. The former is used to produce lengths up to 60 m (197 ft) from slabstock PU, whereas the latter is capable of producing longer lengths, depending on the thickness of the sheet. Profiles are made with CNC cutting (page 316), die cutting (page 336), or manual cutting (page 312).

The mechanical properties of the PU can be designed to suit the application. It may be rigid or flexible, and applied as a coating (page 226) or even adhesive (page 370). The density can be adjusted from 20 kg/m³ to over 400 kg/m³ (1.25–25 lb/ft³) and the proportion of air can be less than half or more than 95%.

The elasticity of the base material is emphasized as foam, because there is less material for the same bulk. Elastic foams may be extended up to 350% and remain stable.

Flame lamination (page 191) is the most efficient method of producing a foam sandwich. When this is not practical, adhesive lamination methods are used. It is possible to include functional films within the lamination, such as impermeable, breathable or waterproof.

MATERIALS

Sheet foams used in textile applications are commonly PU. The polymer is a product of the reaction between polyol and isocyanate with water. The polyol is either polyester- or polyether-based. Foams produced from polyester have more uniform cell distribution and so look smoother. They have superior closed-cell properties and higher tensile strength. Polyether foams are typically softer with an open-cell structure and are used for cushioning and protection. Polyester types are preferred for dry applications and polyether types for products that will come into contact with moisture.

EPP and EPE consist of expanded, primarily closed cells. They are molded into parts used for engineering applications and technical products for sports and leisure. The foam is semi-rigid to rigid and high strength, and has good energy absorption.

COSTS

Foaming uses relatively less material for the same bulk. Slabstock will be less expensive than molded products, because it is manufactured in large blocks cut to size, rather than individual components.

ENVIRONMENTAL IMPACTS

Conventional foaming agents are being phased out because of their negative impacts on the environment. Chlorofluorocarbons (CFCs) were effective

and easy to handle but were linked to ozone depletion. They are banned worldwide and have been replaced by alternatives, such as inert gases. Carbon dioxide (CO_2) and nitrogen (N_2) blowing agents are efficient and inexpensive with high foaming capacity. As a result, they are being widely adopted and used in place of both CFCs and partially halogenated CFCs, which are also forbidden in many countries.

The isocyanates that are off-gassed during the reaction are harmful and known to cause asthma. The polyol/MDI system produces fewer isocyanates than the TDI method.

Scrap from thermosetting plastics cannot be melted and so is recycled by mechanical means. Waste foam is ground up into small pieces and bonded together into multicoloured sheets. The recycled material, also known as rebond, is used as carpet underlay, cushioning in upholstery and sports mats, for example. Alternatively, the foam is ground into a fine powder and added to virgin material, suitable for molding.

Plastics contain relatively high embodied energy compared to other waste materials and so may be incinerated to recover this.

→ Manufacturing Slabstock

The production of PU starts with the raw ingredients. Liquid isocyanate and polyol are stored in preparation for production (**image 1**). The slabstock line is prepared with a fresh roll of paper, which is fed into the metal-sided tunnel (**image 2**).

Production commences and the ingredients are fed to the mix head (**image 3**). As they are mixed together in the presence of a catalyst the reaction begins. The liquid is poured out onto the paper, which passes underneath, transporting

the foam into the paper-lined tunnel (**image 4**).

The mixed ingredients expand and react to form PU foam. The reaction is one-way and cannot be undone. Each 60 m (197 ft) long block is craned into storage (**image 5**), where it is allowed to finish curing before slitting into sheets (see case study on next page).

1

2

3

4

5

→ # Slitting Foam into Sheets and Strip

The long blocks of foam are cut into thinner sheets. This is carried out on a loop slitter, which passes the long length in a giant loop, cutting thin sheets from the top with each pass (**images 1** and **2**).

If thin strip, or tape, is required then the sheets are slit along their length. Multiple rotary blades are lined up at a set distance and cut strips in a continuous process (**image 3**).

To produce narrower widths, the long lengths of pre-cut strip are loaded onto a reel-to-reel slitter (**image 4**). This allows for greater dimensional tolerances and is capable of producing reels that hold up to 35,000 m (114,830 ft).

Each length of strip is passed over a series of tensioning rolls, to avoid breakages and snagging, and cut to width with a rotary blade pressed against a rubber roll (**image 5**). The finished strip (**image 6**) is wound onto reels ready for further processing by lamination or coating, for example.

1

2

3

→ Peeling Sheets from Round Blocks

The slabstock process is used to produce high quantities of foam. Smaller quantities, and longer lengths (slabstock is limited to 60 m/ 197 ft) , are produced from molded round blocks, called logs (**image 1**).

The molded log is loaded onto a spinning axis and the skin is removed to reveal the uniform foam beneath (**image 2**). Sheets are cut using a bandsaw with a straight-edged blade, which passes in a continuous loop around two wheels (**image 3**). The height of cut determines the thickness

of the sheet. Thinner sheets are therefore much longer.

The long length of foam is taken up on a roll (**image 4**) in preparation for further processing, such as laminating between a half-tricot (net) backing layer and tricot (green) outer layer (**image 5**).

1

2

3

4

5

→ Recycled Foam Production

Scrap PU foam is not wasted. It is collected and compressed into bales for storage (**images 1** and **2**). These are transported to a separate facility, where they are ground into smaller pieces, along with post-consumer waste. They are blended together to produce a homogenous mix of different colours and densities, and sprayed with a PU binder. The coated foam is placed into the mold and compressed to the desired density. Once fully cured, the pre-form is peeled into thin sheets (**image 3**).

The sheets are processed like regular foam. In this case, it is laminated with an impermeable membrane backing (**image 4**).

1

2

3

4

Textile Technology
Plastic Film Extrusion

Thermoplastics, elastomers and bioplastics are extruded as film or sheet. Slit into thin strips, films are used to make yarn suitable for weaving, knitting, braiding and lace. With co-extrusion, two or more materials are combined in a single sheet. This is used to combine colours, or to combine dissimilar materials seamlessly.

Techniques	Materials	Costs
• Extrusion	• Thermoplastics	• Low unit costs
• Co-extrusion	• Elastomers	• Rapid cycle time
• Blown film	• Bioplastics	

Applications	Quality	Related Technologies	
• Fashion accessories	• High-quality colour	• Braiding	• Natural rubber
• Packaging	• Poor drape (if any)	• Laminating	and latex
• Apparel	• High strength to weight	• Fluid coating	• Welding

INTRODUCTION

Extrusion is the process of melting plastic and squeezing it through a shaped die to produce long lengths of material with a constant cross-section. Long thin dies produce film, thicker materials are called sheet, while profiled dies are used to make pipes, panels and trim, among other products.

Blown film extrusion (also referred to as tubular film) is the predominant method for making thin film (see page 186). A tube of plastic is extruded, inflated, and cooled to form thin-walled film. The mechanical properties are more uniform than straightforward extrusion, because the film is drawn along its length and around its circumference during production. It is used to make tubular products or slit into sheets.

Film ranges from 10 microns to 0.3 mm (0.012 in.) in thickness. Sheets are thicker. They are used as plain sheets, laminated (page 188) onto fabric, or combined with reinforcement to make light, strong composite film. They are slit into strips to make yarn for weaving (pages 76–105), knitting (pages 126–51), braiding (page 68) and lace (pages 106–19).

A wide range of materials may be extruded as film, including polyethylene terephthalate (PET), thermoplastic polyurethane (TPU), polypropylene (PP), polyethylene (PE), polyvinyl chloride (PVC) and polytetrafluoroethylene (PTFE). Each material has its own characteristics and is used for its specific properties, such as breathability or impermeability, or a combination of these. Films destined for technical applications and packaging are stretched during production to increase mechanical properties. A film is known as oriented or biaxial-oriented depending on whether it was stretched lengthways or both ways.

APPLICATIONS

Plastic film and sheet are available in a wide range of formats and used in diverse applications. Tubular extruded film is mainly used to make packaging, such as for medical applications, shopping bags and agricultural sacks.

Sheet materials are used for a wide range of applications. They are used as plain films – welded (page 376) or stitched (page 354) together – to make packaging and garments. PET and PE films are laminated with a supporting fibre matrix to make reinforced films (see also 3D Thermal Laminating, page 454). These are used for demanding applications that require the performance of plastic film with the strength of fibre reinforcement, such as geotextiles and inflatables.

PE and PP film are split into strips to make yarn that is woven into several products including industrial packaging (see Loom Weaving, page 76) and artificial grass (see Tufting, page 286). PET film is slit into strips – when applied as yarn it is generically referred to as polyester – and used as a novelty yarn in fancy woven and knitted fabrics. As well as being used in its naturally transparent or coloured state, PET is metal coated for functional and decorative purposes, such as to create a highly reflective surface.

Very thin films of TPU and expanded PTFE (ePTFE) are extruded and laminated onto textiles and net to make composite

The Extrusion Process

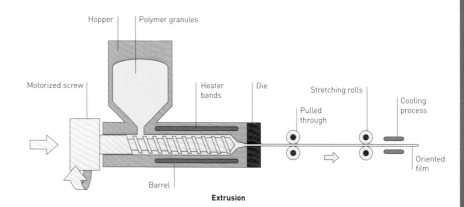

Extrusion

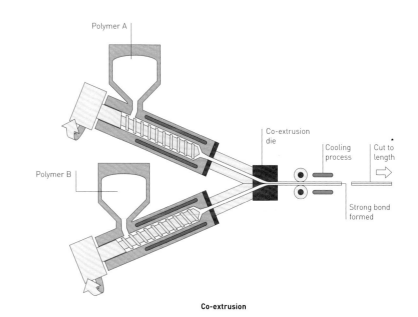

Co-extrusion

TECHNICAL DESCRIPTION

Granules of plastic are fed from the hopper into the barrel, where they are simultaneously heated, mixed and moved towards the die by the rotating action of the Archimedean screw.

The melted plastic is forced through the profiled die and into a cooling chamber or water tank to solidify the hot polymer. Film and sheet extrusions are cooled as they pass between sets of cooling rollers rather than a water tank. This helps to control the wall thickness and apply texture.

Oriented films are stretched in line with the machine direction with two sets of rolls operating at different speeds. By turning the second set slightly faster the material

is drawn out. To stretch the film widthwise a set-up similar to tentering (page 210) is required, which pulls the film outwards by its edges. This technique is used to produce most types of packaging films and tapes such as biaxial-oriented polyethylene terephthalate (BOPET), which is commonly known under the trademark name Mylar.

In co-extrusion, two or more plastics are extruded using conventional equipment and mixed in the co-extrusion die. They are hot and plastic (softened), which means that as they are brought together the polymers combine to form a strong bond. The joint is controlled by the design of the die. The process is suitable for combining

materials of different colours or with different properties. Materials that do not bond well to one another are joined with an intermediate layer of special adhesive, called a tie layer.

The extruded film may also be oriented by stretching, or may simply be cooled and cut to length.

or coated fabrics to provide a breathable and waterproof membrane (microporous). PE is laminated onto paper to make waterproof packaging.

The Italian company Mazzucchelli has devised many innovative ways to co-extrude multicoloured sheets of cellulose acetate (see case study on pages 184–85). These materials are used to make eyewear frames, jewelry and other fashion accessories. Originally developed to mimic tortoiseshell and ivory, cellulose acetate has become a desirable material in its own right and is used to make the highest-quality eyewear.

RELATED TECHNOLOGIES

Lightweight loom-woven fabrics and nonwovens (spun, page 153) tend to be stronger and the processes are more flexible than extrusion. As a result of drawing synthetic filaments during spinning (page 50), they tend to have higher tensile strength than films. While the properties of woven, knitted and nonwoven fabrics are tailored to the requirements of different applications by blending materials and changing density, fibres and films are laminated into composites that take advantage of the properties of both types.

Synthetic leather is manufactured by laminating woven or knitted fabric with a thermoplastic film, such as TPU. This material is used to imitate leather in a range of products, from handbags to footballs. There are several other techniques employed to create synthetic leather, such as needle punching (see Nonwoven, page 152), plastic coating (see page 226) and embossing (page 220). By contrast, genuine leather is protected by calendering a thin film of plastic onto its surface. It may be produced with a gloss, matt or textured finish.

Strips of yarn cut from widths of PP or PE film are used in place of filament-spun yarn (page 50), such as in the production of circular-woven sacks and windbreakers. Such items are low-cost and available in a wide range of vivid colours. Alternatively, very fine strips of PET are metal coated to produce highly decorative yarn suitable for apparel and hosiery.

QUALITY

Films are strong and uniform. The surface finish is semi-gloss and depends on the substrate and cooling process. Calendering the material between polished cylinders straight out of the die produces a high-gloss and high-quality finish on sheet materials, and is used to ensure exact dimensions.

With blown film, the combination of extrusion and inflating creates a film with good mechanical properties. As the film is drawn upwards, the molecules align. As the properties around the diameter of the film (hoop strength) are increased, the properties in line with draw (longitudinal strength) will decrease, and vice versa. The properties of the film such as tensile strength, opacity and tear strength are greatly affected by this molecular alignment.

Stretching films during production significantly increases mechanical properties. There are several different techniques, depending on whether it is extruded as sheet or blown film. Examples include biaxial-oriented PET (BOPET) used to make laminated composite films (see 3D Thermal Laminating, page 454) and packaging, and biaxial-oriented PP (BOPP) used to make packaging.

Surface finish can be gloss or matt, but the quality will not be as good as cast films, because the material is drawn through a die and inflated, which affects haze, transparency and gloss. It is not possible to control film thickness exactly (variation is typically in the region of 4–5%). As it is a continuous, in-line process, imperfections will be visible as straight lines, which can be quite pronounced. This will be emphasized in coloured films and less visible in transparent plastics. These imperfections will usually disappear with secondary processing, such as laminating or coating.

DESIGN

Extrusion is used to produce continuous lengths of film and sheet. Blown film produces tubular film, which may be slit into sheet. Co-extrusion is the process of extruding two or more films simultaneously to take advantage of the properties of both. It is possible

to extrude films consisting of up to 12 separate layers, although it is rare to go above five. Such complex multilayered films are used for the most demanding food and packaging applications. It is slower than monolayer film production, but saves time later in the construction of multilayered films. Qualities such as colour, air permeability and print are enhanced with combinations of different materials. Incompatible materials are joined together with adhesive films, known as tie layers. Other types of material, such as metal, may be encapsulated directly in the extrusion process. This technique is used to make plastic-coated wire, for example, used as a low-cost alternative to braided jackets (page 68).

Colour is added to the polymer and thus throughout the film, or it is printed afterwards (pages 256–79). Adding colour to the film itself is more cost-effective, but not as flexible. The type of printing technology is determined by the type of material, but is most often flexographic printing, because it is high-speed and compatible with PE-type materials.

Coating plastic with a very thin film of metal produces bright, metallic films. The metal is protected with a synthetic coating on top, or laminated between two films, to stop the coating wearing away. This technique is used for both technical (conductive and reflective) and aesthetic (reflective and colourful) purposes.

Reinforced films are fabricated by laminating fibre reinforcement, or other suitable material, between two layers of film. Flexibility ranges from stiff to elastic, depending on the mix of films and fibres. Each material in the layup can be a different colour. These materials are used to make modern lightweight sailcloth, as well as uppers for shoes, garments, geotextiles and industrial fabrics.

Films are bonded onto fabrics as a coating, such as for high-performance breathable and waterproof textiles. Combined with calendering and embossing, a range of surface patterns and textures can be achieved. Embossing is used to create unique designs or imitate other materials and finishes, such as leather-grain TPU.

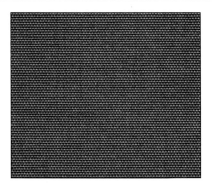

Waterproof Breathable Textile
Material: ePTFE and nylon
Application: Outerwear
Notes: Microporous ePTFE is laminated onto nylon to produce a waterproof and breathable textile.

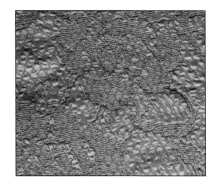

Laminated Lace
Material: PVC and cotton
Application: Apparel
Notes: Lace is bonded onto a plastic film to produce a waterproof material with a decorative surface.

Reinforced Laminate
Material: PVC and polyester fibre
Application: Construction
Notes: Films laminated with fibres are strong and lightweight, taking advantage of the properties of both.

Metallized
Material: Metal-coated PET
Application: Apparel
Notes: Coating plastic with a thin film of metal creates reflective yarn used for decorative and functional fabrics.

Iridescent Film
Material: Metal-coated PET
Application: Packaging
Notes: The vibrant colour is caused by light interference, and changes with viewing angle, an effect known as colour shift.

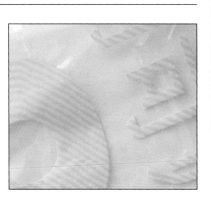

Printed Film
Material: CA
Application: Packaging
Notes: It has good impermeability to air, water and oils, and so is used for a wide range of thin packaging.

Fibrillated
Material: PE
Application: Artificial grass (see Tufting, page 289)
Notes: Strips of extruded sheet are fibrillated to create expanded mesh. It is twisted into yarn for artificial grass.

Strip
Material: PE
Application: Artificial grass
Notes: Extruded plastic film is cut into strips and stretched into strong yarn suitable for fabricating into textiles.

Plied
Material: PE
Application: Artificial grass
Notes: Just like conventional yarn, strips are twisted with other filaments to produce plied yarn.

Textured
Material: PE
Application: Artificial grass
Notes: Heating and compressing strip forms twists and buckles that create a textured and more bouncy yarn.

Knit De-Knit
Material: PE
Application: Artificial grass
Notes: Strip is knitted and heat-set, thus permanently forming it into loop shapes. It is de-knitted back into yarn.

Braided
Material: Metal-coated PET
Application: Apparel
Notes: Metal-coated strip is used flat and wrapped around a filament in the jacket of this decorative braided cord.

Many types of geometry and different colours are possible with plastic extrusion as long as the wall section is constant and the profile is continuous. Large cross-sections are hollowed out using a pin or mandrel in the extrusion die.

Fluorescent pigments glow brightly because they absorb light in the ultraviolet portion of the spectrum and emit light in the visible portion. Therefore tinted materials will appear brighter than their surroundings because they emit more visible light. This quality is emphasized by edge glow: a phenomenon caused by absorbed light being emitted mainly through the edges and thereby intensifying colour. By scoring the surface, laser cutting (page 328) produces a similar effect.

MATERIALS

Most thermoplastics, elastomers and bioplastics can be formed by extrusion. Thermoplastics are polymers that become soft and pliable when heated. The temperature depends on the type of plastic. Elastomers are similar to thermoplastics, except they exhibit similar mechanical properties to rubber.

PVC and TPU are used as a film, coating or laminate in garments and footwear. PET film, which is often referred to by its trade name Mylar, is a lightweight high-performance material used in a wide range of applications, from racing sails to strip yarn. It is commonly used for metallized film applications too, such as balloons, packaging and embroidery. PE thermoplastics, including high-density (HDPE), low-density (LDPE) and linear low-density (LLDPE), are used to make packaging and geotextiles.

Bioplastics are derived from renewable sources, unlike synthetic plastics, which are derived from petroleum. They require less energy to produce as raw materials and they can be fully biodegradable at the end of their useful life.

COSTS

This is a very cost-effective process for large volumes of plain film and sheet. Even so, it is only practical to run substantial volumes, because set-up is time-consuming and costly.

Cycle time is rapid – with modern machines capable of producing 500 m (546 yd) or more per minute – and the process is continuous. In many cases, production runs 24 hours per day. Labour costs are low, because once set up the process is relatively stable and so does not need too much attention. Expensive raw materials, such as PTFE used in microporous fabrics, will have a significant impact on unit cost.

ENVIRONMENTAL IMPACTS

Plastics are either derived from crude oil (and so referred to as synthetic) or from biological ingredients, such as starch (and so referred to as bioplastics).

The environmental impact of each depends on many factors, such as source, production technique and by-products. Taking the full life cycle of the material into consideration, in relation to the application, will ensure that the end product has the lowest possible negative impact on people and the environment.

With regard to eco-efficiency (overall environmental impact), energy recovery by incineration is often more effective than recycling plastics. Even so, thermoplastics can be recycled, but owing to the processes and contamination their strength and quality will be slightly reduced each time, limiting the range of suitable applications.

Plastics are often criticized for containing harmful ingredients. Phthalates and bisphenol A are subject to restrictions in many countries. Phthalate-free PVC is produced using a specially developed plasticizer (free from phthalates). It is used for sensitive applications, including toys and medical devices.

Waste is created at start-up and when transitioning between colours or dimensions of film. The flexibility of the processing equipment helps to reduce waste. If a single type of material, or single colour, is being produced then all the waste can usually be directly recycled: it can be put back into the hopper and re-extruded. Mixed materials cannot be recycled as readily and so are often downcycled into dark-coloured products.

→ Co-Extruding Multicoloured Sheets

Mazzucchelli has learned how to take advantage of this straightforward process to create decorative plastic sheet materials. Mainly used to produce high-end eyewear, these colourful materials are suitable for making jewelry, accessories and even artwork.

Plastic granules are fed into the hopper (**image 1**), heated, melted and extruded. In co-extrusion, the colours are extruded separately and brought together in the co-extrusion die (**image 2**). The material is drawn through polished calendering rolls to smooth and homogenize the hot plastic into an accurate profile (**image 3**).

Each strip is cut to length and checked for visual defects (**image 4**). The cross-section shows how the different colours have been combined to create a precise visual gradient on the surface (**image 5**).

An almost limitless range of colours can be achieved in plastic. Transparent plastics, such as polymethylmethacrylate (PMMA), cellulose acetate (CA) and polystyrene (PS), colour well (**image 6**).

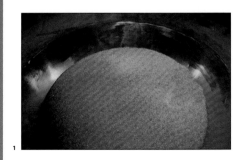

1

2

3

4

6

5

Featured Company

Mazzucchelli
www.mazzucchelli1849.it

The Blown Film Process

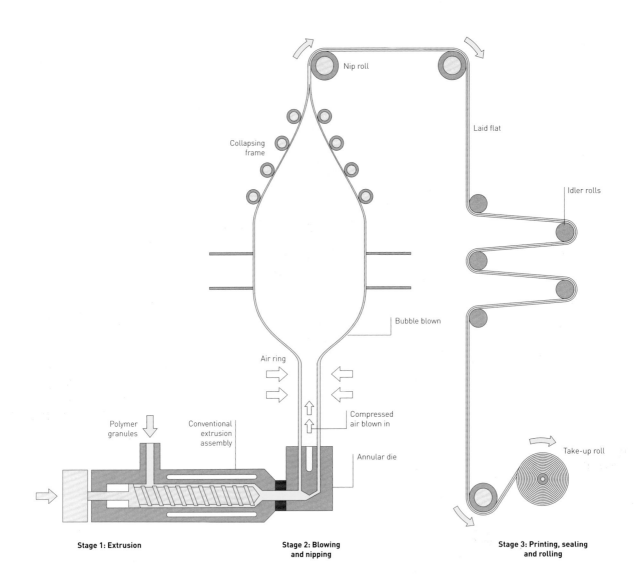

Nip roll

Laid flat

Idler rolls

Collapsing frame

Bubble blown

Air ring

Compressed air blown in

Polymer granules

Conventional extrusion assembly

Annular die

Take-up roll

Stage 1: Extrusion

Stage 2: Blowing and nipping

Stage 3: Printing, sealing and rolling

TECHNICAL DESCRIPTION

In stage 1, a continuous tube is formed by conventional extrusion. Molten plastic passes through the extrusion die, known as an annular die, and is simultaneously pulled upwards by the nip rolls. A single die can be used to make several different sizes and thicknesses of film, because the dimensions can be controlled in the blowing phase.

In stage 2, the tubular extrusion is cooled by the air ring around the outside and inflated from the inside by compressed air. As the diameter increases, the wall thickness decreases. The bubble is drawn through a series of collapsing rolls and into

the nip roll as a flat sheet. By this time, the polymer has cooled and solidified and will not stick to itself. For high-output lines the air inside may also be cooled to speed up the process.

The height of the nip rolls can be 20 m (66 ft) and the bubble may be very small or up to 9 m (30 ft) in diameter. Increasing the amount of stretching at a defined temperature produces oriented film. In operation, the initial bubble is extruded, cooled and then reheated as air is pumped in and the take-up speed increased, thus stretching the film simultaneously in both directions.

In stage 3, the blown film is drawn through a series of idler rolls laid flat. These even out the tension in the film, to ensure consistent properties in the finished product. Printing and sealing, to form plastic bags for example, is carried out before the plastic film is finally wound onto the take-up roll.

→ Blown Film Production

Typically, blown film extrusion is carried out vertically. A continuous tube of HDPE is inflated with compressed air as it rises upwards from the extrusion die (**image 1**). Air is blown inside through a nozzle within the die. The diameter of the tube is carefully controlled to ensure a uniform wall section. The diameter is increased, which further reduces the thickness of the film, as it rises upwards (**image 2**).

Once solidified, the length of tube is gradually collapsed to form a flat sheet with two layers. It continues down the side of the extrusion tower and passes through an in-line printing process at ground level (**image 3**). The tube is sealed, perforated, folded and wound onto a take-up roll (**image 4**).

Production is continuous, only stopping to change colour, or design. Rolls of HDPE are stored in the warehouse (**image 5**).

1

2

3

4

5

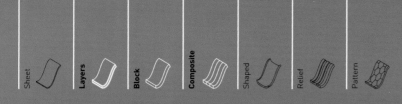

Textile Technology
Laminating

Layers of material are bonded together using heat, adhesive or a combination of the two. A broad range of materials may be combined, including flexible and rigid, thick and thin, and natural and synthetic. Primarily used to enhance the performance of commodity materials, laminating is also utilized in the production of technical and high-performance textiles.

Techniques	Materials	Costs
• Adhesive bonding • Heat bonding	• All types of sheet material	• Low (but depends on materials) • Rapid cycle time

Applications	Quality	Competing Processes	
• Apparel • Footwear • Technical textiles	• Waterproof • Breathable • Lightweight	• Adhesive bonding • Fancy loom weaving	• Fluid coating • Stitch bonding

INTRODUCTION

Laminating is carried out as a continuous process, roll-to-roll, and so is utilized for moderate to high volumes. Alternatively, it may be used for one-offs and low volumes by pressing sheets together on a flat bed. This provides flexibility for designers to experiment and try out ideas before committing to large orders.

Material selection will determine the most suitable method of application. For example, thermoplastic films (see Plastic Film Extrusion, page 180) are bonded onto fabric with heat and pressure, while adhesive may be used to combine all manner of flexible and rigid materials, including sheet, fibre and filament.

APPLICATIONS

Laminating is used in a very wide range of applications. In the automotive industry it is used to produce the upholstery, insulation, headliners (interior roof liners), carpet and leather trim. Interior applications include high-wear upholstery, curtains, underlay and backing for carpets and mattress covers.

Apparel and lingerie make use of laminating to eliminate sewing in molded parts (see Adhesive Bonding, page 370), waterproof and breathable outdoor gear, children's clothing, sportswear and padded footwear. Medical products include plasters and bandages.

There is a range of waterproof and breathable fabrics produced by laminating thermoplastics between a textile and liner. Probably the most well known is manufactured by Gore under the trade name Gore-Tex. The thermoplastic layer is expanded polytetrafluoroethylene (ePTFE), produced by rapidly stretching PTFE. The microporous thin film has unique properties, such as being waterproof, windproof and breathable (allows evaporation). These high-performance materials are utilized for outerwear, footwear and gloves. Multiple different fabric constructions are available, such as two-ply and three-ply, to tailor performance to any given application.

Thermoplastic composites are laminated and utilized in high-performance racing sails (see 3D Thermal Laminating, page 454), protective clothing and lightweight footwear.

Waterproof and impermeable laminates are used to make, for example, inflatables, buoyancy aids, table covers and aprons.

COMPETING PROCESSES

A range of complex properties is achieved with fancy weaving (page 84) and technical knitting (see Weft Knitting, page 126) alone. But with the addition of laminating, properties such as breathability, waterproofing and insulation may be added without detracting from the lightness or handle.

Stitch-bonding (page 196) techniques, such as quilting, are used to combine the properties of two or more materials into a single construction. However, the visual and physical properties of such a combination produce a quite different composite structure and performance.

Similar to laminating, coatings (page 226) are applied to fabrics to enhance the visual and functional performance. Liquid coatings penetrate into the fabric structure and can be impregnated right through, depending on the method of application. Depending on the coating material the look and handle of the base textile may be completely transformed or unaffected.

Thermoplastics may be extruded directly onto the surface of the textile. One of the major advantages of this direct coating method is that the plastic is kept very pure and there is no need for added chemicals or solvents. Otherwise it is very similar to thermoplastic film lamination. One drawback is that it is justifiable only for large orders, because setting up the extrusion line is a lengthy process.

QUALITY

Laminating produces materials that are lightweight and less prone to ravelling. The performance will depend on material selection and bonding process. Plastic foams (page 172) and films are engineered or modified to perform specific functions, such as provide insulation, breathability, abrasion resistance, impermeability to water or air, self-cleaning properties, shape-memory, electrical conductivity and so on.

These properties are combined with the performance and handle of sheets of woven (pages 76–105), knitted (pages 126–51), nonwoven (page 152), leather (page 158), lace (pages 106–19), embroidered (page 298), paper and veneer materials.

Bonded sheet materials individually retain their inherent properties, but only the construction as a whole will provide the full complement of required properties. This gives a high level of control over specific properties, such as cover, breathability and waterproofing, because each layer is independent and not reliant on the base material.

The properties of laminated material will be different from the base materials.

Bonding increases the stiffness of material. However, the increase is typically much lower than if a material with equivalent properties was produced by any of the other more conventional fabrication technologies.

DESIGN

Laminating provides a wealth of opportunity for designers. Materials with the required properties, but that would otherwise be impractical to use, may be combined with a conventional fabric to make them usable. For example, thermoplastic films have many desirable properties, but they are thin and stretchy and do not have good handle. Combining them with a woven or knitted textile produces a more durable and practical material.

Materials can be quite different on the face and back, because the layers retain their inherent properties. The colour, texture, touch and performance can all be tailored to meet specific requirements.

Functional layers may be included, such as self-cleaning; hook and loop fasteners (such as Velcro); non-stick (such as PTFE); shape-memory to produce materials that physically respond to changes in temperature; phase-change materials (PCM) that provide cooling when it is hot and warmth when it is cool; and reflective and colour-changing (light interference, thermochromatic and photochromatic) films.

Rigid and flexible sheets may be combined into a single material. Alternatively, high-performance fibres may be combined with flexible sheet materials to provide strength along predetermined lines of stress, such as in the case of sports shoes and thermoplastic composites (page 446).

Combining multiple materials into a single layer may reduce secondary processes, such as eliminating the need for a liner (page 392).

Depending on the ingredients, laminated sheets may be processed like conventional fabrics, such as printing (pages 256–79), pattern cutting (page 314) and machine stitching (page 354).

Including thermoplastics in the construction means that these materials

The Laminating Process

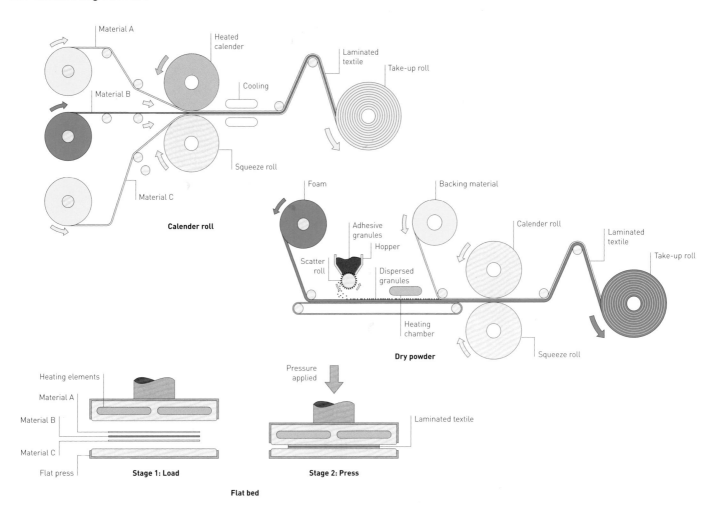

TECHNICAL DESCRIPTION

There are many different processes used to laminate fabrics. All of them may be used to combine two or more layers of material. Two-layer laminate is called two-ply (or bi-ply), three-layer is called three-ply (or tri-ply) and so on. Laminates are bonded with adhesive, or heated to activate the adhesive properties of one or more of the component layers.

Heat-bonding techniques use thermoplastic, either as preformed film or by directly extruding into the laminating process. It is known as hot-melt when thermoplastic is used as adhesive.

Most thermoplastics used in laminating become soft between 80 and 220°C (176–428°F). They are combined with two or more layers of material and form a strong join as they cool. The most common are ethyl vinyl acetate (EVA), nylon and TPU.

Calender-roll laminating presses the layers together between a heated roll and a squeeze roll. There are two main ways this process is used: either the two outer materials (A and C) are fused by the encapsulated thermoplastic film (B); or a fabric (B) is coated with two layers of thermoplastic film (A and C). Alternatively, calendering is used to combine just two materials (in a process similar to coating).

Materials prone to stretch may be laminated with a blanket roll in a process similar to transfer printing (page 272). With this technique, the layers are held firmly together, ensuring there is no slippage as they travel around the heated cylinder under pressure.

In the dry-powder process, granules of thermoplastic hot-melt are spread onto fabric. The adhesive granules range in diameter from 0.002 to 0.2 mm (0.0008–0.008 in). They are spread across the entire width of fabric and melted in a heating chamber; the backing material is then introduced on top and pressed firmly between calender rolls. The powder bonds the layers in evenly dispersed spots. This helps to maintain the inherent flexibility of the materials and is commonly used to bond foam to backing fabric.

The amount of powder adhesive is adjusted to suit the technical requirements of the application. Dispersion is more accurately controlled by replacing the scatter roll with a grooved roll that deposits an exact quantity in a defined pattern.

Flat-bed laminating is used to bond samples, prototypes and low volumes. Sheets of material are cut to size and assembled on the flat bed. With heat and

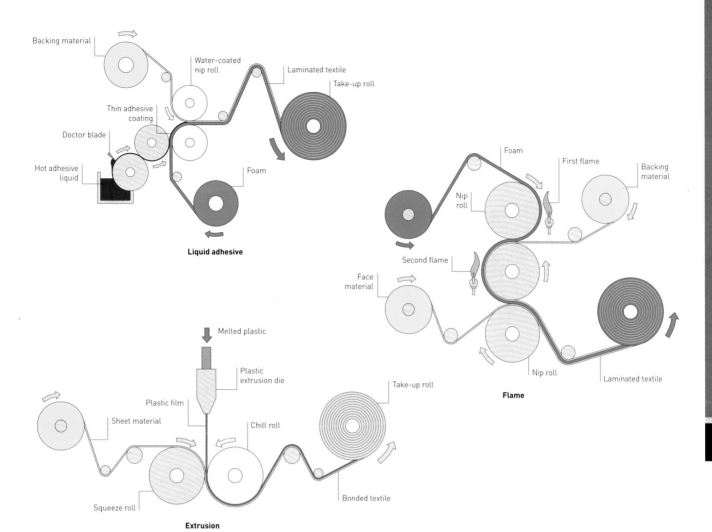

Liquid adhesive

Backing material
Water-coated nip roll
Laminated textile
Take-up roll
Thin adhesive coating
Doctor blade
Hot adhesive liquid
Foam

Flame

Foam
First flame
Backing material
Nip roll
Second flame
Face material
Nip roll
Laminated textile

Extrusion

Melted plastic
Plastic extrusion die
Plastic film
Sheet material
Chill roll
Take-up roll
Squeeze roll
Bonded textile

pressure or pressure alone (depending on the adhesive system), the layers are permanently bonded together. After a predetermined period the press is opened and the laminated material removed.

Liquid adhesive (solvent- or water-based) is applied as a coating to one of the webs of material. A 'doctor blade' controls the film thickness. Alternative means of application include spraying and printing.

The coated material is combined with a second web between two nip rolls, which squeeze the layers together. Curing depends on the type of adhesive used. Some systems can take up to 12 hours.

Adhesive coating is used for materials that are sensitive to heat or pressure, or otherwise incompatible with hot-melt. Adhesive systems can run at very high speed, because heating is not required for

adhesion to take place (although some adhesive systems require heating to speed up curing). Thermoset adhesives, such as polyurethane (PU), which form permanent cross-links and so do not break down at high temperatures, are applied in this way.

An alternative method of bonding foam to fabric, or multiple layers of foam, is flame lamination. This technique is used to bond thermoplastic polyurethane (TPU) or polyethylene (PE) foam to fabric and sheet materials. In operation, the foam is passed over an open flame, which causes the surface to melt immediately before it comes into contact with the backing material.

The layers are pressed together by the nip rolls. The second flame causes the other side to melt, which is subsequently bonded to the face material. Thus the two layers of

material (fabric, film or additional foam) are bonded to the foam layer in a single process.

Extrusion converts plastic granules into a thin plastic sheet or film (page 180). The thin film is bonded directly onto fabric, or between layers of fabric, around a chill roll. The surface of the chill roll may be gloss, matt or textured, and will be reproduced exactly on the surface of the plastic.

Coating directly from the extrusion die works well for polypropylene (PP) and PE, because they are viscous enough to be extruded as thin films. Other thermoplastics are not so well suited to this approach, especially for coatings less than 0.5 mm (0.02 in). Therefore, materials such as polyvinyl chloride (PVC) and TPU are more commonly formed separately and laminated with calender rolls.

are also suitable for welding (page 376) and thermoforming (creating three-dimensional shapes with heat and pressure). These techniques are used to produce one-piece trainer uppers and stitch-free underwear, for example.

Standard laminating machines range from 0.6 to 3 m (2–10 ft) wide, with some up to 5.5 m (18 ft).

MATERIALS

There is a range of techniques, which means that most materials are suitable, from plastic films, lace and sheer fabrics, to wood and rubber.

Commonly used plastic films include polyvinyl chloride (PVC), polyurethane (PU), thermoplastic polyurethane (TPU) and polyethylene terephthalate (PET) (also known under the trade name Mylar).

Commonly used plastic foams are PU, polyethylene (PE) and ethylene propylene diene monomer (EPDM). There are many variables with foam, including flexible to rigid with small to large cells, which may be open or closed.

As well as combining sheet materials, it is possible to laminate filaments between layers of plastic film. For example, North Sails' three-dimensional laminating process combines laminating with thermoforming. Lightness is critical for Grand Prix sailboat racing and with this approach they have been able to reduce weight by 20%.

COSTS

The laminating process itself is relatively straightforward and so low-cost. However, material selection will affect the price, because some materials are easier to laminate than others. Conventional materials and set-ups run at high speed. Bespoke developments, such as combining functional materials, can slow down the process and so can increase cost significantly.

ENVIRONMENTAL IMPACTS

Permanently combining multiple different types of material presents a challenge end-of-life. Only single types of material can be recycled with any efficiency. Therefore, combining foam, sheet or fabric made from a single thermoplastic means it can be recycled at the end of its useful life. Even so, regarding total environmental impact, energy recovery by incineration is often more effective than recycling plastics.

Adhesives should be carefully considered, because they can have negative environmental impacts. Solvent-free systems, such as water-based or thermoplastic (hot-melt) adhesive, are available that have no volatile organic compounds (VOCs), are non-toxic and free from lead, phthalates, and bisphenol A (ingredients that are subject to restrictions in many countries).

VISUAL GLOSSARY: EXAMPLES OF LAMINATED FABRICS

Front · Back

Waterproof and Breathable
Material: Woven nylon face, TPU film and polyester half-tricot backing
Application: Apparel outerwear
Notes: A dry-powder adhesive-bonded three-ply construction used for high-performance outerwear.

 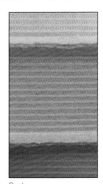

Front · Back

Foam Laminating
Material: Textured nylon face, PU foam and nylon half-tricot backing
Application: Bathing
Notes: Foam is laminated between a knit and a textured weave to produce a soft and comfortable bath scrubber.

Nonwoven
Material: Nylon nonwoven and PU foam
Application: Cleaning
Notes: Abrasive-coated nonwoven nylon is bonded onto PU foam to create a gentle and absorbent abrasive pad.

Fibre-Reinforced Laminate
Material: Various fibres and TPU film outer
Application: Footwear
Notes: Fibres are adhesive bonded with TPU films to make shoe uppers with superior strength to weight.

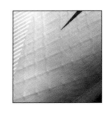

Lightweight
Material: UHMWPE and polyester
Application: Construction
Notes: High-strength filaments are bonded between PET films and used as a lightweight vapour barrier in buildings.

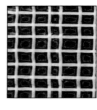

Reflective Film
Material: Plastic film and woven tape
Application: Apparel
Notes: A prismatic structure in the PVC film retroreflects light. The colour is enhanced by fluorescing dye.

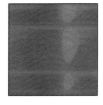

Edge Finish
Material: Nylon weave and TPU film
Application: Apparel
Notes: The plastic film backing stabilizes the weave and stops the edges ravelling.

Three-Dimensional Fabric
Material: Wood veneer and polyester organza
Application: Interior
Notes: Rigid materials are laminated to fabric to provide structure, such as in Elaine Ng Yan Ling's Naturology collection.

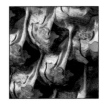

→ Dry-Powder Laminating Foam to Nonwoven

In this case, foam is bonded to nonwoven polyester for use in mattresses. Granules of hot-melt adhesive are loaded into the hopper above a rotating brush (**image 1**). A sheer polyester nonwoven fabric is passed below on a conveyor (**image 2**). As it passes underneath, the surface is given a uniform coating of adhesive particles (**image 3**).

The adhesive is heated, causing it to melt and fuse onto the nonwoven fabric. Foam is laid on top. With the application of pressure, the foam is bonded to the nonwoven layer.

The finished laminate is taken up on a roll (**image 4**). Complete rolls are transferred to the cutting department, where they will be cut into shaped panels on a plotter.

1

2

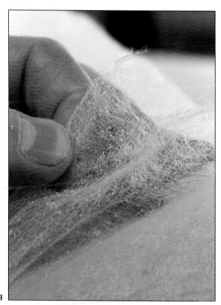

3

4

Featured Company

John Holden
www.john-holden.co.uk

→ Adhesive Bonding Foam to Film

In this case, foam is bonded to film. A roll of polyurethane (PU) foam (**image 1**) and a roll of thermoplastic polyurethane (TPU) film (**image 2**) are loaded onto the laminator (**image 3**). This technique is predominantly used to make materials for upholstery, automotive interiors, acoustic panels, carpet, shoes and office furniture.

The foam is introduced from below and coated with adhesive (**image 4**). Acrylic, latex, silicon and solvent-based liquid adhesive systems are all feasible. Selection is based on the requirements of the application. Nip rolls press the thin film down onto the foam. The laminated material is taken up on a roll (**image 5**). At this point the adhesive is still wet and will take roughly 12 hours to fully cure. The fabric is left under

pressure on the roll to ensure that a strong bond is formed across the surface (**image 6**).

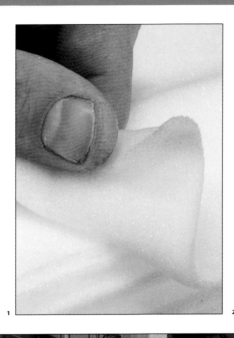

1

2

3

4

5

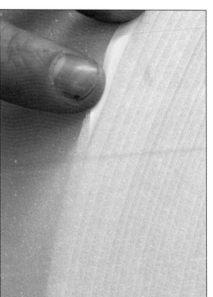

6

Featured Company

John Holden

www.john-holden.co.uk

Textile Technology
Stitch Bonding

This includes a range of mechanical processes such as machine stitching, knit-through and warp knitting with inlaid weft. Sheet materials, fibres or both are locked together with a series of stitches. Thus the properties of a knitted or stitched structure are combined with the base materials to create a composite with unique properties.

Techniques	Materials	Costs
• Knit-through • Quilting • Inlaid yarn	• All sheet materials • Fibres and filaments	• Low-cost (bespoke patterns are higher-cost) • High production rate

Applications	Quality	Related Technologies	
• Apparel • Interiors • Automotive	• Depends on materials and construction • Lightweight • Dense cover to open structure	• Nonwoven • Laminating • Machine stitching	• Weft knitting • Warp knitting

INTRODUCTION

Stitch-bonded fabrics include knit-through nonwovens, quilted laminate constructions and knitting with inlaid yarn. They are primarily used to produce lightweight and low-cost materials with enhanced performance as a result of the combined yarns, fabrics and construction techniques.

Several different machines are used to stitch bond nonwovens, including Arachne (Czechoslovakia, 1949), Malimo (East Germany, 1958) and the variants, which are Malivlies, Maliwatt, Kunit and Multiknit.

The process was conceived as a low-cost and high-speed alternative to weaving (see Loom Weaving, page 76) and knitting (see Weft Knitting, page 126, and Warp Knitting, page 144). Converting webs of fibre directly into fabrics cuts out the lengthy preparation processes required to transform fibres into yarns suitable for weaving.

The machines consist of beds of needles working in reciprocating motion. Joining several layers together produces a composite textile. It is known as quilting when the layers are combined with lines of machine stitching (page 354).

As well as combining nonwovens and other sheet materials, stitch bonding is used to produce biaxial and multiaxial constructions. Inserted warp or weft yarns are combined with pillar stitches and used to produce composite laminates (page 188), pile, terry and fabrics with controlled stretch (the inlaid yarn is straight and so has no more stretch than the yarn properties; see case study on page 201).

APPLICATIONS

Nonwovens are used in a wide range of applications (page 152). Stitch-bonded types in particular are used to make

The Stitch Bonding Process

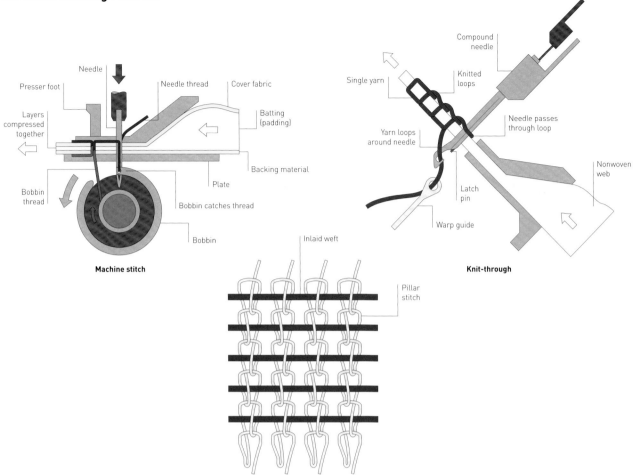

Machine stitch

Knit-through

Inlaid yarn

(Diagram labels — Machine stitch:) Needle; Presser foot; Needle thread; Cover fabric; Batting (padding); Layers compressed together; Backing material; Plate; Bobbin thread; Bobbin catches thread; Bobbin

(Diagram labels — Knit-through:) Compound needle; Single yarn; Knitted loops; Yarn loops around needle; Needle passes through loop; Nonwoven web; Latch pin; Warp guide

(Diagram labels — Inlaid yarn:) Inlaid weft; Pillar stitch; Inlaid yarn

TECHNICAL DESCRIPTION

Knit-through is based on warp knitting (page 144). Many of the same stitches are used, including tricot and raschel, on machines equipped with one or two bars.

Knit-through mechanically bonds sheets of nonwoven fibre webs to form a coherent material with the benefits of a knitted and a nonwoven structure combined.

The pointed compound needle pierces the nonwoven web (and any other sheet material used in the stack, such as film and woven textile). The latch pin of the compound needle is retracted and the yarn is passed around the open hook. The hook catches the yarn, drawing it back through the web of fibres and the existing loop, to form the next stitch.

Most knit-through machines work in a similar way, except the Malivlies machine, which does not use a warp thread. Instead,

the hooked needle catches fibres from within the nonwoven web and stitches those together.

Warp knitting is combined with inlaid weft to produce a fabric with unique characteristics. The inserted yarns run between the technical back and face of the pillar stitch loops during loop formation. By contrast, insertion warp knits utilize the yarn in the structure of the fabric: the inlaid yarns pass around the loops.

Machine stitching joins layers of material together to form a composite (for a more detailed description of machine stitching, see page 354).

Sewing machines produce a lock stitch (illustrated) or chain stitch. A lock stitch consists of two or more threads interlaced from either side. As the needle pierces the layers of material a rotating bobbin catches

the thread. As it travels around the bobbin it is interlaced with the bobbin thread. This process can be repeated up to 4,000 times every minute.

Chain stitches are formed from a single thread, interlooped in a chain-like formation similar to a warp-knitted pillar stitch. The downside of this method is that if the thread is broken at any point it readily comes apart.

Multi-thread chain stitch overcomes the vulnerability of chain stitching with a single thread by interlacing and interlooping the threads on the underside of the join. This creates a stronger stitch than equivalent lock stitching.

filtration, insulation and padding. The laminates are used in furniture upholstery, linings, curtains and mattress ticking, for example. Quilting is used to produce interior fabrics, apparel, footwear and sports equipment.

Fabrics constructed from filaments are used to make geotextiles and the fibre reinforcement for high-performance composites.

RELATED TECHNOLOGIES

Stitch bonding was adapted from warp knitting. The advantage of this technique is that several different types of material are combined to make a single new one. The range of different materials that can be used expands the possibilities of warp knitting for designers.

Stitch bonding is used to produce unidirectional, as well as multiaxial, fabrics for composite laminating. An alternative technique uses adhesive-bonded filaments across the back of the fabric to hold the fibres in place. Eliminating the stitching produces a more uniform appearance. Therefore, this approach is used to produce fabrics that will be used as a visible face layer, such as in sports equipment.

Laminating (page 188) is used to produce composite textiles with heat or adhesive bonding. One of the advantages of mechanical bonding over these alternatives is that a durable interlocked layered construction is produced without heat or adhesive.

A skilled operator may produce quilted fabric on a conventional sewing machine. This allows for greater design freedom, because the path of stitching is not restricted by the set-up of the machine. However, this will be considerably more expensive than stitch bonding with fixed beds of needles.

QUALITY

Mechanically bonding fibres, or layers of material, spreads the load efficiently across the material. The knitted or stitched thread provides cover on the surface, which helps to lock the fibres in and provide a smooth finish. The knit or stitch will always be visible on the surface. Using transparent yarn reduces the visual

impact (this technique is used to cover printed textiles, for example).

The appearance and mechanical properties will depend on the fibres, sheets and construction technique. Many of the visual characteristics and mechanical properties of knitting and machine stitching can be applied to stitch bonding.

The knitted interlocked structure will not affect the handle as adhesive or thermal bonding can. In fact, it may be used to impart some of the advantages of a knitted structure to nonwoven and other sheet materials.

DESIGN

With stitch bonding, the yarn, fabric, foam and film are all interchangeable. In each case, the adjustments that can be made to colour, weight and type of material depend on whether it is woven, knitted, nonwoven, paper, plastic film (page 180), or other sheet material. In addition, all types of fibre may be used, including natural, man-made, staple (page 56) and filament (page 50). There is an unlimited number of combinations.

The method of mechanical interlocking may be adjusted according to the requirements of the application. Conventional set-ups are the least expensive; bespoke patterns and stitches will increase cost.

Quilting joins the materials over a relatively small area. This leaves room for the batting (foam, fibre or down) to expand and provide insulation. The mechanical bonds create cells that prevent the padding from shifting.

The type of stitch, such as lock, pillar or tricot, will determine the appearance of the mechanical join on the face and back.

A wide range of finishes can be achieved including smooth, pile and terry (from hook and loop fasteners to towels). Further finishing processes include printing (pages 256–79) and dyeing (page 240).

MATERIALS

Any material that can be pierced by a needle is suitable for stitch bonding, including for example sheet materials, fibres and filaments.

Batting (also called wadding) is a fibre-based filling used to provide insulation in apparel and bedcovers. Many different weights and thicknesses are used, including nonwovens and foam. Nonwoven fibres are typically in the range of 3.3–17 dtex: polyester, polypropylene (PP) and cotton fibres are all commonly used. Synthetic fibre-based batting is also called fibrefill.

Feathers and down taken from the undercoat of waterfowl, such as ducks and geese, is also used as the fill in quilted items of clothing and bedding.

COSTS

These are high-speed processes capable of producing up to 2,000 stitches per minute in a wide range of fabric densities. The high production rate and low maintenance makes these very cost-effective processes.

However, unconventional set-ups will reduce cycle time considerably and so increase cost.

It is typically more cost-effective to combine the properties of materials in a laminated composite than to try to produce the same properties in a solely knitted, woven or nonwoven textile.

ENVIRONMENTAL IMPACTS

These are dry processes free from heat and adhesive. This means they tend to have significantly fewer negative impacts than laminating.

Material selection is critical, because composites present a challenge at the end of their useful life. Combining different types of material means that they are impractical to recycle. Choosing low-impact materials, such as from renewable or recycled sources, is therefore preferable.

→ Quilted Fabric with Nonwoven Batting

Lock stitches require a bobbin, because two threads are used to make this stitch. The bobbins are prepared with coloured thread (**image 1**). They are loaded into the lower bed (**image 2**). A limited amount of thread can be wound onto a bobbin, so they often need replenishing.

The cotton fabric outer layers and nonwoven polyester batting (**image 3**) are loaded onto the quilting machine. The three layers of material are brought together and stitched through in a continuous process (**image 4**). The needles work in a straight line and the fabric is moved from side to side, which creates the diamond pattern.

The quilted fabric is taken up on a roll (**image 5**). High-quality quilted fabrics are joined with lines of lock stitching, because this stitch is more secure than chain stitching.

The thread is colour-matched with the twill fabric background colour and so blends with the weave (**image 6**). The backing layer is napped prior to quilting to provide a soft finish in application (**image 7**).

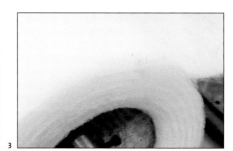

Featured Company

John Holden
www.john-holden.co.uk

Knit-Through

Material: Polyester
Application: Interior
Notes: Stitching creates a smooth surface for printing and improves lengthwise tensile strength.

Front

Back

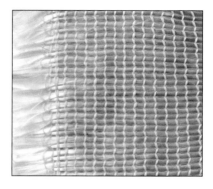

Performance Fabric

Material: Glass tow and polyester thread
Application: Composite laminating
Notes: Stitch bonding is used in the fabrication of high-strength (but brittle) fibre reinforcement for composite laminating. Here, unidirectional glass filament tows (untwisted yarns) are secured with pillar stitches.

Inlaid Weft

Material: Cotton
Application: Kitchen
Notes: Inserting bulky yarn into rows of pillar stitches creates a large surface area for water absorption.

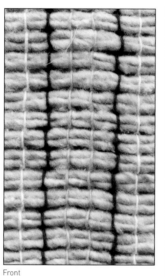

Front

Back

Decorative Fabric

Material: Polyester
Application: Interior
Notes: Filaments used for decorative effect are held between a very fine knitted face (black) and heavier knitted backing (orange) intersected in a diamond pattern.

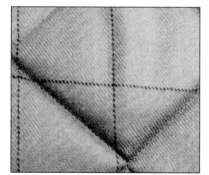

Diamond Quilt

Material: Cotton face and polyester backing
Application: Blanket
Notes: Lock stitches secure a layer of nonwoven batting between twill-woven face and a fabric liner. Web quilting is high-speed.

Linear Quilt

Material: Nylon face with polyester nonwoven batting
Application: Apparel
Notes: Unlike feather and down, foam and nonwoven batting need only to be held in place, and so do not need to be enclosed in cells of stitching.

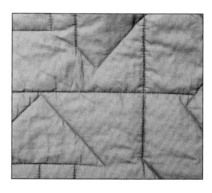

Pattern Quilt

Material: Cotton with polyester nonwoven batting
Application: Apparel
Notes: It is feasible to make bespoke patterns on individual panels, rather than a web, but this will increase cost.

Case Study

→ Warp Knitting with Inlaid Weft

In the same way as weaving latex (page 168), the raw material is supplied in strips, which are split into individual yarns in preparation for knitting (**images 1** and **2**).

Held under tension, the latex yarn is warp knitted. Within each course four non-stretch yarns are inlaid across the wales (**image 3**). As it is drawn out beneath the two needle beds the structure appears open, because it is held under tension (**image 4**). When relaxed, the stitches pull tight to produce a dense narrow fabric (**image 5**). The latex will stretch lengthwise, but will not stretch widthwise thanks to the widthwise-inlaid yarn.

Stitch-bonded fabrics can be easily confused with insertion warp-knits (page 145). The difference is that the inserted yarns in stitch-bonded fabrics are laid in the course of loops, as opposed to around them.

1

2

3

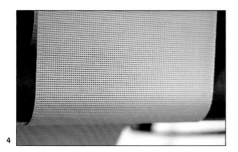

4

Featured Company

H. Seal & Company
www.hseal.co.uk

5

Scouring and Mercerizing

Newly constructed textiles pass through a series of essential treatments in preparation for finishing. Singeing, scouring (washing) and bleaching, collectively referred to as cloth preparation, affect the look and handle. Mercerizing is used to improve the strength and brightness of cellulosic fibres by swelling them in high-strength caustic acid while under tension.

Techniques	Materials	Costs
• Bleaching • Washing • Mercerizing	• Cotton, hemp, flax and other natural fibres • Yarn and textile	• Low

Applications	Quality	Related Technologies
• Apparel • Interiors	• Bright white • Improves strength and appearance	• Dyeing • Printing • Resist dyeing

INTRODUCTION

These processes make up the essential preparatory steps needed to make high-quality coloured, printed and coated fabrics for apparel and interiors.

As well as being essential preparatory treatments, washing is one of the final processes all textiles pass through before they are considered finished. The mechanical action of washing is used in finishing processes such as abrasive, chemical and enzyme washes, reductive techniques that permanently change the colour and handle of the material.

Cotton and other cellulosic fibres are mercerized by soaking in high-strength caustic acid (sodium hydroxide). This

The Singeing and Desizing Process

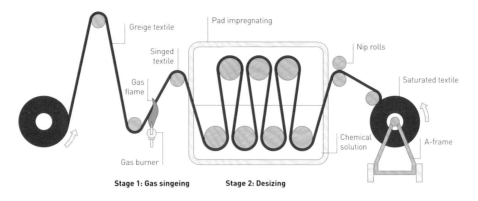

Stage 1: Gas singeing   Stage 2: Desizing

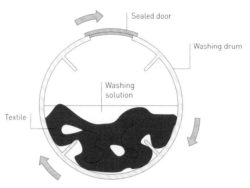

Drum washing

chemically alters the fibre, causing it to swell. Holding the fabric under tension means that it cannot shrink as it swells and contracts. This leads to increased strength and improves the light reflectance of the surface. It is nearly always carried out before vat and reactive printing (pages 256–79) and dyeing (page 240), because the colour appears brighter and more saturated. It is not necessary for pigment printing, because in this case the colour is applied on top as opposed to dyeing the fibres.

APPLICATIONS

In combination with calendering (page 220), these processes are used to make fabrics ready for printing, dyeing, coating and laminating (page 188). They are used mainly for fabrics that will be used in applications that require high-quality appearance and handle, such as apparel and interiors.

RELATED TECHNOLOGIES

These are essential treatments that cannot usually be avoided.

When washing is used to apply the finished appearance and handle it may start to cross over with certain printing, dyeing and resist-dyeing (page 248) techniques. These are additive processes, whereas washing is an abrasive and mechanical process, so the outcomes will be quite different even if they look similar at first. For example, stone washing and chemical washing used to finish denim jeans may be mimicked with printing or dyeing. Over time, the print or dye may fade or wear away. By contrast, washed finishes permanently alter the properties of fibres – only cellulosic fibres are compatible with the chemical and mechanical action – and so the ageing effects, known as patina, will look and feel different.

Using naturally colourful fibre that does not require additional colour from printing or dyeing reduces the amount of washing required and eliminates bleaching. This significantly reduces the water and energy required and so minimizes the environmental impacts. However, natural materials are available in a relatively limited range of colours,

TECHNICAL DESCRIPTION
Freshly woven textile that has come straight from the mill is known as greige or grey goods. In stage 1, both sides of the fabric are singed to produce a smooth finish.

In stage 2, the textile is passed through the desizing solution around a series of rollers under tension. Starch-based sizes are best removed in a solution of enzymes; non-soluble sizes and starch are removed using caustic acid and hydrogen peroxide, detergent and silicate; and water-soluble types are washed out with water and detergent. The textile is passed between nip rolls that apply mechanical pressure, forcing the solution to penetrate the fibres.

This is a continuous process; the fabric is fed through the gas burner at 80 to 150 m (88–164 yd) per minute. The textile is taken up on a bulk roll, supported by an A-frame. Once full, the A-frame is rotated to allow the solution to break down the size.

→ Singeing and Desizing Cotton Fabric

Lengths of fabric are delivered in various formats. Standfast & Barracks of Lancaster, England, grade the fabrics according to their origin (**image 1**), because each batch will finish slightly differently. Therefore, it is important to process batches separately.

The fabric is folded, rather than on a roll, because it takes up less space for transportation. The length is unravelled and inspected (**image 2**). Joins sewn between the lengths of woven fabric are torn out and overlocked to ensure the fabric is straight along its entire length (**images 3** and **4**). This is critical to ensure even finishing and accurate control during stretching.

The fabric is passed over a gas flame at high speed to singe the surface and burn away fluff and protruding fibres (**image 5**). This ensures a smooth surface, which is especially important for printing. The fabric is transferred directly into the first scouring treatment (**image 6**) and immersed in the chemical desizing solution (**image 7**). The first step is to remove size, dirt and contamination that have built up in production. Size is applied to warp yarns prior to loom weaving (see Loom Weaving, page 76, and Fancy Loom Weaving, page 84) to increase strength.

The chemical desizing solution is left on the fabric and the roll is wrapped in plastic (**image 8**). It is left for up to 24 hours, depending on the density and type of cloth, in what is called the 'pig farm' (**image 9**). The roll is rotated continually to ensure the desizing solution is uniformly distributed.

1

2

3

4

5

6

7

8

9

Featured Company

Standfast & Barracks
www.standfast-barracks.com

VISUAL GLOSSARY: WASHED COTTON FABRICS

Chemical Washing
Material: Cotton denim
Application: Apparel
Notes: Chemicals such as alkalis and bleach are added to the wash solution to permanently alter the material.

Abrasive Washing
Material: Cotton denim
Application: Apparel
Notes: Abrasives (such as pumice) saturated with chemicals are added to the wash and tumbled with the fabric.

Organic Cotton
Material: Cotton
Application: Apparel
Notes: Using the natural colour of cotton eliminates bleaching and dyeing and so reduces environmental impacts.

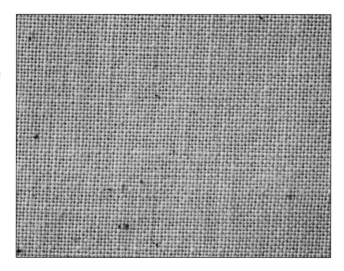

carried out between each finishing step, depending on the processes.

Mercerizing is limited to cellulosic fibre-based textiles, such as cotton, flax and hemp.

COSTS

These finishing processes are relatively low-cost and are an unavoidable part of the processing cycle when converting greige goods into finished products. They may actually reduce the cost of subsequent finishing processes, by improving efficiency. For example, mercerizing cotton reduces the consumption of dyestuff by up to 30%.

ENVIRONMENTAL IMPACTS

Large quantities of water, heat and chemicals are required. Some bleaching techniques produce hazardous acidic or alkali waste that must be carefully disposed of. Many steps have been taken to reduce consumption and waste, such as replacing hazardous ingredients, or using a closed-loop recycling system.

Replacing chemicals with enzymes – such as amylases, cellulases, catalase, pectinase and protease – can help to significantly reduce the environmental impacts. Relatively small concentrations of enzymes are used to replace alkalis in chemical washing and reduce the damage caused to cotton fibres during such treatments (known as bio-stonewashing); to replace acids, alkali or oxidizing agents in scouring and desizing (known as bio-scouring); to produce a smoother and glossier appearance on cotton (known as bio-polishing); and even to clean up spent bleach after whitening.

and synthetics must be produced in very large volumes to include colour in the raw fibre, and this reduces flexibility in production.

QUALITY

Singeing, bleaching, washing and mercerizing are fundamental steps in the production of high-quality fabrics for a variety of applications. Singeing provides a smooth finish free from stray fibres. Bleaching removes the colour and stains, which is essential for achieving high-quality colour in dyeing and printing. In the process of whitening, bleaching damages the fibres, weakening them slightly. Washing removes dirt and contamination, which affect the cleanliness and uniformity of the finished material. Mercerizing produces a higher-strength material with a lustrous surface.

DESIGN

Designers cannot affect the feel and appearance these processes give to fabrics. However, there may be scope to challenge the number of cycles and the

exact ingredients within each process. Reducing water consumption and the use of chemicals will reduce their environmental impacts.

MATERIALS

Used in various configurations, these processes are compatible with a wide range of materials. Almost all plant fibre fabrics are singed, except those that are napped (page 216). Materials that melt or shrink when singed, such as wool and synthetics, are not treated in this way, as it would produce an unsightly surface.

Natural materials require bleaching to improve whiteness. They may require several cycles with a variety of chemicals to produce a suitably bright white base for dyeing and printing. Synthetics do not require such heavy treatment, because the whiteness may be controlled during extrusion. Mild forms of bleaching may be used to improve whiteness.

All materials are washed, either in preparation or to remove manufacturing contaminations. A washing cycle is often

The Washing and Bleaching Process

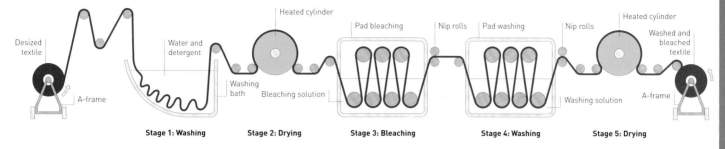

The Mercerizing Process

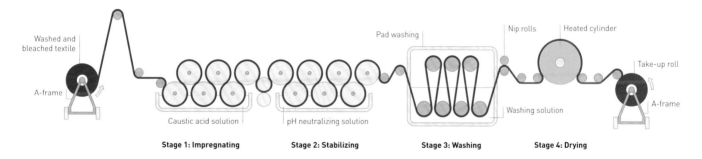

TECHNICAL DESCRIPTION

The impregnated textile, which has been soaking for 24 hours or so, is loaded onto the bleaching line. In stage 1, the textile is washed in mild detergent to remove the desizing solution. In beck washing the textile is run into the bath slack and pulled out of the other side under tension. This allows it to dwell for a short period in the hot wash solution.

In stage 2, it is dried around a series of heated cylinders. All of the heating, including maintaining the solutions at temperature, is typically carried out with steam. This means factories have to produce large amounts, which is expensive.

In stages 3 and 4, the dried textile is bleached and re-washed. The pad process involves running the textile through the solution over a series of rollers, which apply mechanical pressure to the surface, forcing the liquid to penetrate. There may be several stages to bleaching and washing, including different concentrations and mixes of chemicals, to slowly neutralize the pH. After each bathing nip rolls squeeze excess liquid from the fibres.

In stage 5, the textile is dried over a series of heated rolls and taken up on an A-frame at the end of the line. Fabrics may be passed through these processes more than once to achieve a desirable level of whiteness. But finished whiteness also depends on the fibre; for example, flax will never be as bright white as cotton.

There are alternative washing techniques, such as those used to carry out chemical and abrasive washes for aesthetic finishes. These are typically carried out in rotating drums, which range from tens to several hundred litres, depending on the volume of fabric to be processed. Finished garments are loaded into the drum with the abrasive solution and tumbled for a short period, until the desired effect is realized.

TECHNICAL DESCRIPTION

Mercerizing is carried out on cellulosic fibres, including cotton, flax and hemp, for three main purposes: to strengthen the fibres, to improve lustre and to increase dyeing efficiency.

In stage 1, the bleached and washed textile is impregnated with high-strength caustic acid. The solution causes the fibres to swell. In stage 2, the textile is held under tension and pressure. The swelling fabric cannot shrink and so the fibres take on a rounder profile with a smoother surface. The process is combined with tentering (page 210) to maintain lengthwise and crosswise tension. Alternatively, in a process known as slack mercerization, the textile is held under moderate tension and allowed to shrink slightly. This produces a fabric with improved stretch and recovery.

In stage 3, the fabric is washed to remove the caustic acid and rebalance the pH. Finally, the fabric is dried over a steam-heated roll (stage 4).

This process increases strength by as much as 20% and reduces the consumption of dyestuffs by up to 30% (it may be cost-effective to carry out low-concentration mercerization simply to improve dyeing efficiency).

→ Washing, Bleaching and Drying

The fabric is washed to remove the desizing chemicals, bleached and dried in preparation for subsequent finishing operations. The roll is loaded onto the line and the fabric is fed into a beck washer (**images 1** and **2**). It is thoroughly cleaned before passing through the bleaching chemicals (**image 3**).

Many different types of chemicals are used, depending on the type of fabric and desired effect. Examples include caustic acid, peroxide and silicate. Bleaching whitens the fabric with each pass. Some fabrics require more whitening than others, either because they respond less quickly to the whitening effects of the bleaching solution, or because a brighter white is required. However, with each bleaching cycle the fibres are weakened slightly, so there is a limit to how much a fibre can be exposed to the aggressive solution.

After bleaching, the fabric is passed through a hot water wash (**image 4**), steam dried (**image 5**) and taken up on a roll.

1

2

Featured Company

Standfast & Barracks
www.standfast-barracks.com

3

4

5

FINISHING

208

→ Mercerizing Cotton

Mercerizing can be carried out at various stages during pre-treatment: after desizing, scouring or bleaching. Each of these sequences has its benefits. Mercerizing is most commonly used for cotton but is suitable for all cellulosic fibres, including flax, hemp and cellulose.

The fabric has been singed, desized and bleached – although these processes are not essential for mercerizing – and is fed into the line (**image 1**). Production is continuous; the fabric runs over the rolls and is impregnated with high-strength caustic acid (**image 2**). The solution chemically alters the cellulose, causing the fibres to shrink and swell, and increases crystallinity of the molecular structure lengthways. Once treated it is thoroughly washed. The fabric is dried as it travels up a tower of steam-heated rolls, before being taken up on an A-frame (**image 3**) ready for dyeing and printing (see case study, page 265).

1

2

3

Featured Company

Standfast & Barracks
www.standfast-barracks.com

Textile Technology

Stretching and Shrinking

During production, tension builds up in the yarns of woven and knitted fabrics. Controlled stretching and shrinking are employed to relieve this tension and so create a more stable material that will maintain its shape and performance throughout its lifetime. They are often combined with other finishing processes to reduce rewinding.

Techniques	Materials	Costs
• Compressive shrinking • Tentering (natural yarns) and heat setting (synthetic yarns)	• Woven textiles • Knitted textiles	• Low • Rapid cycle time

Applications	Quality	Related Technologies
• Apparel • Interior textiles • Upholstery	• Consistently on-grain and high-quality fabric • Very low shrinkage	• Nonwoven • Plastic film extrusion

INTRODUCTION

These mechanical finishing processes adjust the dimensions of fabrics. Tentering (or stentering) is used to straighten and set fabrics to the desired length and width as they dry. This ensures that the fabric is on-grain (warp and weft yarns are perpendicular and aligned), meets the length and width specification, is ready for further finishing processes and will be stable in use.

For synthetic fabrics, a process called heat setting is used. This is essentially the same as tentering, except that the heating processes must be controlled so that the material is taken close to its softening point, but not above. Once

The Tentering, Drying and Heat Setting Process

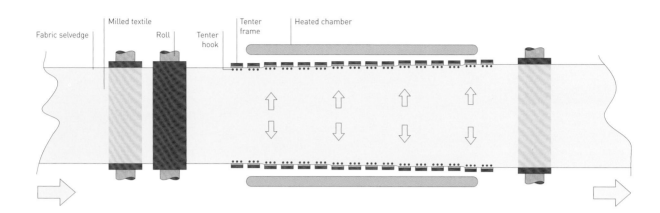

Fabric selvedge | Milled textile | Roll | Tenter hook | Tenter frame | Heated chamber

cooled, the slightly stretched fabric will be on-grain.

Fabrics are pre-shrunk by compressive shrinking, known widely under the trade name Sanforizing. This is carried out to ensure that the fabrics do not shrink during further processing or washing when in use.

Some mills carry out their own finishing, while others specialize in specific types of finishing. Alternatively, the fabrics are purchased unfinished, known as grey or greige goods (regardless of colour) – meaning that no wet or dry finishes or treatments have been applied. Factories that carry out all the necessary processes to convert fibres into finished goods are called vertical mills. The advantage is that they can control every step of production, ensuring the highest-quality products.

APPLICATIONS
Most woven (pages 76–105) and knitted (pages 126–51) fabrics are tentered, while compressive shrinking is a fundamental step to ensure high-quality cotton and cotton blends. Ultimately, these fabrics are

TECHNICAL DESCRIPTION

During production and processing, fabrics are subjected to stresses, which cause misalignment in the warp (and weft). Stress and imbalanced stresses cause materials to shrink when washed and they will not hang properly. Tentering, also known as stentering, straightens the grain of woven and knitted fabrics by stretching the cloth across its width. The small holes found in the selvedge of finished fabrics are evidence of this.

Hooks on the tentering frame pick up the selvedges. As it progresses through the heating chamber, the distance between the hooks is gradually

increased, causing the warp (and weft) to be aligned (and perpendicular to one another). An alternative set-up uses clips instead of pins.

Synthetics, such as polypropylene (PP) and polyester, are tentered by heat setting. This is essentially the same process, except that as fabric is straightened it is heated to close to the softening point of the thermoplastic. Once cooled, the slightly stretched fabric will be on-grain.

used in all types of application, including garments, upholstery and interior textiles.

RELATED TECHNOLOGIES
These mechanical finishes cannot really be avoided, nor replaced with other finishing steps, although using alternative materials and manufacturing techniques

may reduce or eliminate the need for finishing in this way. For example, nonwovens (page 152), extruded plastic sheet (page 180) and leather (page 158) do not require tentering or compressive shrinking. However, they have their own unique finishing requirements.

QUALITY

Tentering is fundamental in the finishing of woven and knitted materials. During operation, the warp and weft are straightened and set to produce a more stable textile, known as on-grain. This is an important step before subsequent finishing operations, such as dyeing (page 240) and printing (pages 256–79).

Synthetics are tentered by heat setting, which takes the material close to its softening point. Under tension, the fabric is straightened and the stresses evened out. The fibres cool and set in place to produce a material less prone to shrinkage during subsequent finishing processes and washing.

Compressive shrinking produces cotton and cotton blend fabrics that have a shrink-proof value of less than 1% of overall dimensions. This is essential for high-performance fabrics, such as those used to construct outer garments.

DESIGN

There is little impact a designer can have on these processes, except to specify their requirements. All modern mechanical production processes are monitored by sensors, which ensure that fabrics are on-grain and pre-shrunk to the designer's requirements. It may be necessary to pass fabrics through a process more than once to achieve the desired quality.

MATERIALS

Tentering is used to stretch and dry woven and knitted fabrics made up of natural yarns, including wool, cotton, flax and hemp. Heat setting is used for synthetics, such as nylon and polyester, or blends involving 50% or more synthetic.

Compressive shrinking is used to finish cotton and cotton blends.

COSTS

Unit costs are low and cycle time is high. These are fundamental steps in the production of high-quality textiles. Specifying strict requirements and small tolerances will likely increase the cost of finishing.

ENVIRONMENTAL IMPACTS

The environmental impacts can be significant, because finishing uses large quantities of water and chemicals. These processes can use millions of litres of water per day for mass production. Owing to the high cost and environmental impacts, there are many developments under way to reduce consumption, such as closed-loop systems to reclaim and recycle water and chemicals. Moving from liquid-based to foam-based treatments, and other less water-intensive processes, reduces consumption further still.

Hazardous chemicals and additives are being phased out wherever possible. However, the location of finishing is critical, because many countries have less stringent pollution controls in place.

→ Tentering Cotton

This cotton is being prepared for screen printing. Prior to tentering, the fabric was scoured and mercerized (page 209) to increase the fabric's strength, lustre and affinity for dyes.

The cotton is fed into the tentering line via a water bath to slightly dampen the fabric (**image 1**). Fluid coatings (page 226) and lamination (page 188) may also be applied during tentering.

The selvedges are picked up by clip frames (or pins) and crosswise and lengthwise tension is applied to the fabric as it is dried (**image 2**). This ensures the weave is on-grain.

Skew, bow and width are monitored as the fabric is processed (**image 3**). A vacuum cleaner works over the rollers to remove any fluff and contamination that has collected on the surface (**image 4**). And the tentered cotton is taken

up on an A-frame ready to be transferred to the printing line (**image 5**).

1

2

3

4

5

Featured Company

Standfast & Barracks

www.standfast-barracks.com

The Compressive Shrinking Process

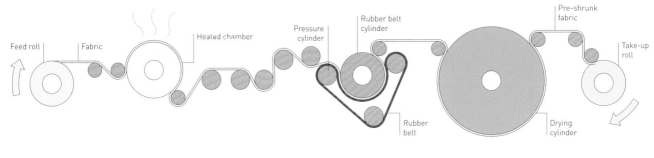

Feed roll | Fabric | Heated chamber | Pressure cylinder | Rubber belt cylinder | Pre-shrunk fabric | Take-up roll | Rubber belt | Drying cylinder

Stage 1: Moistening   Stage 2: Tentering   Stage 3: Rubber belt unit   Stage 4: Drying chamber

TECHNICAL DESCRIPTION

This is the process of pre-shrinking fabrics, known under the trade name Sanforizing, to virtually eliminate further shrinkage, such as during washing cycles.

In stage 1, the fabric is moistened with water or steam. This lubricates the fibres, allowing them to slip over each other more easily and shrink. Moistening the fabric to around 15% throughout is usually enough to get sufficient shrinkage.

In stage 2, the fabric is passed through tentering, to set the fabric at the correct width. Then, in stage 3, the moistened fabric is passed between the rubber belt and cylinder. The pressure cylinder squeezes the rubber, causing it to stretch. As the fabric passes between the belt and the cylinder, the belt contracts and so compresses the fabric, causing shrinkage and packing the fibres and yarns tightly together. The higher the pressure at the nip point (between the pressure cylinder and the rubber belt cylinder), the greater the shrinkage.

In stage 4, the fabric is passed through a drying chamber to remove the moisture and lock the fibres in their pre-shrunk state.

Case Study

→ Pre-Shrinking Cotton

Here the Sanforizing technique is being used to pre-shrink cotton to exact requirements. A roll of woven cotton, which has been dyed on the roll, is fed into the moistening chamber (**image 1**). The fabric passes around a cylinder that pumps steam through the fibres (**image 2**), raising the moisture level and thus allowing the yarns to move more freely against one another.

The moistened fabric is fed into a tentering line where the selvedges are gripped with clips and the fabric is drawn out widthwise and lengthwise. This ensures the fabric is on-grain prior to compressive shrinking. Sensors measure the dimensions as the fabric passes underneath (**image 3**). Comparing this to the dimensions of the fabric as it emerges from the rubber belt and cylinder determines the process parameters.

The fabric emerges from the end of the tentering line and the selvedges are released from the clips. It passes through a series of tensioning rolls and is drawn between the pressure roll and rubber belt (**image 4**). It is shrunk by the compression of the belt and fed into the drying chamber.

The process operates at high speed. The finished material, which has been pre-shrunk, is taken up on a roll on an A-frame (**image 5**). At this stage the cotton is ready to be passed on to the next finishing process, which in this case is fluid coating with wax (see page 232).

1

2

3

4

5

Textile Technology
Napping and Sueding

These mechanical finishing techniques are used to raise the nap on the surface of textiles to create a softer finish, increase cushioning and improve insulation. Sueding is an abrasive process and is also known as emerizing and sanding. Napping lifts fibre ends from the face of the fabric by brushing, creating a longer pile.

Techniques		Materials		Costs
• Shearing	• Sueding,	• Woven textiles	• Nonwovens	• Low to moderate
• Napping	emerizing and	• Knitted textiles	• Leather	• Rapid cycle time
	sanding			

Applications	Quality	Related Technologies
• Garments and footwear	• Soft finish	• Flocking
• Interior textiles	• Good insulation	• Pile fabric
• Upholstery	• Reduced strength	

INTRODUCTION

These are cost-effective techniques used to raise the nap on the face of woven, knitted and nonwoven textiles. This is different from weaving pile (page 96), which employs a separate set of yarns for the purpose.

In sueding, also referred to as emerizing or sanding, fabric is drawn over an abrasive surface, which breaks the fibres and so produces a softer-feeling surface. This can significantly reduce the strength of fabrics.

In the past, napping was done with dried teasel pods. Teasels are still used today to produce some of the finest woollen products, such as scarves and

The Napping and Sueding Process

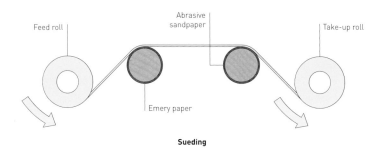

Feed roll | Abrasive sandpaper | Take-up roll

Emery paper

Sueding

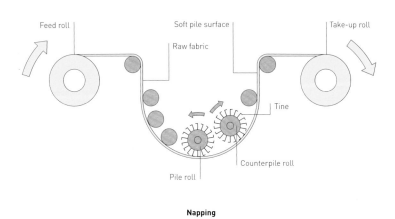

Feed roll | Soft pile surface | Take-up roll

Raw fabric

Tine

Counterpile roll

Pile roll

Napping

TECHNICAL DESCRIPTION

SUEDING

Also known as emerizing or sanding. Drawing fabrics at high speed over abrasive paper opens up the yarns to expose fibres, resulting in a low pile and soft touch. The first roller is covered with emery paper, followed by rollers covered with successively more abrasive sandpaper. This finishing technique may be applied to the face, or to both sides of the fabric.

NAPPING

Napping produces a more pronounced pile on the surface of fabrics. It may be carried out on the face or on both sides.

The fabrics, including woven and knitted, are fed past counter-rotating rolls covered with tines. Dried teasel pods were used in the first napping machines and are still considered superior for high-quality soft woollen fabrics, such as cashmere. Most modern napping machines use metal tines.

The pile roller rotates faster than the fabric is fed in, and in line with the direction of travel, to raise the nap. In double-action napping, the counterpile roll rotates counter to the direction of travel and more slowly than the fabric. This produces a smoother surface.

blankets; however, the majority of napping is now done with wire brushes, on what is sometimes called a Mozer. The two processes are essentially the same, except the brushing is done with either a natural material or a wire brush.

Shearing crops the nap to a set length. It is also used to sculpt three-dimensional patterns into fabric. Most pile fabrics are finished in this way, as well as napped fabrics.

APPLICATIONS
The soft suede-like nap produced by emerizing (or sueding) is used to finish denim, felt (see Nonwovens, page 152) and fabrics for garments and interiors.

Like fabric, suede leather is finished with emery paper. However, the properties of leather compared to woven, knitted and nonwoven materials are quite different, and abrading the surface will produce a different feel and softness in split leather compared to other textiles. Fabric may be preferred to suede leather thanks to better liquid repellency and stain resistance.

Napping is used to finish products that require a high level of insulation as well as increased softness. It is mostly used for clothing, such as wool scarves, wool blankets, flannel (wool or cotton) and polyester fleece. It is also used to create some fake furs.

RELATED TECHNOLOGIES
Pile fabrics have a separate set of yarns to create the pile. They are more expensive and are used to produce fabric with pile up to 12.5 mm (0.5 in.) long. The pile may be consistent all over, or created in ribs (corduroy) or other patterns.

QUALITY
All of these techniques can be applied to one side or both sides. In shearing, the nap is cut to a uniform length, to produce a more consistent surface feel and appearance. Sueding produces a softer-feeling surface, referred to as a 'soft hand'.

Napping produces a longer pile. The soft napped pile traps air and so provides

Twill Weave Pattern
Material: Lambswool
Application: Rug
Notes: Napping produces a soft hand and fuzzy appearance on patterned weaves.

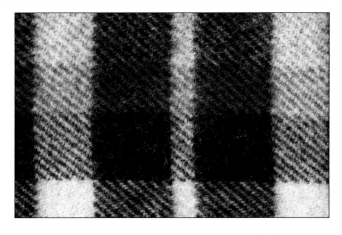

Fleece
Material: Polyester
Application: Outerwear
Notes: Fleece is weft knitted with an inlaid yarn, which is napped to produce a soft and insulating outer layer. After napping, the knitted structure is hidden below the layer of raised pile.

Front

Back

Long
Material: Wool and cashmere
Application: Apparel
Notes: A long pile is raised with heavy napping to produce a soft fabric. The subtle weave pattern remains visible.

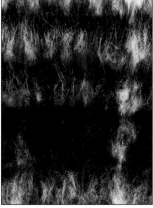
Front

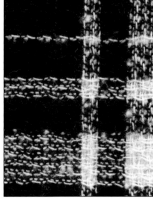
Back

words, they will look different when viewed from the top or bottom. Brushing will further enhance this visual effect. This must be taken into account when assembling cut panels from sheared, suede or napped fabric. Likewise, it can be used to distinguish panels and materials.

There are many variables in the napping process, such as fabric tension and speed, and pile and counterpile roll speed. These factors, along with combining napping with other finishing processes, give scope to the designer to tailor the material's appearance and performance to their application. For example, the pile may be combed, raised, semi-felted (pulled out and tucked in), or completely felted.

As well as producing a consistent finish, shearing may be used to cut strips and other three-dimensional patterns into the pile.

MATERIALS

Cotton, wool and leather are commonly finished with sueding. Cotton and wool are also suitable for napping, as are staple synthetic yarns, such as polyester and nylon. However, all types of materials can be finished with abrasives and most woven (pages 76–105) and knitted (pages 126–51) textile materials are suitable for napping.

COSTS

Unit costs are low for standard finishes. In some cases, sueding may be carried out in-line with weaving, eliminating further finishing steps.

This is a more cost-effective way to produce a soft hand than pile weaving.

ENVIRONMENTAL IMPACTS

Even though these processes do not require water, chemicals or other additives, the dust produced by abrading textiles can be irritating and harmful. Of course, this depends on the selected materials and exact processing.

The offcuts produced by shearing nap can be recycled to make products that utilize short fibres, such as insulation and automotive parts.

good insulation. The additional thickness may increase water repellency, depending on the material. A longer nap will produce a fuzzier appearance, softening the edges of knitted and woven patterns.

Abrading the surface of fabrics reduces the strength, and will affect the wear and abrasion resistance. Reinforcing the fabric with additional yarns, laminating (page 188), blending synthetic and natural fibres or using core ply yarns helps to maintain tensile strength.

DESIGN

Owing to fibre alignment, these finishes tend to have a direction of view. In other

→ Various Techniques for Raising the Nap on Wool

Abrading the surface of the fabric causes the fibres to separate and open up into a soft pile, and so produces a softer hand. The first roller is covered with emery paper, followed by rollers covered with successively more abrasive sandpaper (**image 1**).

Delicate and fine wool, such as cashmere, is teaselled to raise the nap and create a very soft pile. The dried pods are still used to finish the finest-quality wool (**image 2**). They are hand-loaded into metal carriers, which are in turn mounted onto rotating centres (**images**

3 and **4**). The fabric is drawn past the rotating teasels at high speed (**image 5**). As it does so, the tines of the teasels pick up fibres from the surface, raising them into a soft pile with a distinctive wavy pattern (**image 6**).

Wire napping is considered too harsh to use on soft materials such as cashmere, but hardier fabrics such as lambswool and synthetics are suitable (**image 7**). Wire rolls do not produce the same wavy pattern in wool pile.

1

2

3

4

5

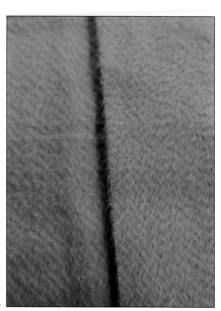

6

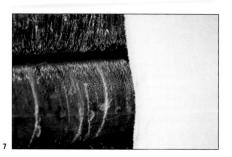

7

Featured Company

Johnstons of Elgin
www.johnstonscashmere.com

Textile Technology
Calendering and Embossing

These mechanical finishing techniques change the appearance and feel of fabrics. They are used to smooth the surface and improve lustre, or to reproduce a three-dimensional pattern. Depending on the material and technique, they may be semi-permanent or last the lifetime of the fabric. Finishes include glazed, chintz, moiré, schreiner, embossed and debossed.

Techniques		Materials		Costs
• Calendering (glazed, ciré, schreiner and moiré)	• Embossing • Punching	• Woven textiles • Knitted textiles • Nonwovens	• Leather • Plastic film • Natural, synthetic	• Low-cost for 'standard' products • High-cost for specialist finishes or low volumes

Applications	Quality	Related Technologies	
• Apparel • Interior textiles • Upholstery	• Improved lustre on the face • Repeatable pattern or texture	• Flocking • Fluid coating • Laminating	• Napping • Printing • Laser etching

INTRODUCTION

Calendering and embossing are straightforward processes. Fabric is passed between metal cylinders under high pressure. As a result, the surface profile of the roller is imparted onto the surface, or right through the structure, of the fabric.

Smooth-surfaced rolls produce a flat finish on fabrics. The pressure applied will compress the fabric, making it slightly thinner. Friction calendering (and hot calendering) uses heated metal rolls, which are rotated faster than the fabric is passed between them, polishing the surface. This produces the glazed finish. For example, chintz is glazed calico

Calendering, Embossing and Punching

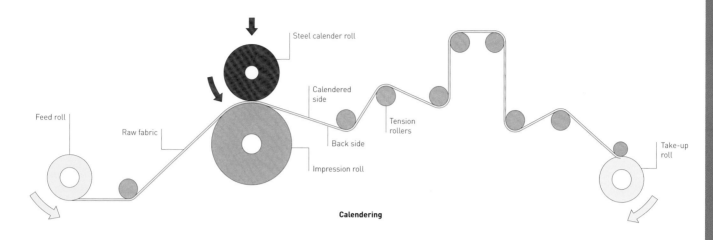

Calendering

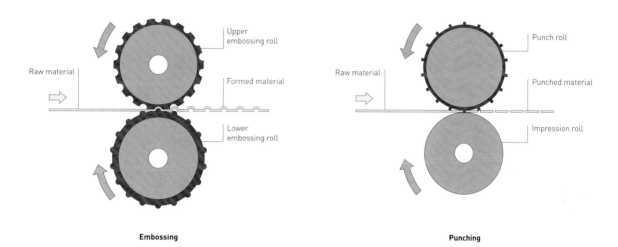

Embossing

Punching

TECHNICAL DESCRIPTION

In calendering, a highly polished steel roll flattens and compresses the surface of the fabric. Cold calendering produces a semi-permanent finish that will disappear with the first wash, whereas calendering with a heated steel roll produces a long-lasting polished finish.

Glazed finish is formed by friction calendering, whereby the calender roll rotates at higher speed than the impression roll. The glazed finish is produced on the front side. Synthetics require only heat, pressure and friction, whereas cotton and silk benefit from the addition of wax (to produce ciré finish or fabric).

Calender rolls engraved with patterns are used to create embossed sheet materials. In schreinering, the calendering roll is covered with very fine diagonal lines. The embossed effect enhances the lustre of the front side.

It is possible to reproduce deep-profile patterns by passing the sheet material between two embossed rollers with heat and pressure: the top and bottom match and so deform the sheet across its entire thickness, not just the front side.

Punched designs are reproduced using a steel roll engraved with a sharp-edged pattern. Either it cuts against a metal roll

with the reverse pattern (for cutting away large areas), or it pierces through the material into a sacrificial plastic roll (for punching small holes).

Deep Emboss
Material: Weft-knitted PP
Application: Apparel
Notes: Deep profiles are formed in synthetic knitted, woven and nonwoven textiles with heat and pressure.

Moiré Finish
Material: PVC
Application: Footwear
Notes: With very high pressure, a fine wavy pattern is embossed into the surface of synthetic fabric, creating a moiré effect.

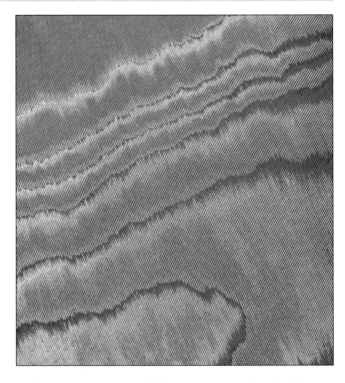

Embossed Leather
Material: Leather
Application: Footwear
Notes: Split leather is embossed with natural skin pattern to look like high-quality full-grain leather.

Embossed Weave
Material: PE
Application: Furniture
Notes: Hot-press embossing flattens the weave to create a permanent and durable finish suitable for high-wear applications.

Embossed Laminated Coating
Material: TPU film-coated polyester knit
Application: Outerwear
Notes: Embossing laminated coatings produces a durable and well-defined pattern finish.

Embossed and Profiled
Material: Paper
Application: Doily
Notes: Some materials may be cut and embossed simultaneously to reduce cycle time and eliminate the potential for misalignment.

(plain-woven unbleached cotton fabric). Ciré is a very smooth and highly glazed finish produced by hot calendering with wax. Ciré fabrics are used in apparel, such as to make waterproof jackets.

Patterns are reproduced in the surface of fabrics by calendering with shallow-engraved rolls. This technique is used to produce finishes such as moiré, which has a wavy, watery appearance, and schreiner (named after Ludwig Schreiner, a German textile manufacturer), which consists of fine diagonal lines that create a soft, silk-like lustre.

Embossing uses metal rolls with a deeper surface profile. Debossing is when the pattern is depressed into the surface,

rather than raised up. With heat and pressure, patterns are pressed into the surface or right through the structure of the textile, to create a long-lasting and hard-wearing three-dimensional finish. Materials must be sufficiently malleable (moldable) to be effectively embossed and debossed. For example, elastic material will spring back, unless it is thermoplastic, in which case it can be permanently reformed with heat and pressure; whereas natural fibres such as cotton and wool cannot be stretched far without breaking and so will not emboss in a satisfactory manner.

Finishing processes may be carried out in the mill, where the fabric is

produced, or at a specialist converter. Alternatively, the fabrics are purchased unfinished, known as 'grey' or 'greige' goods (regardless of colour), meaning that no wet or dry finishes or treatments have been applied. In some cases, the fabricator has unique requirements, such as punching, which are done in-house to maintain flexibility during production.

APPLICATIONS
Calendering is used on most fabrics during production, but to varying degrees. Low-temperature calendering produces a smooth finish with a slight sheen, similar to ironing, which will disappear with the first wash. Glazed calendered finishes are

used to produce fabrics with a polished finish, such as chintz and chino. Ciré fabrics have a highly polished finish, such as lightweight performance fabric used to make outerwear. It is usually referred to as ciré or ciréd finish, because it is suitable for a wide range of fabrics and not one in particular.

Embossing and punching are used to mimic expensive and decorative materials, such as full grain, ostrich and peccary leather. Embossing is utilized in the production of wallpapers and interior fabrics to reproduce floral and patterned designs. Schreinering is used on cotton sateen, to increase the visual appeal, such as for damask and apparel. Textiles coated with plastic film are embossed to improve performance and used to make parts of shoes and apparel.

RELATED TECHNOLOGIES

These mechanical finishing processes may be used in conjunction with, or reduced by, the use of other finishing techniques, such as coating (page 226), laminating (page 188) and proofing.

Patterns can be reproduced in the surface of fabrics by shearing, laser etching (see Laser Cutting, page 328) or dissolving the pile. Alternatively, by flocking (page 280) a stencilled pattern, a design may be reproduced without changing the cross-section or cutting into the surface of the fabric.

QUALITY

Calendering improves the lustre of fabrics, because the surface will be smoother and so more light will be reflected. The finish will be semi-permanent in natural fibre-based fabrics and permanent in synthetic fabrics. Semi-permanent finishes will fade or disappear with the first wash.

Calendering reduces the porosity of fabrics, because it flattens the yarn slightly, thus reducing the gaps between warp and weft. This gives improved properties when coating and proofing. Highly polished and glazed finishes are achievable with friction calendering, or by adding wax to natural fibres.

Embossed designs will be permanent and durable in synthetic fabrics, because thermoplastics – such as polypropylene

(PP), polyethylene (PE) and nylon – soften when heated and so will change shape. Leather embosses equally well, retaining a long-lasting and durable finish. Mixing resin with natural fibres will produce a similar effect, because it adds stiffness to the material.

Calendered and embossed fabrics will have a face and a back, which appear and feel different.

DESIGN

There is an almost unlimited number of variables that can be achieved with these mechanical finishing techniques. In many cases, the optimum set-up has been established to create 'standard' materials. Because of the set-up and testing required, deviating from these standard finishes will likely require minimum-volume production runs, such as in the case of dyeing (page 240), so is typically reserved for high-volume production.

The rolls used to apply an embossed or punched pattern are designed so that with each rotation, the pattern aligns. Any pattern or design can be reproduced, and it is possible to create a bespoke design if the volumes justify the investment costs.

Depending on the shape of the rollers, either a single side of the material is formed, leaving the reverse smooth (paper, leather, metal and similar malleable materials), or both sides are formed simultaneously (embossing). Alternatively, very small production runs, samples, or experimental embossing may be achieved using a flat press with textured tools.

Pattern embossing is used for both decorative and functional applications. For example, embossed paper and card have improved strength and stiffness compared to flat sheets. Textured surfaces disguise wear and tear.

Embossed patterns tend to have soft edges, due to the nature of the materials. However, synthetic materials may be heated and formed, so sharper designs can be reproduced. Punched patterns may have a very well-defined edge, but this depends on the hardness of the material: softer materials will deform more easily and so will not retain a sharp edge.

MATERIALS

Calendering and embossing are applied to a wide range of materials, including knitted and woven fabrics made up of natural or synthetic yarns. Conversely, punching is suitable only for materials that will not unravel, such as warp knits, nonwovens, sheet plastics, leather, rubber and paper. Holes cut into woven or weft knitted fabrics that are prone to unravel must be secured in some way. For example, laser-cut holes in synthetic materials will not fray, because the ends are simultaneously cut and melted together. Likewise, cutwork employs needle lace (see Embroidery, page 298) techniques to secure the exposed edges and create an openwork structure.

Textiles coated with a plastic film, such as thermoplastic polyurethane (TPU), are embossed with patterns and textures for both functional and decorative purposes. Gloss leather (also called glazed) is coated with plastic and calendered to produce a dense smooth surface.

COSTS

Unit costs are low for standard finishes. Bespoke finishes, patterns and effects are high-cost and usually limited to high-volume production.

Cycle time is very short for standard finishes and roll-to-roll processes.

ENVIRONMENTAL IMPACTS

Calendering and embossing have a minimal environmental impact, because they do not add or remove any material. However, heating may be required, which increases the amount of energy used. If resins are used to stiffen natural fibres (see Spread Tow and Fabric Prepreg, page 234), this may affect the recyclability or biodegradability of the material at the end of its useful life. Using bioplastics reduces the environmental impact of composite constructions.

→ ## Calendering Cotton

Lengths of cotton are calendered between a steel roll and a nylon roll (**images 1**, **2** and **3**). The steel roll is heated to 80°C (176°F): cold rolling produces a temporarily smoother finish, whereas rolling at high temperature imparts a more permanent polished finish.

In this case, calendering is carried out for two reasons. First, it reduces the porosity of the woven fabric by flattening the fibres and thus reducing 'hole size' between warp and weft. This helps with down-proofing, for example, and so is used to make pillowcases.

Second, it is carried out for aesthetic reasons. Calendering produces a shiny surface. Washing reduces the glazed finish, but cottons that have been proofed with wax (see Fluid Coating, page 226) naturally repel water and so maintain the glazed finish.

1

2

3

→ Punching Leather

Peccary leather is easily distinguished from other types of pigskin, because it is marked with a distinctive pattern of hair follicles. It is recognized as ideal for glove making, because it is strong, very soft and supple.

The hair follicle pattern is reproduced in less expensive leathers by punching. The steel roll is decorated with a raised profile (**image 1**). As leather is passed between the steel and the nylon rolls, holes are punched right through (**images 2**, **3** and **4**), reproducing the pattern exactly. It is designed in such a way that there is no start or finish to the pattern, so that with each rotation the pattern is seamlessly integrated.

The leather is now ready to be assembled into gloves by machine stitching (page 354).

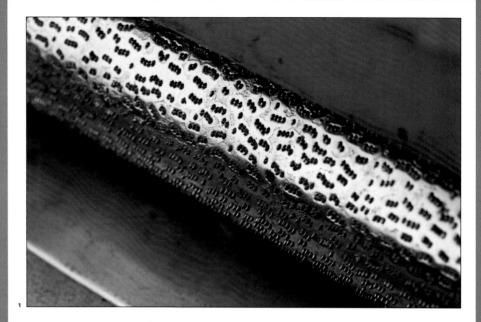

1

2

3

4

Featured Company

Dents
www.dents.co.uk

Textile Technology
Fluid Coating

Coatings are applied to enhance the functional and visual properties of woven, knitted and nonwoven textiles. They are applied using a range of techniques, including dipping, spraying and printing. Depending on the coating material, the look and handle of the base textile may be completely transformed or unaffected.

Techniques		Materials	Costs
• Direct	• Spray	• Knitted, woven and nonwoven textiles	• Low to moderate for standard coatings
• Dip	• Transfer	• Leather	• High unit cost for specialist and bespoke
		• Natural and synthetic	

Applications		Quality	Related Technologies	
• Interiors	• Automotive	• Surface coating or fully impregnated	• Prepreg	• Laminating
• Apparel	• Technical	• Improved strength and tear resistance	• Dyeing	• Vapour deposition
		• Proofing, such as waterproof or flameproof		

INTRODUCTION

Liquids, pastes and foams are applied as coatings to improve specific properties. Proofing gives textiles resistance to something, as in waterproofing and flameproofing; liquid plastic coating is used to make textiles water-repellent while maintaining breathability, and to improve tear resistance and shape retention; functional materials are microencapsulated; and nanocoatings provide exceptional resistance to dirt and staining. Alternatively, coatings are applied by laminating fabric with a plastic film (page 180).

The correct coating method is determined by the required properties

The Direct Coating Process

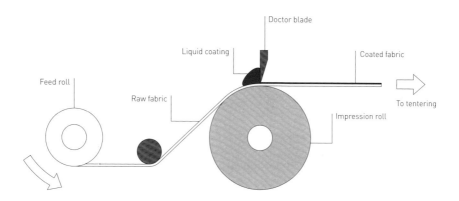

Doctor blade

Liquid coating

Coated fabric

Feed roll

Raw fabric

To tentering

Impression roll

TECHNICAL DESCRIPTION

There are three main types of set-up for direct coating: roll, blanket and air (or floating knife). The diagram shows the doctor blade (knife) over an impression roll, the most versatile and widely used method. The doctor blade spreads the coating over the surface of the textile as it passes beneath, held under tension. The distance between the blade and the textile determines the coating thickness. As the coating is spread over the textile it is drawn into the surface. The amount of penetration depends on the

absorbency of the base textile and rate of evaporation of the coating.

The blanket method uses a short conveyor, suspended between two rolls, to support the textile as it passes underneath the blade. This ensures that uniform pressure is applied, which means delicate and unstable textiles may be coated.

A floating knife applies the coating without any support underneath the textile. Therefore, compressive pressure is applied to the coating, forcing it to penetrate deeper.

Direct coating usually precedes tentering (page 210). Immediately after the coating has been applied the textile is picked up by hooks or clips on the tentering frame. As it progresses through a heating chamber, the distance between the hooks is gradually increased, causing the warp and weft to be aligned perpendicular to one another.

During direct coating the textile is held under tension. Therefore, the process is only suitable for woven and knitted textiles that can withstand the applied loads.

The Dip Coating Process

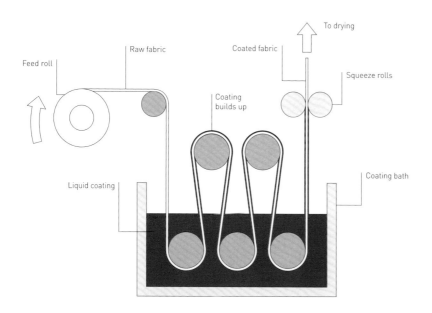

To drying

Raw fabric

Coated fabric

Feed roll

Squeeze rolls

Coating builds up

Liquid coating

Coating bath

TECHNICAL DESCRIPTION

The dip coating process is also referred to as impregnation or saturation. The textile is drawn though a coating bath for a set period of time, known as dwell time.

No pressure is applied to the coating during dipping, so the amount of penetration depends on the ability of the textile to take up the liquid. This process is mainly used to apply coatings that are too liquid, or otherwise unsuitable, for the direct coating method.

As the textile is drawn upwards and away from the coating bath, excess material is removed by a pair of squeeze rolls. Alternatively, excess material is removed with doctor blades.

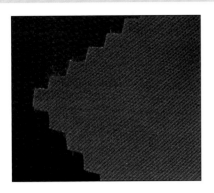

Waxed Cotton
Material: Wax-coated cotton
Application: Apparel
Notes: The fabric develops a handle and appearance unique to the individual. The wax coating is reapplied periodically to extend the useful life of the garments.

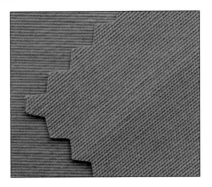

Waterproof Cotton
Material: PU-coated cotton
Application: Apparel
Notes: PU coatings are applied for functional reasons such as waterproofing. The liquid coating penetrates the fabric and so can build up on the back side.

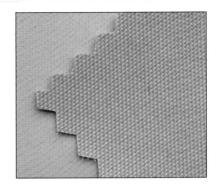

Durable Rubber Coating
Material: Synthetic rubber-coated nylon
Application: Technical
Notes: Hard-wearing synthetic rubbers, such as polychloroprene (CR), or neoprene, are used to coat fabrics for demanding applications such as inflatables.

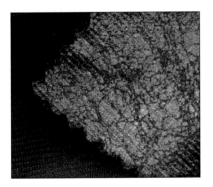

Fake Leather
Material: PU-coated nylon/elastane
Application: Apparel
Notes: Coated onto a four-way stretch fabric, the bright and glossy finish cracks and wears with use to mimic leather.

Slip-Resistant Coating
Material: Screen-printed flax
Application: Packaging
Notes: Open weaves are coated to reduce movement between yarns and so help retain the textile's structure.

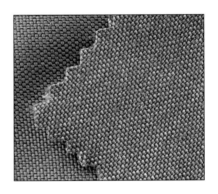

Thin Film Coating
Material: Silicone-coated nylon
Application: Outdoor equipment
Notes: Silicone provides a lightweight, waterproof and breathable coating. Only silicone will stick to silicone, making it easy to clean.

and chosen materials. Direct coating with a doctor blade (known as knife coating or spread coating) applies a thin coating to one side of the textile as it is drawn underneath. The blade moves up and down to control film thickness.

Dip coating is a straightforward process. The textile is submersed in a bath of coating for a set period of time. The textile does not need to be under high tension as with direct coating, so high-stretch and lightweight fabrics may be treated in this way.

Spray coating is a versatile process suitable for both one-offs and mass production. The coating must be liquid and compatible with the spray nozzle, which typically limits this technique to very thin coatings (although multiple layers can be applied). No pressure is applied to the coating, so it tends to sit on the surface and in the gaps, rather than penetrating into the fibre structure.

The transfer method tends to be the most expensive, because it requires multiple application steps and the use of an expendable transfer film. It uses the same principle as transfer printing (page 272) and prepregging (page 234) and is suitable for lightweight, elastic and unstable textiles, because the fabric does not need to be held under tension during application.

APPLICATIONS
Coatings are used for a range of property enhancements and so are utilized in diverse industries, such as apparel, upholstery, automotive, architectural and geotextiles.

Apparel fabrics ranging from 55 to 950 g/m² (1.3–22.5 lb/ft²) are coated with wax or polyurethane (PU) for application in outdoor apparel, sportswear, shoes, hats and luggage. Heavier textiles coated with PU and silicone, such as cotton, linen and polyester over 275 g/m² (6.5 lb/ft²), are used to make tents, sails and luggage.

RELATED TECHNOLOGIES
Liquid coatings penetrate into the fabric structure and can be impregnated right through, depending on the method of application. By contrast, laminating bonds a thin layer of thermoplastic to the surface of a textile, or multiple layers of textile together. With this

approach the layers of textile retain their inherent properties, and only the entire construction will provide the required properties. This creates more control over specific properties, such as cover, breathability and waterproofing, because each layer is independent and not reliant on the base material.

Laminating can run at higher speeds, because a thermoplastic film is easier to control than a liquid coating and the exact amount is applied regardless of the speed of the machine.

Additives may be used in conjunction with dyeing to achieve certain properties, such as flameproofing and resistance to ultraviolet (UV) light. Likewise, coatings may be used to apply colour to the base textile. In this way, two processes are combined into one operation. However, the range of properties will be limited by the chemical composition of the dyeing or coating materials.

Fluoropolymer coatings such as polytetrafluoroethylene (PTFE) may also be applied as a powder form and sintered. This requires that the base textile have a high melting point or the fluoropolymer a low sintering temperature.

Nanotechnology is either embedded into the fibres during production (page 50) or applied as a coating. Other than liquid-based coating, vapour deposition and impregnation are both suitable means of application. Process selection depends on the nanotechnology and its properties.

QUALITY
Coatings are applied to give textiles specific properties such as making them water-repellent (resists absorption and penetration of water and similar liquids at a specified pressure and for a given period of time); waterproof (completely impenetrable by water); flameproof (will not actively flame or burn); antibacterial (repels or inhibits the growth and reproduction of bacterial, mold and mildew); shape-retaining (improves the resilience of textiles by reducing wrinkling and wear); slip-resistant (warp and weft are coated to make them less prone to moving in relation to one another); breathable (allows moisture to evaporate through); printable (improves print quality); and improved durability (such as resistance to staining, heat, chemicals and UV).

Applications can be very demanding in terms of both mechanical function and visual performance. For example, textiles used to cover seats on a boat have to be able to withstand extended periods of cold temperatures, midday sun, UV, salt water and heavy use. By contrast, coatings used in medical applications must be hygienic (such as antibacterial and easy to clean) and resistant to abrasion to avoid scuffs that create dirt traps. And all textiles used in public spaces need to be durable and flame-retardant.

Depending on the coating material, the look, feel and handle of the base textile may be unaffected or may be completely transformed. Plastic coatings, such as PU and polyvinyl chloride (PVC), are available in a range of hardness, from soft to rigid. The choice of coating will affect the handle of the fabric accordingly. One quality does not always compromise another. For example, coated textiles may be very soft and resilient, such as lightweight PU-coated fabrics used to make paragliders.

The coating may be on the surface of the textile or permeate right through. This depends on the type and viscosity of material. Coatings that lie on the surface, such as PVC, are prone to delamination. Coatings that permeate right through, such as PU, silicone and wax, produce a consolidated composite textile with balanced properties. These types of fabric are less prone to fraying along the cut edge and will not delaminate.

Calendering (page 220) reduces the gaps between warp and weft by compressing the fabric and flattening the fibres. Liquid coatings are used to plug the gaps and make waterproof and airtight textiles, such as used to build inflatables and dry suits.

Transparent coatings do not affect the colour of the textile. And clear abrasion-resistant coatings will ensure that textiles retain their inherent appearance for longer, especially in high-wear applications, such as upholstery. The gloss level is determined by coating material.

PU is available matt or glossy, depending on the chemistry. It may be improved or altered by calendering and embossing (page 220).

DESIGN
The huge range of existing and as yet unexplored material combinations gives designers a wide scope of opportunity. Such diversity means it is important to be clear about the required properties and their priority so that the correct coating and base material are developed, or selected from existing materials.

The textile being coated – whether weave (pages 76–105), knit (pages 126–51), nonwoven (page 152), leather (page 158), natural or synthetic – determines property limitations, such as tensile strength and elasticity. The coating contributes to this, enhancing specific properties such as resistance and resilience. So, it is possible to increase a textile's resistance to UV with coating, but it is not possible to increase its elasticity, for example, without changing the base material.

Coatings may be built up in layers to increase thickness, or incorporate more than one coating material and thus multiple properties. Coating thickness depends on the material being used. For example, PVC is typically 0.3 to 1.5 mm (0.01–0.06 in.).

Each layer can be transparent, translucent, coloured, fluorescent or even reflective. Plastic coatings may be matched to a colour reference, such as Pantone.

As well as the functional and visual benefits of coating, it may be utilized in preparation for secondary processes, such as calendering or embossing, and printing (pages 256–79). Coating textile with a weldable plastic, such as styrene butadiene rubber (SBR), means that textiles that would normally have to be machine stitched (page 354) can be welded together and formed into three-dimensional shapes. This technique is used to make the upper part of trainers, for example.

Printing techniques such as screen and rotogravure may be used to apply a coating for functional and decorative purposes. This method is utilized for articles that require defined areas of performance, such as additional waterproofing, breathability and stiffness. It is slower than applying a coating to the entire surface area and so is more expensive to set up and run. However, the added benefit outweighs the cost for some applications, such as sportswear and trainers.

MATERIALS

The most commonly coated materials include woven cotton, flax, polyester, nylon and glass fibre. These are the most common because they are the ones typically used in applications that demand the enhanced performance offered by coating.

Most materials can be coated with liquid- and foam-based systems. With developments in knitted and nonwoven fabrics for high-performance applications, coating systems suitable for these materials are becoming more widespread.

The most common liquid coatings are PVC, wax, latex, silicone, PU, acrylic and SBR.

Paraffin-based wax coating, also known as wax proofing, was originally utilized in sail making. Today it is synonymous with country sports and biking jackets, such as Barbour and Belstaff. Wax provides a flexible weatherproof barrier that ages with a natural appearance, and it can be reproofed to extend the life of the garment.

Latex coating produces a stiff and waterproof fabric. It provides good grip, and moderate cut and abrasion resistance (see also Dip Molding, page 442). It makes a comfortable material, with good shape retention and resilience, and so is widely used in carpet, upholstery and footwear fabrics and padding.

SBR exhibits similar properties to latex and natural rubber, and so is sometimes used as a replacement. It is available in a limited colour range.

PVC is the most widely applied coating. It is low-cost, versatile and available in a range of densities and colours. It is used for applications in construction, agricultural, marine and inflatables.

PU coatings are used to improve the visual appearance of textiles for high-performance and technical applications, as well as providing waterproof (or water-repellent), breathable and anti-abrasion properties. PU used in coatings is thermoset, as opposed to thermoplastic (laminating uses thermoplastic polyurethane, which can be welded). Being thermoset means it can be applied as a liquid at room temperature. The resin cures in a chemical reaction, caused by mixing two parts together. Permanent cross-links are formed between the polymer chains to form a durable and protective coating.

Silicone is a tough and resilient high-performance material. As a coating it is resistant to abrasion, water, chemicals and heat, while maintaining breathability. It provides good grip with low surface friction, making it practical and easy to clean. In fact, very few things will stick to silicone, including adhesives. Like PU and PVC, silicone is cured in a one-way reaction, and so is permanent. Therefore, three-dimensional shapes applied to textiles during coating, such as creases and pleats, will be permanent.

Acrylic is a water-based coating often applied by spraying. It exhibits good all-round performance in demanding environments including prolonged exposure to UV, such as tents, awnings and outdoor covers.

Fluoropolymer coatings – principally PTFE, which is commonly known under the DuPont trade name Teflon – provide high repellency to water and dirt and high resistance to heat and chemicals, and so they are utilized for medical and industrial applications.

Microencapsulated functional materials are micrometre-sized particles made up of a liquid or gas encapsulated within a shell. Dispersed within a liquid coating they are applied by direct, dip and spray coating. Functional microencapsulated materials include phase-change material (PCM), thermochromatic pigment and fragrance for example. Alternatively, the microcapsules may be incorporated directly in the yarn during extrusion (page 50).

Nanocoatings are applied in solution or by vapour deposition. They are typically 2 to 4 nanometres thick and so invisible to the naked eye. Nanocoatings are mainly used to make the surface of textiles hydrophobic. Also known as easy to clean, or self-cleaning, this means the surface will repel water, dust, oil and dirt.

COSTS

High-performance coatings are more expensive than commodity types. Silicone rubber (also known as SI-rubber) is one of the more expensive, followed by PU, acrylic and SBR. PVC and thin acrylic coating are the most cost-effective.

Unit price may be reduced and performance maintained by applying a thin topcoat of high-performance material to a more cost-effective basecoat, such as combining PU and PVC.

Cycle time is rapid, although set-up can be a lengthy process because the processing parameters are critical for a consistent and reliable product. Minimum order quantities tend to be quite high for bespoke solutions. Stock materials will be much more cost-effective.

ENVIRONMENTAL IMPACTS

Coatings are used to make longer-lasting and higher-quality products. Coatings that can be replaced or repaired, such as waxed cotton, increase the lifespan of the product further still. And coatings can reduce the amount an item needs to be washed, saving significant water and energy over the lifespan of apparel. However, coating produces a composite textile, which can make it more difficult to recycle or compost at the end of its useful life.

Depending on the coating technique, large quantities of water and heat may be required. Some techniques produce hazardous waste that must be carefully disposed of. Many steps have been taken to reduce consumption and waste, such as foam finishing, replacing hazardous ingredients, using a closed-loop system and recycling.

Water-based coatings are mostly latex or acrylic polymer emulsion dispersed in water. They are cost-effective, less toxic and reduce hazardous emissions

→ Direct Coating Polyurethane

Polyurethane (PU) is prepared by mixing (**image 1**). It is a versatile material and there are several different types to choose from, including non-breathable, breathable and hydrophobic, and microporous.

The dry fabric is drawn underneath the doctor blade over a roll, and a thin film of PU is applied to the top surface (**image 2**). The process operates at high speed and the textile is transferred directly into the tentering frame (**image 3**). Pins grip the textile along each selvedge and hold it tight as it is dried

and set in a series of ovens (**image 4**). The fibre composition, weight of fabric and coating applied determine the temperature of the ovens, which ranges from 80 to 200°C (176–392°F). The series of ovens become gradually cooler, by around 20°C (69°F) per stage, to ensure uniform drying.

The coated textile emerges from the last oven fully dried and set. It is passed over a copper wire to take away the static that has built up through the drying process and taken up on a roll (**image 5**).

1

2

3

4

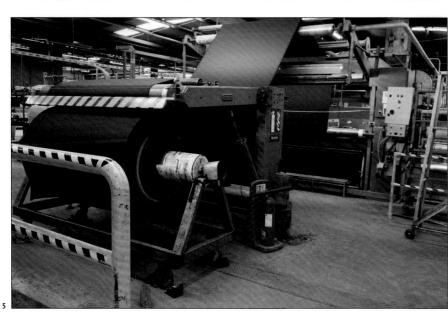

5

→ Dip Coating Wax

Originally developed for sailmaking, oil-treated weatherproof fabrics began to be produced for apparel, luggage, footwear and the military in the 1880s. The techniques have changed very little over the years and the fabrics continue to be used in familiar applications, such as the famous Barbour jacket (**image 1**).

Silicone and other high-performance coatings have developed rapidly over recent years, providing colourful and lightweight weatherproofing. Even so, wax remains an important material in the fashion industry thanks to its unique appearance and natural handle.

The exact ingredients of the wax solution vary for each cloth (**image 2**). Wax is applied to coated or uncoated textile by dipping. The textile is submerged in the hot wax for a set period of time. Squeeze rolls remove excess coating as the fabric is drawn out of the bath of wax (**image 3**).

The coated fabric is taken up on a roll (**image 4**) and left for a period of time to condition. This allows the wax to fully migrate into the cotton fibres.

In the past, waxed fabrics were available only in dark colours, such as black and olive-green. Now waxed fabrics are available in a range of colours, and it is possible to make new colours to order (**image 5**).

1

2

3

4

5

→ Testing Coated Textile

Many tests are used to assess the performance of the coated textile, such as flame resistance and air permeability.

The Mullen-type hydrostatic tester (**image 1**) is used to assess the resistance to water penetration. A sample is held between two circular clamps and hydraulic water pressure is applied to the underside of the fabric. As the pressure is increased the amount of visible water is assessed.

Alternatively, a sample of coated textile is folded to make a cone (**image 2**). It is filled with water and the amount of water that drips into the container below is measured after a set period of time.

Tear strength is measured by placing a sample of predetermined width and length between two clamps, which are slowly moved apart (**image 3**). This provides a measure of stretch as well as of the load required to tear the textile.

1

2

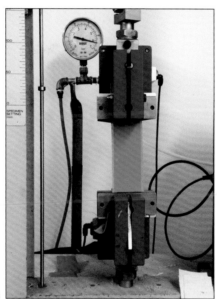

3

compared to solvent-based types. Traditional solvent-based systems have a significant impact on the environment including volatile organic compound (VOC) emissions during application and toxic additives that are harmful in use (such as indoor air quality). However, there are many different types of water-based types, so care should be taken to assess the full list of ingredients and make sure that the coating will perform well in the environment and be less harmful than solvent-based alternatives.

Plastics are often criticized for containing harmful ingredients. For example, phthalates and bisphenol A are subject to restrictions in many countries. Phthalate-free PVC is produced using a specially developed plasticizer (free from phthalates). It is used for sensitive applications, including medical devices and food wrapping.

Textile Technology
Spread Tow and Fabric Prepreg

In this process, used to fabricate composite materials for molding applications, a precise quantity of resin matrix is combined with fibre reinforcement. In this way, the laminator does not have to mix and handle the resins and curing agents during molding. Spread tow is made up of filaments laid in parallel, which reduces the reinforcement to the thickness of a few filaments.

Techniques	Materials	Costs
• Hot-melt prepregging • Fibre tow spreading for ultra-thin filament tapes	• Synthetic and natural high-performance fibres • Thermoset, thermoplastic and bioresins	• Moderate to very high cost depending on resin matrix and fibre selection

Applications	Quality	Related Technologies
• Automotive, aerospace and marine • Sports products	• Precise material distribution • Very high strength and stiffness to weight	• Composite laminating • Coating and laminating • Loom weaving

INTRODUCTION

Pre-impregnating with resin, abbreviated as prepregging, is the process of embedding continuous fibre reinforcement into a resin matrix. The choice of fibre, textile construction, resin and laminating technique will ultimately determine the properties of the material.

Prepregging produces very consistent materials. Precise quantities of each ingredient are combined to produce reliably high-strength parts. It is therefore utilized in the production of high-value materials, such as carbon-fibre-reinforced epoxy, for high-performance and critical applications. This means that it tends to be high-cost. However, it is possible

The Hot-Melt Prepreg Process

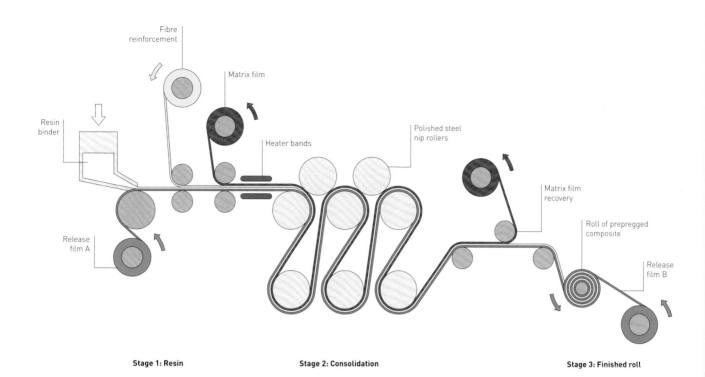

to combine many types of material, from commodity to high-cost and from synthetic to biobased, making it suitable for a broader range of applications.

High-performance yarns are typically made of multiple filaments of carbon, aramid, glass or ultra-high-molecular-weight polyethylene (UHMWPE) twisted together (page 60). Each filament is typically between 5 and 19 microns in diameter, depending on the fibre type. Unidirectional (UD) fabrics are made up of multiple yarns running in parallel. By contrast, spread tow tape is made up of filaments laid in parallel, reducing the fibre reinforcement to the thickness of a few filaments with conventional prepregging, or in the case of North TPT (page 237), to a single individual filament.

In the pursuit of ultimate strength to weight, composite laminates are designed with spread tow running along the lines of predetermined stress. In spread tow fabrics the filaments all run in the same direction. Thus, combining four layers (each a quarter the weight of equivalent woven fabric) in different orientations will produce superior performance, because

TECHNICAL DESCRIPTION

This is the conventional prepreg process (also called precasting a resin film), which differs from the process North TPT uses to produce thin-ply UD tape (see case study, page 237).

In stage 1, an exact measure of resin is applied to the release film A, ensuring a controlled resin-to-fibre ratio. The resin contains many ingredients, such as adhesive and hardener. It is precatalysed, which makes it viscous and tacky.

The fibre reinforcement is laid on top, locked in by the matrix film and compressed together. Types of fibre reinforcement include knitted, woven, nonwoven, unidirectional, multiaxial, tow and spread tow.

In stage 2, the fibre or woven textile passes through a series of polished steel nip rollers, which force the resin into the reinforcement. In stage 3, the backing substrate carrying the resin film is removed and the prepregged material is wound onto a roll with release film B on the top side.

The partially cured resin is fully hardened when heat and pressure are applied during molding.

the direction of filaments is determined by design rather than by the weave.

Most composite parts continue to be laminated by hand. Each layer of prepreg is individually placed into the mold. Large structures take several months to prepare, making them very expensive. To overcome the high cost and quality control issues

associated with manual fibre placement, North Sails developed a novel approach called TPT (Thin Ply Technology). Grown out of the company's 3Di sailmaking process (page 458), it has combined very light spread tow with automated tape laying (ATL) to produce parts with superior strength, stiffness and lightness.

TPT and 3Di are effectively the same prepreg tape production and bespoke fibre placement technology, but employing differing resin matrixes for varying end uses: epoxy resin matrixes for rigid composite structures and high-molecular-weight polyester thermoset adhesives for flexible structures.

APPLICATIONS

The performance of prepreg composites is unrivalled for many applications, such as in the construction of racing cars, boat hulls, masts and spars. They are used for their lightness and resilience to make wind turbines, aircraft (aerospace is the greatest consumer of prepreg), safety helmets, road cars and sports products (fishing poles, tennis rackets). They are applied for their thinness, strength and perceived value in consumer products.

RELATED TECHNOLOGIES

Prepregging is precise and versatile and allows for a wide range of resins to be combined with all types of fibre reinforcement. There are alternative methods of application, such as coating (page 226) and laminating (page 188), which depend on the materials and molding technology.

Thermosetting resins may be applied directly during molding as a liquid, thus cutting out pre-impregnation. This technique is rarely used for high-value composites, such as those using carbon and epoxy, but is still possible. Resin infusion sits between prepreg and wet lay-up. It is more expensive than wet lay-up but has much better resin control and so produces a superior finished part.

Certain types of resin matrix are woven (page 76) or knitted (page 126) together with the fibre reinforcement. This is relatively inexpensive, because it cuts out many preparation steps, but is limited to materials that can be molded with heat and pressure (see Thermoplastic Composite Molding, page 446). Several types of thermoplastic can be combined in this way, but thermosetting resins (used for the most demanding applications) cannot.

QUALITY

Performance depends on three key factors: yarn, construction and resin selection. The two broad categories of prepreg materials are fabrics (woven) and unidirectional spread filament tape (UD tape).

For fabrics, several different types of weave can be used including plain, twill and satin. Plain fabric (one over, one under) is very stable, but difficult to drape around sharp profile changes. Using a heavy balance of fibres in the warp direction produces a near unidirectional format. Twill fabrics (two over, two under to create a diagonal pattern) have an open weave, readily draping and conforming to complex profiles. Satin weave (for example four over, one under) is a much flatter fabric that can be easily draped to a complex surface profile. However, owing to this construction, such weaves are unbalanced.

Unidirectional fabrics and tapes are made up of parallel yarn or tow. This provides the highest tensile strength in line with the fibres, but reduces strength

Case Study

→ North TPT

North TPT (Thin Ply Technology) combines ultra-thin prepreg unidirectional composite with its unique fibre placement technology developed in Minden, Nevada, and Cossonay, Switzerland. Different resin matrixes are employed to create either flexible or rigid composite structures. Whereas carbon fibre is combined with epoxy to make rigid structures, such as for racing boats and Formula One cars, ultra-high-molecular-weight polyester adhesive is used to make flexible textiles (sail membranes, page 437).

As a result of the specific requirements for yacht sails, North TPT had to develop its own prepreg process that could easily manufacture tape down to 25 g/m² (0.08 oz/ft²) and as light as 16 g/m² (0.05 oz/ft²). Rather than compacting carbon-fibre tow, the thousands of filaments that make up the tow are physically spread apart and laid down side by side after passing through a resin bath. The lower limit on thin and light unidirectional tape weight and thickness is now the size of the individual filament, rather than the size of the tow you start with.

The process begins with reels of fibre reinforcement, such as carbon, aramid and UHMWPE. Each of the untwisted yarns (tow) is opened up and spread out as it enters the prepregging line. The fibres are coated with thermosetting resin and placed onto backing paper (**images 1** and **2**).

Since 2011, several Formula One R&D teams have evaluated thin ply for various applications. Each of the teams has its own take on how the technology might be of most use. Utilizing the TPT automated tape-laying (ATL) head, 18 g/m² (0.06 ft²) prepreg is plotted to make a body panel (**image 3**). A series of two-, three- and four-layer complexes were fabricated to compare final surface finish and weight. Using the new thin-ply skins reduced the weight of the panels by around 25%.

A Formula One racing car wishbone is formed into a three-dimensional shape by press forming (**image 4**). Other applications include skis, helicopter blades, fishing rods, sail masts and racing bikes. The first 100% TPT yacht mast, produced by Southern Spars for the TP 52 Matador, was up to 19% stiffer along the length of the tube than previous generations.

The advantage of being able to place each layer of fibre at a different orientation is being utilized in the production of ultralight boat hulls. Fibre placement is calculated with computer modelling software and exploded to illustrate the various layers (**image 5**). The result of this development is lighter, stronger and more competitive racing designs.

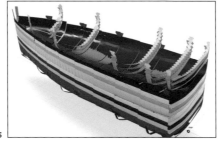

Featured Company

North Thin Ply Technology
www.thinplytechnology.com

in other directions. Therefore, multiple layers of unidirectional fibre are often combined in different orientations.

The matrix supports and bonds the fibres, transferring applied loads and protecting the fibres from damage. It also governs the maximum service temperature of a composite. Thermoplastics, thermosets and bioplastics are all suitable matrices. The defining difference is that thermosetting plastics form permanent cross-links between the polymer chains. This creates a more durable molecular structure that is more resistant to heat and chemicals. The formation of cross-links means that the curing process is one-way: they cannot be remolded. Thermoplastics do not form permanent joins and so can be remolded with heat and pressure.

North TPT combines the advantages of prepreg with the rapid and precise placement of UD tape, thus reducing cost and quality issues associated with hand lay-up. Its automated tape-laying system means that the designer is no longer limited to the traditional 'easy' fibre orientations of 0°, 45° and 90°, which yield a quasi-isotropic part. Using a computer-guided plotter, the designer can fan out the fibres every few degrees to produce an isotropic structure tailored to the load-bearing requirements of the application.

DESIGN

Prepregs are molded into lightweight and high-strength parts by composite laminating or press forming. Composite laminating tends to be for lower-volume products, while press forming is used for smaller parts, high volume and mass production.

Each production technology has different requirements. Across all processes material selection plays a key role in determining the design opportunities. For example, more intricate mold profiles are reproduced with pliable fibres and textiles, whereas bulkier textiles can be built up more quickly.

There are significant design and engineering advantages to using multiple thin layers (plies) of spread tow as opposed to fewer thick layers of fabric.

Increasing the number of plies, while optimizing filament placement, allows greater design freedom and improved structural optimization. This is known as the 'thin ply effect'.

MATERIALS

As long as the fibres and resin are compatible there is a wide range of material options. Typical fibre types include carbon, aramid, glass and UHMWPE.

The proportion of different materials depends on the application. Carbon fibre is very stiff, but this makes it a little brittle; aramid is strong and more flexible than carbon, but is sensitive to sunlight; and UHMWPE is durable and flexible.

Natural fibres, such as hemp and flax, are also suitable for prepregging. Combined with the appropriate resins, natural fibres are used to make lighter-weight composite parts (compared to glass-fibre-reinforced material), which are already being utilized in automotive applications and sports products.

Commonly used thermosets include epoxy, phenolic and polyester; thermoplastics include polyamide (PA), polypropylene PP and polyetheretherketone (PEEK); and bioplastic includes thermoplastic polylactic acid (PLA) and thermosetting polyfurfuryl alcohol (PFA).

COSTS

The amount this adds to the base cost of the fibre will depend on the resin being pre-impregnated and whether the fibre is processed into textile or UD tape format for impregnation. High-performance plastics such as epoxy are significantly more expensive than commodity types like PP and polyester.

Very thin and light carbon-fibre-reinforced unidirectional tapes have not been widely available, owing to the high cost of production. This is partly due to the legacy of the composites industry and the bulk of the development being focused on large-scale aerospace and military applications, and partly to the technical challenge of making very thin materials. This is gradually changing with innovations in the production techniques.

ENVIRONMENTAL IMPACTS

Conventional composites, such as carbon-, glass- and aramid-reinforced plastic (CFRP, GFRP and AFRP), are very light, stiff and strong. They are changing the way high-performance products are designed and built, improving efficiency and enabling structures that would otherwise be impractical. However, carbon and aramid (also known under the trade names Kevlar and Twaron) are very energy-intensive to produce; and at the end of their life composites are very complex to recycle.

Emerging technologies include recyclable resins and the equipment needed to process composites back into raw materials at the end of their useful life. Therefore, it is likely that the impact of high-performance composites will be reduced in the foreseeable future.

Biocomposites are made up of natural-fibre reinforcement (see Plant Fibres, page 478) and bioplastic, also called bioresin. Alternatively, they are biobased and consist of a synthetic resin matrix or non-bio fibre reinforcement. Thermoplastics are highly recyclable; thermosets are less easily recycled.

Bio ingredients require less energy to manufacture than synthetic equivalents. The recyclability has also been studied: flax/polylactic acid (PLA) laminates were shredded, regranulated, injection molded and tensile tested – this cycle was performed five times. The strength of the recycled material reduced by 10% with each round of recycling, but the modulus was unaffected. Flax/PLA is also biodegradable and compostable in the right conditions.

→ Preparing Flax Fibre Composite

This is the production of flax–epoxy biobased composite. In this case, the fibre reinforcement is twill flax (**image 1**). Similar to hemp, flax has good mechanical properties and can be grown in a European climate, near to the factory. The epoxy resin is mixed in preparation (**image 2**). An exact measure is applied to the lower release film (**image 3**). The flax weave is laid on top and it is passed through the hot-melt process (**image 4**). It is consolidated with resin and taken up on a roll between release films (**image 5**).

The process is carefully controlled to ensure the highest-quality material (**image 6**). This is essential because these materials are used in demanding applications where predictable mechanical performance is critical.

This prototype part by Composites Evolution is made up of woven flax and PLA Biotex and formed by composite laminating (**image 7**). This example is a rear-door module for the Jaguar XF. It is 35% lighter than the current glass-filled (GF) polypropylene (PP) component of the same thickness.

1

2

3

4

6

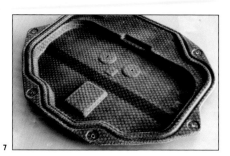
7

Featured Companies

Umeco
www.umeco.co.uk
Composites Evolution
www.compositesevolution.com

Textile Technology

Dyeing

Colour is added at one or more stages during production, from raw fibre to finished item. All types of material and fabric can be dyed, but the chemicals and techniques will differ, and the quality depends on the raw material. Brightness is improved with fluorescing agents, which absorb invisible ultraviolet light and re-emit blue light that is visible to the naked eye.

Techniques	Materials	Costs
• Stock (fibre and sliver) • Yarn • Piece (roll)	• All types of raw material except those unaffected by dyeing	• Low to moderate • Natural dyes are the most expensive

Applications	Quality	Related Technologies
• Apparel • Interiors • Technical	• Durable solid colour • Resistant to abrasion • Colour fastness varies	• Bleaching • Printing • Inherent coloured • Resist dyeing fibre

INTRODUCTION

Depending on when colour is introduced in the production of the textile, a wide range of patterns and effects can be achieved with dyeing. It produces durable and high-quality colour because the colourant penetrates into the structure of the fibre. Patterned and feathered colours are formed by the dyeing technique or by combining different-coloured fibres or yarns in the structure of a textile.

The unique character of indigo denim (used to make typical blue jeans, jackets and other apparel) is achieved by dyeing only the warp. Twill-woven with a white-coloured weft, traditional denim has a blue front while the back is white. The indigo-coloured dye penetrates only the outside surface of the hard-wearing cotton warp, leaving the core uncoloured. Therefore, as the surface is gradually worn away by wear and tear, the denim fades to white. This patina has become so desirable that techniques have been developed to artificially wear away the surface in a controlled fashion, such as by abrasive or chemical washing (see Scouring and Mercerizing, page 202).

Dyeing uses the same chemicals as printing (pages 256–79); the difference is that dyeing involves immersing the material in dye solution to achieve a solid colour throughout, while printing is used to reproduce multicoloured patterns and designs on the surface.

APPLICATIONS

Virtually all textiles are coloured in some way. If not dyed, then the colour will come from the base material, or is added during filament extrusion (see Filament Spinning, page 50) or printed. Dyeing is used to add a base colour for printing. Known as ground, a dyed background is either printed over or removed by discharge printing (page 270).

The Dyeing Process

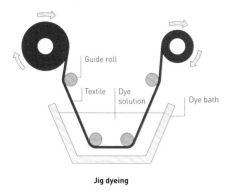

Jig dyeing

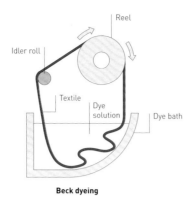

Beck dyeing

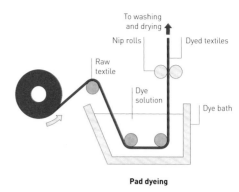

Pad dyeing

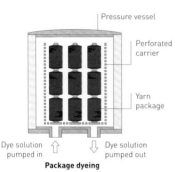

Package dyeing

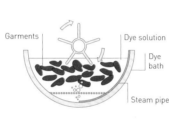

Paddle dyeing

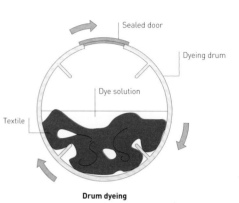

Drum dyeing

TECHNICAL DESCRIPTION

To achieve uniform dyeing the material is circulated within the dye solution. There are three principal systems: moving the material through the dye, circulating the dye solution around the material, or circulating both.

Systems that circulate the material are referred to as batch dyeing because they are suitable for colouring short as well as long runs of yarn and textile in batches. Jig, pad and winch dyeing are all batch processes. In package dyeing, the solution is forced through the fibres. Processes that circulate both the material and the dye include jet, drum and paddle dyeing.

Jig dyeing uses a stationary dye bath. A roll of fabric is drawn through the dye solution and onto a roll on the other side. When the second roll is full, the process is reversed. The fabric is drawn back and forth until the dye has fully penetrated the fibres. The textile is held under tension and so is suitable for woven materials and those prone to creasing. Long lengths may be dyed in a single process.

Pad dyeing also uses a stationary bath. The textile is passed through the dye solution at high speed and the dye solution is forced into the fibres by the pressure of the nip rolls. It is a continuous dyeing process utilized in mass-production dyeing applications. For high volumes, pad dyeing typically makes up the first stage of a modular process that also consists of washing (page 202) and drying. By the time the fabric is taken up on the roll at the other end of the line it is finished.

Beck dyeing, also called winch dyeing, is an economical batch process. A length of yarn or textile is bound together to make a loop and is then fed through the dye bath in a continuous rotating cycle. The material is not held under tension, which makes this process suitable for knitted and lightweight textiles. It is a versatile process used to colour short and long lengths.

Jet dyeing uses a similar principle to winch dyeing. The difference is that the dye solution is circulated at high pressure, so the operating temperature can be above

boiling point. Therefore, it is used to colour synthetics such as polyester, which are dyed at 130°C (266°F).

Package dyeing is used to colour batches of yarn on cones, or yarn or textile on beams (beam dyeing). The material is loaded into a perforated carrier so the dye solution can pass freely. Dye is pumped through the fibres at high pressure. Capacity varies from a few litres to very large machines. Bulky yarns, such as combed wool wound into long coils (skeins), are dyed at atmospheric pressure to avoid compacting the fibres.

Paddle and drum dyeing are typically used for finished products or garments. As in jet dyeing, both the material and the dye are circulated, which provides good uniformity of colour. The items are submersed in a dye bath and circulated by a rotating paddle. As with the other technique, the solution is heated with steam. Drum dyeing uses similar equipment to leather dyeing (page 164). The dye solution is pumped in and the drum is rotated. Drums are used to apply unique finishes.

RELATED TECHNOLOGIES

Certain materials are inherently colourful and so dyeing can be avoided. For example, natural fibres such as wool, hemp and nettle come in a variety of colours ranging from dark greys and browns to pale greens and off-white. Natural coloured cotton (page 480) grows in shades of green, brown and beige. Eliminating dyeing by using naturally coloured materials means that fewer chemicals are used in production, if any at all, and subsequently water consumption is reduced.

Some materials are not suitable for dyeing, and so are available only in their inherent colour. For example, aramid fibre comes only in yellow and black and carbon fibre is available only in black.

White and off-white colours are produced by washing and bleaching. Natural fibres, such as flax, require repeated bleaching to produce bright white. Washing and bleaching often precede dyeing because the quality of dyed colour depends on the cleanliness and whiteness of the raw material.

All of the printing processes may be used in place of dyeing to apply solid colour, including transfer, screen, block and digital printing. Printing is more versatile than dyeing because colour and pattern can be chosen much later in the production cycle, especially when using digital techniques. However, printed colour only penetrates the surface of textiles and does not go right through like dyeing, unless printing openwork or sheer fabrics. Printing is used to apply pigment ink. Pigments coat the surface of the raw material, as opposed to chemically bonding to its structure like dyes. Therefore, it is much more straightforward to colour-match with pigments than it is with dyes. The drawback is that they are more vulnerable to abrasion and rubbing.

QUALITY

Fastness is measured in terms of resistance to washing, ironing, steaming, light, perspiration (salt water), gas fading and rubbing (also called crocking). Dye molecules penetrate deep into the fibre and bond with its structure.

No dye has high durability in all areas, so compromises are made depending on the application. For example, interior fabrics require a high level of light fastness and low wash fastness because they are unlikely to be washed frequently but they are exposed to prolonged daylight. Apparel fabrics, on the other hand, require a high level of wash fastness so that they can be cleaned often without losing colour.

Colour is enhanced by adding optical brighteners (or fluorescent agents), which absorb light in the ultraviolet (UV) part of the spectrum and re-emit it in the visible blue portion. Thus, the colour will appear brighter than its surroundings. Washing powders often contain optical brighteners, which help to maintain colour saturation in frequently washed apparel.

When two items appear the same colour in one lighting condition but different in another this is known as metameric failure. Careful dye and fibre selection is critical to ensure a good colour match in all lighting conditions.

DESIGN

The design opportunities vary according to the techniques used to apply the colourant, through to the construction of the textile or finished item.

Dyeing at the fibre stage is known as stock dyeing or top dyeing (which refers to dyeing combed wool sliver). It is the least versatile because all the fibre is dyed one colour. Any variation in colour is removed, or reduced, by the blending action of spinning the fibres into staple yarn (page 56) or entangling fibres to make a nonwoven. Yarns made up of different-coloured yarns produce a heathered (flecked) effect.

Yarns may be dyed at various stages and may be dyed in much smaller quantities than raw fibre while maintaining the option to combine different-coloured yarns during weaving (pages 76–105) or knitting (pages 126–51). This makes it a versatile technique.

Yarns may be dyed as skeins (skein dyeing), on cones (package dyeing) or on the warp beam (beam dyeing).

Case Study

→ Dyeing and Blending Wool Fibres

In preparation, the wool is cleaned and scoured (see Scouring and Mercerizing, page 202). Dyeing takes place in large vats of boiling water (**image 1**). The wool is hoisted in bags (**image 2**), which are used to transfer it to the drying processes.

The wool is blended in a Fearnought machine, which breaks down the clumps of wool through counter-rotating action (**images 3** and **4**). Various shades of colour are blended to create a rich colour in application. The fibres pass through a synthetic coiling unit to reduce dryness and are deposited into a hopper (**image 5**).

The dyed and blended wool is now ready to be carded and spun into staple yarn (page 56). Mixing various shades produces a yarn with slightly irregular colour or heathered (flecked) effect. Viewed up close, the different colours are clearly visible, but when seen from a distance, the colours will merge into a single shade. This assimilation is called the Bezold effect, after the German meteorologist who discovered it, and gives heathered yarn its depth and character.

1

Rolls of fabric are coloured in a process known as piece dyeing. High-quality colour can be achieved in small and large orders, making this a versatile approach because colour selection does not need to be made until late in the production cycle.

Finished products or garments may be dyed when complete or at any stage during construction. This is the most flexible technique of all as the colour is selected at the end of the process.

By combining two types of fibre that are compatible with different dyestuffs, two colours may be achieved with dyeing alone. This technique is known as cross dyeing and the patterns and colours that can be created depend on the method of production. For example, it is more cost-efficient to warp knit (page 144) seamless garments in plain white because setting up the machines with different colours each time would take too long. Two types of yarn that respond to different dye systems, such as nylon and polyester, are knitted with jacquard effects. Once complete, the colours are selected and the finished item is dyed in two stages, highlighting the jacquard pattern. In this way, a single design may be dyed a wide range of colours without having to change the yarn.

Alternatively, finished items may be dyed a single colour even if they are made up of multiple fibre types, in a process known as union dyeing. Care must be taken to ensure that all the elements of the garment will appear the correct colour (or remain white), especially if it is complete with trim and labels.

Applying dye to the finished product or garment means other effects may be reproduced, such as a pre-faded appearance, normally created with abrasive washing techniques.

MATERIALS

The appropriate dye system is determined by the raw material. Each type of dye may be used on more than one material.

Acid dyes are used mainly for wool, silk, acetate and nylon. They are bright and

2

3

4

5

Featured Company

Mallalieus of Delph
www.mallalieus.com

relatively inexpensive, but they do not have good fastness to washing and so are not usually used for apparel. If they are used then the garment will be designated dry clean only. They are applied in acid solution such as vinegar.

As with some other dye systems (including direct and reactive), metal ions are added to improve the wash and colour fastness. Referred to as complex metal dyes, chromium, cobalt, nickel and copper are used.

Azoic dyes are available in a bright and broad colour range, with especially good reds, yellows and blacks with very good colour fastness. They are mainly used to colour cotton but are also compatible with nylon and polyester.

Basic dyes are also known as cationic, because they contain positively charged ions that combine with acids in the raw material. They may be water-based or solvent-based and are available in a range of brilliant colours including fluorescents. They are compatible with wool (although they are not colour fast on wool), silk and acrylic. Combined with a mordant (fixing agent) they may be applied to other types of fibre, such as cotton. The mordant is applied beforehand. It reacts with the dye causing it to fix to the base material.

Direct dyes are widely used and are available in a full and bright colour range, often referred to as the commercial colours, because they are relatively inexpensive and not the highest quality. They have fair resistance to light and poor wash fastness. As the name implies, they are relatively straightforward to apply directly to cellulosic materials, using hot water and common salts to regulate dyeing. Similar to acid dyeing, treating the dyed fabric with copper salts improves fastness.

Disperse dyes are equally widely used and available with a good colour range. They were originally developed to colour CA and are now used with most popular synthetics including nylon, polyester, polyvinyl chloride (PVC) and acrylic. Fibres such as polyester are disperse dyed at elevated temperatures and so require additional pressure during dyeing. They are ideal for transfer printing.

Natural or vegetable dyes are derived from plants, animals or minerals and are used to colour natural fibres. The majority are derived from vegetable matter, such as leaves, seeds, bark and berries. The techniques have been used for thousands of years and are still practised in many countries, mainly by very skilled craftspeople, which makes vegetable dyeing relatively expensive. The process is straightforward: the dyestuffs are added to hot or cold water, and the yarn or textile is immersed in the solution until the dye has soaked into the fibres. The dye is fixed with mordant, such as alum, ash, oxidized iron or limestone. A range of vivid colours is possible, though this may depend on availability.

Reactive dyes come in the full colour range and have very good colour fastness. They are a relatively new development compared to the other dye systems, having been created in the 1950s. There are several types compatible with dyeing cotton, wool, silk, acrylic and nylon. After dyeing, the material is steam-heated to develop the colour and then washed thoroughly.

Sulphur dyes are used to colour cotton and flax with a range of dark colours, including black, brown and blue. They are relatively inexpensive and have good colour fastness. Dyeing is carried out at high temperature and there are several separate dipping stages, including application, oxidizing and washing.

Vat dyes are available in a wider colour range than sulphur types (although still lacking in reds and oranges) and have superior colour fastness. They are relatively expensive and are used to colour cotton, flax and CA. Using a mordant they may also be applied to wool, nylon, polyester and acrylic.

Pigments, which are sometimes referred to as pigment dyes, although this is not technically correct, are applied by coating (page 226). The colour is applied to the surface of the fibre or fabric and is held in place by a resin binder. It is not as durable to abrasion as dyeing, but a wide range of colours and effects (such as metallic) can be achieved that are not possible with dyestuffs.

COSTS

Unit cost depends on the combination of dye system and raw material. Stock dyeing is the most cost-effective and direct dyeing is the least expensive. Setting up bulk production for new colours is expensive, so is only suitable for high volumes. Small orders are batch processed.

ENVIRONMENTAL IMPACTS

Dyeing uses a great deal of water – up to 150 litres (40 gal.) per 1 kg (2.2 lb) of textile – while some processes use harmful chemicals, heavy metals and salts. Great effort has been made to make sure they are safe and purely a colouring agent. Declarations for health and safety and tests for fabric before going into apparel are necessary to avoid any incidents arising from manufacture or use. Significant improvements have been made and pollution to air and water has been reduced. None the less, this is not the case in all countries or factories, so care should be taken to select a responsible dye house.

Using natural-coloured fibres, such as wool and cotton, eliminates dyeing and the associated chemical and water consumption. Cutting out the dyeing process also reduces the amount of water and chemicals required for the preparation of the fibre. In addition, natural-coloured cotton is produced without the need for pesticides and fertilizers. But the colours are limited to shades of white, green, brown and beige. Natural and vegetable dyes contain less harmful ingredients. But they too have some colour limitations.

The process of dyeing with carbon dioxide as the carrier significantly reduces water consumption and the associated energy and waste. DyeCoo in the Netherlands has developed the first commercial production equipment capable of dyeing with recycled carbon dioxide. The gas is heated and pressurized to make it supercritical, at which point it has similar density to liquid. A specially developed disperse dye is dissolved into it and applied to fabric. It is quicker and requires much less energy than conventional wet processes.

→ Skein Dyeing Wool

Skein dyeing, also called hank dyeing, is relatively slow and expensive but produces very good colour penetration. Many different types of fibre are dyed in this way, and each type comes in a variety of natural colours (**image 1**). Wool ranges from off-white to brown. It is has been scoured, carded (page 27) and twisted into yarn. In preparation, the skeins are bleached and washed (**image 2**) to ensure uniform colour for dyeing.

The mix of ingredients required for dyeing is prepared (**image 3**) and added to the dye bath. The wool is loaded on a carrier and dipped in the hot dye solution (below boiling point). After repeated dipping the desired colour is reached and the skeins are lifted from the tank (**image 4**). Each of these tanks is capable of holding 1 ton of fibre. Neighbouring tanks are linked to increase colour consistency. This is especially tricky for long runs of uniform colour. To ensure consistency, fibres from different batches are carefully checked in a colour-corrected light box (**image 5**).

After dyeing, excess dye is washed out and the yarn is spin dried and hot-air dried. The finished material is stacked ready for use. Package dyeing has replaced skein dyeing in many cases, because it is more controllable and efficient. However, skein dyeing produces yarn with superior bulkiness, and so continues to be utilized for high-quality wool and acrylic yarn used in knitwear, and pile yarn used to make woven carpets. In addition, wool is dyed below 100°C (212°F); any hotter and the fibres would be damaged.

1

2

3

4

5

DYEING

245

Featured Company

Calder Dyeing
www.caldertextiles.co.uk

→ Package Dyeing Silk

Silk has a smooth and reflective surface, so when dyed the colour appears very bright and lustrous.

In preparation for dyeing, silk is wound onto perforated plastic cones, wrapped in cloth and loaded onto the carrier (**images 1** and **2**). Package dyeing takes place in a pressurized vessel (**image 3**). Multiple cones are dyed the same colour at one time. Vanners, of Sudbury, England, uses both acid and reactive dyes on silk, mainly acid on wool and cashmere.

The dyed yarn is thoroughly washed to remove excess colourant (**image 4**) and dried. The packages of yarn are loaded onto a creel (**image 5**), ready to be transferred to a warp beam (**image 6**) in preparation for weaving.

Vanners has been weaving silk for more than 250 years. Its fabrics are used in the neckwear, couture, furnishing and bridal markets. In this case, the brightly coloured warp and weft are converted into a fancy jacquard-woven fabric (**image 7**). The blue-dyed warp determines the underlying colour and the pink weft creates the pattern.

1

2

3

4

5

6

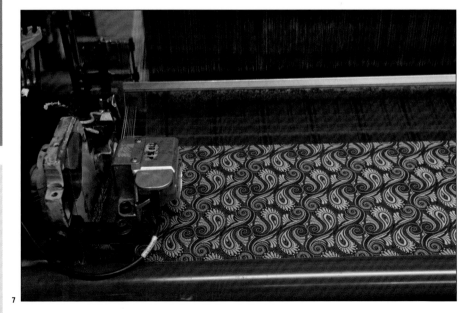

7

Featured Company

Vanners
www.vanners.com

→ Piece Dyeing Cotton Textile

In preparation for dyeing, plain-woven fabric (**image 1**) is washed, bleached and mercerized, followed by tentering. These steps are critical to ensure high-quality colour in the finished material.

The jig-dyeing process is used to colour a specific length, which is wound onto a roll on one side of the bath (**image 2**). The fabric is passed through the steaming green dye solution below and taken up on the roll on the other side. When the full length of fabric has passed through the dye the process is reversed. After dyeing, the fabric is washed and re-tentered to produce the finished material (**image 3**), ready for printing or conversion into finished items by cutting and sewing (page 354). Alternatively, two types of yarn are combined in the fabric construction, such as polyester and nylon. These yarns are compatible with different dye solutions. Either both yarns are coloured by carrying out two dyeing operations, or one of the yarns is dyed in a single operation (**image 4**).

1

2

3

4

Featured Company

Standfast & Barracks
www.standfast-barracks.com

Textile Technology
Resist Dyeing

A resist is applied to yarn or fabric to stop dye absorption in those areas, while exposed areas continue to absorb colour. It is also known as resist printing, because multiple colours may be built up. The designs are made by hand and so range from accurately registered to unique and random. Commonly used techniques include ikat, batik and tie-dye.

Techniques	Materials	Costs	
• Ikat • Batik • Tie	• Natural fibres and natural or vegetable dyes	• High	
Applications	**Quality**	**Related Technologies**	
• Apparel • Interiors • Ceremonial	• Soft outline • Variable • Colour limited by dyes	• Block printing • Digital printing	• Dyeing • Screen printing

INTRODUCTION

Resist dyeing techniques are hand processes that are still widely practised in many countries. As well as apparel and interior fabrics, resist-dyed materials are used for important traditional ceremonies, such as weddings and burials, and so are held in high regard.

The crafts are ancient and are practised by highly skilled craftspeople. The exact technique will depend on the location and perhaps the individual, but the basic principle is that the dye is prevented from penetrating areas of yarn or fabric by a resist. The exposed areas make up the pattern and the resisted areas form the ground.

The Resist Dyeing Process

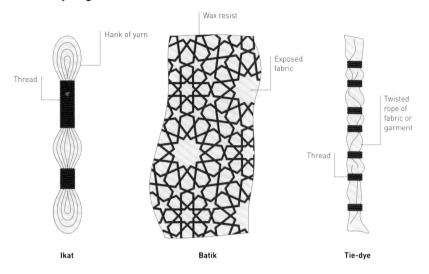

Hank of yarn

Thread

Ikat

Wax resist

Exposed fabric

Batik

Twisted rope of fabric or garment

Thread

Tie-dye

TECHNICAL DESCRIPTION
Resist dyeing processes are manual and so can be accurate or totally random, depending on the design. Ikat involves dyeing hanks of yarn before weaving. The yarn is bundled together and tied, sometimes with the addition of wax, to control where the dye can permeate. Exposed areas of yarn absorb dye and so produce bands of colour on the yarn. After dyeing it is loaded onto the loom in such a way that cloth will be patterned when woven.

Tie-dyeing is a similar approach applied to woven textile. Areas of the sheets are bound tightly with thread to stop the dye permeating. As the textile is dyed, the liquid soaks into the textile and underneath the thread, creating gradients of colour. The effect is reversed by soaking the threads in colour before tying them onto the fabric and soaking in hot water.

Batik is the reverse of printing. Sheets of woven fabric are drawn on with wax resist. The wax soaks into the fibres and so stops any dye reaching those parts. The fabric is dyed and the wax is removed with hot water or chemicals. Several colours are built up in stages, to make complex and elaborate patterns.

The three main techniques are ikat, batik and tie-dye. Ikat involves dyeing hanks of yarn with bands of colour. The patterns are formed in the way the warp and weft are combined during weaving. Batik is a wax-resist process carried out on woven fabric. Complex and colourful patterns are built up with multiple waxing and dyeing stages. The distinctive pattern of tie-dye is formed by tightly binding the fabric with thread before dyeing. The binding stops the dye penetrating and so produces radiating patterns around the binding.

APPLICATIONS
These finishes tend to be relatively expensive and so are limited to few applications. They are used for headwear, neckwear, shirts, blouses and gowns. Interior applications include covers, drapes, tapestries and rugs.

RELATED TECHNOLOGIES
Printing techniques, such as screen printing (page 260) and digital printing (page 276), are used to reproduce batik, because they are much more cost-effective. These printing techniques are set up to reproduce identical parts; they cannot mimic the unique quality of individual batik.

Block printing (page 256) has similar qualities to batik. The techniques are reversed: in batik the areas to be left are blocked with wax, whereas in block printing the colour is applied directly with shaped blocks. Block-printed patterns tend to have a more clearly defined edge, due to the direct application of the dye, rather than soaking areas with wax.

Warp printing (see Screen Printing, page 268, and Digital Printing, page 278) produces very similar effects to ikat. This printing technique involves printing the warp before the final weaving. As a result, the finished pattern has undefined edges and the quality is largely dependent on the skill of the weaver, making it virtually indistinguishable from ikat.

Winch dyeing (see Dyeing, page 240) regular fabric in rope form can produce similar types of patterns to tie-dyeing, because the creases and wrinkles are dyed less evenly.

QUALITY
There are many variables when using natural materials and it is unlikely that two sets of dyed fabric will look the same, unless they are from the same batch.

These are hand techniques and the quality will be largely dependent on the skill of the craftsperson. The unique qualities of each process may be considered imperfections elsewhere, but in resist dyeing they give the materials their desirability.

Batik uses hot wax, which soaks into the fabric slightly unevenly, and is prone to cracking. The dyeing process is the same for every piece, depending on colour.

Ikat uses even less precise dyeing techniques. The yarns soak up the solution and the extent of dyeing may depend on the duration as well as the depth of container. Dyeing is stopped by washing out remaining colourant.

Tie-dyeing faithfully shows all of the creases and bundles as they were made. The dye soaks in to produce a gradient around the edge of the pattern.

DESIGN
These processes are based largely on tradition and ceremony. Therefore, the materials, dyes and colours have changed very little. Traditional colours include white (ground), brown, indigo and black, thanks to the use of natural dyes. However, any colour can be applied

as long as it is compatible with the resist technique. For example, batik is constrained by the melting temperature of the wax, because the dye system must be effective at a lower temperature than the wax will soften.

Traditional styles are associated with the country of origin. Malaysian batiks tend to be bright and floral, African batiks are more geometric and Chinese batiks contain complex figured motifs.

There are many variations of ikat, including dyeing the warp, weft or both. Dyeing the weft is the most flexible for the weaver, because the pattern is formed and tweaked with each pick.

These processes lend themselves to experimentation, although creating new patterns is very challenging if you are not an expert in the technique.

The processes are slow and labour-intensive. Multicolour patterns are built up in layers. Depending on the process, each colour requires the same amount of work, increasing cost as a result.

MATERIALS

Natural dyes are derived from biological materials. A wide range of rich colours can be achieved, including blue (indigo), black (ebony), red and purple (stick lac, cochineal, madder and alkanet), grey (teak), pink (sappan and brazilwood), yellow (ochre, turmeric and marigold), brown (betel, almond and rosewood) and orange (annatto). Varying the mordant (fixing agent) and overlaying colours opens up a broader range of colours. While these techniques are traditionally carried out on natural-fibre materials, such as cotton, linen and hemp, virtually any material can be used. As with natural dyes, the quality will vary according to the source and may be limited by the season.

COSTS

All of these techniques rely on highly skilled labour and so tend to be high-cost in production. Organically cultivated natural materials are the most expensive.

Blocks are used to print wax for batik, which greatly increases output and consistency. The blocks may be costly to make, especially for complex designs.

ENVIRONMENTAL IMPACTS

Cultivated organically and sourced ethically, natural dyes have a much lower environmental impact than synthetic equivalents. They are derived from vegetable matter, insects or animals, and minerals. They do not contain, or require the addition of, any harmful ingredients, and they can be disposed of safely at the end of their life. Many natural dyes are used in food colouring and cosmetics.

Natural fibres can have very low environmental impacts too. However, all of the preparation and processing should be taken into account. Depending on the source and processing, cotton can have significant negative impacts before dyeing is even considered, while hemp is fast-growing and high-yield without the need for any fungicide, pesticide or herbicide. Certification is key to monitoring the overall environmental impact of the production processes.

Case Study

→ # Dyeing Yarn for Ikat

A range of natural ingredients is used to dye the yarn, including annatto seeds (**image 1**). Each ingredient is prepared in a different way and it takes a very skilled craftsperson to harbour good colour. For example, annatto seeds are separated from the inedible fruit by boiling water and the solution is ready to be used as dye immediately. By contrast, indigo leaves must be soaked for several days before limestone is added to turn the dye blue.

Hanks of silk are prepared by hand and bound tightly with thread (**image 2**). They are carefully placed with a mix of indigo leaves and limestone in boiling water and left to soak up the colour (**image 3**).

When the desired colour has been achieved, the hanks are removed and soaked in cold running water to wash out the excess dye (**image 4**). Then they are hung to dry in preparation for handloom weaving (**image 5**).

Ikat weaving uses coloured warp, coloured weft, or both. In this case, coloured weft is hand-woven with plain warp (**image 6**). The weaver aligns the threads with each pick (weft insertion) to create the pattern.

1

2

3

4

5

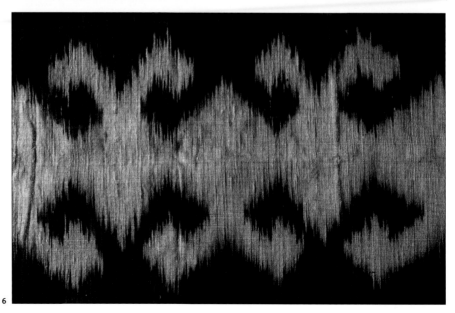

6

→ Hand Drawing Batik

Batik is practised as art as well as craft. The tools and techniques make the process suitable for free-drawn one-offs (known as batik tulis) or for block printing repeating patterns (batik cap, see case study on page 254).

The fabric is prepared by soaking in sodium silicate for four to five hours. This increases the pH and in doing so activates the molecules in the cellulose-based fibres to enable a strong chemical bond to form between them and the reactive dye.

The hand tool used to draw on the fabric with wax, known as a tjanting or canting, has a copper pot and wooden handle. It is dipped in hot wax, which may be a combination of beeswax, paraffin wax and natural plant extracts (**image 1**). The wax flows through the spout as the artist draws on the cotton fabric (**image 2**).

The fabric is held on a frame throughout the process, so that the wax is not disrupted. The dye is applied with a brush (**image 3**), rather than the fabric being placed in a dye

bath. Once dry, the wax is removed or added to, and indigo dye is applied over the top (**image 4**). When dyeing is complete the wax is removed to reveal the finished design and the fabric is placed in boiling water to remove any remnants (**image 5**).

1

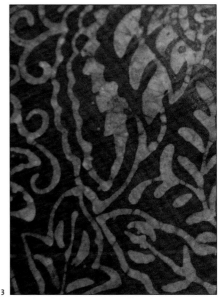

Featured Company

Penang Batik
www.penangbatik.com.my

→ Block Printing Repeating Patterns

When producing multiple copies of batik, or repeating patterns, a copper or wooden block is used to apply the wax (**image 1**). Using a similar principle to block printing, the batik block is dipped in hot wax (**image 2**) and then placed onto the fabric, transferring the wax to the fabric in the process (**image 3**).

The printer gradually works across and up the fabric, aligning the pattern with each impression (**image 4**). It is very important that the pattern is accurately aligned. The fabric is laid on a bed of split banana tree stems. This helps to cool and solidify the wax more quickly.

Less wax is applied by this method, so heavier fabrics may require printing on both sides. The exposed areas of the fabric, which remain white after printing, will be dyed with the first colour.

DECORATING

254

1

2

3

Featured Company

Penang Batik
www.penangbatik.com.my

4

→ Dyeing Batik

With the fabric held over a frame, only the outlines need to be made in wax, because each area is carefully filled in with dye by hand (**image 1**). The uneven coloration, caused by the dye absorbing into the fibres at different rates, gives batik its unique appearance.

Large areas of the design are filled in (**image 2**). Smaller brushes are used to work in and around the edge of the pattern to complete the design (**image 3**). Afterwards, the wax is removed with hot water and the dyed fabric allowed to dry in the sun.

1

2

3

Featured Company

Penang Batik
www.penangbatik.com.my

Textile Technology
Hand Block Printing

An ancient technique used to decorate fabrics, with evidence in China dating it to more than a millennium ago, hand printing with wooden blocks is used to produce repeating patterns of solid colour. Since then it has changed very little; it continues to be practised as a traditional craft characterized by the delicate idiosyncrasies of a highly skilled artisan's handwork.

Techniques	Materials	Costs
• Woodblock printing	• High-quality natural fibres such as cotton, linen (flax) and silk	• Very high labour costs

Applications	Quality	Related Technologies
• Covers, curtains and bedspreads • Apparel, scarves and sarongs	• Every piece is handmade and so unique • Intricate patterns wet on wet or wet on dry	• Digital printing • Resist dyeing • Screen printing

INTRODUCTION

Block printing is a relief-printing process: ink or dye is applied to a carved wooden – or cast-metal – block, which is pressed onto paper or textile. Colour is applied wet on dry, or wet on wet. Printing wet dye onto dry print, known as fall-on, means that more colours can be produced with fewer printing dyes and the edges of the print are sharply defined. Block printing is the ultimate execution of this technique (see also Screen Printing, page 260), with colours allowed to dry before the next layer is applied. Dark colours are typically applied first, with lighter colours printed on top. Thus, green is achieved by printing yellow over blue. In this way, four dyes may be used to produce a design consisting of eight colours or more.

Block printing may also be used to print wet on wet, like other printing techniques. This is quicker of course, but the print quality will not be so high.

Each colour requires a separate block, as do different parts of the pattern or text. In some cases, several blocks may be used for a single colour. Mechanized relief-printing techniques have evolved – letterpress and more recently flexographic printing – but these are mainly used for paper and board, not fabric.

APPLICATIONS

Block printing has remained a traditional craft and continues to be practised in India, China, Thailand, South Africa and a handful of other countries. Some fabrics that have been produced in this way for decades, or even centuries, have different names. For example, in India there is ajrak (resist and mordant technique from Gujarat), kalamkari (traditionally from Andhra Pradesh), and Sanganeri and Bagru (the former is on a white and the latter on a black-and-red ground, and both are from Rajasthan).

It may be used to decorate all types of fabric, but is reserved mainly for high-value interior textiles such as bedspreads, upholstery, quilts, drapes, curtains and pillowcases. It is used to decorate items of clothing too, as has been the case for hundreds of years, such as dresses, scarves and shawls. It is also used to print single-colour logos and motifs on fabric packaging, such as for food and other high-value items.

RELATED TECHNOLOGIES

Block printing crosses over with resist dyeing (page 248) techniques, such as when blocks are used to apply the wax resist used in batik. Reversing the process in this way – applying the non-printed areas as opposed to the dye – allows larger areas of colour to be filled in more easily. The quality is still largely dependent on the skill of the craftsperson.

Block printing cannot compete with rotary-screen printing (page 265) for speed and efficiency. However, for low-volume and high-value applications it may be interchangeable with flat-bed screen printing (page 267).

Digital printing (page 276) is the most versatile of all the decoration techniques. By scanning woodblock-printed artwork it is possible to reproduce something that looks very similar at first glance. On close inspection the difference in quality will be noticeable, but digital printing has other advantages, such as significantly reduced set-up costs.

QUALITY

The quality of reproduction depends on the skill of the printer, because each colour is applied by hand with a separate block. The blocks are aligned with a matrix of registration pins or located by eye to a pre-printed outline. The disadvantage of using registration pins is that they have to be checked before printing to ensure they align; this requires a higher degree of skill. The slight inconsistencies, which would be viewed as imperfections in other processes, give block printing its unique and desirable appearance.

Techniques have evolved to accommodate finer details, such as outlining with bent metal strip. Infilling

The Block Printing Process

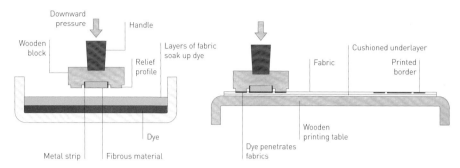

Stage 1: Tiering and dipping

Stage 2: Positioning and printing

TECHNICAL DESCRIPTION

Printing starts with the preparation of the solid wood block. Cut from dense hardwood, the block is usually no more than 50 mm (2 in.) in depth to reduce the likelihood of warping. Any material can be used for the block as long as it will sustain a relief profile. For example, linoleum (rubberized cork) has a smooth printing surface and is easy to carve for printing large areas of colour, while potato is used in workshops and homes alike, as it is so easily carved.

The design is transferred with tracing paper to reverse the image (so that it will be the right way around when printed). The areas of the pattern not printed are cut away, producing a relief profile. Large areas may be hollowed out and filled with fibrous material, such as felt, to soak up the dye and ensure more uniform colour.

The quality of print obtained will be different from the metal strip (thin lines and edging), fibrous infill and wooden relief profile. The subtle contrast is used to enhance the design of the pattern.

Printing itself is a straightforward process. Dye is poured into a bath and covered with several layers of textile. In stage 1, the liquid is soaked into layers of textile spread over the top, known as tiering. This creates an ideal surface from which to ink the wooden block.

The block is dipped onto the dye-soaked textile and the relief parts are saturated with a layer of colour. In stage 2, the block is transferred to the fabric and carefully positioned. Pressure is applied using a special mallet, called mauling, and the ink is forced into the surface against the wooden table.

large areas of colour on the block with felt improves the uniformity of dye saturation. Even so, relief printing produces a distinctive halo effect as a result of the slightly higher concentration of dye around the perimeter of each area of colour; this is due to the behaviour of liquid dye on a surface of the block.

DESIGN

Using hand-sized blocks means this technique is typically most practical for repeating patterns with motifs no more than around 300 mm (12 in.) across. An exception to this is Turnbull Prints' unit in Thailand, the last hand-printing workshop to be producing complex furniture designs (see case study on page 259), where blocks may be up to 650 mm long and 450 mm across (25 x 18 in.).

If the areas of colour are larger than is practical to print, then the fabric is dyed first and printed over; resist dyeing techniques are employed; or mordant

Colour
Material: Cotton
Application: Interior
Notes: Each colour requires a different block. In this case, the colours are used to create a three-dimensional effect.

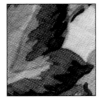

Fall-on
Material: Cotton
Application: Interior
Notes: Blocks are overlapped to create a third colour and thus more complex design with fewer printing dyes.

Registration Points
Material: Cotton
Application: Interior
Notes: Small dots in the design, known as pin marks, are used to align the various blocks.

Texture
Material: Cotton
Application: Interior
Notes: In a relief-printing technique, smoother textiles produce more uniform colour.

Lines
Material: Cotton
Application: Interior
Notes: Fine lines are reproduced using metal strip embedded in the wooden block.

Outlines
Material: Cotton
Application: Interior
Notes: Dark outlines help to define the pattern and overcome any slight misalignment.

Imitation
Material: Cotton
Application: Interior
Notes: The fine quality of block printing is here used to mimic the visual quality of embroidery yarn.

Nonwoven
Material: Nonwoven
Application: Packaging
Notes: A range of different materials is printed in this way, including woven, knitted and nonwoven fabrics.

Front   Back

Show-Through
Material: Cotton
Application: Interior
Notes: With manual pressure and relatively viscous dyes, the colour does not penetrate right through the fabric.

Wooden Block
Material: Hardwood
Application: Interior
Notes: The most commonly used woods are sycamore and fruitwood (such as pear). Modern CNC cutting is used to produce very precise blocks.

Fibrous Infill
Material: Wood, metal and wool felt
Application: Interior
Notes: Large areas of colour are filled in with fibrous material to produce more uniform colour reproduction.

(fixing agent) is printed that reacts to different dyes. By using different mordants that absorb dissimilar dyes, a multicolour print may be built up with only a single dyeing step.

Colours tend not to be butt-fitted; instead they are overlapped slightly to prevent flashes of white in the design. Following the shape of the pattern, as opposed to finishing with a straight line, is preferred, because it is very difficult to achieve a touch fit between two blocks. Very fine details, such as text, tend to be avoided (even though the first use of block printing was likely to have been literature), because in relief printing, thin lines and sharp corners are easily damaged.

Straight lines and geometric patterns, which typically show up the smallest misalignment, do not seem to be such a problem in block-print design. Complex patterns are aligned with reference points, known as pin marks.

Owing to the complexity of printing multiple colours, block printing is most commonly used to add one or two, and rarely more than four, printing colours (the ground colour may be dyed before printing). Small areas of additional colour are sometimes added by hand, depending on whether block printing is convenient. Dotted designs may be formed by metal wire, saving a considerable amount of carving.

MATERIALS
The most commonly used fabrics are the traditional ones, which are cotton, linen (flax) and silk. Choice of fabric is determined by whether the dye will fix to the material.

In the past, vegetable dyes would have been used. In most cases these have been replaced by aniline dyes. Paper printing uses ink instead of dyes, and pigment dyes are used to decorate leather.

COSTS
A high degree of skill is required in every part of the process. Preparing the wooden block by hand may take weeks, or even months, depending on the complexity. CNC cutting (page 316) makes this much

→ Block Printing Cotton Fabric

Turnbull is a unique company, whose expertise ranges from advanced digital printing (page 276) to hand block printing using traditional techniques that have remained unchanged for generations. This is the last remaining fully operational production unit for complex block printing.

In this case, furnishing fabrics are being produced (**image 1**). Turnbull has an archive of blocks dating back more than 150 years. Alternatively, a new block may be produced, either by hand or by CNC cutting.

In a process known as tiering, dye is soaked into layers of fabric to provide just the right amount of liquid each time the block is pressed on top. It is spread over the surface with a brush (**image 2**).

Once the dye has been prepared, the block is pressed down onto the soaked fabric (**image 3**). A thin layer of dye coats the relief parts of the block. Very quickly, the block is aligned on the pattern, using dots to ensure accuracy, and the printer presses it onto the surface (**image 4**).

Pressing the block down firmly, the printer 'mauls' it with a special lead mallet – the weight is selected according to the size of the block – to ensure the dye is evenly transferred (**image 5**). Slight misprints are touched-in with dye using a copper implement (similar to a plectrum), and the print is complete (**image 6**).

1

2

3

4

easier. The patterns are aligned by eye, so as not to slow down the process, although rulers and guides may be employed by less experienced printers. It is therefore expensive to commission.

ENVIRONMENTAL IMPACTS

There are all the usual chemicals associated with fabric preparation and dyeing, such as bleaches, colourants and mordants (see also Screen Printing, page 260). But as an entirely manual process, the environmental impacts tend to be relatively low.

5

6

Textile Technology

Screen Printing

Also known as silkscreen printing, this wet-printing process is used to apply graphics and coatings to all types of textiles and sheet materials. Whereas rotary screen printing is used to decorate rolls of textile with repeating patterns, flat-bed is used to apply graphics and multicoloured designs to batches of readymade items, as well as rolls of textile.

Techniques	Materials	Costs
• Colour • Discharge • Burnout (devoré)	• All sheet materials • Colourants include pigment ink, discharge ink and dye	• Low to moderate unit cost • Moderate set-up costs
Applications	**Quality**	**Related Technologies**
• Apparel • Bags and accessories • Interiors	• High quality and repeatable • Sharp definition • Dyed colour is hard-wearing	• Block printing • Digital printing • Transfer printing

INTRODUCTION

This technique is used to apply print to almost any type of flat material or surface. The principle is straightforward, but the process variabilities are subtle and complex. A mask is created on a flat-woven mesh screen, through which colourant (collective name for inks and dyes used in screen printing) is squeezed. Areas unprotected by the mask are printed. It is not just used to apply ink; any substance the right consistency can be printed. For example, glue may be screen printed for flocking (page 280), discharge ink printed to dissolve fabric dye in selected areas (page 270), or chemical gel applied to create devoré pile fabric; and by applying foaming ink (also called puff ink) a three-dimensional effect is produced.

Screen printing is a reasonably cost-effective process used to reproduce high-quality graphics with sharp definition of detail. This makes it suitable for one-offs, such as artworks and prototypes, through to mass-production applications.

Flat-bed is suitable for one-offs and batch production. Rotary screen is capable of much higher printing speeds than flat-bed and so is used for high-volume production. The process is constrained by the diameter of the rotary screen and the length of the cylinder. Therefore, the design of overlapping repeating patterns will be different for rotary than for flat-bed.

APPLICATIONS

Screen printing is the principal technology used to mass-produce patterns and graphics on textiles that go on to be constructed into apparel, bags, packaging, carpet and interior textiles. Rotary screen is used to print the paper for transfer printing (page 272). Applying individual prints to fully assembled items

The Rotary Screen Printing Process

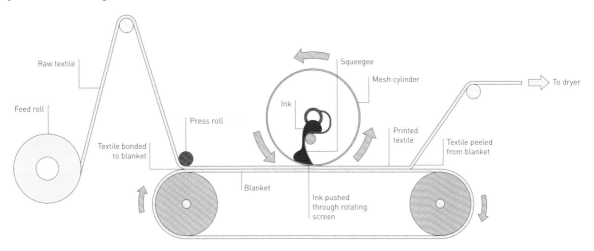

is a flexible means of prototyping and production. Apparel, bags, accessories and packaging are commonly finished in this way. It is also used to print garments made in one piece, as opposed to cut and sewn from flat sheets that have been printed on a roll, such as circular-knitted t-shirts.

Screen printing is used to apply coating (see Fluid Coating, page 226) to designated areas to enhance specific properties, such as waterproofing and breathability.

RELATED TECHNOLOGIES
Digital printing (page 276) has evolved as an alternative to screen printing for a range of materials. It is inexpensive for low volumes because print plates are not required and there is little or no set-up cost. Full-colour graphics are reproduced in a single process, unlike screen printing, which requires a separate, registered screen for each colour. Digital printing inks are limited, because they must be the correct consistency for the inkjet or laser-printing process. Therefore, it is less suitable for experimenting and for developing the tactile and visual quality of a print or graphic, such as by combining different inks and media, compared to screen printing.

Inkjet printing is used to make the films for transfer printing. Alternatively, sublimation inks can be applied directly to the substrate by inkjet printing, cutting out the heat transfer process. Sublimation means that the dye transitions between

TECHNICAL DESCRIPTION
Rotary screen printing is used to reproduce graphics and patterns on continuous lengths of fabric and is capable of printing up to 70 m (230 ft) per minute. The principle is the same as flat-bed: colourant (ink or dye) is squeezed through a stencil onto the surface of the textile, except that in rotary screen, the stencil is wrapped into a cylinder. The screen rotates in time with the textile passing underneath. Ink is fed onto the screen and squeegeed through the stencil in a continuous process.

The textile is temporarily bonded to the blanket, which ensures it stays flat and stable as it is printed. After printing they are peeled apart. The blanket continues back to the start and is washed

and fresh adhesive is applied along the journey. The textile is transferred to drying and taken up on a roll.

Modern rotary screens are made by electroforming the mesh on a mandrel (cylindrical metal mold). This is necessary to make a seamless screen. With electroforming is it possible to make bespoke meshes that are solid where print is not required, cutting out the stencil process. These screens cannot be made with such fine mesh (to reproduce intricate patterns), but the mesh is more stable and so can be used to make wider cylinders for printing wider fabrics.

If more than one colour is required then multiple screens are lined up over the same blanket.

the solid and gaseous state without becoming liquid. In textile production, dye sublimation is used to decorate polyester garments, such as t-shirts, sportswear and interior fabrics. The dye penetrates into the fibres, rather than coating them, to provide excellent colourfastness.

Block printing (page 256) is an ancient manual process used to decorate textiles and paper. Each colour is applied with a separate block. Rotary-block printing (flexography) competes with rotary screen in mass production, but is becoming less common owing to the high set-up costs.

Paper is printed with a variety of techniques such as offset lithography, rotogravure and flexography, because they can be set up quickly and can run at very high speed – up to 14 m (46 ft) per second – in the mass production of newspapers, magazines and wallpaper, for example. Technically it is possible to use these processes to print graphics on textile, such as nonwovens used in place of paper (Tyvek for example), but it is not very common owing to the high volume required to justify the set-up cost.

QUALITY

Quality depends on many factors, principally the textile, colourant and method of application.

The absorbency and smoothness of the textile will determine the viscosity and quantity of colourant required. Each material receives colourant differently and the quality relies on finding a balance in production between speed and finesse.

Colour is applied by pigment, dye or discharge. Pigment inks do not penetrate into the fibre and so must have good coverage of the surface so the background colour is not visible. Bright and fast colours are possible and there tends to be very little show-through of the ink on the back side of the cloth. However, this type of ink is not good for high-wear applications, owing to poor rub-fastness. And the elasticity of the ink will affect the handle and drape of the textile.

The amount of colour applied depends on the fineness of the mesh, the viscosity of the paste and the speed of printing. These in turn affect the quality of colour matching. The base colour of the substrate may affect the final appearance of the printed graphics, depending on the colourant. Water-based pigment inks tend to be more transparent, but recent developments include thicker, more opaque inks. Drying the ink between colours improves the quality of print.

Dyeing depends on the base material because natural and synthetic fibres require different dyeing techniques. Generally, brighter and more durable colours can be achieved on synthetics. Care must be taken to ensure the dyeing system fits the application, because the requirements are different. For example, upholstery must be resistant to heavy wear, whereas t-shirts require excellent colourfastness when washed.

Registration refers to the alignment of the print on the textile. It is especially important when printing multicolour repeating patterns. Generally, colours that sit side by side are printed with a small overlap to reduce the likelihood of an unsightly light-coloured gap. However, depending on the colour, the overlap may result in an undesirable colour effect. When flat-bed printing a continuous web of textile, the motion must stop and start for each colour to be applied. This presents a challenge for achieving very accurate registration, especially with lightweight, high-stretch and delicate textiles. Bonding to an expendable backing cloth improves stability and thus quality of registration.

Flat-bed and rotary screen do not apply high pressure to the surface of the textile, unlike block printing. This means that the surface profile of the textile is less affected, while the print maintains better 'bloom'.

Screen printing produces graphics with clean edges. Heavier-gauge screens will deposit more colourant, but have lower resolution of detail. These tend to be used in the textile industry, which requires copious amounts of colourant, whereas light mesh screens are used to print on paper and other less absorbent materials.

DESIGN

Each colour is applied with a different screen, which will require registration. Depending on the colourant, each colour may have to cure, or dry, between applications, but this can be accomplished in a matter of seconds with UV curing systems. Two colours may be achieved by printing onto a dyed background colour, or ground. When the background colour is printed, it is known as 'blotch'. Printing large areas of colour, such as blotch, becomes tricky because if a screen comes into contact with wet colourant it can leave an imprint.

Continuous repeating patterns are best printed with rotary screen. Since screen making techniques that utilize electroforming have been developed that eliminate the seam in the mesh cylinder, the pattern does not have to try to hide it. Electroformed screens are produced on a mandrel, which means they are made to set dimensions. Repeating patterns must therefore fit into the circumference an equal number of times. Alternatively, longer repeats may be printed with alternate lifting screens (whereby part of the repeat is printed with one screen and another part by an additional screen). Diameter ranges from 50 cm to 100 cm (20–39 in.).

The width of rotary screen is constrained by the stability of the mesh. Wide fabrics require a sturdier screen, which will have a coarser mesh. Maximum width is 220 cm (87 in.).

Rotary screens have uniformly spaced hexagonal holes for maximum strength. Screen fineness is measured as holes per linear inch and ranges from coarse (60) to fine (100). Flat-bed does not have the same constraints. The flat-bed frames can be much larger and be made up of fine mesh.

Mesh size will affect the fineness of pattern that can be reproduced, and the quality of edge finish. Fine meshes are utilized to print intricate designs. However, finer meshes require lower-viscosity colourants. Coarser meshes make it possible to lay down high quantities of viscous paste, but they have lower resolution and there may be a visible stepped pattern along the edge. Mesh size is determined by the number of threads (38–380 imperial and 15–150 metric) and thread diameter: 31, 34 or 40 (small, medium or heavy).

It is possible to print very fine details down to 0.12 mm (0.005 in.).

Flat-bed printing lines typically have space for 15 or more screens. Carousel printing lines used to finish items such as t-shirts and bags have space for six or so screens.

Gradients are produced by reducing the area of colourant to dots, known as 'halftone'. As the concentration of dots for each colour is reduced, the visible colour changes. This technique is used in all printing. The dots will be visible close up. Alternatively, tonal designs are achieved by printing one colour on to another. This is known as 'fall-on'. To ensure reproduction of the highest quality, colour overlapping should be avoided.

Screen printing may be combined with other finishing techniques to achieve a wide range of effects. For example, screen printing adhesive allows flock, gold leaf or glitter to be applied. The finishing material will stick to the coated areas, leaving the uncoated areas bare.

Repeating patterns must be designed with consideration for the process, such as whether it is flat-bed or rotary. In addition,

Front face   Back face

Show-Through
Material: Cotton
Application: Interior
Notes: The amount a dye will show through will depend on the weight and density of the fabric being printed.

Before   After

Dye Activation
Material: Reactive dye on flax
Application: Upholstery
Notes: Reactive dyes are activated by steam and a mild alkali solution. The colour intensifies and fixes to the fibre.

Front   Back

Devoré
Material: Silk and flax
Application: Apparel
Notes: Flax yarn is satin woven onto silk sheer, printed and then selectively dissolved with devorant paste, which is applied by screen printing.

Flat-Bed Mesh
Material: Polyester
Application: Various
Notes: Flat-bed screens are made up of woven filaments and so have square holes.

Rotary Screen Mesh
Material: Nickel
Application: Various
Notes: Electroformed screens are produced on a mandrel. Hexagonal holes are the most stable.

Registration
Material: Cotton
Application: Apparel
Notes: Colours that sit side by side are slightly overlapped to remove unsightly gaps in the design.

Halftone
Material: Ink on paper
Application: Poster
Notes: As the concentration of dots for each colour is reduced, the visible colour changes.

White on Black
Material: Black-dyed cotton
Application: Bag
Notes: Plastisol-based ink is sufficiently opaque to cover the black completely.

Nonwoven
Material: PE
Application: Packaging
Notes: Textured surfaces, such as nonwoven fabrics, are coated well by screen printing.

Foaming Ink
Material: Plastisol-based ink on cotton
Application: Cotton glove
Notes: Foaming inks are used in thin layers to create a soft-touch ink, or built up for added texture.

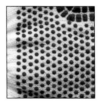

Warp Printing
Material: Silk
Application: Interior
Notes: The motif has a soft edge owing to the unavoidable misalignment of the warp during loom weaving.

there are several techniques employed to overlap the edges of the pattern, used to disguise the transition from one screen to another. Following the edge of a motif, as opposed to finishing along a straight line, is preferred, because it is very difficult to achieve a touch fit between two screens. For the same reason, straight lines that run across more than one screen should be avoided, because any error in registration will be accentuated.

In a process known as warp printing, the yarn is printed and rewoven to produce a graphic with blurry definition (see case study, page 268). The warp is woven into an open textile, printed and unpicked. It is then rewoven at the desired density. It is used mainly for curtains, screens and other fabrics that will be backlit, because the print is through the entire thickness and so eliminates any chance of misalignment (if the fabric is printed separately on both sides it is difficult to align the pattern exactly).

Colourant comes in a wide range of colours and effects including plain, clear, metallic, pearlescent, fluorescent, thermochromatic and foam. The type of colourant that can be used is determined by the performance requirements of the application combined with the material being printed.

MATERIALS
Almost any material can be screen printed, including paper, plastic, rubber, metal, ceramic and glass. Pigment printing inks are water-based, plastisol-based, solvent-based or ultraviolet (UV) curing. The pigment does not have any affinity with the type of fibre, but not all inks are compatible with all textiles.

Plastisol inks are based on polyvinyl chloride (PVC). They are also known as solid ink. They are heated after printing to between 150°C and 165°C (302–329°F) to polymerize and solidify the PVC. They have varying levels of flexibility, determined by the quantity of plastisol, which can cope with stretching fabrics. For multicolour

→ Design for Screen Printing

Design for production is critically important when screen printing. Standfast & Barracks employ a team of pattern and colour designers to ensure that the most suitable print process, design repeat, screen mesh, textile and colourant are selected.

Both handwork and digital techniques are employed. When drawing patterns by hand, a resist is used to ensure a crisp edge (**images 1** and **2**). Drawing by hand gives the final print a softer appearance than if it was drawn from scratch on computer. The designs are finessed and colours separated with computer-aided design (CAD) software (**image 3**).

CAD files are then used to produce sample colourways (**image 4**) by digital printing. These are used to assess the design on the correct fabric. The printers are colour balanced to provide an accurate match to what can be achieved by screen printing in bulk production.

The CAD files are also used to laser engrave the rotary or flat-bed screens (see case study opposite).

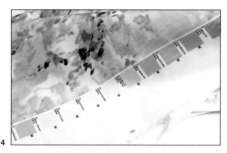

Featured Company

Standfast & Barracks
www.standfast-barracks.com

prints they can be applied wet-on-wet, greatly accelerating production speeds. Very good colour reproduction is achieved on both light and dark colours.

Water-based ink systems are based on evaporation, but may contain co-solvents to help accelerate drying time. As well as being less harmful to the environment, they are the ink of choice when a soft-feeling finish is desirable. In other words, they do not affect the handle (feel) in the same way as plastisol inks. They are also well suited to high-nap fabrics, such as towel, because the ink wicks deep into the fabric.

In the past, the use of water-based inks has been limited to printing on light-coloured textiles. Recent developments include more opaque inks, thicker consistency, intense colours and fluorescents suitable for printing on dark colours. Lighter colours affect the handle least of all.

Similar to water-based inks, solvent-based inks rely on evaporation and are air-dried or heated to accelerate drying time. UV inks contain chemical initiators, which cause polymerization when exposed to UV light. These inks have superior colour and clarity, but are also the most expensive.

Dyes are absorbed into the fibre and compatibility is the same as dyeing (page 240) plain-colour textile. Combined with the physical requirements of the application, the textile determines dye selection: natural fibres are compatible with direct, acid, mordant, reactive and vat dyes; wool and other protein fibres are compatible with acid, basic, direct and reactive dyes; polyester and acrylic are compatible with disperse dyes, and nylon is compatible with acid and direct dyes.

Dyes do not affect the handle in the way that pigment inks do. Therefore, high-cover and blotch designs are typically reproduced with dye rather than with pigment inks.

Discharge inks can be used to simply remove the dyed ground and leave behind the natural colour of the textile, or they can be combined with a dye that is impervious to the discharging agent to add colour to the area that has been discharged. As a result of removing dye

→ Rotary Screen Printing

This is the most widely used mass-production technique for printing textiles. The mesh cylinders come pre-formed and are coated with a blue emulsion. The emulsion is etched away by computer-guided laser to produce a stencil (**image 1**). Each cylinder is very carefully checked to ensure that there are no pinholes or other imperfections that will affect the quality of printing (**image 2**).

The fabrics are carefully prepared by controlled stretching and shrinking (page 210), mercerizing (page 202) and dyeing (page 240) to ensure the highest-quality material and surface finish prior to printing (**image 3**).

The finished cotton fabric is loaded onto an A-frame and fed into the rotary screen printing line (**image 4**). A separate cylinder is required for each colour. The water-based reactive dye is printed onto the surface wet-on-wet (**image 5**). As the fabric progresses down the line the pattern is completed (**image 6**). To fix the dye, the fabric is passed through steam and bicarbonate of soda (mild alkali) solution. This causes the dyes to react and penetrate deep into the fibre for a durable and permanent finish.

The printed fabric is inspected (**image 7**). Then it will be re-tentered to produce a stable, flat fabric suitable for cutting and sewing into apparel.

1

2

3

4

5

6

7

Featured Company

Standfast & Barracks
www.standfast-barracks.com

TECHNICAL DESCRIPTION

A charge of colourant (dye or ink) is deposited onto the screen, and a rubber squeegee is used to spread the ink evenly across it. An alternative system uses a bar, which is rolled across, guided by a magnet underneath. The method is selected according to the application, as the rolling bar tends to lay down more colourant, while more pressure can be applied using a squeegee.

Those areas protected by the impermeable film (stencil) are not printed. The screens are made up of a frame, over which a light mesh is stretched. The mesh is typically made up of nylon, polyester or stainless steel.

The screen is placed slightly above the surface of the textile. This is known as off-contact printing. As the squeegee is drawn across the surface it is pushed downwards, forcing the screen onto the surface of the textile. This has two important functions: the screen pops away from the surface after printing to ensure a crisp edge and does not press onto existing wet colourant on the textile.

Each colour requires a separate screen. A full-scale positive image of each colour is printed onto separate sheets of acetate. The areas to be printed with colourant are black and the areas not to be printed are clear. The full-scale positive is mounted and registered on the screen, which is coated with photosensitive emulsion. This is exposed to UV light, causing it to harden and form an impermeable film. The areas that were not exposed, under the opaque areas, are washed away to produce the stencil.

Roll printing, also known as A-frame, is set up in-line and each colour is applied in sequence. The textile is temporarily glued to a blanket, which transfers it accurately underneath each screen.

Individual items are printed on a carousel, where the bed rotates through six or so printing operations. Drying stations are placed in between if required.

The Flat-Bed Screen Printing Process

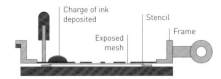

Stage 1: Load

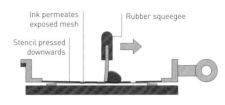

Stage 2: Print

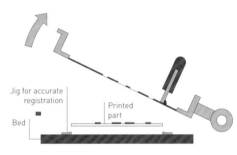

Stage 3: Unload or transfer

it may not always be possible to match colours, or produce bright whites and other light colours.

Burnout is the process of selectively dissolving areas of yarn from a pre-woven fabric. It relies on using two chemically dissimilar yarns, such as protein, cellulose or man-made. After weaving, a chemical solution is screen printed onto the areas of fabric to be dissolved. For example, sodium bisulphate is used to dissolve cellulose-based or viscose fibres, while leaving the protein yarns unaffected. The effect is known as devoré when applied to pile or satin fabrics. The surface is often printed or dyed before selectively dissolving areas away.

COSTS

Pigment printing is the least expensive and most versatile, owing to the simplicity of the procedure and the compatibility of the inks. It is therefore commonly used for low-volume and one-off applications as well as for high-volume production.

Screen-making costs are low, but a separate screen is required for each colour to be printed. Therefore single colour is the least expensive.

Rotary screen is the fastest, but the screen-making costs are higher than flat-bed, so it is most economical for high volumes. Printing speeds can be up to 70 m (76 yd) per minute although it is generally necessary to print at much slower speeds to ensure good-quality reproduction of detail.

Labour costs are relatively low for automated processes and moderate to high for manual techniques.

ENVIRONMENTAL IMPACTS

Printing uses large amounts of water and chemicals. Therefore it is essential to assess all the options available for a given application. With consideration the environmental impacts may be neutralized or even eliminated.

PVC-, formaldehyde (UF)- and solvent-based pigment inks contain harmful chemicals, but they can be reclaimed and recycled to avoid water contamination.

Water-based inks contain less harmful ingredients than petroleum-based equivalents. They can be printed onto paper and metal, and – in the case of textiles – are used to replace plastisol ink, which contains PVC. After printing, the ink is washed from the screen with water and the solids are separated before it enters the sewage system. Whether water-based or PVC-based, inks are chemical compounds that require proper handling and disposal.

Dyes tend to be more problematic than pigment inks, owing to the additional washing off and solvents required. Natural dyes are steadily developing as an alternative in some applications, although they cannot compete across all applications because of the relatively high cost and limited colours.

Screens are recycled to save money and this helps to reduce the environmental impacts. The impermeable film is dissolved away from the mesh so that it can be reused.

→ Flat-Bed Screen Printing

As with rotary screen printing, the fabric must be carefully prepared to ensure a high-quality surface for printing. The roll is loaded onto the print line (**image 1**). Flat-bed lines can accommodate 15 or more screens to make complex and colourful designs (**image 2**).

The fabric is temporarily bonded to the blanket, which transfers it accurately between each print station. With each progression the screen is automatically lowered and the squeegee is drawn across, applying dye through the stencil (**images 3** and **4**). The fabric moves forward between screens and is kept flat and straight on the blanket (**image 5**).

After the final colour has been applied the fabric is peeled from the blanket (**image 6**). The same reactive dye system is used here (see case study on page 265), so the fabric is finished with steam and a mild alkali solution. The printed cotton is inspected prior to re-tentering (**image 7**).

1

2

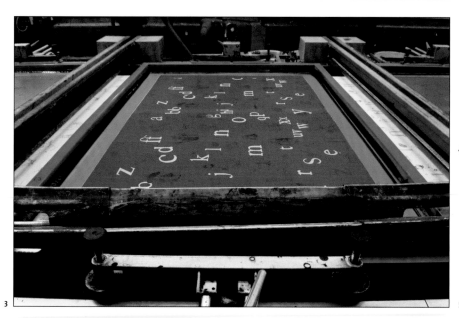

3

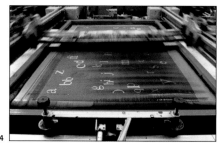

4

5

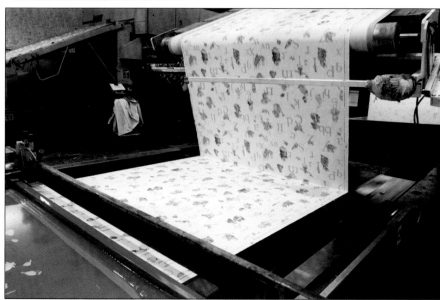

6

7

Featured Company

Standfast & Barracks
www.standfast-barracks.com

→ Warp Printing Silk

Warp printing is a technique used to produce soft-edged motifs. The pattern will be perfectly aligned between front and back, because the fabric is printed only once and the colour penetrates right through the yarns. It involves many steps, which are loom weaving (page 76), screen printing, unpicking and reweaving.

The first weave produces a textile with well-spaced weft yarns (**image 1**). This gives more space for the colourant to penetrate around the warp yarns during screen printing. The fabric is printed and the weft is unpicked to reveal the printed warp (**images 2** and **3**).

The warps are re-beamed and rewoven on a conventional loom (**images 4** and **5**). Plain-woven with a fine and densely packed weft, the printed pattern looks the same from both sides.

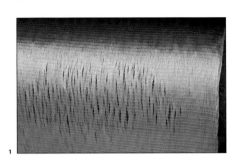

1

2

3

4

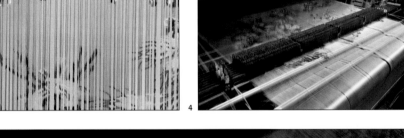

5

Featured Company

Prelle
www.prelle.fr

DECORATING

268

→ Multicolour Flat-Bed Printing a T-Shirt

When printing garments and other readymade items, there is a wide range of base colours available as standard (**image 1**). When applying ink onto a dyed ground, the colour will affect the quality of the print: dark colours require thicker and more opaque inks to achieve the desired colour.

In this case, a plain white knitted t-shirt is used. It is loaded onto the screen-printing bed (**image 2**). It is smoothed flat and held in place by a low-tack adhesive. The registered screen is lowered down in line with the t-shirt and the first colour of the leopard-print design is printed (**image 3**). The colours are printed in sequence and laid down wet-on-wet. After violet, the yellow is printed (**images 4** and **5**) followed by the black outline (**image 6**).

The black outline is applied last, and so covers up any slight misalignment in the registration (**image 7**). This helps to speed up the printing process and reduces waste. The ink is cured and the t-shirt is finished (**image 8**).

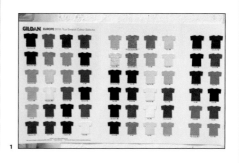

1

2

3

4

6

5

7

8

Featured Company

K2 Screen
www.k2screen.co.uk

→ Discharge Printing a Black T-Shirt

Also known as extract printing or colour discharge printing, discharge printing is an excellent way to produce light-coloured prints on dark-coloured natural textiles (**image 1**). Screen printing is used to apply water-based discharge ink (**images 2, 3** and **4**). The ink (discharge chemical mixed with pink dye) is barely visible on the surface of the black cotton t-shirt (**image 5**). Activated with heat, the ink undergoes a chemical reaction to remove the black dye (**images 6** and **7**) and leave behind a pink-dyed

graphic. This produces light-coloured prints with a softer feel than conventional light-coloured inks, which is known as 'soft hand'.

Discharge printing is utilized for gaming tables because, unlike with conventional pigment inks, there is no thickness, and thus no edge, to the print. It is also used for large areas of print because it does not affect the handle of the textile as plastisol-based ink does.

1

2

3

4

Featured Company

K2 Screen
www.k2screen.co.uk

5

6

7

DECORATING

270

→ Preparing a Screen

The screens are made up of a frame over which a light mesh is stretched. The mesh is typically made up of nylon, polyester or stainless steel. The type and density of mesh depends on the colourant, design and level of detail sought.

Screens can be used to reproduce several thousand prints over their lifetime, or they can be reused many times for shorter print runs (**image 1**). The ink and stencil are carefully dissolved and washed away to leave the bare mesh, which is recoated with emulsion (**image 2**).

Once the emulsion is dry the screen is loaded onto a vacuum table along with the negative (**images 3** and **4**). The design is printed in opaque black onto transparent acetate. A separate negative is produced for each colour and screen. It works by blocking the selected parts of the screen from exposure to UV light (**image 5**).

The areas of the emulsion exposed to UV light harden. Unexposed areas are washed away with water (**image 6**). This creates the durable and precise stencil screen ready for printing (**image 7**).

1

2

3

4

Featured Company

K2 Screen
www.k2screen.co.uk

5

6

7

Textile Technology
Transfer Printing

Also known as sublimation printing, this process involves transferring dyes from printed transfer paper onto textile. It is carried out as a continuous web-fed process, or on a flat press. Paper is less expensive and more straightforward to print than textile. This creates opportunities for design and makes transfer printing suitable for low-volume to mass-production applications.

Techniques	Materials	Costs
• Sublimation	• Polyester • Disperse dyes	• Low

Applications	Quality	Related Technologies
• Apparel • Interiors • Upholstery	• Bright colour • Accurate registration	• Digital printing • Loom weaving • Screen printing

INTRODUCTION

Printing onto an expendable material and later transferring the dye onto textile has many benefits. Transfer paper is more straightforward to print onto, so very high-quality images and bright colours can be produced at relatively low cost. Paper is less expensive and easier to store than textile. Therefore, a large quantity may be printed in a single operation but only used in small batches, as the printed textile is required. The transfer process itself is compact and low-cost compared to wet-printing presses.

Sublimation transfer printing is based on the principle that at elevated temperatures disperse dye will pass from solid to gas and condense in the substrate material (polyester) without becoming liquid. With heat and pressure the print is reproduced on the polyester with clear defined edges.

APPLICATIONS

Transfer printing is used for a wide range of applications, including sportswear, nightwear and other garments, footwear, lining, banners, signage, gaming tables, upholstery, curtains, drapes and floor coverings. As well as printing fabrics, transfer printing is utilized to print onto polyester-coated products, such as high-end packaging and sports equipment.

RELATED TECHNOLOGIES

The transfer method is limited to printing mainly on polyester, although other materials are susceptible to sublimation dyes, but in varying amounts. By contrast, block printing (page 256), screen printing (page 260) and digital printing (page 276) are compatible with a wide range of materials, inks and dyes.

A significant advantage of heat transfer is that it can be used to print

The Transfer Printing Process

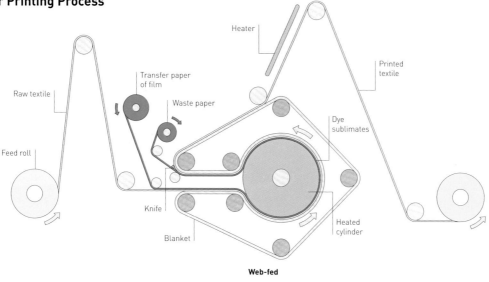

Web-fed

Stage 1: Load   Stage 2: Heat transfer

Flat bed

on three-dimensional surfaces, including flexible substrates and rigid objects.

QUALITY

The quality of the image will be determined by a combination of printing technique and textile substrate. The colour will appear bright and lustrous, depending on the substrate. It is absorbed into the structure of the textile and so cannot be worn away like surface coatings with pigment inks.

The temperature of sublimation affects the durability of the dye in application. Higher-temperature sublimation dyes tend to be more durable, but are harder to process, whereas low-temperature dyes tend to have poor colourfastness and wash durability.

Since transfer printing is used for synthetic fibres, temperature has to be carefully controlled, because if the material starts to soften the surface will become glazed and the fibres may bond together, affecting the handle.

TECHNICAL DESCRIPTION

Sublimation dye goes from solid (printed transfer) to gas (with heat) and condenses in the substrate without becoming liquid. Vapour is more rapidly taken up by the substrate than liquid dye, making this a clean and efficient process.

Transfer printing is carried out as a web-fed process or a flat-bed operation. In web-fed printing, the raw textile, which has been washed (page 202) and tentered (page 210), is fed in between the blanket and transfer paper. The image side of the paper faces the textile.

The textile and paper are held firmly together, ensuring there is no slippage, and travel around the rotating heated cylinder. The temperature must be maintained very accurately, because too low and the dye will not sublimate; too high and the substrate will start to soften.

Held under pressure between the blanket and the heated cylinder (205°C/401°F), the dye converts into gas at 200°C (392°F) and sublimates from the paper into the textile.

A knife separates the paper and textile. The waste paper is taken up on a roll. The textile progresses into the heating chamber, where the disperse dye is set at 200°C (392°F), becoming fully colour fast by the time the temperature has dropped to below 180°C (356°F). After inspection it is taken up on a roll.

Flat-bed printing is used to transfer print onto textiles and rigid materials. The shape does not have to be flat; it can be curved to fit a three-dimensional profile. The polyester substrate is either textile or rigid material coated with a suitable polymer.

The image on the transfer paper or film is reversed, because it will be mirrored during the transfer process to be 'right reading' once printed. In stage 1, the transfer medium and textile are loaded into the press. In stage 2, the press is closed and pressure (or vacuum) is maintained for two to five minutes. Heat and pressure cause the sublimation dye to transfer into the textile or coating.

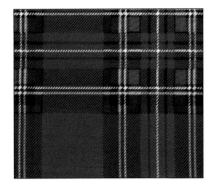

Colour
Material: Polyester
Application: Interior
Notes: Disperse dyes produce bright surface colour. The colours are built up on the paper transfer one by one, and applied to the polyester substrate in a single step.

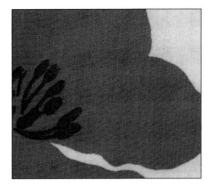

Edge
Material: Polyester
Application: Interior
Notes: The high pressure during printing ensures a crisp edge and clean transitions between colours.

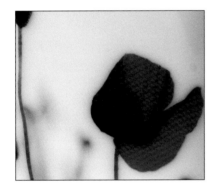

Packaging
Material: Coated glass
Application: Packaging
Notes: Transfer printing can be applied to virtually any material, as long as it can be coated with a suitable plastic.

Gingham
Material: Polyester
Application: Interior
Notes: Large areas of colour are applied; this is useful for designs that cover the entire width.

DESIGN

Print design opportunities are governed by the processes used to create the transfer paper. All of the commercial paper-based printing techniques may be used, depending on the volume. These include flexography, rotogravure, offset lithography and screen printing. All of them are web-fed (used to print reels of paper); offset lithography and screen printing are also used to print sheets. All may be used to print directly onto certain textiles, in particular nonwovens. Screen printing is mainly used to print onto textiles, rather than transfer printing.

Offset lithography is used in a wide range of applications. It is often referred to simply as offset litho and is capable of very high-quality printing. It relies on the basic principle that oil and water do not mix. The non-image areas of the printing plate absorb water, whereas the image areas repel water. The oily ink therefore sticks to the image areas and is transferred to the paper via a blanket cylinder. This indirect method of printing is known as offset. It is mainly used for high-volume production, because the set-up costs are high.

Flexography, also known as flexo, is a relief-printing technique that works on a similar principle to block printing. Ink is transferred from a relief-printing plate onto the transfer paper. The printing plates are less expensive than for offset lithography, making it more cost-effective for lower volumes. Modern presses operate at 100 m (109 yd) per minute.

Rotogravure, also known as gravure, is an intaglio process. The print cylinder is engraved with tiny recesses, called cells. Ink is picked up in the cells and excess is removed by a doctor blade. The ink left in the cells is transferred to the paper as it is pressed against the cylinder. It is capable of producing continuous-tone images. However, image reproduction is not as sharp as offset lithography and flexography, because the ink dots blend together. The cylinders are expensive and the process is capable of operating at 800 m (875 yd) per minute. So it is typically reserved for high-volume production.

Offset litho, flexo and gravure are all based on the four-colour process: cyan, magenta, yellow and key (black), collectively known as CMYK. Colours that cannot be produced by mixing CMYK during printing may be introduced as a spot colour, increasing cost.

Screen printing is not based on CMYK; solid colours are mixed according to a design's requirements. It is a versatile process with relatively low set-up costs. The amount of ink deposited depends on the size of mesh. Heavier gauges will deposit more ink, suitable for absorbent textiles, but have lower resolution of detail. Rotary screen paper printing runs at 80 m (87 yd) a minute, which is quicker than printing direct to textile.

As well as these bulk printing processes, digital printing is utilized in the production of transfer paper. There are no set-up costs, so one-offs and proofs are often done in this way. Digital printing can produce very good image quality.

In other industries transfer printing is used to print onto three-dimensional surfaces, such as protective covers for mobile phones and sports equipment. The same principle is applied to printing seamless graphics onto knitted t-shirts and other three-dimensional items.

MATERIALS

Polyester fabric is used in isolation, or blended with other materials, such as cotton. Around 70% synthetic content

→ Printing Faux Tartan onto Polyester

The tartan pattern has been screen printed onto a web of lightweight paper. It is loaded onto the front of the machine face down in contact with the mid-weight woven polyester (**image 1**). The disperse dye appears dull on the paper transfer. Its colour gets brighter and saturated when activated by heat.

The two materials are pressed together against the blanket and drawn through the heating process at 205°C (401°F). In less than a minute the dye has sublimated into the polyester and the two materials are separated

by a blade (**image 2**). It is critical that they are split cleanly to avoid disturbing the fresh dye.

The footprint of the transfer printing press is small (**image 3**). The fabric is drawn through a final heating step to set the disperse dye (**image 4**). The print appears bright and lustrous on the surface of the polyester compared to the transfer paper (**image 5**).

Waste transfer paper is collected for recycling (**image 6**).

1

2

3

4

5

6

Featured Company

Sharps
www.sharpsfabricprinters.co.uk

---

is required to maintain quality and colour fastness. The weave will affect the proportion of polyester on the surface and so affect dye pick-up. New fabrics and derivatives of polyester are tested to check print quality.

Woven (pages 76–105), knitted (pages 126–51) and nonwoven (page 152) fabrics from openwork to sheer and stretch to rigid and laminated (page 188) are compatible.

The transfer paper varies according to the application. Lighter-weight papers are less expensive but only suitable for web printing, whereas heavier papers are printed with sheet processes, including offset lithography and screen printing.

COSTS

Transfer printing is relatively low cost for high-volume production, because printing onto paper is both rapid and economical, as is the transfer process. Once the paper is printed, small orders of printed textile are processed on demand.

Web printing operates at up to 35 m (38 yd) per minute.

ENVIRONMENTAL IMPACTS

Printing onto paper generally requires less water and fewer chemicals than textiles, because the whole process is more efficient. The transfer process itself does not require any additional chemicals or water. However, using an expendable

material means there is waste at the end of the process. Recycling the paper minimizes the environmental impacts.

Water-based inks are non-toxic and can reduce harmful emissions, such as volatile organic compounds (VOCs). However, there are many different types so care should be taken to assess the full list of ingredients.

For efficiency, relatively large quantities of paper are printed and stored. Only the required amount of textile is printed, reducing the unnecessary consumption of relatively expensive textile. However, a drawback of preparing the paper beforehand is that late changes to design, colour and material become impractical.

Textile Technology
Digital Printing

Initially developed for proofing, prototyping and one-offs, this is the fastest-growing area of printing used in full-scale production. It is less expensive to set up than conventional printing processes, because the design is transferred directly to the fabric without the need for plates, stencils or blocks. The quality and speed depends on the technology.

Techniques	Materials	Costs
• Thermal (inkjet)	• All types of sheet material	• Reduced set-up costs
• Piezoelectric	• Dyes depend on textile	• Moderate unit costs

Applications	Quality	Related Technologies
• Apparel	• Very high definition of detail possible	• Screen printing
• Interiors	• Continuous-tone images	• Transfer printing
• Functional coatings		

INTRODUCTION

With these techniques, digital files are outputted directly to the printer. Precise dots of colour are sprayed onto the textile at high speed. The colour gamut (range of printable colour) depends on the system. Digital printing is primarily based on cyan, magenta, yellow and key – black (CMYK), a legacy from the paper-printing industry. Additional ('spot') colours are used to enrich the reds, greens and blues.

With textile-printing machines, it is not uncommon to have 12 colours or more in the printhead, including grey, green and gold, among others. By mixing the colours, continuous-tone photorealistic images can be reproduced exactly.

The Inkjet Printing Process

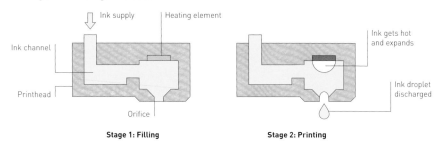

Stage 1: Filling   Stage 2: Printing

Thermal inkjet printing

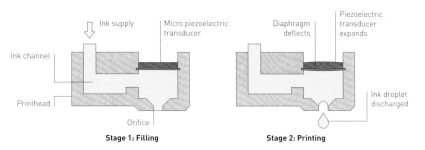

Stage 1: Filling   Stage 2: Printing

Piezoelectric inkjet printing

TECHNICAL DESCRIPTION
These techniques are also referred to as drop on demand (DOD). This describes how only the required amount of each colour is deposited at any point. Ink is squeezed through tiny orifices around 10 microns in diameter directly onto the substrate, which is 1 mm (0.04 in.) below.

Pressure is applied to the ink by either thermal or piezoelectric means. In thermal inkjet printing, the heater causes the ink in contact with it to expand rapidly, forcing the ink in the orifice onto the textile below. It cools and the cycle is repeated many times per second.

Piezoelectric materials deform when an electric field is applied. This forces the diaphragm downward and subsequently an ink droplet is discharged from the orifice.

Both processes form roughly circular dots of ink on the surface of the textile. By combining different coloured dots virtually any colour can be reproduced. After printing the dyes are fixed with steam or heating, depending on the dyes used (see Screen Printing, page 260).

The major advantage of digital printing is that set-up costs are reduced. However, with textile printing it is impractical to completely remove set-up costs, owing to the need for colour proofing and ensuring consistency between batches of fabric – natural fibre-based fabrics will inevitably have some variability.

APPLICATIONS
Digital printing is widely used to proof designs in preparation for mass production. This involves printing samples, known as strike-offs, which are used to check colour, quality and design. It is also used for prototyping, such as producing small batches that will be used to launch a new collection or demonstrate an idea at a show.

It is used as a standalone production technique for low to medium volumes in the fashion and furnishing industries. Digital printing requires investment, but once set up it is by far the most cost-effective method of printing short production runs in all manner of applications. In a process known as direct to garment (DTG), digital printing is used to reproduce designs on knitted t-shirts and woven bags.

It is also particularly well suited to producing one-off large-scale items including carpet, wall coverings, banners (such as for advertising) and covers (such as temporary building façades and lorry sidings).

Functional materials, in the form of nano- or microparticles, are applied by digital printing. Exact doses of reactive, reflective and conductive materials are deposited onto fabrics.

RELATED TECHNOLOGIES
Digital printing is rapidly becoming the dominant manufacturing technology for low- to medium-volume production. Rotary screen printing (page 265) continues to be the technique of choice for high-volume production, while flat-bed screen printing (page 267) provides the highest-quality reproduction – wet-on-dry printing produces colour with the sharpest definition of detail (see also Hand Block Printing, page 256) – for high-end and expensive fabrics.

Screen printing is capable of applying higher-performance inks. Those developed for digital printing have to be able to pass through the precise and very small orifice in the printhead. Nor do the fabrics need to be as smooth and uniform when using screen printing.

Digital printing has provided designers and manufacturers alike with the ability to try out designs very quickly, thus providing a faster way to respond to changing trends and customer demands. Transfer printing (page 272) may be combined with digital printing to take advantage of both systems, because transfer paper can typically be printed more quickly than fabric. Reducing the resolution further increases output. The range of materials is determined by the transfer process.

QUALITY
Layers of dots of dye are built up gradually to achieve very high-quality prints. It is not uncommon for each swathe of the image to take six or more passes of the printhead. Resolution is typically between 600 to 1200 dots per inch (DPI) and is determined by the requirements of the application.

The quality and colour of the textile being printed will affect the printed

Colour

Material: Linen (flax)
Application: Furnishing
Notes: The entire surface of the fabric is covered with dye, except the white areas, which remain unprinted.

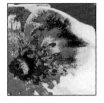

Warp Printing

Material: Silk
Application: Interior
Notes: Similar to screen printing, warps may be printed before weaving, producing a unique appearance.

Mesh

Material: Polyester
Application: Advertising on buses
Notes: Meshes and other openwork structures are suitable for digital printing.

Pile

Material: Cotton velvet
Application: Interior
Notes: Fancy fabrics are suitable for digital printing as long as the surface is uniform and they can be held flat.

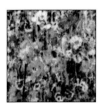

colour. For example, very bright and lustrous colour can be achieved on woven silk, whereas sheer fabrics will appear washed out. Bright white is preferable as a base colour, prepared by bleaching (page 202) and coating (page 226), but any colour can be printed on. A test print is carried out to assess the edge definition and quality of colour reproduction.

There is no contact between the nozzle and the surface of the material, unlike in screen and transfer printing. High-quality prints may be reproduced on delicate, napped (page 216) or pile (page 96) fabrics without changing the surface profile.

The quality of colour matching between the computer-aided design (CAD) file and printed fabric depends on the calibration of the equipment, compatibility of the dye system and preparation of the fabric.

DESIGN

Digital printing is extremely versatile and is capable of reproducing very fine details up to large areas of colour. Designs are created with digital software. Colours should be split into layers to make changes quick and easy. Patterns can be designed as a repeat, similar to screen printing, or alternatively the design can fill the entire width and length of fabric. The size of repeat is governed by the size of the CAD file and the processing capability of the printer, rather than by

mechanical means as is the case with other printing technologies.

It is important to consider the colour gamut of the printer. A greater number and diversity of colours in addition to CMYK will provide the broadest colour range. Care must also be taken to ensure that the quality of colour seen on screen can be achieved with printing colour. This is because computer displays are based on adding red, green and blue (RGB) light. The quality of colour that can be achieved is different in these two colour models.

Digital printing is used to mimic other printing techniques, such as block and screen. It is capable of producing continuous-tone images on virtually any flat surface. Artworks are even printed onto canvas to mimic paintings and reproduce photographs.

Set-up and changeover is less between prints than with other printing technologies. Therefore, digital printing is cost-effective for very small runs down to 1 m (or 1 yd) or less and batch production, through to the mass customization of standard items. This flexibility has proved very useful in the fashion industry, where it is not uncommon to produce short runs of the same design in several colours.

MATERIALS

Virtually any type of textile and sheet can be printed. Other than paper, the most commonly used are woven (pages 76–105) and nonwoven (page 152) types, because

they are the most stable and uniform. The fabric is stuck down during printing, but if prone to edge curling, like knitted fabric, this can cause problems.

For textile printing the dyes are selected according to the base material. Reactive dyes are compatible with natural fibres, disperse dyes are compatible with polyester, and acid dyes are used to print wool, silk and nylon.

Pigment inks are used to print onto paper, wood and several other types of sheet material.

COSTS

Digital printing has less downtime, which reduces the cost compared to screen printing for short production runs. For example, screen-printing lines may be utilized around 50% of the time owing to set-up, whereas digital print equipment can be 90% or more efficient.

Some machines are capable of printing up to 200 m^2 (240 yd^2) per hour. Adding more jets to the printhead increases potential output, but will increase the cost of the equipment significantly.

The dyes are expensive, as they have to be very high quality and consistent.

ENVIRONMENTAL IMPACTS

Applying only the required amount of ink is energy-efficient, reduces waste and means these are low-impact processes. Another significant advantage of digital printing is that for low-volume production, only the required amount of fabric is printed, thus reducing waste compared to processes that require run-in and run-over (excess printing at start and end of run) to achieve the desired quality.

Much less water is consumed than in screen printing, because there are no screens to wash. The transfer paper used in combined digital transfer printing is expendable and used only once. Recycling reduces environmental impacts.

The humidity and temperature is carefully controlled in the print room. As with other printing methods, this contributes to the energy consumption. The base material, preparation and dye system should be carefully considered, because they contribute significantly to the total environmental impact.

Case Study

→ ## Inkjet Printing Linen Fabric

A roll of plain-woven linen (flax) is prepared by bleaching, washing and mercerizing (page 202). It is critical that the material is uniform and stable prior to printing.

Directly before printing it is finished with a specially formulated coating to ensure the best-quality surface on which to lay down the dye.

The fabric is still damp from coating when it is bonded to the blanket by the impression roll (**image 1**). The blanket is coated with an adhesive that becomes tacky when it is warmed up. So just before the fabric is pressed on the surface it is heated.

The surface must be as flat as possible. Any undulations in the surface stop the machine (**image 2**), because they could damage the printhead.

The CAD file is transferred to the printer via raster image processor (RIP) software that converts the graphics into code compatible with the printer software.

The printhead is made up of several sets of nozzles. With each pass strips of dye are built up in layers (**image 3**). The dye is carefully filtered and the air removed before being transported to the printer nozzle (**image 4**). Bright spot colours, such as orange, are added to the standard CMYK.

The printer is fully computer controlled. An operator keeps an eye on the process to ensure everything runs smoothly (**image 5**).

The printed fabric is peeled from the blanket under tension (**image 6**) and loosely folded, rather than taken up on a roll, ready for finishing (**image 7**).

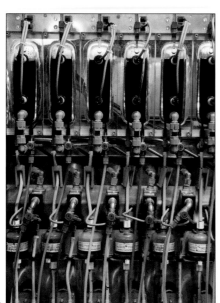

Featured Company

Turnbull Prints
www.turnbull-design.com

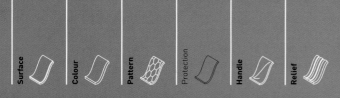

Textile Technology

Flocking

Uniform and densely packed fibres are applied to the surface of fabric, paper, board and three-dimensional surfaces by electrostatic flocking. Combined with masking or screen printing, flocking is used to create patterned and multicoloured pile finishes with a velvety appearance. A range of fibres is used, the most common being nylon and cotton.

Techniques	Materials	Costs
• Electrostatic • Mechanical • Patterned	• All materials may be used as the substrate • Natural and synthetic flock fibres	• Low to moderate, depending on the complexity

Applications	Quality	Related Technologies
• Apparel • Interiors • Accessories	• Uniform and dense • Soft finish • Vivid colours	• Napping • Pile weaving • Screen printing

INTRODUCTION

Originally used to decorate interiors, flocking was not adopted as a finishing technique for a broader range of materials and applications until the development of modern adhesive systems in the 1970s.

Nowadays, this versatile technique is used to produce pile on all types of material, from knitted t-shirts to molded leather. It is used to imitate luxury material, such as velvet, as well as to create colourful, decorative and functional finishes that cannot be achieved in any other way.

The flock fibres are adhesive bonded to the surface of materials in one of

The Electrostatic Flocking Process

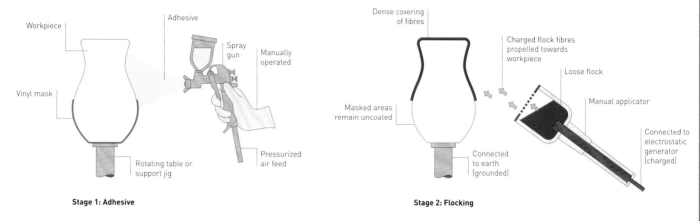

Stage 1: Adhesive

Workpiece
Adhesive
Spray gun
Manually operated
Vinyl mask
Rotating table or support jig
Pressurized air feed

Stage 2: Flocking

Dense covering of fibres
Charged flock fibres propelled towards workpiece
Loose flock
Masked areas remain uncoated
Manual applicator
Connected to earth (grounded)
Connected to electrostatic generator (charged)

Manual flocking

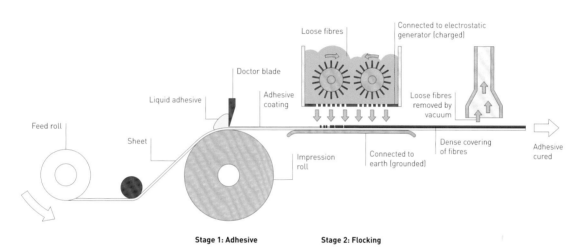

Stage 1: Adhesive   Stage 2: Flocking

Doctor blade
Liquid adhesive
Loose fibres
Connected to electrostatic generator (charged)
Adhesive coating
Feed roll
Sheet
Loose fibres removed by vacuum
Impression roll
Connected to earth (grounded)
Dense covering of fibres
Adhesive cured

Automated flocking

TECHNICAL DESCRIPTION

Flocking is either manual or automated. Both techniques may be used to produce patterns and multicolour finishes. The process is carried out at close to 20°C (68°F) and 50 to 60% humidity. Environmental conditions are important, because higher moisture content causes the fibres to stick together, and lower moisture content means that the fibres will not accept the electrostatic charge.

Manual flocking is more versatile and the set-up costs are minimal. In stage 1, a vinyl mask is applied. The process is the same for flat and three-dimensional items. Of course, it is easier to apply the mask accurately to flat sheets. Alternatively, selected areas of the workpiece are coated with adhesive by screen printing (page 260). The adhesive is conductive and built up to a sufficient thickness to support the fibres once they are applied. The fibres are coated to increase their conductivity.

In stage 2, the flock is applied with a manual applicator. The item to be coated is grounded and the fibres are charged with high voltage (40,000–80,000 volts depending on the application). This difference in electrical potential draws the fibres towards the surface of the workpiece, where they penetrate the layer of adhesive and stand perpendicular to the surface. Automated flocking is continuous and so is much more cost-effective for high-volume production. However, as a result of the complexity of set-up it is far less versatile and large minimum orders are usually required.

In stage 1, the sheet is coated with a layer of adhesive. When applying a single colour across the entire surface, direct coating is used (page 227). Patterns are created by screen printing (flat-bed or rotary) the adhesive.

In stage 2, the loose fibres are dispensed from an overhead hopper. They are positively charged as they pass through the metal mesh and are attracted to the grounded sheet by the difference in potential. They penetrate the layer of adhesive and as the surface is gradually covered the electrically charged fibres are drawn to the areas that are at earth, which ensures that a dense and even coating is built up. Next, loose fibres are removed by vacuum and the adhesive is cured. Each square metre (10.8 ft²) of the flocked surface consists of tens of millions of individual fibres.

two ways. Electrostatic flocking charges the fibres with high voltage while the adhesive-coated substrate is grounded. The potential difference draws the fibres to the surface and causes them to stand upright, thus creating a pile finish.

Alternatively, the fibres are applied by gravity, in what is termed mechanical flocking. With this technique the flocked substrate is vibrated to drive the fibres into the adhesive coating. However, this does not produce such durable flock and is seldom used.

A range of finishes and effects can be achieved, such as combining different colours, lengths or types of flock. Graphic patterns are applied by masking the adhesive during application.

APPLICATIONS
Flocking is used for both decorative and functional reasons. Applied to woven (pages 76–105) and knitted (pages 126–51) textiles, leather and other flexible sheet materials, it is used to provide soft pile finishes for apparel (logos on t-shirts and sportswear), accessories (bags and jewelry), interiors (flocked wallpaper), inflatables (cushions), toys (teddy bears), stationery (relief motifs) and packaging (particularly cosmetics).

Flocked surfaces are not very hard-wearing, because only the end of each pile fibre is secured by adhesive, so care should be taken when specifying this finish for potentially high-wear applications. Even so, flocking is used to finish fabrics for upholstery (page 404) and floor covering, such as for schools and hospitals, because the finish is low-cost, relatively hygienic and easy to keep clean. Rigid items, such as hats and shoes, are flocked with a soft and luxurious surface.

Functional uses include automotive applications (for example sound insulation in glove compartments and interior linings) and military parts (for its anti-reflective properties).

RELATED TECHNOLOGIES
The cut-pile finish is similar to what can be achieved with weaving (see Pile Weaving, page 96), knitting and napping (page 216). Flocking is more cost-effective and so is used as an alternative in many applications, such as carpets, upholstery and apparel. However, its versatility makes it suitable for a much broader range of products than these relatively restricted fabrication techniques. For example, it may be applied to plastic film that is welded to make inflatable cushions; injection-molded panels that line the interior of a car; or cuddly toys.

Relief decoration may also be achieved by screen printing foaming ink (page 263), appliqué (see Embroidery, page 298) or embossing (page 220).

QUALITY
The densely packed fibres produce rich, uniform and matt coloured pile. Synthetic flocked surfaces are quite durable and damage may be readily repaired. Thicker fibres tend to be more durable and resistant to abrasion, but this depends on material type. Intensity of colour, light-fastness and other similar qualities are also dependent on material.

The softness varies according to the type and length of fibre. Combined with colour and pattern, this is used to reproduce life-like animal fur.

Flocking is applied to reduce friction, such as on buffing and polishing surfaces. It is used for sound and thermal insulation purposes. Standing on end, the fibres increase surface area, a quality used to promote liquid retention or evaporation, for example. Also, the fibres may be treated with a range of finishes, before or after application, to provide specific added functional properties.

DESIGN
Flocking is suitable for flat and three-dimensional surfaces. In both cases, masks may be used to create patterns and graphics. These are applied either by screen printing (page 260) the adhesive, or applying the adhesive over a vinyl mask (see CNC Cutting, page 316).

Both techniques are quite versatile. However, there are limits to the extent of the shape. Whereas screen printing is suitable for curved surfaces (one direction) and cylindrical items, vinyl masks may be applied to more challenging three-dimensional forms. The mesh size affects the fineness of pattern that can be reproduced when screen printing. Adhesives are viscous and a thickness of 100 to 250 microns is required to provide sufficient holding strength. To achieve this a relatively coarse mesh of 20 to 40 threads per cm (0.4 in.) is required. Therefore, fine lines and details are best avoided. Different restrictions apply when using vinyl masking and a new one is required for each flocking cycle, making this method more expensive for high volumes.

Several different techniques are utilized to create multicolour patterns. Either the colours are applied on top of one another using masking, the fibres are applied to selective areas, flocked designs are transferred whole, or the pattern is printed (by screen printing or digital printing, page 276). Each method has its merits and drawbacks. Applying multiple coloured fibres, one on top of the other, is the most expensive, because an entire flocking cycle is required for each colour. Masking during flocking is the most complex to set up but the most cost-effective in production, and very good-quality results can be achieved. Printing adds a step in the production cycle, but is capable of reproducing all types of pattern (with little or no difference in price between designs) and with saturated colour (depending on the fibre).

Three-dimensional effects are produced by embossing or shearing the flocked surface once the adhesive has fully cured.

MATERIALS
Virtually any type of monofilament can be used ranging from 0.2 to 10 mm (0.0079–0.4 in.) in length. Fibre weight ranges from very soft and fine 3.3 dtex to 22 dtex for high-wear applications. Nylon is the most common flocking fibre; it is durable and used for functional as well as decorative applications. Other commonly used man-made fibres include polyester (used for many of the same applications as nylon, including automotive interiors), acrylic (decorative packaging) and viscose (wallpaper). The most frequently used natural fibre is cotton, which produces a soft suede-like finish.

Colour
Material: Nylon flock
Application: Various
Notes: Nylon fibres are bright and durable. Thomas & Vines holds more than 600 colours in stock.

Precision-Cut
Material: Nylon flock
Application: Various
Notes: Fibres are cut to an exact length and so produce a uniform and densely packed pile.

3D Colour
Material: Nylon flock
Application: Interior
Notes: Using electrostatic manual flocking any shape may be coated with uniform colour.

Insulation
Material: Nylon flock
Application: Automotive
Notes: Used in the automotive industry for functional reasons such as to reduce vibration, noise and glare.

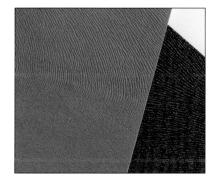

Texture
Material: Viscose flock
Application: Packaging
Notes: Soft textured surfaces are reproduced with a combination of patterned flocking and short fibres.

Vinyl Masking
Material: PVC film
Application: Various
Notes: PVC film is CNC cut to produce single-use masks, which are used to apply adhesive to selected areas.

2D Pattern
Material: Nylon flock
Application: Stationery
Notes: The fibres stick only to the adhesive-coated areas of the material (see Vinyl Masking above).

3D Pattern
Material: Nylon flock
Application: Various
Notes: Patterns are reproduced on 3D surfaces using vinyl masking or by screen printing the adhesive.

The fibres are cut to length during production. Whereas random-cut fibres vary in length, precision-cut fibres produce a uniform pile. Deeper pile is typically produced from precision-cut fibres and random-cut types are used to make very short pile, such as found on decorative packaging and toys.

Fibre cross-section will affect the colour and finish. Trilobal yarn, created from synthetic types such as nylon and polyester, produces brighter and punchier colour, because the triangular shape reflects more light.

All materials can be flocked onto, depending on the application. This includes textiles, leather, plastic and paper. The adhesive system, which may be one- or two-part, depends on the substrate.

COSTS
Flocking is a relatively low-cost method of forming cut pile, suitable for one-offs to mass production. Even so, it will add significantly to the cost of the base material. In other words, a sheet of paper or leather hide will be several times more expensive when flocked.

There are no set-up costs, unless a pattern is being applied. Even then, creating masks does not have to be expensive. Labour costs are moderate for low volumes, owing to the high level of skill required to ensure a quality finish. High-volume production is automated.

ENVIRONMENTAL IMPACTS
Chemicals are used and dust is produced during flocking, so suitable facemasks and breathing equipment are required (see also Adhesive Bonding, page 370). Flock materials are unsuitable for recycling.

The surface finish can last for many years, depending on application, and it is possible to retouch damaged coatings, thus extending their useful life.

→ Flocking on Leather

Thomas & Vines, based in London, UK, is a unique and innovative company. Its work spans numerous technical applications, from flocking dashboards to eliminate reflections for Subaru Prodrive and Mitsubishi Ralliart rally cars, to decorative items for fashion and textiles.

This case study illustrates flocking goatskin with brightly coloured animal pattern. The skin is tanned (page 158) and dyed. No other preparation is required prior to flocking (**image 1**). The pattern is created by screen printing the adhesive. First of all, the adhesive is mixed with colour (**image 2**), similar to the final colour of the flock fibres. It is applied through the screen so that only selected areas of the skin are coated (**image 3**).

The loose fibres are loaded into the applicator. They are negatively charged and so propelled towards the surface of the electrically grounded workpiece as they leave the applicator (**image 4**).

A dense covering of fibres is achieved within about a minute. Excess fibres are removed by blowing air across the surface (**image 5**). The pattern is faithfully reproduced on the surface and the flock is set aside for 24 hours or so to allow the adhesive to fully cure (**image 6**).

1

2

3

4

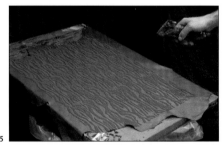

5

6

Featured Company

Thomas & Vines
www.flocking.co.uk

→ Flocking a Plastic Stool

The composite-laminated stool is flocked with blue trilobal nylon fibres (**image 1**). Unlike conventional fibres that have a round profile, trilobal fibres are triangular and so shimmer in the light, resulting in brighter and more intense colour.

The surface is prepared by spray coating (**image 2**) with a suitable adhesive, which in this case is polyurethane (PU).

The nylon fibres are loaded into the applicator (**image 3**). The fibres produce a uniform coating, even on the underside, because they are drawn to the electrically grounded stool (**image 4**). Excess fibres are gently blown away and the finished stool is set aside to allow the adhesive to cure (**image 5**).

1

2

3

Featured Company

Thomas & Vines
www.flocking.co.uk

4

5

Textile Technology
Tufting

Pile is formed through a base fabric by hand, or with hundreds of needles on an industrial tufting machine. Hand techniques are extremely versatile and capable of producing intricate patterns and three-dimensional effects. Machine tufting is a cost-effective method for producing pile fabrics for applications that range from carpet to artificial grass.

Techniques	Materials	Costs
• Pile, cut pile, sculpted • Machine tufting or gun tufting • Hand knotted	• Cotton, jute or synthetic base fabric • Natural or synthetic pile yarns	• Low for machine-made • Moderate to high for gun tufting • High for hand knotting

Applications	Quality	Related Technologies
• Carpets and rugs • Linings • Fake fur	• Variable height of pile possible • Variable density	• Nonwoven • Pile weaving

INTRODUCTION

Hand knotting is an ancient technique used to make pile fabrics for decorative and functional purposes. It is laborious, with individual yarns being inserted into the base fabric and cut off at the correct length. Nowadays, a tufting gun enables practical weavers to make up to 10 m² (108 ft²) per day. Only time and money limit the number of colours and different types of pile that may be combined in a single piece.

Machine tufting was developed as a cost-effective alternative to hand techniques and is now used to produce around 90% of carpets worldwide. It consists of long beds of needles, typically 5 m (16.4 ft) wide, which punch yarn through the base fabric to form loops, and may be subsequently cut to form pile. Patterns are recreated using multiple coloured yarns and sideways-sliding needle beds.

APPLICATIONS

Tufting is predominantly used to make fitted carpets, rugs, bath mats and wall hangings. It is also used to make fake fur and imitation sheepskin. Artificial grass (also called synthetic turf) is tufted using strands of extruded polypropylene (PP) or polyethylene (PE) yarn (page 183).

It is less commonly used to make items of clothing, but can be used to decorate fabrics or create layers of insulation.

RELATED TECHNOLOGIES

Tufting was developed in the early 20th century in the United States, well after pile weaving (page 96) techniques such as Axminster and Wilton, and it was not until the 1940s that it was developed for use in the carpet industry. Its cost effectiveness contributed to the trend of carpeting wall to wall. Since then, it has dominated the flooring industry.

Woven pile continues to be used thanks to its high quality and hard-wearing features. Axminster carpets (page 104) consist of dense cut pile, and it is possible to combine up to 12 colours. This process is costly to set up and so typically only suitable for large orders. Wilton is considered superior quality pile, which may be loop (also known as Brussels carpet) or cut, and multiple different colours. Wilton is used to cover the floors of commercial buildings, aeroplanes and public transport. The type of yarn determines the overall cost. For example, even though traditional English Wilton is a more expensive process than Axminster, synthetic Wilton would be cheaper than wool Axminster. Both are considerably more expensive than tufting.

Nonwovens (page 152), such as thermal-bonded loops of polyvinyl chloride (PVC) filament and needle-punched PP, provide a very durable and cost-effective floor covering suitable for many applications including commercial, industrial and sports.

QUALITY

The quality of tufting depends on whether it is machine or handmade, as well as the type of materials and density of pile.

Yarn selection will affect the aesthetics and performance of the product. Hard-wearing materials, such as PP, nylon, polyester and wool, are used for commercial applications where the surface will be exposed to high wear. A mix of wool and synthetic yarn (80/20) is chosen for the places that will get the hardest wear, because the wool ages gracefully; 100% wool is used only for the most expensive domestic applications. Synthetics are more cost-effective and are utilized for applications that are expected to be changed more frequently.

Artificial grass is produced with PP, PE or a mixture of the two. These materials are resilient and lightweight, providing long-lasting surfaces even in the most demanding sports applications.

DESIGN

Machine tufting is a low-cost mass-production technique. A wide range

The Machine Tufting Process

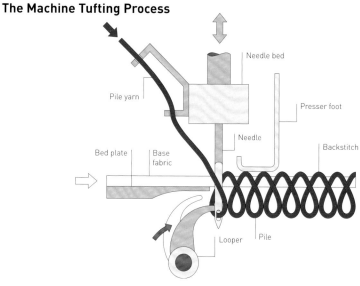

Machine tufting uncut pile

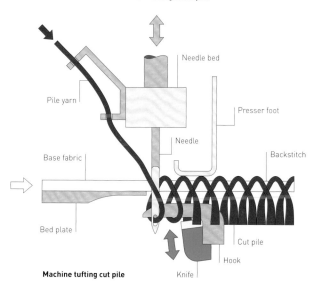

Machine tufting cut pile

TECHNICAL DESCRIPTION

Tufting is done either by hand with a tufting gun, or by machine. In both cases, a base fabric is used – also called a backing or carrier – which is typically woven cotton, jute, polyester or nylon. This provides the structure onto which the tufting yarn is sewn.

Machine tufting uses rows of needles on beds, typically 5 m (16.4 ft) wide, to apply the pile yarn. Each needle is fed with its own yarn, or multiple plied yarns. As it penetrates the base material a looper catches the yarn before the needle is retracted. It is a high-speed process capable of producing up to 2,000 stitches per minute.

When producing cut pile, the hook is reversed, so that as it catches a new loop a previous one is cut. Pile height is typically in the range of 5 to 70 mm (0.2–2.7 in.), determined by the height of the bed plate and machine stroke.

The part of yarn that remains on the back side of the base fabric is called a backstitch. Thus the length of yarn is twice the length of pile plus the backstitch. The yarn is temporarily held in place by the tension of the base fabric. After tufting is complete, the back side is coated with latex or polyurethane (PU) foam to lock the yarns in place.

Loop Pile
Material: Wool
Application: Interior
Notes: Tight loops create a dense and hard-wearing surface.

Cut Pile
Material: Wool
Application: Interior
Notes: Cut pile is softer than loop and less hard-wearing. This example uses two-ply yarn for a dual-colour effect.

Straight Line Pattern
Material: Wool
Application: Interior
Notes: Using different-coloured yarns side by side makes linear patterns in line with machine direction.

Staggered Pattern
Material: Wool
Application: Interior
Notes: Sliding the needle bed from side to side forms a simple repeating pattern.

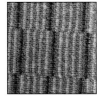

Space-Dyed Yarn
Material: Wool
Application: Interior
Notes: Space dyeing utilizes resist dyeing or digital printing in combination with different mordants to produce yarn with two or more colours repeating along its length.

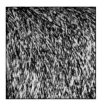

Space-Dyed Yarn Pile
Material: Wool
Application: Interior
Notes: Using space-dyed yarn results in a visually random coloured surface on tufted pile.

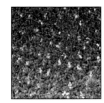

Space-Dyed Yarn Back Side
Material: Wool
Application: Interior
Notes: Multicoloured yarn is combined with a single sliding-needle bed to produce a visually random pattern.

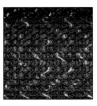

Printed Pile
Material: Wool
Application: Interior
Notes: Digital printing overcomes the inherent pattern limitations of colour tufting.

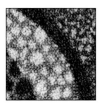

Dye Penetration
Material: Wool
Application: Interior
Notes: A vacuum is applied to the fabric during printing to draw the dye deep into the pile.

Dyed Pattern
Material: Wool
Application: Interior
Notes: Subtly coloured patterns are made by combining two-ply and three-ply yarn dyed the same colour.

Jute Backing
Material: Jute
Application: Interior
Notes: Hardy woven jute backing is applied to the fabric with latex to lock the pile yarns in place.

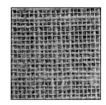

Embossed Foam Backing
Material: Latex coating
Application: Interior
Notes: Machine-tufted bath mats are coated with embossed latex to provide a slip-resistant backing.

of different types of yarn is compatible. The pile may be looped, cut, or a combination of the two at varying heights.

In machine tufting, patterns are reproduced using multiple different coloured yarns, or the pile is digital-printed (page 276). The quality will be very different for these two approaches: using coloured yarns means the pattern runs right through. However, printing is more versatile and the complexity of design and number of colours is less restricted.

Fixed single-needle bed (FSNB) machines are used to create linear patterns. Deviating from a straight-line pattern requires sideways-sliding needle beds. Single sliding-needle beds (SSNB)

can produce simple patterns by moving a single bed up to three steps each row. More complex patterns are produced using a double sliding-needle bed (DSNB). The needle beds are computer guided, moving independently or together.

Hand tufting with a gun is versatile and is used to produce samples, artworks and highly decorative pieces. There is no limit to the number of colours, patterns and pile heights that may be combined although every change will increase the length of time it takes to make.

Pile height can be up to 70 mm (2.7 in.) or more, depending on the requirements of the application. Multiple different heights of pile may be combined, or it

can be sheared afterwards to produce a flat or relief profile.

Tufting may run up to the edge, or the fabric may be trimmed afterwards. Areas of the base material may be left without pile, leaving the backing fabric exposed.

Hand-knotted carpets, made without a tufting gun, may take months or even years to produce. Warp yarns are suspended in a frame and the weaver uses weft yarn and individually knotted pile yarns to build up the texture and pattern row by row. Traditional techniques are still practised in many places, especially India and the Middle East. These are labour-intensive processes and the fabrics tend to be expensive.

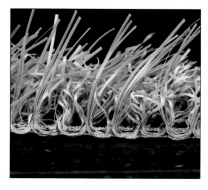

Profiled Yarn
Material: PE
Application: Exterior
Notes: Strips of extruded plastic yarn are profiled to mimic natural variation in the structure of the grass.

Infill
Material: PE with sand and rubber particles
Application: Exterior
Notes: The infill supports the long pile, keeping it upright, while providing soil-like cushioning underfoot.

Fibrillated Yarn
Material: PE
Application: Exterior
Notes: Plastic sheet is slit into mesh and twisted to make dense yarn that is a durable and cost-effective surface.

Textured Yarn
Material: PE
Application: Exterior
Notes: Permanently rippled plastic yarn produces a short dense pile ideal for tennis courts and hockey pitches.

Textured Yarn Cross-Section
Material: PE
Application: Exterior
Notes: The yarns twist and interlock to produce a dense pile that provides a reliably bouncy surface.

Mixed Yarn Cross-Section
Material: PE
Application: Exterior
Notes: Mixing strip, profiled and textured yarn produces long pile supported by a more compact underlayer.

Back Side
Material: PE
Application: Exterior
Notes: Neat rows of pile yarn are penetrated through the woven backing layer.

Latex Coated
Material: PE and latex
Application: Exterior
Notes: The tufted yarn is locked in place with a latex coating.

MATERIALS
A wide range of different yarns, including staple (page 56), filament (page 50) and plied (page 60) are suitable for the pile. Extruded plastic (page 180) is cut into strips (strands) or mesh (fibrillated) and tufted to make artificial grass.

Synthetic pile yarns include PE, PP, nylon and polyester. Natural materials include wool, silk and cotton.

The base fabric is typically plain-woven cotton, nylon, polyester or PE. Machine-tufted carpets are reinforced with jute. After tufting it is coated with latex or polyurethane (PU) foam to secure the pile yarns and prolong its life.

COSTS
Machine tufting is a low-cost process and operates at high speed. Very large areas of carpet may be produced in a single piece.

Hand tufting is more expensive owing to the high labour costs and length of time it can take. Intricate patterns and multiple colours make the process slower and consequently more expensive. Hand knotting, such as still practised in India, Iraq and Afghanistan, is the most expensive of all.

ENVIRONMENTAL IMPACTS
The most important factor to determine the sustainability of tufted fabrics is material selection. Tufted fabrics typically contain several different types of material, including a latex or PU backing, which makes them impractical to recycle. Care must be taken to ensure the raw materials are from renewable sources and do not contain any harmful ingredients.

Very little carpet ends up in landfill at the end of its useful life. Almost all waste carpet is recycled to make underlay (for new carpet). Artificial grass is recyclable. However, with regard to eco-efficiency (overall environmental impact), energy recovery by incineration is often more effective than recycling plastics. Natural materials are biodegradable and so are suitable for composting.

→ Machine Tufting Commercial Carpet

Machine tufting is carried out equally well with synthetic and natural fibres. In this case Cavalier Carpets of Lancashire, England, is using high-quality worsted wool yarn (page 27), supplied on a creel (**image 1**). The yarns are fed to the loom by tube (**image 2**). This helps to avoid snagging, because the yarn is being pulled through very quickly and has a tendency to twist. It is important to use high-quality yarn, because it is pulled under considerable tension and so prone to breaking, which slows down the process.

The needles are densely packed, requiring a high density of yarn feeds (**image 3**). Pre-woven polypropylene (PP) fabric is fed through the loom beneath the bed of needles. As it is stepped through, the needles punch the bulky pile yarn through from the back side (**image 4**). To form cut pile the loops are cut underneath.

The tufted carpet progresses from the back of the loom upside-down and fully formed (**image 5**). It is carefully inspected to check for any missed stitches (**image 6**) and gaps are filled in with a handheld tufting gun to ensure the carpet has full pile density (**image 7**). It is mended easily, because the pile yarn is not woven into the structure of the fabric.

The carpet is taken up on a roll with the pile facing inward (**image 8**). Finishing involves steaming, coating with latex, backing with fabric and oven-setting.

Uniformly coloured carpet is produced on a loom with a sliding-needle bed. Tracking the rows back and forth during tufting produces a wavy line of stitches (**image 9**). This helps to avoid visible lines forming in the finished pile and helps to produce more uniform colour (**image 10**).

DECORATING

290

1

2

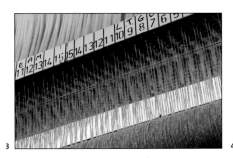

3

4

5

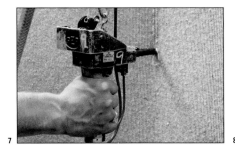

6

7

8

9

10

Featured Company

Cavalier Carpets
www.cavaliercarpets.co.uk

→ Machine Tufting Coloured Patterns and Finishing Carpet

Using either single sliding-needle beds (SSNB) or double sliding-needle beds (DSNB), a range of patterns and different colourways is possible (**image 1**). More complex design may be realized with DSNB. In operation, yarn is supplied to two rows of needles. The front row consists of light-coloured yarn (**image 2**), while the back row is supplied with dark-coloured yarn (**image 3**). As the primary backing is passed under the needle beds they work in tandem to create the pattern (**image 4**).

There are many alternative techniques used to form patterns. For example, multicolour-dyed yarn will produce colour variation even if the pile is stitched in straight rows (**image 5**).

After tufting, the carpet is sheared to produce a homogenous surface (**image 6**). The underside is coated with latex and a secondary backing fabric is applied. Jute is preferred (**image 7**), but synthetic alternatives are sometimes required, such as if there has been a bad harvest.

This locks the pile in place and stiffens the construction overall.

1

2

3

4

5

6

7

Featured Company

Cavalier Carpets
www.cavaliercarpets.co.uk

→ Tufting Artificial Grass

The colour is selected and lines worked out prior to manufacturing the turf. Selecting the appropriate feed yarn can produce areas of colour and straight lines in the direction of the machine. Patterns that do not follow the direction of tufting, such as curves, are inserted by hand.

Tufting machines used to make artificial grass are adapted from carpet tufting (page 290) to accommodate the longer length of pile (**image 1**). Rollers apply tension to the extruded plastic yarn, which is fed into each needle according to the colour requirements laid out in the design (**image 2**). The yarn passes through a row of tufting needles that penetrate the backing fabric to a predefined depth (**image 3**). At this stage the yarn is held in place by the tension of the woven backing layer (**image 4**). The yarns unravel to create a uniform pile of specific density (**image 5**).

When a new colour is required, such as a marking line, the location is checked by measuring the material on the loom and the relevant ends are replaced with a new colour yarn (**image 6**).

The entire surface is carefully checked to ensure that all the pile has been cut (**image 7**). Loops must be removed, because otherwise the strong yarn could become a hazardous snagging point (**image 8**). If the yarn has broken to leave a gap or where colour is required that is not in line with machine direction, yarn is added by hand (**image 9**) and cut to length on the front side (**image 10**).

Once tufting is complete, the fabric is coated with latex (**image 11**). Held under tension on a tenter frame, it is passed through a series of bar heaters (**image 12**) to bring the material up to temperature before passing through an oven to fully cure the backing. This creates a durable material that will perform well even under the most demanding playing conditions.

With careful installation and maintenance the pitch can last seven years or more (**image 13**). At the end of its useful life it is recyclable.

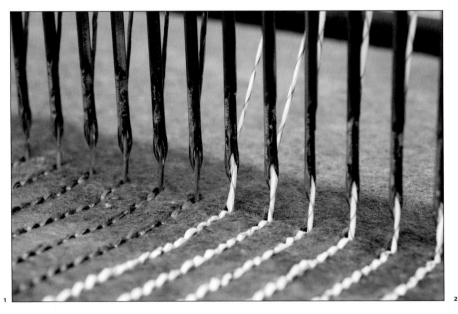

1

2

3

4

5

6

7

8

9

10

11

12

13

Featured Company

TigerTurf
www.tigerturf.com

The Gun Tufting Process

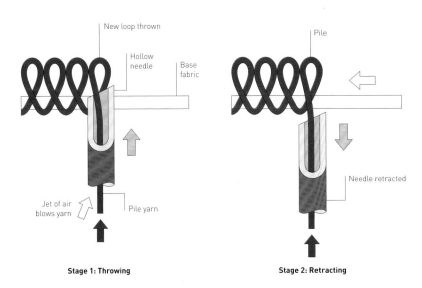

Stage 1: Throwing

Labels (left diagram):
New loop thrown
Hollow needle
Base fabric
Jet of air blows yarn
Pile yarn

Stage 2: Retracting

Labels (right diagram):
Pile
Needle retracted

Gun tufting uncut pile

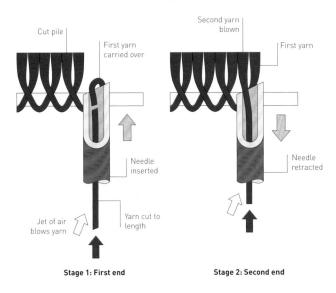

Stage 1: First end

Labels (left diagram):
Cut pile
First yarn carried over
Needle inserted
Jet of air blows yarn
Yarn cut to length

Stage 2: Second end

Labels (right diagram):
Second yarn blown
First yarn
Needle retracted

Gun tufting cut pile

TECHNICAL DESCRIPTION

Gun tufting uses a hollow needle to apply one loop at a time, or multiple loops in the same location. The yarn is fed through a hollow needle in the tufting gun. In stage 1, as the needle penetrates the base fabric, the yarn is blown down the needle with a jet of air. The length of yarn fed in determines the height of the loop.

In stage 2, the needle is retracted and with a stepping motion progresses to the next row. In this way, a row of interconnected loops are made along a predetermined path, held in place by the base fabric. Handheld tufting guns can operate at up to 1,400 cycles per minute, but are rarely used at this speed for long periods.

Cut pile is formed by cutting the yarn with a rotary blade, which is mounted behind the hollow needle. The yarn is cut and blown down the hollow needle. The length is determined by the height of pile required.

Each time the hollow needle penetrates the fabric it carries out two operations: in stage 1, the second half of the previous yarn is thrown and in stage 2, the first part of the next length of yarn is blown through.

Similar to machine tufting, the yarn is temporarily held in place by the tension of the base fabric. Once complete, the back side is coated with latex to lock the yarns in place.

Hand Knotted

Material: Wool and silk
Application: Rug
Notes: Each of the pile yarns is individually hand knotted and so any number of colours may be used.

Pile

Back side

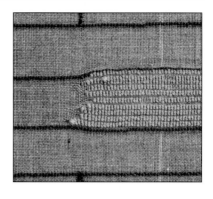

Guides

Material: Wool
Application: Rug
Notes: A design is copied onto the base fabric using guidelines, which help to ensure a crisp edge to the patterns when gun tufting.

Gun Tufted

Material: Wool
Application: Rug
Notes: Colour is filled in in the areas between the guidelines using continuous lengths of yarn.

Pile

Back side

Graphic

Material: Wool
Application: Rug
Notes: As a manual process, gun tufting is versatile and all types of pattern, graphics and artwork are reproducible.

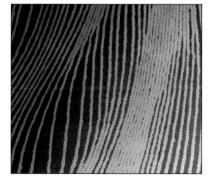

Colour

Material: Wool
Application: Rug
Notes: The type of yarn determines the minimum line weight that can be achieved.

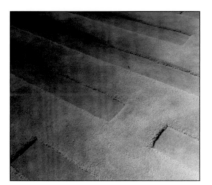

Relief

Material: Wool
Application: Rug
Notes: Relief surface patterns are made by tufting different heights of pile, or afterwards, using shears.

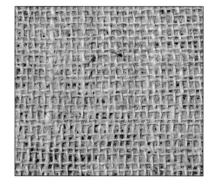

Latex Coated

Material: Wool
Application: Rug
Notes: The back side is coated with latex and a second layer of fabric is applied to firmly secure the pile yarns.

→ Hand Making a Rug with a Tufting Gun

A tufting gun is equipped with a hollow needle. In this case, four two-ply yarns are being used to fill areas of blue in the rug (**image 1**). The head of the gun is equipped with a rotating blade, which cuts the yarn to the correct length.

The base fabric is stretched over a frame (**image 2**). The weaver works from the back side of the fabric (**image 3**). As he presses the gun against the fabric and pulls the trigger, lengths of yarn are cut and blown through around the weft, which runs horizontally

(**images 4** and **5**). Next, rows of loop pile are added. Instead of cutting the yarn to length, it is blown through in a continuous line (**images 6**, **7** and **8**). The needle determines the length of the pile.

Colours are changed relatively quickly and the finished pattern takes shape (**image 9**).

Featured Company

Area Rugs and Carpets
www.arearugs.co.uk

→ Shearing and Backing

Once the tufting is complete, rugs may go through a series of finishing steps. In the case of gun tufting the guidelines used to mark out the design are removed (**image 1**). The underside is coated with latex and backed with an additional layer of fabric to hold the pile yarns in place and maintain the dimensional stability of the rug (**images 2** and **3**).

The height of the pile is equalized with a shearing machine (**image 4**). The sharp blade rotates at high speed to produce a flat and uniform surface (**image 5**). The edges are sculpted with clippers (**image 6**) to produce a neat chamfer.

1

2

3

4

5

6

Surface Colour Pattern Protection Handle Relief

Textile Technology
Embroidery

Threads are stitched onto fabrics to produce decorative imagery, patterns and motifs. Natural and synthetic materials are used in accordance with selected techniques. Hand embroidery is used in the production of couture, high-end and bespoke embellished fabrics for applications including apparel, theatre and film. Machine embroidery is employed for batch and mass production.

DECORATING

298

Techniques	Materials	Costs
• Hand embroidery • Vertical looms • Machine and digital embroidery	• Fabric and thread • Ribbons, sequins, beads, purl, pearl, bullion, plate and spangles	• High for couture, bespoke and high-end • Moderate for machine embroidery and digital embroidery

Applications	Quality	Related Technologies
• Ecclesiastical • Interiors • Fashion • Military • Film • Theatre	• Flat, raised or openwork • Hand, machine or digital embroidery • Metallic, matt, iridescent and so on	• Lace • Printing • Weaving

INTRODUCTION

Embroidery is highly skilled and is practised in many countries around the world. For centuries it has been used to make elaborate pieces, and the style and application varies according to tradition.

The structure of embroidery is unlike any other process. The thread is oriented in all directions, sewn onto layers of padding or couched (laid down on fabric and fixed in place by stitching with another thread) to create three-dimensional surface structure and decoration. There are many types of stitch and factors that can be adjusted according to the design, such as length, direction and colour.

Machine-made embroidery is designed to give the appearance of hand embroidery, but the stitches and materials are different. There are two main categories: vertical looms (embroidery made by more than 1,000 needles as one continuous piece); and horizontal machines, which includes machine embroidery (hand-guided sewing machine) and digital embroidery (computer-guided machine).

Vertical looms are equipped with beds of horizontal needles several metres wide, which move forwards and backwards to form stitches as the fabric is moved in all directions. It is a cost-effective alternative to lace (pages 106–19). The thread is interlocked and self-supporting, and so is suitable for decorating net, sheer and sacrificial fabrics (to make lace). It is used to produce wide fabric, or several narrow strips, suitable for apparel, upholstery, interiors and trim.

Several types of patterns can be applied by a skilled operator using a conventional sewing machine. For example, satin fill stitches are created with tightly packed zigzags, and chain stitches may be used to create columns or areas of colour. This type of embroidery is labour-intensive.

Computer-guided embroidery machines are single- or multi-head with a needle for each colour. The needle moves relative to the fabric to produce a wide range of stitches. Mass production is carried out on embroidery machines with 30 or more heads reproducing the same design on different garments simultaneously.

APPLICATIONS

Hand embroidery is used to embellish textiles in all manner of applications including footwear, garments and headwear. It is labour-intensive and carried out by a highly skilled embroiderer, so tends to be reserved for high-status applications such as monogramming a handkerchief, shirt or towel; couching gold thread onto a military ceremonial jacket, slipper or hat; or adorning artwork, rugs, covers, bedspreads and wall hangings.

Multi-head machines are most effective when producing multiples of the same design, such as batch or mass production of company logos, branding and monograms.

RELATED TECHNOLOGIES

Pattern, graphics and colour are produced with digital (page 276), screen (page 260) and block (page 256) printing. Inherent and relief designs are reproduced with flocking (page 280), pile (page 96) and jacquard weaving (page 84). All of these processes may be used to embellish the surface of fabrics and finished garments, but the surface quality and appearance are very different.

Embroidery crosses over with lace and openwork fabrics in many cases, especially when embroidering onto cutwork, net, sheer or sacrificial fabrics. There is a structural difference between embroidered and woven, knitted and handmade lace; but in some cases they are difficult to tell apart based on appearance.

Schiffli lace, also known as chemical lace, is a mass-production technique that uses more than 1,000 needles. The front and back threads are sewn into complex structures on wide rolls of fabric. It resembles handmade lace and embroidery, but is much more cost-effective.

Tambour is a technique used in both lacemaking and embroidery. It is the most effective method of beading – with beads or sequins – onto a fabric or net ground. With hand embroidery there is a higher chance of making mistakes; this is not the case with tambour (see case study on page 307).

QUALITY

Embroidery adds thread to the surface of fabrics and therefore can be flat or relief in profile. Stitching onto additional layers of material (appliqué) or couching over thread increases the height that can be achieved and so emphasizes the embroidery.

Density is measured as number of stitches per square centimetre or square inch, or number of rows per centimetre or inch, and varies according to stitch type. It is also referred to as stitch spacing. Higher density results in greater reproduction of colour, but will affect the handle of the fabric more than lower-density designs.

Quality of colour and appearance depends on the stitches. Fabrics that will move and flex in application (stretch) are treated different from those that will remain static in application (such as stiff materials, or fabric that will be under tension).

Dense designs are suitable for heavier fabrics with good drape. Sparsely placed stitches are preferable for lighter-weight, flowing fabrics, because dense designs will increase stiffness and weight and so affect handle.

Napped (page 216) fabrics, such as fleece, are typically embroidered with additional layers of thread (underlay), or topping (fabric), to ensure adequate cover and that the embroidery does not disappear into the pile.

Fabrics with stabilizing qualities are utilized to reinforce the back of fabrics (ground) to avoid distortion and puckering in the finishing stages. They are either left as part of the construction or removed after stitching. There are several means of removal, depending on the stabilizer, including tearing, cutting and dissolving in water.

Stabilizers are important for maintaining the integrity of fabrics that will be marked by tensioning over a hoop (known as ring burn) during stitching. This includes velvet, suede, leather, satin weave and other delicate fabrics.

DESIGN

Embroidery typically starts with a draft (artwork). When using digital embroidery techniques, the draft is converted into computer-aided manufacture (CAM) data by the digitizer (also called a puncher). The file is overlaid with different types, lengths and colours of stitches to faithfully recreate the design.

It is important to consider the size of the hoop or reach of the head of the sewing machine. Larger designs will require re-clamping, which increases the likelihood of misalignment in the finished work.

Compensation must be made for the pull of the thread tension to ensure

Common Embroidery Stitches

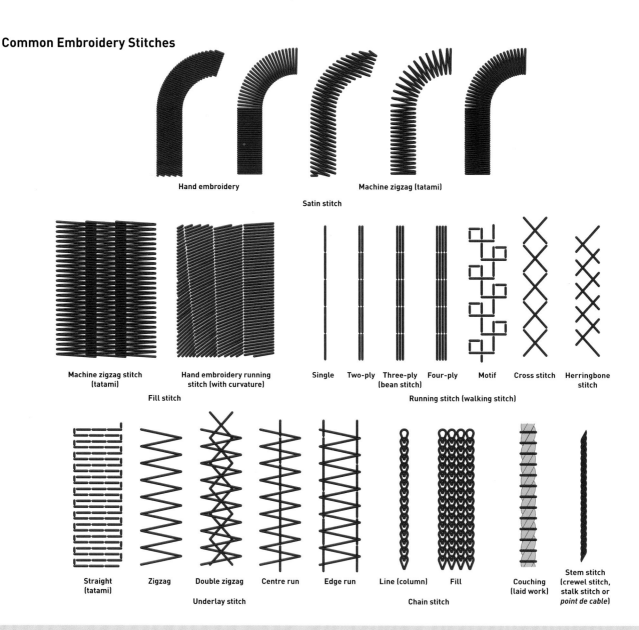

Satin stitch
Hand embroidery   Machine zigzag (tatami)

Fill stitch
Machine zigzag stitch (tatami)   Hand embroidery running stitch (with curvature)

Running stitch (walking stitch)
Single   Two-ply   Three-ply (bean stitch)   Four-ply   Motif   Cross stitch   Herringbone stitch

Underlay stitch
Straight (tatami)   Zigzag   Double zigzag   Centre run   Edge run

Chain stitch
Line (column)   Fill   Couching (laid work)   Stem stitch (crewel stitch, stalk stitch or *point de cable*)

TECHNICAL DESCRIPTION

Stitches are made by hand or machine. The principal difference is that with hand stitching the needle can be inserted into the fabric in one location and brought back through somewhere else, while this is not possible with machine stitching, because the needle punches into and out of a single hole.

Machine stitching has been developed to mimic the appearance of hand embroidery, but the structure of the stitch will be different and not all stitches are possible with machine.

Digital (computerized) embroidery uses lock stitches (page 355) to produce tatami, satin (compact zigzag) and running stitches, and machine embroidery (hand-guided) uses chain stitches (page 357).

Satin stitches are used to create letters and borders and to fill small areas typically no wider than 9 mm (0.4 in.). Hand embroidery utilizes a combination of short and long stitches to cover areas greater than that. In hand stitching, the thread is passed behind the fabric to produce a clean linear appearance on the face. Machine stitching utilizes a zigzag stitch, whereby the backstitch is visible on the front face. The density and angle are adjusted according to the design.

Fill is used to cover larger areas with colour. It is created with multiple satin stitches (vibrant colour) or a series of overlapping zigzag stitches (tatami produces a matt finish). Fill does not have to follow straight lines: curvature, open areas and changes in density are all feasible.

Machinists prefer shorter overlapping stitches to avoid unnecessary travel (machine movement) and reduce cycle time.

Large areas of fill are often finished with a satin stitch or running stitch border to ensure a neat edge.

Running stitches are applied as single, two-ply, three-ply (also known as a bean stitch) and four-ply, depending on the size of thread and depth of line required. Multi-ply stitches may be formed in a single pass, by stitching each of the plies one after the other before moving onto the next run. This avoids making multiple passes to build up line thickness.

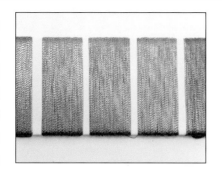

Metallic Thread

Material: Polyester thread and PET film
Application: Apparel
Notes: A strip of metallized PET film is wound around polyester filament, producing a bright and reflective thread.

Gold-Wrapped Thread

Material: Gold and cotton
Application: Apparel
Notes: Gold plate is wrapped around a cotton core to produce a lightweight flexible thread with a bright gold finish.

Check Purl

Material: Gold
Application: Apparel
Notes: Sparkle is achieved by wrapping round metal wire around square (check) and round (rough) mandrels.

Metallic Coloured Thread

Material: Polyester thread and PET film
Application: Apparel
Notes: The metallized coating is tinted with colour to create a shimmering visual effect.

Smooth Purl

Material: Gold
Application: Apparel
Notes: Flat gold wire is wound over a mandrel, which is removed to produce lengths of smooth hollow thread.

Gold Plate

Material: Gold
Application: Apparel
Notes: Precious metals are cut into strips and sewn over padding to create smooth-surfaced relief designs.

that the fabric does not pucker and good registration is achieved in the final design. Otherwise there is a danger that, for example, circles will appear elliptical and squares rectangular. Alternatively, this property may be exaggerated with elastic thread, which is stitched under tension and so pulls the fabric into three-dimensional patterns when relaxed.

In hand and machine embroidery a wide range of stitch types is employed in combination with different colours and styles of thread. The choice depends entirely on the design. The same design at different scales will likely require different stitches, because enlarging a design will increase the stitch length while reducing the coverage, so the stitches will need to be adjusted.

Stretchy items, such as knitted garments, are embroidered under tension. This ensures there is no undue stress applied to thread or fabric during use, and the design will look as intended. Elastic and lightweight fabrics work best with designs that have plenty of open space; dense stitching affects the handle.

The widest range of design possibilities is possible with hand techniques, and depends on the skill of the embroiderer. Hand and machine embroidery is used in combination with other techniques, such as lace, cutwork and appliqué, further expanding the range of possibilities.

When applying multiple machine-made patterns on a single piece, each one must be digitized separately to ensure it is completed before the machine moves on to the next. Hand stitches can jump on the back side of the fabric, enabling more complex and intricate patterns such as herringbone.

Underlay stitches are commonly used to produce a foundation for further machine embroidery. They are used to stabilize the fabric – to reduce puckering and stretch – and hold the surface of napped and pile fabrics flat so that the top layer of embroidery will not sink in and disappear. Centre-run and edge-run underlays provide extra lift for satin stitching. Straight, zigzag and double zigzag tatami-like stitches give cover and stability to large areas of fill.

Chain stitches are made by machine (known as chenille or cornely) or by hand (known as tambour). They are named after the machines or equipment used. Unlike lock stitching, which is the principal machine-embroidery technique, chain stitching forms interlooped stitches from a single thread. This means it can be used to apply beads and sequins. The embellishments are loaded onto the thread and moved into place when they are needed.

Machine-made chain stitches are used to make straight and curved lines of colour as well as to fill large areas. There are many more possibilities with hand embroidery, such as braided and knotted chain stitches. Chain stitches are prone to ravelling if they are broken at any point, because they are constructed with a single thread.

Hand couching is also known as laid work. It is used for borders as well as large areas of colour. Heavy inflexible thread, and other types of material that are not practical for stitching, is held onto the fabric by stitching a finer thread over the top. The covering thread may be virtually invisible.

Stem stitch (also known as stalk stitch or *point de cable*) is a popular and straightforward hand stitch. It is commonly used for applying narrow lines of colour, such as found in monograms, flower stems and borders.

Front   Back

Freehand
Material: Cotton
Application: Apparel
Notes: Satin stitches make up the flowers and petals, and green stem stitches are used to make the stems.

Front   Back

Paper Template
Material: Cotton
Application: Apparel
Notes: Working over a card template, known as carding, makes complex patterns straightforward to produce. Shaded areas indicate different colours.

Silk
Material: Silk
Application: Interior
Notes: Silk has a natural lustre and is utilized in hand embroidery to produce realistic imagery, such as renditions of flowers.

Front   Back

Appliqué
Material: Fabric
Application: Interior
Notes: This is a cost-effective method for filling in shape with colour. Beads are added by hand.

Front   Back

Thread on Net
Material: Cotton and nylon
Application: Apparel
Notes: Echoing the design effect of lace, this technique is utilized in the production of fabrics for apparel, theatre and film.

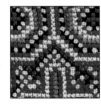

Cross Stitch
Material: Cotton
Application: Bag
Notes: This is a popular and straightforward hand stitch formed with counter whipstitches.

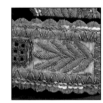

Goldwork
Material: Gold
Application: Apparel
Notes: Precious metals including gold and silver are embroidered onto fabrics for high-end applications, such as this collar on a military dress uniform.

Embroidering onto a sacrificial fabric that is dissolved away afterwards to leave just the stitched threads produces an openwork structure (macramé and guipure lace). Where the sacrificial fabric is dissolved away by chemicals, the lace is referred to as chemical lace; this technique has been phased out in many countries because the chemicals are harmful, and has been replaced with water-soluble systems.

Size is determined by process selection. Hand embroidery can be any size and shape, and multi-head embroidery machines are typically limited to the size of the hoop, or maximum travel of the head.

MATERIALS

Virtually any fabric can be embroidered, as long as it is suitably stable. Knitted (mainly jersey) and woven textiles are the most common. Other examples include paper, board, veneer, leather and plastic.

All types of natural and man-made thread can be used. The most common are shiny decorative threads such as silk, polyester, viscose, gold and silver. Beads and sequins are used for decorative effects.

Threads include filament (page 50), staple (page 56), twisted and plied (page 60), ribbon (see weaving narrow widths, page 82, and Plastic Film Extrusion, page 180) and plate (strips of precious metal such as gold and silver).

Metal wire wrapped around a metal rod (removable mandrel) to produce a hollow wire coil is known as purl. There are several types including smooth (flat wire coiled around a circular mandrel), rough (round wire coiled around a circular mandrel), check (wire wrapped around a mandrel with rivets or pearls embedded in it) and pearl (larger diameter wire coiled and slightly stretched to open up the wire). The diameter of these threads ranges from around 0.1 to 3.5 mm (0.004–1.4 in). They are couched by hand onto fabric and used to make flat or relief designs. Technical versions, known as round-wire coil, or flat-wire coil, are used for medical devices, such as endoscopes.

Front   Back

Satin Stitch
Material: Cotton
Application: Apparel
Notes: The tough white bobbin thread is stitched under just enough tension to pull the needle thread tight.

Chain Stitch
Material: Cotton
Application: Interior
Notes: Lengths of chain stitch are hand guided on a chenille or cornely machine to produce patterns and graphics.

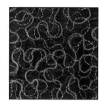

Sequins
Material: Cotton and polyester
Application: Apparel
Notes: Used to make fashion fabrics, a variety of sequins and beads are attached with colour-matched thread.

Front   Back

Two-Colour Dyeing
Material: Polyester and cotton
Application: Apparel
Notes: Large thread is couched around the perimeter. The two materials respond to different dye systems.

Front   Back

Macramé Lace Effect
Material: Polyester
Application: Apparel
Notes: Embroidering onto a sacrificial fabric, which is dissolved away after stitching, creates openwork fabrics. Heavy lace effect is known as macramé.

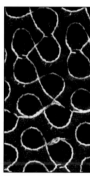 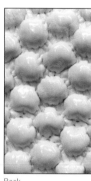

Front   Back

Relief Patterns
Material: TPU-laminated fabric and elastane thread
Application: Apparel
Notes: Elastic thread is stitched under tension and so pulls the fabric into a three-dimensional shape when relaxed.

Front   Back

Guipure Lace Effect with Gold Finish
Material: Polyester and metallized foil
Application: Apparel
Notes: The foil is pressed onto embroidered fabric and bonded to the surface, emphasizing the relief profile. Finer lace effect is known as guipure.

COSTS

Hand techniques require a high degree of skilled labour and so are typically high- to very high-cost, depending on size and complexity. The materials may be very expensive too, such as silk, gold and silver.

Machine techniques are less expensive, although the cost of drafting, digitizing and preparing the artwork can be moderate to high depending on the scale and complexity. Production price depends on the number of stitches. As with hand embroidery, the cost is calculated according to the amount of time it is likely to take, in addition to the value of the materials used.

High volumes of the same design may be produced in a single colour and dyed different colours to increase flexibility without increasing cost. To achieve this, two or more types of thread are used that respond to different dye systems.

ENVIRONMENTAL IMPACTS

Embroidery is an additive process. However, there is very little waste – especially when using precious materials – and the outcomes typically have greater personal and economic value than plain textiles, which means they will be used and maintained for longer.

Carefully selecting where the materials are sourced from will help to reduce the

Cutwork
Material: Cotton
Application: Apparel
Notes: Pre-woven fabric is embroidered and cut out in a single operation to create partial openwork.

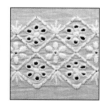

environmental impact. Natural materials should be certified and synthetics derived from recycled or biobased ingredients. Fully biodegradable stabilizers ensure that washed- or torn-away material does not produce harmful by-products or contribute to landfill.

→ Marking a Pattern for Embroidery

Designs that are embroidered by hand need to be transferred onto the fabric ground prior to production. Several techniques are used, including drawing with a thin pencil (the line will be covered by the thread), transfer paper (heat, carbon or wax) and pouncing (demonstrated here). Temporary or permanent methods are used depending on the type of fabric and whether the surface can be marked or dampened.

Hand & Lock, of London, UK, has been producing fine hand embroidery since 1767.

It continues to practise traditions and techniques that date back to Roman times. Pouncing is an example of one of the oldest techniques. It is centuries old and yet remains well suited to large patterns and marking onto nonwoven, pile and napped fabrics.

The design is copied onto tracing paper (**image 1**). A needle is used to pierce small holes along the outlines (**image 2**). Fine chalk powder (pounce) is applied with a soft cloth (**image 3**). The powder

penetrates the holes to transfer the pattern to the fabric below (**image 4**).

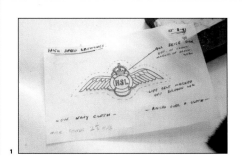

1

2

3

4

Featured Company

Hand & Lock
www.handembroidery.com

→ Embroidering a Monogram

Monograms are frequently applied to garments, accessories and linens. The plain-woven cotton shirt is held onto a sturdy backing fabric with tack stitches. The backing fabric is held under tension between two clamps to ensure a sturdy base for embroidery (**image 1**).

Three letters are marked onto the white cotton with water-soluble pencil. A thread is threaded onto the needle and folded to make a pair of threads. A knot is tied at the end of the thread, which is pulled through until the knot stops on the back of the fabric. Stitching typically starts at the lower left-hand corner of the first letter.

Stem stitches are used to make fine letters. Passing the needle from left to right in successive lapping stitches forms a smooth round profile (**image 2**). For broad letters and large flat areas of colour satin stitches are used.

Each letter is individually embroidered (**image 3**). The whole monogram takes no more than a few minutes to create.

Excess thread is then cut away from the back of the fabric (**image 4**) and the monogram is complete (**image 5**).

1

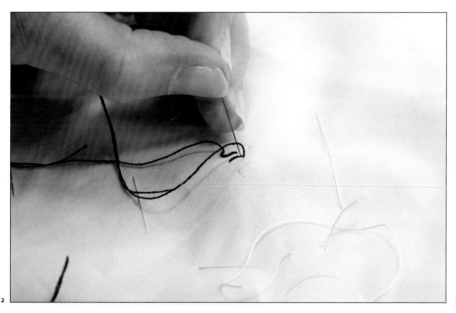

2

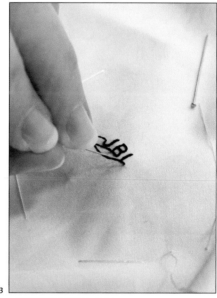

3

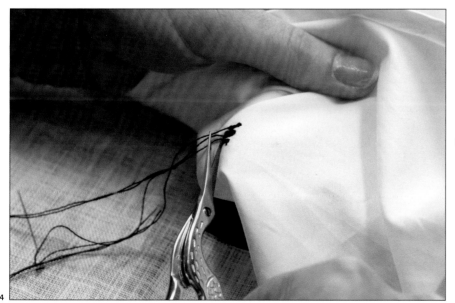

4

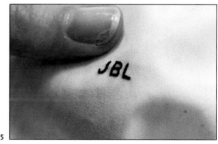

5

Featured Company

Hand & Lock
www.handembroidery.com

→ Goldwork for Ceremonial Accoutrements

Jackets and other garments worn by the military at ceremonial events are traditionally adorned with fancy goldwork. Embroidery, lace and braiding are used side by side in the construction.

Embroidery is used to make relief patterns with precious metal thread, such as gold and silver purl (**image 1**).

The pattern is marked out on the fabric by pouncing (page 304). Depending on the shape of the relief profile the metal thread is stitched over additional layers of fabric, or alternatively over twisted threads (**image 2**).

Gold purl is cut into small lengths and fed onto a needle (**image 3**). It is placed onto the fabric: one end is located by the needle and the other is held by the thread (**image 4**). The needle is pushed through the fabric and the thread is pulled tight, forming the purl into a neat arc over the fabric topping (**images 5** and **6**).

Each piece is cut by hand. The embroiderer skilfully calculates how each piece will fit together to make the finished pattern. The centre line where the purls meet is covered with a line of sequins.

1

2

3

4

5

6

Featured Company

Hand & Lock
www.handembroidery.com

→ Tambour Bead Embroidery

Also called *broderie chaînette* and *broderie de Lunéville* (named after the French town where the hook technique first gained renown in Europe), this form of embroidery is used to make highly embellished pieces incorporating beads and sequins with a range of threads.

Unlike conventional embroidery, patterns are created on the back side of the fabric. In other words, the tambour hook allows the embroiderer to form chain stitches, by working upside down.

The fineness of the hook depends on the fabric and threads being embroidered. The fabric is stretched over a circular tambour frame and held taut (**image 1**), while the thread is loaded with beads and suspended below.

The chain stitches are formed in much the same way as on a sewing machine (page 357). The hooked needle is passed through the fabric (sheer in this case so you can see through) to catch the thread (**image 2**). The thread is pulled back through the fabric

to form a loop, which stays on the needle as it is passed back through the fabric to catch the next loop (**image 3**). Each time a bead is required it is slid into place by the embroiderer and held with a stitch (**image 4**).

1

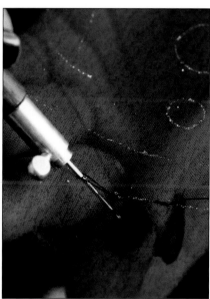

2

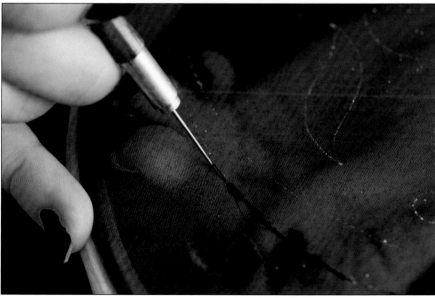

3

4

Featured Company

Hand & Lock
www.handembroidery.com

→ Machine Embroidery

Machine-embroidered designs, such as this martial arts club logo (**image 1**), have a lustrous appearance similar to hand embroidery, but the stitches are usually different. The process begins with the original artwork (**image 2**), which is translated into a pattern of stitches in a process known as digitizing (or punching). This can be time-consuming for large or complex designs.

A hoop is prepared with backing material, which stabilizes the fabric during embroidery (**image 3**). The fabric is aligned on the hoop using a grid (**image 4**) and held onto the stiff backing material with tack stitches.

The whole assembly is loaded into the embroidery machine. The digitized design has been transferred to the machine and guides the stitching, including the order of colours (**image 5**). The head is equipped with six colours and it is possible to stitch up to 15 colours on some machines.

The blue ring is embroidered in two layers: the underlay of running stitches stabilizes the fabric and the second covers the area with fill stitches (**image 6**). The fill (or tatami) stitches are shorter than the depth of the ring, so multiple stitches are combined to cover the full width, producing a more stable fill with greater flexibility.

The white inner ring is applied with the same two layers of stitching (**image 7**). Subsequently the rest of the design is filled in and the stitching is complete (**image 8**). The operator removes extraneous stitches accumulated during embroidery, such as jumps between areas of the same colour.

1

2

3

4

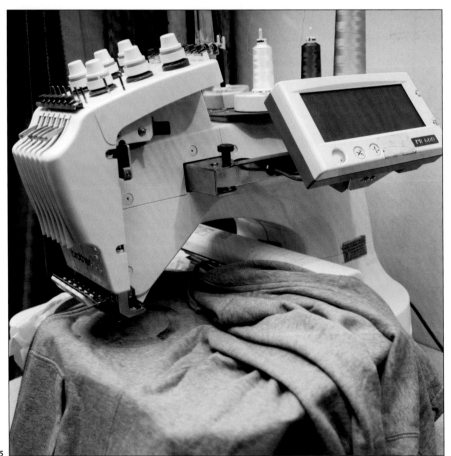

5

6

7

8

Featured Company

Hand & Lock
www.handembroidery.com

Part

3

Construction Technology

Construction Technology
Manual Cutting

Pattern cutting is the art and science of measuring and designing patterns. Patterns are designed by a cutter, created by hand or using computer software, or by a combination of the two. Once complete, the patterns are used as templates to mark where to cut fabric, or alternatively the data is used to drive computer-guided cutting machines.

Techniques	Materials	Costs
• Pattern cutting • Hand shears • Mechanized knife	• Woven, knitted, nonwoven, leather, rubber and plastic film	• Low to high, depending on location (wages) and complexity

Applications	Quality	Related Technologies
• Apparel • Accessories • Upholstery	• Shape depends on skill of pattern cutter • Edge depends on knife	• Laser cutting • CNC cutting • Die cutting

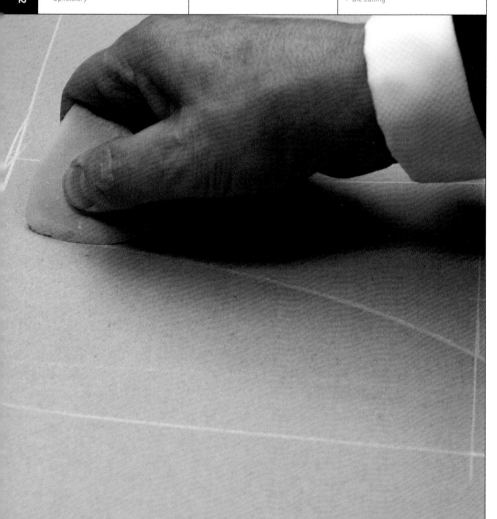

INTRODUCTION
Patterns are used to create bespoke tailored items through templates used in mass production. Importantly, as well as creating the right shape for a garment (see case study on page 314), shoe (see Leather Molding, page 430), upholstered furniture (page 404) or other textile item, patterns ensure that the grain is correctly aligned when the panels are constructed.

Hand-cutting techniques include shears and mechanized knives. The choice of cutting technique depends on the volume and requirements of the application.

Hand shears, such as those used by tailors, produce a very neat edge. Textiles are abrasive, so the blades must be sharpened regularly to maintain their quality. Mechanized knives operate at high speed and are capable of cutting through several plies at once. Straight blades are used to cut around curved profiles, while rotary blades are better suited to cutting straight lines and large-radius curves (see Slitting and Crosscutting, page 322).

APPLICATIONS
Pattern cutting in one form or another is utilized in all textile-related industries, for example in the design and production of garments, shoes, headwear (see Blocking, page 420), upholstery and composite laminates.

RELATED TECHNOLOGIES
Manual cutting is generally reserved for samples, bespoke items and low-volume production, although this does not mean computer-guided cutting is limited to high volume production. CNC (page 316) and laser cutting (page 328) are versatile processes utilized in the production of one-offs as well as in batch and mass production. They are employed in the

The Manual Cutting Process

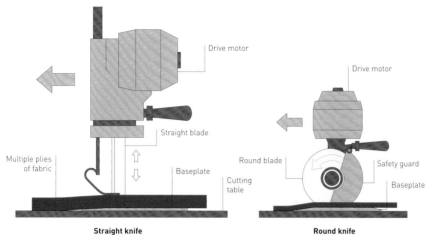

Straight knife

Drive motor
Straight blade
Multiple plies of fabric
Baseplate
Cutting table

Round knife

Drive motor
Round blade
Safety guard
Baseplate

Rotary shears

Handheld motor
Round blade

production of panels that can range from small to very large.

Die cutting (page 336) requires the production of a metal tool. Therefore sufficient volumes are required to justify the additional set-up costs. But once up and running, this is the fastest method of profiling small- to medium-sized items and is highly repeatable.

To accelerate the process from taking measurements to finished garment, the patterns may be cut out directly (freehand cutting) without the use of a paper template. A skilled cutter can accommodate changes in body shape and size when marking up the fabric. Committing cutting instruction to memory, as opposed to following a set of instructions, increases the speed and flexibility of the cutter. However, there is then no record of the design, except the garment itself.

QUALITY
The quality of the pattern design depends on the competency of the cutter and the accuracy of the measurements. The pattern is typically laid out on lightweight board, which is turned into the template for cutting fabric. Mock-ups may be used to assess the fit and hang before going into production. In garment production, these are known as muslins (USA) or toiles (UK).

Jigs or guides are not used, so the quality of hand cutting depends largely on the skill of the operator and their

TECHNICAL DESCRIPTION
Motorized cutters have either straight or round blades. Straight blades move up and down at 1,200 to 4,000 rpm, cutting through the fabric with a sawing action. They come with a range of finishes. Straight-edged blades are used to cut softer materials, such as cotton, wool and knitted fabrics. Angle-tipped blades are used to cut tougher materials, such as denim, canvas, aramid and fibreglass. Wavy-edged blades are used to cut materials that are prone to fusing along the cut edge, for example nylon, polyester and polypropylene (PP).

Round blades rotate at very high speed. They are either perfectly round and so cut with a sawing action, or shaped (square or hexagonal) and so cut with a chopping action. Round blades are used for light to heavy-duty cutting. They are ideal for cutting long straight lines and large-radius curves. Handheld shears are used for lighter work, one-offs and on-site cutting (such as carpet fitting).

Knives are sharpened to varying levels of coarseness, which is paired to the material being cut. A fine edge finish is ideal for cutting knitted, high-pile and loosely woven fabric; medium finish is used to cut synthetics, cotton and wool; a coarse edge finish is used to cut heavier-weight woven materials; and a rough edge is ideal for heavy fabrics such as denim.

ability to reach all parts of the pattern being cut. Sharp blades ensure a clean and snag-free edge.

The sawing action of straight knives generates frictional heat. This can cause problems with thermoplastics because if they heat up to beyond their softening temperature then the edges may fuse together. In this case a wavy-edged straight knife is used, which reduces the build up of heat.

Precision-engineered straight knives are capable of maintaining cutting tolerances of 0.25 mm (0.001 in.).

DESIGN
In production, patterns are transferred to a marker, which is typically dotted or plain white paper. It is placed on top of the stack of fabric plies and held in place. With each cutting cycle the marker is consumed in the process. Alternatively, the pattern may be drawn directly on the top layer of fabric. In this case, it is best to cut upside down, so any residual marks are left on the inside after construction.

Complex shapes may be measured with a three-dimensional scanner to provide very accurate statistics to calculate the pattern cutting.

→ Pattern Cutting a Bespoke Suit

Bespoke suits are made to measure to ensure the best fit possible. In this case, the cutter measures the client, as well as creating the patterns and cutting out the fabric panels.

Measuring the client is the first step in the process. It is very important that the measurements are accurate to ensure a high-quality end product. A series of reference points are used around the body to make sure the cutter has all the information he needs to draw the patterns (**images 1** and **2**).

Here, the suit jacket is symmetrical (although that is not always the case). The measurements for the right side front and back panel are taken down in a notebook ready to be transferred to the cloth (**image 3**).

Measurements may be taken at the tailor's office, or on location. Anderson & Sheppard regularly work in the United States, even though they do all the tailoring in Savile Row, London. The measurements are used to create the outline of the pattern on the fabric. Using a series of established rules and guides, learned as an apprentice, the cutter skilfully draws the shape of the front panel using a triangle of tailor's chalk (**image 4**). Lines that indicate three-dimensional seams on the jacket, such as where the arm joins the body, are drawn freehand (**image 5**).

The tailor cuts out the pattern with a pair of shears (**image 6**). Usually, the client is not present at this point, but in this case the tailor can demonstrate how the pattern will fit around the front of the chest (**image 7**). The pattern does not have seam allowances. These are added during tailoring.

Bespoke patterns for each customer are archived so that a new suit can be made to order without having to take new measurements (**image 8**). This pattern forms the basis for a bespoke jacket, which is expertly constructed by hand (see page 346).

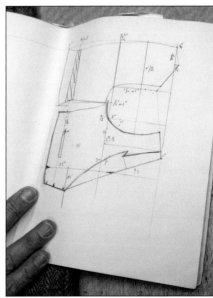

1

2

3

Once a pattern has been designed and cut it is scaled up and down by grading. For production pieces that require more than one size to be made, a pattern cutter will typically start with the middle size and work upwards and downwards from there.

The pattern and materials will determine the most appropriate cutting equipment. Hand shears offer the most precise cutting, but are the most time-consuming to use. Motorized straight knives are suitable for straight as well as shaped profiles, whereas round knives are unsurpassed for straight lines and large-radius curves. Cutting thickness ranges from a single ply with a pair of shears to tens or even hundreds of plies with a motorized straight knife. Straight knives range from around 100 to 300 mm (3.9–11.8 in.) cutting depth. Fabrics that are prone to snagging or pulling, such as delicate synthetics or pile, are best cut with a round knife.

Woven fabrics, where the warp thread runs parallel to the selvedge (edge), have the least stretch. Fabrics tend to be unrolled and pulled out lengthways (along the grain) to ensure the most accurate cutting. Crosswise grain has more stretch; and bias grain (45°) has the most stretch and is easily distorted. Therefore cutting on the bias is the most difficult and least accurate.

MATERIALS

All types of sheet materials are suitable for manual cutting, ranging from delicate nonwoven to heavy-duty fabric.

COSTS

The demand for more consistent cutting, larger fabrics, reduced labour costs and higher throughput has driven the adoption of computerized cutting systems (page 316) in large parts of the garment and upholstery industries. Even so, many manufacturers still rely on manual cutting techniques when making samples. This is the dominant method of cutting for bespoke and tailored items.

4

5

6

7

8

Pattern cutting is highly skilled and manual cutting techniques are labour-intensive, which means these processes are moderate- to high-cost.

ENVIRONMENTAL IMPACTS

The amount of waste produced during cutting depends on the optimization of the marker. Computerized nesting maximizes yield and reduces placement errors. Cutting a single item in one size will be the least efficient; cutting multiple sizes mixed to give the least waste is preferable; and the most efficient method is to arrange the number of pieces from all sizes and styles depending on the order quantities.

All textiles have grain, to a greater or lesser extent, as a direct result of their manufacture. In addition, woven and knitted fabrics have a right side, or technical face, and a grain that runs lengthways. The marker is designed accordingly. Fabrics with pile, or nap, must be oriented so the surface fibres always point or flow in the same direction.

Much more waste is produced when cutting natural materials, such as leather and fur, which is partly why it is more expensive. When using computerized nesting, flaws in the hide are marked with a light pen and the patterns are arranged around them in the most efficient manner.

Featured Company

Anderson & Sheppard
www.anderson-sheppard.co.uk

Construction Technology
CNC Cutting

CNC knife cutting is also known as automated cutting and multi-ply cutting. It is a high-speed computer-guided process that cuts panels without the need for dies or paper markers. The knife travels on x- and y-axis tracks, following a path defined by the CAM file. CNC routers are equipped with a spindle-cutting tool used to profile thick and rigid materials.

Techniques	Materials	Costs
• Rotary, drag and reciprocating knife • Router	• Woven and knitted textiles • Thin to thick and flexible to rigid	• CNC knife cutting is low-cost • Routing is low- to moderate-cost

Applications	Quality	Related Technologies
• Apparel • Upholstery • Tensile structures	• High quality and repeatable	• Laser cutting • Die cutting • Manual cutting

CUTTING

316

INTRODUCTION

This technology has automated the process of manual cutting (page 312), to make profiling panels in single or multi-ply lay-ups much more cost- and time-efficient. The equipment is often referred to as plotter/cutter, because the cutting head may equally be used to mark out (plot) patterns as well as cut the shapes.

Used in a variety of industries, CNC cutting comes in several formats, ranging from high-speed multi-ply cutters used in the production of garments to cutters up to 3.96 m (156 in.) wide used to mark and profile panels employed in the construction of fabric tensile structures. CNC cutters operate at high speed and

The CNC Cutting Process

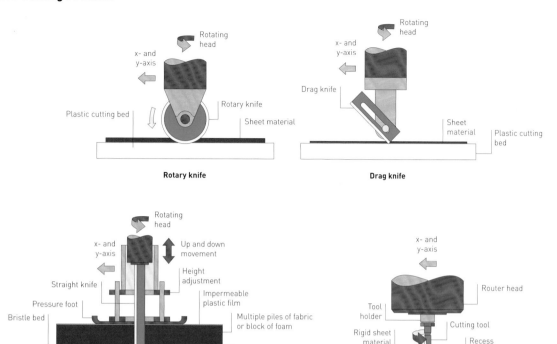

Rotary knife

Drag knife

Reciprocating knife

Router

TECHNICAL DESCRIPTION

CNC knife cutting includes rotary knife, drag knife and reciprocating knife. The head, which holds the knife, moves along horizontal x- and y-axis tracks. The head rotates to keep the blade in line with the cutting direction.

Rotary and drag knives are used for single- and low-ply cutting. Rotary blades cut by pressing down onto the sheet material. Drag blades cut as they are pulled across the surface of the material.

Reciprocating knife cutting is used for cutting textile and foam. A long straight blade is moved up and down at high speed to produce a sawing action. The height is adjusted according to the thickness of material being cut. The head rotates in line with the direction of cutting, to keep the blade pointing in the right direction.

Materials that are easily cut through may be stacked several plies thick. An impermeable plastic film is placed over the top ply, which enables a vacuum to be drawn on the stack, compressing the layers into a dense block. With this technique it is

possible to cut through stacks of fabric up to 70 mm (2.75 in.) thick and foam up to 120 mm (4.7 in.) thick.

Blades used for general cutting applications are typically made from steel. When cutting abrasive and heavier materials, ceramic and tungsten carbide blades are used, which hold a sharper edge for longer.

In addition to carrying knives, the cutting head may be used as a plotter, to apply crease lines with a v-notch knife or to drill holes.

Cutting beds are fixed or a moving conveyor, so that sheets or rolls may be processed. Conveyors automatically move new plies of textile into the cutting area after each cutting cycle, allowing the operator to remove the cut parts.

Routers are used for rigid materials. Routers cut out panels using a cylindrical cutting tool, which is rotated by a spindle motor at very high speed (up to 60,000 rpm). Three-dimensional shapes, such as those produced by composite laminating, are cut

out using a four- or five-axis CNC routers (see Thermoplastic Composite Molding, technical description, page 446). This allows the router to travel around profiles on multiple planes and cut at any angle.

Many different types of cutting tool are used, including flute (side or face), slot drills (cutting action along the shaft as well as the tip for slotting and profiling), conical, profile, dovetail and ball-nose cutters (with a dome head, which is ideal for 3D curved surfaces and hollowing out).

High force is applied by the speed of the cutting bit, so the material must be clamped to stop it moving. A vacuum bed draws air from underneath the sheet so that it is held down firmly.

are capable of cutting up to 1.5 m (60 in.) per second in suitable materials.

A router is used to cut heavy, thick and rigid materials. These are also known as milling machines, a term derived from the metalworking industry. The cutter on the router removes all material that it comes into contact with, unlike a knife, which slices or saws through material with a narrow cut width (known as kerf). Routers operate at slower speeds, up to around 5 m (197 in.) per minute, and are accurate to 0.1 mm (0.004 in.).

APPLICATIONS

Predominantly used to cut out panels for garment construction and upholstery, CNC cutting is also utilized in the production of composite laminates, sails, canopies, tents, swimming pool covers, lorry tarpaulins and side curtains, inflatables, hot air balloons and parachutes.

CNC routing is used in a wide range of applications, from cutting wooden frames for upholstery to profiling foam packaging.

RELATED TECHNOLOGIES

Laser cutting (page 328) is used in place of CNC cutting in many applications, thanks to its ability to cut, score and engrave on such a wide range of materials. One of the downsides of laser technology, and the reason it is not used for many cosmetic applications, is that the heat of the laser can leave scorch marks on the surface and edge of materials. CNC cutting leaves a clean edge, without burning, but exposed yarns may be prone to ravelling in certain types of fabric construction.

Die cutting (page 336) is used to produce shoes, garments and upholstery. Unlike CNC cutting, the blades are fixed. This means that it is more cost-effective when high volumes are required. The pressure needed to force the knives through multiple plies means that it is impractical to produce parts any larger than 1.5 m (4.9 ft) with this technique. CNC plotting is not limited by such constraints. In addition, CNC cutting may make more efficient use of material, because parts may be nested very

closely and even side by side. This is not practical with die cutting.

QUALITY

CNC cutting uses shearing or sawing action to cut through textiles and thin sheet materials. The sawing action of straight knives generates frictional heat. This can cause problems with thermoplastics, because if they heat up to beyond their softening temperature then the edges may fuse together. In this case a wavy-edged straight knife is used, which reduces the build-up of heat.

When using a drag blade, care is taken to ensure the material does not clog up around the blade holder. A lower number of plies may be cut with a rotary blade. Hard or abrasive materials are cut out with ceramic knives (carbide or zirconia).

As a computer-guided process, it is highly repeatable and produces very accurate parts.

DESIGN

Long straight lines, or large-radius curves, in single- and low-ply cutting may be

---

→ Cutting with a Rotary Knife

The movement over the textile rotates the blade. Therefore shaped and patterned blades may be used. In this case, the cutting head is loaded with a scallop-shaped blade (**image 1**).

A single ply of polyester is spread onto the cutting bed (**image 2**). The rotary knife is pressed down onto the sheet, cutting through it as the head is guided along the x- and y-axis cutting path (**image 3**). These small squares will be used as sample swatches (**image 4**). The blade penetrates through the material and into the plastic cutting bed to ensure a clean cut. No sideways force is applied to the sheet so it does not need to be held down during cutting.

The blade is readily replaced in preparation for the next cutting job, which requires a straight-edged cut (**image 5**).

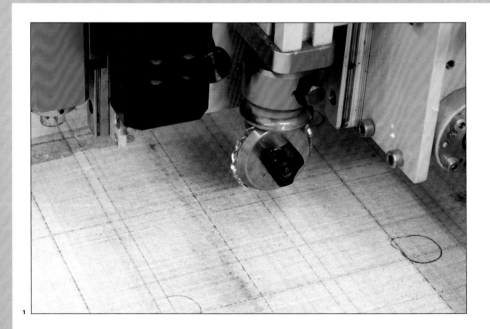

1

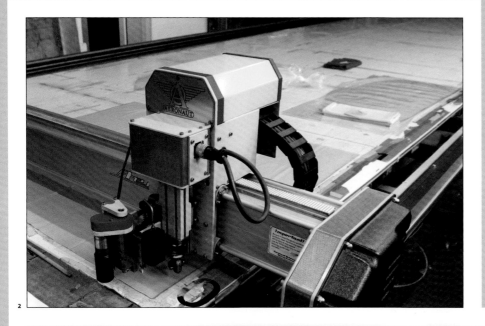

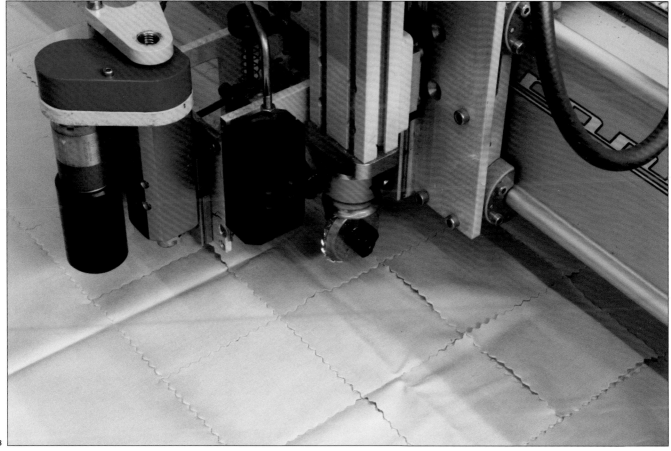

Featured Company

Automated Cutting Solutions
www.automatedcutting.co.uk

→ Multi-Tool CNC Cutter

CNC cutters may be equipped with several different tools to carry out multiple functions in a single pass (**image 1**). In this case, coated nylon fabric is marked, cut and punched in preparation for assembly into covers.

The rotary blade produces clean straight lines. Sharp corners are made by overcutting one edge then lifting the blade out and reinserting it in line with the next cutting path (**image 2**). Holes are formed with a profiled punch and used to mark assembly points (**image 3**).

Once the cutting is finished, the parts are gathered up ready for assembly by machine stitching (page 354).

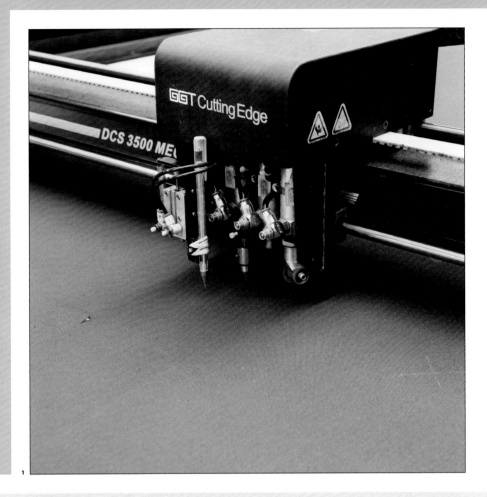

1

carried out with rotary knives. This allows for very high-speed cutting.

Patterns with more intricate details and sharp corners are cut out with reciprocating straight knives, or drag knives. Sharp corners may be cut by taking out the knife, rotating the head and punching back through the fabric.

Tool heads may be equipped with multiple types of cutting knife, as well as markers, drills, punches and a labelling device. This means that multiple processes may be carried out in a single operation and without having to re-spread the fabric each time.

For heavy-duty and thick materials the head may be replaced with a router,

which has a high-speed spindle-cutting tool. This technique is used to profile rigid plastic sheet, foam and molded composite parts (see case study on page 452 in Composite Press Forming), for example.

Advantages of cutting with a router include working on three, five or more axes (which means three-dimensional profiles may be cut out) and the ability to cut recessed areas, as well as right through the sheet material.

The minimum internal radius possible with a router is determined by the size of the cutter. For example, if a 10 mm (0.4 in.) diameter straight-sided cutter is used, then the minimum internal radius will be 5 mm (0.2 in.).

The flexibility of CNC cutting is determined by programming the cutting paths. Therefore, it is suitable for use in prototyping and sampling as well as mass production. The build-up of heat affects certain materials. In this case, overlaps in the cutting path and tight bends should be avoided.

MATERIALS

The large number of different cutting heads available makes CNC knife cutting a very versatile process. Rotary and straight knives are used to cut textile, prepreg, film and foam.

Routers are used to cut foam and rigid materials, including wood, board,

2

3

4

Featured Company

Felthams
www.cplfelthams.co.uk

plastic and composite laminates. They are not suitable for cutting flexible sheets and textiles.

COSTS

These are high-speed and efficient processes used for single-, medium- or high-ply cutting. There is no set-up required for each job, because there is no tooling, although bespoke jigs may be required to hold the workpiece during cutting.

Conveyor systems may be used for high volumes. The fabric is automatically transferred onto the cutting bed by conveyor and moved on once cut out. This reduces handling.

ENVIRONMENTAL IMPACTS

Like all cutting processes, the amount of waste depends on the efficiency of nesting. Computerized nesting and optical recognition systems make the most efficient use of material (see also Manual Cutting, page 312). Straight-sided patterns may be nested alongside one another, without any space between, which greatly reduces waste.

CNC knife cutting is a mechanical process, so there are no fumes or gases given off during operation. CNC routing is a reductive process, so generates waste in operation. Modern CNC systems have very sophisticated dust extraction that collects all the waste for recycling or incinerating

for heat and energy use. Dust that is generated can be hazardous, especially because certain material dusts become volatile when combined.

Construction Technology
Slitting and Crosscutting

Webs of cloth, foam, paper and film are cut into narrow widths in a continuous, high-speed process known as slitting. It is called rewinding when a roll is cut into strips that are taken up on separate rolls. This technique is used to prepare materials for further processing, or to mass-produce items such as scarves or tape. Sheets are produced from lengths of textile by crosscutting.

Techniques	Materials	Costs
• Mechanical • Ultrasonic • Thermal	• Textile, nonwoven and paper • Rubber, film and foil	• Low

Applications	Quality	Related Technologies
• Scarves and blankets • Tape and film • Packaging	• Precise and clean cut depending on technique • Thermal-cut synthetics have sealed edge	• Laser cutting • Pattern cutting • Water-jet cutting

INTRODUCTION

Slitting is a high-speed and precise process used in mass production. Rolls are cut into narrow widths and rewound at up to 120 m (131 yd) per minute. Slit widths may be trimmed to length by cutting across the grain.

Fabric and sheet materials are cut by hand with scissors, shears, guillotine or rotary cutters. These techniques are used to process lower volumes, such as for batch production.

Mechanical and ultrasonic cutters may be automated or handheld. Ultrasonic techniques have replaced mechanical cutting in many applications. As well as being utilized to cut fabric, film and sheet,

The Cutting Process

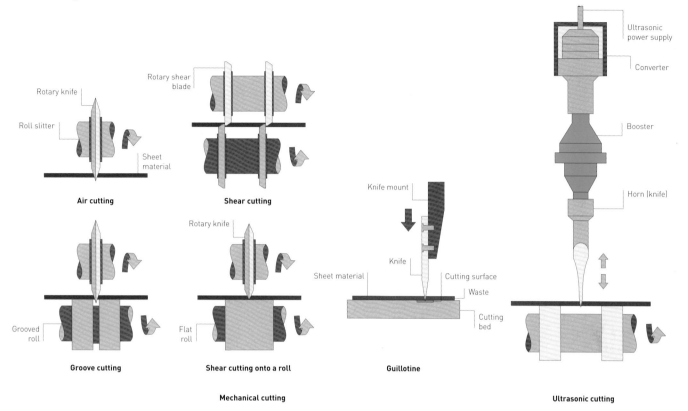

Air cutting

Shear cutting

Rotary knife
Roll slitter
Sheet material

Rotary shear blade

Groove cutting

Shear cutting onto a roll

Grooved roll
Rotary knife
Flat roll

Mechanical cutting

Guillotine

Knife mount
Knife
Sheet material
Cutting surface
Waste
Cutting bed

Ultrasonic cutting

Ultrasonic power supply
Converter
Booster
Horn (knife)

TECHNICAL DESCRIPTION

Mechanical slitting and crosscutting are carried out with sharp blades, which are either straight (razor, scissors or guillotine), or rotary disc (roll cutting or handheld rotary cutter).

Slitting without any support underneath, known as air cutting, is suitable only for materials that can be easily cut through and are stable under tension. Woven fabrics, films and foam may all be cut to width in this way.

Shear cutting is suitable for cutting a wide range of materials, from lightweight nonwovens to metal foil, into strips and sheet. Scissors, dies (see Die Cutting, page 336), guillotines and rotary blades all cut with shear force: when the force exceeds the ultimate shear strength of the material, the material will fail (cut through).

Sheet materials that require support are slit against a grooved roll or flat grooved surface that moves with the rotary disc knife. The most delicate materials, such as lightweight nonwovens, are cut against a flat surface to minimize strain on the material.

Ultrasonic cutting works on the principle that electrical energy can be converted into high-energy vibration (cutting action) by means of piezoelectric discs. Electricity is converted from mains supply (50 Hz in Europe or 60 Hz in North America) into 15 kHz, 20 kHz, 30 kHz or 40 kHz operating frequency. The frequency is determined by the application; 20–40 kHz is the most commonly used frequency for cutting.

The converter consists of a series of piezoelectric discs, which have resistance to 15, 20, 30 or 40 kHz frequencies. The crystals that make up the discs expand and contract when electrically charged. In doing so, they convert electrical energy into mechanical energy with 95% efficiency.

The mechanical energy is transferred to the booster, which modifies the amplitude into micro-vibrations. The vibrations are transferred to the surface of the material by the horn (disc or straight knife), where the micro-cutting motion slits the material without exerting any pressure. Thus the cut is made without burning or compacting the

fibres. The size and length of the horn are limited because it has to resonate correctly.

The same equipment is used to weld thermoplastics in a process called ultrasonic sewing. In this case the vibrations are converted into frictional heat rather than cutting action, which causes the layers of material to plasticize and bond together.

→ Hand Cutting Linen

Lengths of fabric may be cut with hand tools in straight lines along the grain or on the bias (diagonal). Shears (a type of scissors used to cut fabric) have slightly bent blades, which ensure that the shear-cutting surfaces are pressed tightly against one another throughout the cutting cycle.

The correct width is marked with a rule (**image 1**). The seamster picks a warp from the fabric and uses it as a guide to cut a straight line along the grain (**image 2**). Slippery or delicate materials may be held between layers of tissue paper to make them more manageable.

For longer lengths a rotary disc cutter is used (**image 3**). A sharp blade is rotated at high speed. The location of the cut is marked during the construction of the fabric by modifying the weaving arrangement to produce a dashed line.

During operation, the fabric is pressed onto the blade by the grooved foot so that an accurate cut is made and minimal stress is applied to the material (**image 4**).

1

2

4

3

ultrasonic technology has advantages for other industries, such as cutting food (cakes and sandwiches for example).

An ultrasonic cutter separates material with a micro-sawing action. The knife oscillates several thousand times per second to produce a clean cut without compacting the material (it is how cakes are cut without smudging the layers). The technique is limited by the length and shape of the knife, because it has to resonate correctly, as well as by the compatible materials.

Synthetic materials may be cut using a heated knife. With this technique the cut is formed with a sealed edge.

APPLICATIONS

Rolls are slit into narrow widths and rewound to make them easier to handle in secondary processing. The narrow widths may be crosscut to make sheet materials, such as in the production of paper and board.

Slitting may be the final stage in the production of an item, such as cutting scarves, blankets, tape, film, banners and awnings to the correct width. Crosscutting converts items to the desired length, such as rectangular blankets.

RELATED TECHNOLOGIES

Slitting is the most economical method for producing long lengths of narrow

widths from tubular (see Plastic Film Extrusion, page 180, and Circular Knitting, page 138) or full-width sheet materials. In some cases it may be more practical to cut the roll using the log-cutting method rather than slitting and rewinding.

Alternatively, materials may be pattern cut (page 314) directly into the desired size and shape for stitching and finishing.

Laser cutting (page 328) is a high-speed process capable of producing narrow widths (straight lines) or patterns (shapes) directly from webs of material. For certain applications this added flexibility eliminates the need for a separate slitting process. Water-jet cutting is a versatile process. A supersonic

Case Study

→ # Guillotining Cotton

Guillotines are useful for cutting stacks of fabric to a specified length in a single operation. Cotton fabric is hemmed in preparation for sewing to make packaging. Multiple plies are stacked and cut through with a single stroke of the knife (**image 1**).

The knife is heavy and powered by a hydraulic ram. It cuts onto a rubber pad to ensure clean edges from top to bottom (**image 2**). Care must be taken when cutting compressible materials, because they deform under the weight of the knife before cutting and so the edges may not be perpendicular. The cut piles are stacked and ready to be assembled into soft bags (**image 3**).

1

2

3

Featured Company

Felthams
www.cplfelthams.co.uk

jet of water, which is typically mixed with abrasives, is used to cut through almost any sheet material, from soft foam to titanium. It is a versatile process: intricate and complex profiles can be cut out with a water jet, as well as foam up to 200 mm (7.9 in.) thick.

QUALITY

Mechanical cutting produces a clean edge on most materials. It is essential that blades are maintained and kept sharp to avoid over-stressing the material along the cutting line (soft materials have a tendency to stretch and rigid materials form a burr on the back side). Some knitted and woven materials are prone

to ravel if they are not sealed with a hem or selvedge.

Thermoplastic fabrics, such as nylon and polyester, are cut with a hot knife. Melting produces a bead of material, which joins the ends of the yarns, preventing the material from ravelling.

When compared to conventional cutting methods, ultrasonic cutting of soft and thermoplastic materials has many advantages. Cutting force is reduced by 75%, which avoids the defects caused by applied pressure in mechanical cutting. Heat is not required, unlike laser cutting, so the edge is not burned or discoloured. The cutting action simultaneously seals the edge of thermoplastic woven and

knitted fabrics, preventing the material from unravelling. And the edge is tapered, without a bead that would add bulk.

DESIGN

Slitting and crosscutting is limited to straight lines. Shaped pieces are cut out with laser- or pattern-cutting techniques.

Material selection will determine which cutting techniques may be used. All types of material may be slit to width, including flexible to rigid plastics, fabrics and metal foils.

The thickness of material will also affect the choice of process. Ultrasonic cutting is limited to materials no more than around 20 mm (0.8 in.) thick,

→ Hot-Knife Cutting Synthetic Webbing

Nylon webbing is woven to a specified width (page 82). Coils are cut to length to make straps and reinforcement for various applications (**image 1**). To ensure a clean edge that will not unravel, the nylon is cut with a hot knife.

It is pulled through to the correct length and the heated metal knife is pressed down, cutting through and sealing the end in a single step (**image 2**). Without sealing the relatively large-denier nylon would quickly fray.

Stitched onto buckles, the ends do not need to be hemmed, which results in a smoother join and a lighter-weight part over all (**image 3**). However, on outward-facing joins the edges are hemmed to conceal the melted bead of fabric.

1

2

3

CUTTING

326

because the knife is limited to 25 mm (1 in.). Thicker materials are cut with a knife, razor or water jet.

Many parallel cuts may be made simultaneously. The distance between the knives is set up before cutting, and in some cases may be adjusted during processing.

MATERIALS
Mechanical cutting may be used to slit all types of material, from delicate nonwovens to supple fabric, stiff metal foil and plastic film.

Ultrasonic cutting is limited to thermoplastic materials, such as nylon, polyester, polypropylene (PP), acrylic,

certain polyvinyl chlorides (PVC), thermoplastic polyurethane (TPU) and certain blends; rubber; and paper and board. Woven, knitted, nonwoven, foam and film made from these materials are all suitable.

COSTS
Slitting is a low-cost, high-speed process. Abrasive and stiffer materials will require the knives to be changed more frequently.

ENVIRONMENTAL IMPACTS
Mechanical and ultrasonic cutting are efficient processes and do not produce any waste. They do not emit any smoke or fumes, unlike thermal cutting processes.

→ Cutting Scarves From Full-Beam Fabric

Cashmere is blended (page 242), twisted into yarn (page 56) and woven into fabric several scarves wide (page 76).

It is more efficient to process the fabric full-width and cut the scarves out at the end. After weaving, the tassels are formed by pearling (twisting) the warp yarns together at each end.

The fabric is scoured (page 202), tentered (**image 1**) and napped (page 216). These processes ensure a soft and stable fabric that is durable and less likely to shrink when in use.

During weaving, warp yarns were omitted from the joins between the scarves and a selvedge formed to create a secure edge. The disc knives cut the wefts that run between the edges of the scarves (**image 2**). The strips of fabric, which are the correct width (**image 3**), are crosscut to length. Finally, the scarves are pressed flat.

1

2

3

Featured Company

Johnstons of Elgin
www.johnstonscashmere.com

Construction Technology
Laser Cutting

This is a high-precision CNC process used to cut, etch, engrave and mark a range of materials. It uses thermal energy and does not apply force during cutting, which means that delicate materials may be cut into very intricate shapes, large parts cut out in a single operation, and the edge of thermoplastic fabrics such as nylon and polyester sealed as they are cut.

Techniques	Materials	Costs
• Cutting • Scoring • Engraving	• Textile, film, foam and composite • Wood, veneer, board and paper • Foil and sheet metal	• Low to moderate

Applications	Quality	Related Technologies	

• Apparel
• Architecture
• Model making
	• Precise and repeatable
	• The cut edge of synthetics is sealed
	• Some materials will scorch
		• Calendering,
		  embroidery and lace
		• CNC knife cutting
			• Die cutting
			• Devoré (screen
			  printing)

INTRODUCTION

Laser cutting produces a clean cut in materials ranging from thin textiles up to 40 mm (1.57 in.) thick sheet plastic. The quality and finish depends on the type of material. For example, synthetic fabrics melt under the heat of the laser and so the edge is sealed as it is cut, whereas natural materials burn to leave a dark line.

Laser cutting works by focusing thermal energy on a spot 0.1–1 mm (0.0004–0.004 in.) wide to melt or vaporize the material. There are two main types, which are CO_2 and Nd:YAG (also called fibre laser). The main difference between them is that CO_2 lasers produce

The Laser Cutting Process

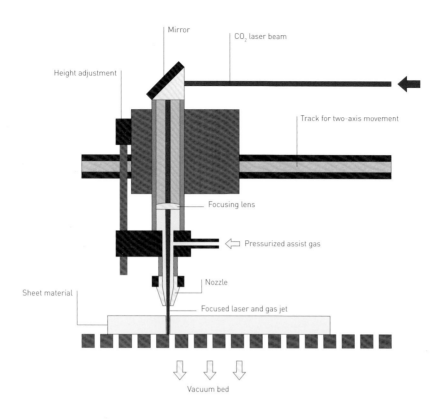

Mirror

CO$_2$ laser beam

Height adjustment

Track for two-axis movement

Focusing lens

Pressurized assist gas

Nozzle

Sheet material

Focused laser and gas jet

Vacuum bed

a 10 micron infrared wavelength and Nd:YAG lasers produce a more versatile 1 micron infrared wavelength.

CO$_2$ lasers are mainly used to process thin sheet materials such as plastic and textile. They cannot cut through metal but are capable of engraving the surface.

APPLICATIONS
Laser cutting is a versatile process used to make one-offs, and batch- and mass-produced items of clothing, upholstery, toys, packaging and labels. It is used to cut out large items too, such as architectural façades, tensile structures and sail panels.

Similar to CNC cutting (page 316), laser cutting is particularly useful in the production of clothing and upholstery when a high level of flexibility is required. For example, when using leather, each panel must be carefully placed on the hide to avoid the flaws while using the material as efficiently as possible. With computerized nesting, flaws in the hide are marked with a light pen and the patterns are arranged around them in the most efficient manner and subsequently

cut out with laser. This approach greatly increases efficiency.

RELATED TECHNOLOGIES
Laser cutting has replaced conventional mechanical cutting in many applications. It is capable of profiling shapes that would be impractical with shears or CNC knife cutting.

For high volumes and mass production, laser cutting will not be as rapid or economical as die cutting (page 336) in most cases. And the burn marks that are visible on certain materials make laser cutting less desirable for cosmetic applications. In this case, CNC knife cutting may be used as an alternative.

Relief patterns are produced on thermoplastic textiles, such as nylon, polyester and polypropylene (PP), by calendering and embossing (page 220). Alternatively, areas of fabric may be dissolved away with chemicals or other suitable solutions. Examples include devoré (see Screen Printing, page 260) and embroidery using sacrificial fabric (page 298).

TECHNICAL DESCRIPTION
CO$_2$ and Nd:YAG laser beams are guided to the cutting nozzle by a series of fixed mirrors. Owing to their shorter wavelength, Nd:YAG laser beams can also be guided to the cutting nozzle with flexible fibre-optic cores. This means that they can cut along five axes because the head is free to rotate in any direction.

The laser beam is focused through a lens that concentrates the beam to a fine spot, between 0.1 mm and 1 mm (0.0004–0.004 in.). The height of the lens can be adjusted to focus the laser on the surface of the material. The high-concentration beam melts or vaporizes the material on contact. The pressurized assist gas that blows along the path of the laser beam removes the cutting debris from the cutting area.

QUALITY

Laser cutting is precise to 0.01 mm (0.0004 in.). It produces smooth and clean cuts. The cutting width, known as kerf, is different for each type of material.

It uses thermal energy to make the cut, so thermoplastics such as polyester and nylon will melt. This means the edges are sealed as they are cut, which reduces fraying. Rigid sheet plastics, such as polymethyl methacrylate (PMMA), are cut with a polished edge, eliminating subsequent edge-finishing processes.

Natural materials, such as cotton, leather and wood veneer, will burn as they are cut owing to the high temperature of the laser beam. Thicker materials will require more energy and slower speeds to cut through, resulting in more pronounced scorching. The visual effects are different for each type of material, determined by its specific properties, such as resin content and density.

Several different techniques are used to reduce the visible signs of scorching. Blowing compressed air onto the cutting zone (pressurized assist gas) cools the area and so reduces scorching. Masking the area with tape further reduces the visible signs of burning. The effects will be least visible on darker materials. Using an inert pressurized assist gas such as nitrogen (N_2) is the most effective method of eliminating scorching. However, a large quantity of N_2 is required to be purged through the nozzle, which increases costs.

Engraving will tend to produce lighter areas on dark materials and darker areas on light-coloured materials, although the quality will vary according to the type of material and thickness of engraving.

When engraving large areas in flat sheet materials, lines may be visible on the surface as a result of the pulsing action of the laser beam. This can be reduced by slowing the laser down.

DESIGN

These are vector-based cutting systems: the lasers follow a series of lines from point to point. The files used are taken directly from CAD data, which is divided up into layers that determine the depth of each cut. It is important that all lines are 'pedited' (joined together) so that the laser cuts on a continuous path. Overlapping and repeated lines cause problems as the laser will treat each as another cut and so increase process time.

Compatible file formats include .dxf and .dwg. Any other file formats may need to be converted.

This process is ideally suited to cutting thin sheet materials down to 0.2 mm (0.0079 in.); it is possible to cut sheets up to 40 mm (1.57 in.), but thicker materials greatly reduce processing speed, depending on the type of material.

Different laser powers are required for different operations. Lower-powered lasers (150 watts) are more suitable for cutting sheet plastics because they leave a polished edge. High-powered lasers (1 to 2 kilowatts) are required to cut metals.

The strength of the laser may be controlled to produce a variety of finishes ranging from surface marking (changing the texture) to removing material to a given depth (engraving). Marking large areas in thin sheet materials can cause distortion, which will reduce the quality of reproduction.

Logos, pictures and fonts may all be engraved on the surface. A layer of material may be removed completely when engraving laminated materials (page 188) thereby revealing a different-coloured core.

Deep engravings are feasible in some materials, but the increase in power and the reduced cutting speed magnify the visual effects of laser cutting, such as melting and scorching. And when the depth exceeds the width, problems with material removal result in a poor-quality finish on the cut edge.

Lasers produce a narrow cut. However, thin sections of material can be fragile and so as a general rule the distance between cutting lines should be no less than the thickness of the material. The minimum width for fabrics will depend on the construction and how few yarns are required to maintain the structure.

Thermoplastic sheets, films and fabrics are well suited to laser cutting. Transparent and tinted material with a polished edge exhibit edge-glow: a phenomenon caused by light picked up on the surface of the material being transmitted out through the edges. Thermoplastic fabrics, such as nylon and polyester, are finished with a sealed edge, eliminating the need for hemming and stitching (page 354) in many cases.

These processes do not stress the material as mechanical cutting techniques do, so small and intricate details can be produced without reducing strength or distorting the part. Therefore, very thin and delicate materials can be cut in this way (care must be taken to ensure that the heat does not distort the material and so reduce cutting quality).

It is also feasible to process cylindrical parts and curved surfaces, although setting up for this is complex and so not as widely practised as sheet cutting.

MATERIALS

These processes can be used to cut a multitude of materials including plastic film and sheet, wood, veneer, paper, flexible magnet, composite laminates, rubber and certain glasses and ceramics.

All types of textile and related material may be cut in this way, including woven (pages 76–95), knitted (pages 126–51), nonwoven (page 152), pile (page 96), fake fur, fleece, polyester and nylon fastening tape (such as Velcro), and lace (pages 106–19). Commonly used plastics include polypropylene (PP), polycarbonate (PC) and polystyrene (PS).

Of the metals, steel cuts better than aluminium and copper alloys, for example, because it is not as reflective to light and thermal energy.

COSTS

Data is transmitted directly from a CAD file to the laser-cutting machine via the manufacturer's CAM software. Cycle time is rapid but dependent on material thickness, because thicker materials take considerably longer to cut.

This process does not apply pressure during cutting and so the sheet material to be cut does not need to be clamped or fixed. This helps to reduce cycle time. Thin and wavy materials may need to be held on a vacuum table to ensure that the material stays flat.

The process requires very little labour. However, suitable CAD files must be

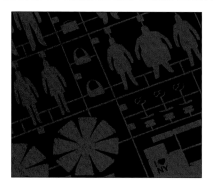

Paper
Material: Paper
Application: Model making
Notes: These 1/100 models by Terada Mokei demonstrate the precision and cleanliness of laser cutting paper.

Plastic
Material: PMMA
Application: Promotion sample
Notes: Laser cutting leaves a polished edge on PMMA sheets without any further finishing required. This sample was produced by Zone Creations.

Edge-Glow
Material: PMMA
Application: Signage
Notes: Scoring acts like an edge and so lights up. The effect is caused by light transmitted out through the edges.

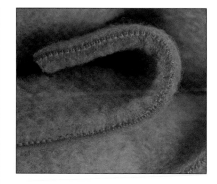

Fleece
Material: Polyester
Application: Padding
Notes: The edge of synthetic fabrics melts as it burns, sealing the edge and eliminating fraying.

Leather
Material: Cow hide
Application: Accessories
Notes: Leather is suitable for cutting, scoring and engraving. Laser cutting produces a clean, dark-coloured edge.

generated for the laser-cutting machine, which may increase set-up costs.

ENVIRONMENTAL IMPACTS
Careful planning will ensure minimal waste, but it is impossible to avoid offcuts that are not suitable for reuse. Computerized nesting and optical recognition systems make the most efficient use of material (see also Manual Cutting, page 312). Thermoplastic scrap, paper and metal can be recycled.

Some materials will burn or vaporize and give off fumes during cutting. These must be carefully extracted and filtered to minimize exposure.

Wood
Material: Birch
Application: Invitation
Notes: Laser cutting produces a dark edge, caused by burning, which in this case is used to the designer's advantage.

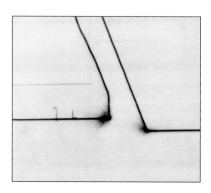

Scorching
Material: Polyester
Application: Yacht sail
Notes: Tails are caused by the laser firing at the beginning of the path. Some materials burn more than others.

→ Laser Cutting Fleece Padding

1

Laser cutting begins with the 2D CAD file (**image 1**). After creating the layout in a suitable graphic design program, such as Adobe Illustrator or CorelDraw, it is converted into a file format for the software that will drive the laser-cutting machine.

In this case, many thousands of parts are being cut out, so a roll of fleece is fed directly into the laser-cutting bed from the back (**image 2**). Alternatively, sheets of material may be placed directly onto the bed.

The laser works its way rapidly across the width of fabric, operated on x- and y-axes (**image 3**). Once the full bed area has been cut out (**image 4**), the operator removes the pieces by hand (**image 5**). The edges of the polyester fabric are sealed by the cutting process and so will not fray or ravel (**image 6**).

Once the parts have been removed the fabric is pulled through from the front and the sequence is repeated (**image 7**).

2

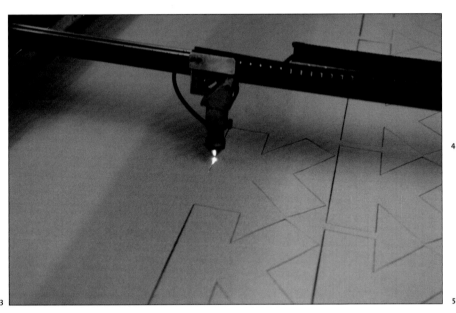

3

4

5

6

7

Featured Company

Automated Cutting Solutions
www.automatedcutting.co.uk

→ Large-Format Cutting

Large parts, such as used to make sails (**image 1**), banners, façades and tensile structures, are cut out on a large-format laser cutter. In this case, a long length of plain-woven polyester (known under the trade name Dacron) is unrolled onto the cutting bed (**image 2**). It is registered along one edge, but does not need to be fixed or clamped in any way, and laser cutting begins, guided by the computer data.

The head is held on a gantry, which travels on x- and y-axes to cut lines in all directions. It is fitted with an extractor, which removes fumes and gases produced by the burning and melting plastic material (**image 3**).

The panels are nested together to reduce waste (**image 4**). The cut parts are removed one by one (**image 5**) and rolled up ready for assembly (**image 6**). It is a reasonably heavy material, which requires high power to cut through. This leads to visible scorch marks along the cut edge.

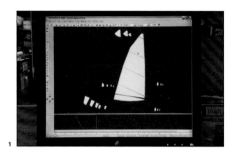

1

2

3

4

5

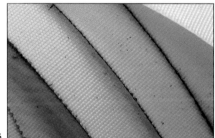

6

Featured Company

Automated Cutting Solutions
www.automatedcutting.co.uk

→ Laser Cutting Fake Fur

Fur-like fabric is manufactured with knitted synthetic yarn (**image 1**). It is laser cut by laying pile-side down on the bed so that the flat technical back faces the laser (**image 2**). The back is smoothed out as much as possible and the laser optically measures the distance with each cutting path to ensure the correct focal length of laser beam.

The laser penetrates through the knitted back (**images 3** and **4**). The fur topside remains intact and appears untouched by the effects of laser cutting. Small holes are made during cutting for attaching the parts by sewing.

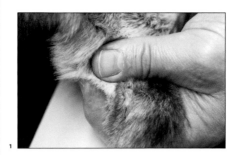

1

2

3

4

Featured Company

Automated Cutting Solutions
www.automatedcutting.co.uk

→ Cutting Intricate Details

Laser cutting is capable of reproducing complex, intricate patterns with fine details as easily as straight lines. This sets it apart from mechanical cutting processes.

The laser is aligned above the surface of this thin board and begins cutting (**image 1**). It works around the pattern, following each path as it is laid down in the CAD file (**image 2**). The order of cutting can affect the efficiency, especially when cutting many small and intricate details across a large area, because the laser has to track back and forth to cut each one.

The width left between cuts is typically no less than the thickness of material. This ensures that the positive part (to be kept) of the board is maintained during cutting. The sheet is supported by a honeycomb bed, which reduces the amount of scorching on the back side of the material caused by the laser hitting the bed. As a result, some small parts of the pattern fall away during cutting (**image 3**).

1

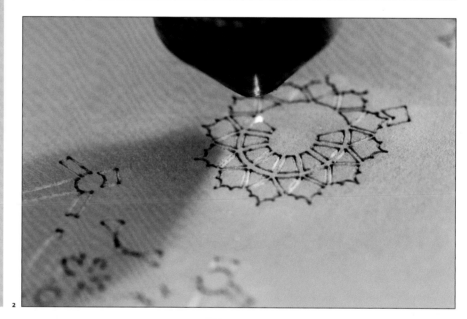

2

3

Featured Company

Automated Cutting Solutions
www.automatedcutting.co.uk

Construction Technology

Die Cutting

This process, also referred to as clicking or blanking, is a high-speed pattern-cutting operation. It is utilized to process suitably soft materials including knitted, woven and nonwoven fabrics, paper and board, and plastic sheet materials. It is rapid and used to cut out small and medium-sized parts, as well as for kiss cutting, perforating and scoring.

Techniques	Materials	Costs
• Through cutting, punching and perforating • Kiss cutting • Scoring	• All sheet materials except glass and metal	• Low for high volumes • Tooling costs are moderate • Rapid cycle time

Applications	Quality	Related Technologies
• Pattern cutting • Packaging • Embellishments	• High quality and repeatable • No sideways pressure	• CNC cutting • Laser cutting • Water-jet cutting

INTRODUCTION

Shapes are cut from sheet materials by steel knives mounted onto dies. It combines many operations into a single stamping action. The depth and shape of each steel rule determines whether it cuts through, kiss cuts (in which the top layer is cut through and the support layer is left intact), scores or perforates the material.

For batches and mass production it is rapid, and multiple layers of material can be cut through in a single stroke, making it a very cost-effective process. However, the use of dies means that it is not as flexible as computer-guided technologies such as laser cutting (page 328) and x–y plotting (see CNC Cutting, page 316).

The Die Cutting Process

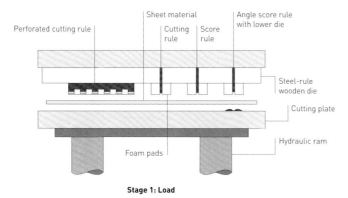

Perforated cutting rule
Sheet material
Cutting rule
Score rule
Angle score rule with lower die
Steel-rule wooden die
Cutting plate
Hydraulic ram
Foam pads

Stage 1: Load

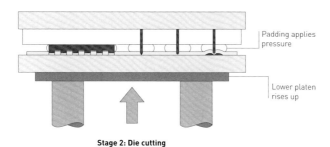

Padding applies pressure
Lower platen rises up

Stage 2: Die cutting

Flat-bed die cutting

APPLICATIONS

Die cutting is predominantly used in the packaging industry to mass-produce folded card packaging. It is a versatile process and used for a wide range of other applications including cutting out shoe uppers, upholstery fabric patterns, gloves, garment patterns, watch straps, padding, hygiene products from nonwovens, jewelry, crafts, toys and sample swatches.

Kiss cutting is used to cut to a specified depth. This is especially useful for cutting out labels, and similar items that are produced on a backing film.

RELATED TECHNOLOGIES

Die cutting is limited to applications with sufficient volume to justify building the dies. Computer-guided processes, such as laser cutting, do not require tooling, which means the cutting path may be changed as quickly as it can be programmed. Therefore these technologies are suitable for one-offs as well as volume applications. Another benefit of laser cutting is that it seals the edges of thermoplastics as it cuts.

However, thermal processes are not suitable for all materials and light-coloured fabrics may show up burned edges. In production, laser cutting is more expensive than and not as quick as using dies, but this is rapidly changing.

Die cutting is limited to materials that can be cut through with steel rules. Brittle materials that cannot be die cut, such as glass and ceramics, may be suitable for water-jet cutting. This computer-guided two-axis cutting process is very versatile and used to cut all types of material from thin foam to sheets of titanium.

Die cutting is only suitable for profiling parts up to the size of the cutting bed. Laser cutting, water-jet cutting and CNC knife cutting may be used to cut much larger parts, because they are not limited by the pressure required to cut out a complete outline in a single operation. Instead, these processes operate on x- and y-axis tracks, cutting only a fraction at a time, and can span large distances.

QUALITY

The tools are laser cut and so are precise. The steel knives wear very slowly and

TECHNICAL DESCRIPTION

Die cutting consists of three main elements: the steel-rule die, the press either side of it and the sheet material to be cut.

The steel rules pass right through the wooden or metal die (a framework to hold the rules in place) and press against the steel support plate. This ensures that all of the energy produced in the hydraulic rams is directed into the cutting or scoring action.

The pressure required for cutting is determined by the thickness and type of material. Multiple sheets may be cut in a stack, though this also depends on the type and thickness of material. Generally, the pressure required to die cut is between 5 and 15 tonnes, but some die cutters are capable of 400 tonnes of pressure.

In stage 1, the sheet is loaded onto the cutting plate, which rises up to meet the steel-rule die. In stage 2, the sharp steel rules cut right through the sheet material, while foam and rubber pads either side of each cutting rule apply pressure to the sheet material to prevent it from jamming. These pads are not always necessary, because, for example, soft and loose materials will not jam. The cutting action is instantaneous, each sheet being processed within a few seconds. However, particularly tight and complex geometries can clog up the die, in which case the operator may have to remove material manually.

It is possible to score certain materials, such as corrugated card and plastic, using different techniques, including perforating and creasing, and by adding a ribbed strip on the cutting plate. The type of steel rule used to score the material is determined by the angle and depth of score required.

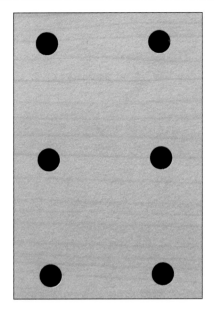

Wood
Material: Birch veneer
Application: Interior
Notes: Wood veneer is die cut to produce a clean edge, but the knives must be kept sharp.

Plastic
Material: PP
Application: Lighting
Notes: Thin plastic sheet produces a clean cut. Some materials will change colour along the cut edge. These butterflies make up the Libellule light by Black & Blum.

Paper
Material: Stock card
Application: Folding packaging
Notes: Packaging is mass-produced with die cutting. The materials are well suited to cutting and scoring.

have a long lifespan. The cutting action is between a sharp cutting rule and a steel cutting plate (or expendable plastic surface), and therefore results in clean, accurate cuts.

Scoring, kiss cutting and perforating are equally precise, and the depth of cut is accurate to within 50 microns. Some materials resist certain cutting techniques, especially kiss cutting. Cutting such materials may result in a halo around the perimeter, or cause laminated materials to delaminate. Testing is essential to eliminate such aesthetic defects.

DESIGN
A multitude of materials can be die cut to specific depths in increments only microns apart, such as in kiss cutting. The type of material will determine the thickness that can be cut.

Possible shape, complexity and intricacy are determined by the production of the steel-rule dies. Steel-rule blades can be bent down to a 5 mm (0.2 in.) radius. Holes with a smaller radius

are cut out using a profiled punch. Sharp bends are made by joining two steel rules together at the correct angle.

The width between the blades is limited to 5 mm (0.2 in.); any smaller and ejection of material becomes a problem. Thin sections should be tied into thicker sections to minimize ejection problems.

Cutting beds can typically accommodate sheet materials up to 1.5 x 2.5 m (5 x 8 ft). Nesting parts is essential to reduce material consumption, cycle time, scrap material and costs in general.

MATERIALS
All non-brittle sheet materials are suitable for die cutting, including paper and card; woven, knitted, nonwoven and unidirectional textile; plastic film, sheet and foam; cork; wood and veneer; and leather.

COSTS
Tooling costs are low to moderate. Cycle time is rapid and automated systems may cut up to 4,000 feeds per hour. Each feed may produce four or more

parts, which means a cycle time of 16,000 parts per hour. Labour costs are low in automated systems.

Manual operations are still quite quick, because profiles are cut in a single operation. Multiple small shapes can be mounted onto a single die, maximizing output. And in addition, several layers can be cut through simultaneously.

ENVIRONMENTAL IMPACTS
Die cutting does produce offcuts. However, scrap can be minimized by nesting shapes together on a sheet. Thermoplastic scrap can be recycled by the manufacturer.

Case Study

→ # Preparing Leather for Cutting

Dents makes some of the finest leather gloves through a combination of high-quality leather, expert cutting and precise sewing (**image 1**). In preparation for die cutting, the leather is dampened to loosen the fibres slightly and allow the cutter to stretch the leather out. This helps to make a better-fitting glove, and enables him to assess the look and feel to determine the best place to take each part of the glove from, because no two pieces of leather are alike (**images 2** and **3**). It requires such a high level of skill that a master cutter's apprenticeship lasts seven years.

The pattern will determine the shape and style of the finished item. Card templates are used to determine the size of leather to be cut (**image 4**). The cutter will try to get all the leather for a pair of gloves from a single piece, ensuring a consistent quality (**image 5**). One pair requires around 0.3 m² (3 ft²). And a one-piece glove will be more expensive, because a larger area of leather without imperfections is required. The selected leather is gathered (**image 6**) in preparation for die cutting (see case study on next page).

Featured Company

Dents
www.dents.co.uk

→ Die Cutting a One-Piece Glove Pattern

The steel-rule die is loaded onto the press (**image 1**). It has been used thousands of times: this type of die is easily repaired, extending its lifetime. Two pieces of leather are placed onto the die (**image 2**). They will make up the left and right hands. The press is brought in line with the die and pressure applied, cutting out the leather patterns in a single stroke (**image 3**).

Each panel is marked with an identification number, which is tracked throughout the process to ensure consistency and quality (**image 4**). The cut edge is clean and precise (**image 5**).

Every part of the pair of gloves is die cut (**image 6**) in preparation for assembly by hand sewing (page 342) or machine stitching (page 354). This includes tranks (palm and fingers), thumbs, gussets (triangles for between the fingers), quirks and fourchettes (sides of the fingers).

1

2

3

4

5

6

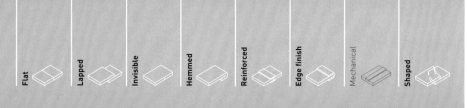

Flat · Lapped · Invisible · Hemmed · Reinforced · Edge finish · Mechanical · Shaped

Construction Technology

Hand Stitching

Hand stitching is one of the oldest construction techniques and continues to be considered superior for constructing and finishing high-value and bespoke items. Traditionally utilized by tailors and milliners, among other professions, these techniques allow designers to experiment with a range of materials, from conventional to cutting-edge.

Techniques	Materials	Costs
• Straight, edge, tack, hem and looped (fancy)	• High-quality fabrics are used, including silk, cotton and wool • Natural, synthetic and novelty thread	• High to very high

Applications	Quality	Related Technologies
• Suit making • Dressmaking • Finishing	• Very high • Depends on skill and experience	• Machine stitching • Welding • Adhesive bonding

INTRODUCTION

While machine processes (page 354) have become a very rapid and efficient means of assembly, hand techniques continue to be utilized in the production of the highest-quality items of clothing, headwear and footwear.

Before clothes became semi-disposable – as a result of continually decreasing prices – hand-sewn items were repaired, maintained and even turned inside out to get the maximum use out of a garment. At the end of their useful life, they were often taken apart and converted into a new fabric piece, such as quilting or cleaning items. Before fabrics were invented, animal skins and fur were

sewn together by hand to make clothing and shelter.

The quality and attention to detail represented by generations of development and refinement is maintained by skilled tailors, dressmakers, embroiderers, shoemakers and milliners.

APPLICATIONS

As well as being used to make tailored suits and dresses, hand stitching is utilized in the production of footwear, headwear and jewelry. It is also used to make decorative pieces, such as wall hangings and quilts.

It remains the most practical way to finish some machine-made items, such as knitted garments (page 390) and woven silk scarves.

Hand stitching is frequently used in conjunction with other handcraft techniques, such as lace (page 112) and embroidery (page 298).

RELATED TECHNOLOGIES

While it is impossible to replace the sensitivity and dexterity of hand techniques, modern sewing machines are capable of making virtually any type of stitch. However, fancy stitches, such as the running stitch and prick stitch, which are commonly made by hand, are more expensive to make than stitches developed specifically for mechanized production.

Machine stitching ensures a consistent and repeatable join at higher speeds than sewing by hand (fancy stitches are produced at a rate of more then 100 per minute and lock stitches can be rattled out at up to 4,000 stitches per minute).

Since the invention of the sewing machine, these two processes have evolved for different types of applications. Hand techniques are reserved for the highest-quality applications and for elements of construction that machines are still not capable of.

Other processes used to join fabrics that may be used in place of hand sewing include welding (page 376) and adhesive bonding (page 370). For example, sails were traditionally constructed by hand. Nowadays, panels are cut using a computer-guided laser (page 328)

and adhesive bonded to form ultra-lightweight racing sails.

QUALITY

Density ranges from tightly packed stitches, typically around five per 10 mm (12 per 1 in.), to loosely stitched hems that are intended to be flexible in use. It is important to select the appropriate thread, needle and stitch for the fabric and task.

There are many types of stitch; each one has evolved for a particular function and appearance. They may be visible and matched or contrasting colour depending on the desired effect, or concealed within the hem. Today, designers and craftspeople work with proven techniques, as well as continually developing new styles and approaches.

The final quality is to a large extent dependent on the skill and experience of the person sewing.

DESIGN

Traditional hand sewing, such as used by tailors and dressmakers, has developed to fulfill particular requirements, such as a strong join (backstitch), allowing stretch and movement (running stitch), concealing the thread within a hem (slip stitch), providing firmness and shape (pad stitch), secured a soft rolled hem on lightweight materials (whipstitch) or stopping raw edges from fraying on thicker materials (whipstitch or blanket stitch).

In the production of one-offs and prototypes, designers can exploit the huge versatility of the process, such as joining flexible and rigid materials that may not be practical or possible to join with machine stitching.

Hand stitching may be used to combine virtually any material, whether it can be sewn through or not (materials too stiff to pierce with a needle may be hole punched in preparation). In addition, a wide range of materials may be threaded onto the needle and used in place of conventional thread, such as novelty yarn, ribbon and metal. Wire that may be too delicate to machine sew, such as illuminating electroluminescent (EL) material, may be sewn by hand.

MATERIALS

All types of sheet material may be utilized. Conventional fabrics include woven, knitted, leather and nonwoven. Other types of sheet material that may be used include plastic film, metal and wood veneer.

All types of thread may be used, including cotton, silk, nylon, metallic, novelty and ribbon, depending on the application.

Designers may choose to combine material not usually associated with fashion and textiles, by adjusting their technique to accommodate the different properties.

COSTS

Hand stitching is a highly skilled process. A simple procedure, such as sewing around the hem of a scarf may take only a few minutes. By contrast, a suit can take several days, or even weeks.

These techniques are often used for high-value items, using fine materials. The highest-quality materials are very expensive and will raise the unit price significantly.

ENVIRONMENTAL IMPACTS

Hand sewing itself has no negative impacts. High-quality items that have been made with care and great attention to detail are more likely to be well looked after and maintained (see also Machine Stitching, page 354).

TECHNICAL DESCRIPTION

The most common types of hand stitching include straight stitching, edge stitching, tack stitching (also called tailor's stitching), hem stitching and looped stitching. Within each of these groups the techniques may be interchangeable, such as using a hem stitch to join two edges.

Most hand stitches are made with a single thread and worked right to left, if the person sewing is right-handed.

Straight stitches are based on making a single line of stitches that follow a predetermined path. They are used to make joins, or for decoration.

The simplest of these is the running stitch (called a basting or tacking stitch when the lines are longer and used to join layers temporarily while tailoring, or for decoration). With this stitch the front and back will look the same. The stitch allows for some give when used to join materials and so is used only when the layers will not be subjected to much stress.

A backstitch is effectively a doubled-up version of the running stitch and so is stronger and sturdier. It is used to join trim and zips onto fabric, for example. When making each stitch the needle travels twice as far on the back side as it does on the front, thus a straight line of evenly spaced stitches is formed on the front and chain-like stitches are formed on the back. The stitch may be pulled tight.

A prick stitch is a variation on the backstitch and often used as a temporary or decorative element parallel to a seam. The short stitches on the front are spaced out by long lapping stitches on the back.

Edge stitches are used to prevent cut edges from fraying and ravelling, as well as for joining materials and for decoration. They include whipstitch (also called overcasting) and buttonhole stitch. Both may be sewn at varying densities, and from either direction, depending on the requirements of the application.

A whipstitch is formed by passing the needle through the materials (layers or hem) perpendicular to the face. With each stitch the thread is pulled at an angle. Applying counter whipstitches on top forms a cross stitch.

A buttonhole stitch is very durable when sewn densely around an edge. As the name suggests, it is most commonly used to finish the edges of buttonholes, but it is also used to attach fasteners or appliqué for example. With this technique, each stitch pulls the previous one tight, to ensure a smooth and sturdy seam. It is known as a blanket stitch when used to neaten the edge of thick materials.

Tack stitches are used to join layers of material, such as temporarily in embroidery and tailoring; to mark an area of fabric; for decoration; and for permanently joining dissimilar materials.

Pad stitches (also called diagonal basting stitches) are essentially running stitches, except the entry and exit points of the needle are offset to produce a diagonal pattern. They may be stitched in parallel or symmetrical formation (chevron).

A cross stitch is formed by sewing two rows of basting in opposing directions, one on top of the other. It forms a strong and flexible join. As well as being used for pleats and tucks, such as in men's jackets, it is used extensively in embroidery.

Hems are formed using one of many techniques, such as catch, slip (blind),

Types of Hand Stitching

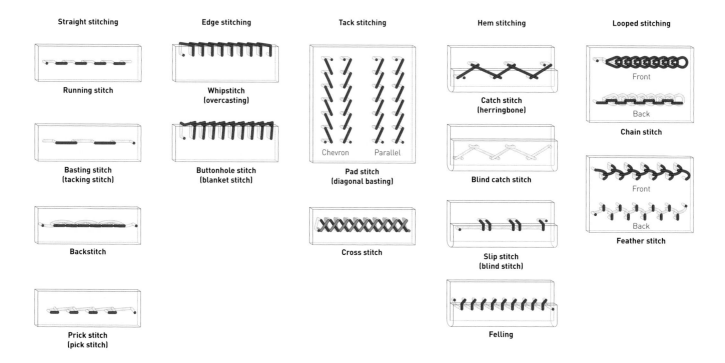

Straight stitching
- Running stitch
- Basting stitch (tacking stitch)
- Backstitch
- Prick stitch (pick stitch)

Edge stitching
- Whipstitch (overcasting)
- Buttonhole stitch (blanket stitch)

Tack stitching
- Chevron / Parallel — Pad stitch (diagonal basting)
- Cross stitch

Hem stitching
- Catch stitch (herringbone)
- Blind catch stitch
- Slip stitch (blind stitch)
- Felling

Looped stitching
- Front / Back — Chain stitch
- Front / Back — Feather stitch

felling or one of the previously mentioned stitches.

A catch stitch provides good support with some flexibility, similar to a cross stitch. This makes it particularly well suited to joining knitted fabrics. It is called a herringbone stitch when used flat or in embroidery. It is worked from left to right, and the needle from right to left, to create overlapping cross-shaped stitches.

A blind catch stitch is invisible on both sides. It is worked into the inside of the hem, which is turned back to allow access, and only passes partway through the material, perhaps catching only a single yarn.

A slip stitch is used to join two materials, or to form a hem. It is also known as a blind stitch, because it can be virtually invisible, or completely concealed by picking up only one or two yarns in the fabric being joined while concealing the long part of the stitch between the layers.

Felling stitches form a short diagonal pattern. This technique is used to form hems, attach appliqué and close seams (such as attaching collars). Short stitches are made right through the material, providing good cover and a strong join.

Looped stitches are used for decorative and functional purposes. A chain stitch (see also Weft Knitting, page 126, and Machine Stitching, page 354) is formed by passing a loop of thread through the fabric, and before it is pulled tight the next loop is pulled through it, which prevents it from pulling back through the needle hole. Thus with each new stitch the previous loop is secured and pulled tight.

Feather stitches are used for functional and decorative purposes. They are similar to buttonhole, in that each stitch is caught by one perpendicular. The difference is that each stitch is offset from the line, and made from alternate sides. The looped pattern is often utilized for embroidery.

VISUAL GLOSSARY: HAND-STITCHED SEAMS AND JOINS

Rolled Hem with Chain Stitches
Material: Silk
Application: Scarf
Notes: This soft edge finish works best on light- to medium-weight woven fabrics, such as silk, voile and organza.

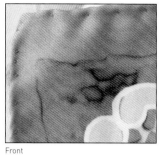
Front

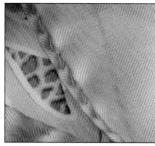
Back

Whipstitch
Material: Silk
Application: Scarf
Notes: Colour-matched thread is whipstitched over the rolled fabric to form a secure and soft hem.

Running Stitch
Material: Cotton thread, wool fabric
Application: Waistcoat
Notes: Long running stitches, known as saddle stitches, are used by tailors to temporarily hold patterns together.

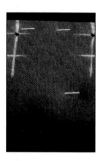

Feather Stitch
Material: Cotton thread on cotton
Application: Apparel
Notes: Feather stitches are used to decorate the hem of this sheer cotton top.

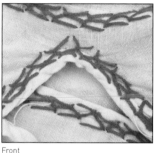
Front

Back

Pad Stitch
Material: Silk thread, wool interlining
Application: Jacket
Notes: Layers of material are held together with long pad stitches to maintain the fabric's natural flexibility.

Wood–Fabric Laminate
Material: Veneer and fabric
Application: Artwork
Notes: Hand stitching is used to join a range of materials, including the wood–fabric laminated Climatology series (see case study on page 353).

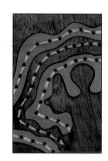

Pre-Punching
Material: Leather
Application: Gloves
Notes: Holes are punched or cut in materials that are impractical to pierce with a needle and to ensure accuracy.

Laser Cutting
Material: Veneer
Application: Artwork
Notes: Rigid materials unsuitable for punching by mechanical means may be laser cut in preparation for sewing.

→ Tailoring a Savile Row Suit

PART 1 – FITTING A COLLAR TO THE BODY OF THE JACKET

Tailoring a well-fitted suit is a very highly skilled craft involving several people, including a cutter (page 314), tailor and seamster. After the first measurements are taken the cutter produces a pattern, which is used to create panels that are placed on a length of fabric so that when cut and sewn together the suit will hang correctly and the woven pattern will be aligned. A well-made suit will look seamless, but may include hundreds of panels, including the outer, lining and several interlinings; and several thousand stitches.

The choice of fabric is largely dependent on the application and desired end result. Wool is one of the most common fabrics used to make suits. It is either woollen or worsted. Worsted is preferred for suit making, because the tightly packed yarns are smooth, strong and wear well. Fibre diameter ranges from superfine to thick: somewhere between 100 and 120 is optimum for a fine-quality and durable suit. Too fine and the suit may look very good but be prone to abrasion.

Alternatively, a suit may be constructed from cotton or linen (casual suits for warm weather, but prone to creasing), silk (light and lustrous but prone to surface abrasion), or man-made fibre (engineered performance, but has visibly synthetic drape and surface finish).

This suit consists of twill-woven worsted wool outer, canvas interlining and jacquard-woven silk liner. The tailor works the fabric and seams by hand, creating the suit in three dimensions so that it fits perfectly.

At this stage, the body and lining have been assembled and tacked together with a series of basting stitches in white thread. The location of the base of the collar is marked around the neck with a triangle of white chalk (**image 1**). Two layers of interlining – melton (heavy woollen fabric with dense nap) and

canvas (strong plain-woven flax or similar fibre) – make up the structure of the collar. They are held together with a series of pad stitches to create a strong join with just enough flexibility, and secured to the jacket along the bottom edge (**images 2** and **3**).

The lapel and collar interlining are carefully pressed. Tack stitches made previously mark the start and end of the crease line (**images 4** and **5**). Next, the tailor draws the shape of the collar onto the interlining over a shaped cushion – this helps

to create the three-dimensional finished profile – and trims it to size (**image 6**).

The wool collar panel is cut to size and carefully pressed into a three-dimensional shape to help it fit with the body of the jacket (**image 7**). It is pressed again (**image 8**) – tailoring is a continuous process of measuring, cutting, sewing and pressing repeated many times – and the excess is trimmed (**image 9**). The fit is checked before the seams are finished with compact diagonal felling stitches (**images 10** and **11**).

1

2

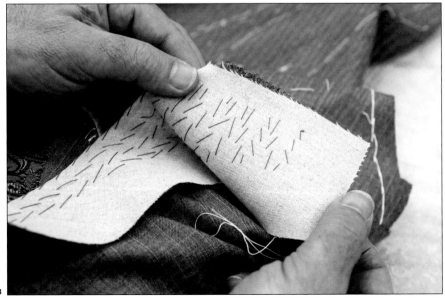

3

PART 2 – SEWING A BUTTONHOLE AND LINING THE POCKET

The location of the buttons is marked in chalk (**image 12**). The inside of the line is hole-punched to create a wider opening while ensuring that a potential tear point is not created at the ends of the buttonhole (**image 13**).

The buttonhole is opened up with a pair of scissors and prepared for sewing (**image 14**). The interlining is visible between the layers of wool (**image 15**). This provides support for the breast and helps to reinforce the buttonholes.

The buttonhole stitches are made with plied silk thread (**image 16**), as well as couching over an additional thread, to create a sturdy edge finish with added body (**image 17**). Finally, the buttonholes are temporarily bound with thread to stop them opening up and affecting the cut of the jacket during subsequent tailoring steps (**image 18**).

The pockets are lined with jacquard-woven silk like the jacket (**image 19**). The lining and wool outer are stitched together with fine felling stitches. The silk thread is colour matched to the grey colour of the wool and so virtually invisible on the face (**image 20**). The weave pattern lines up perfectly between the body and pocket flap (**image 21**).

The small white stitch left behind on the pocket, after all the basting stitches are removed, was left by the tailor to indicate where the sleeve should hang when the jacket is relaxed. The tailor made the mark when the customer was first measured up. It is these seemingly small details that make all the difference with a bespoke suit.

12

13

14

15

16

17

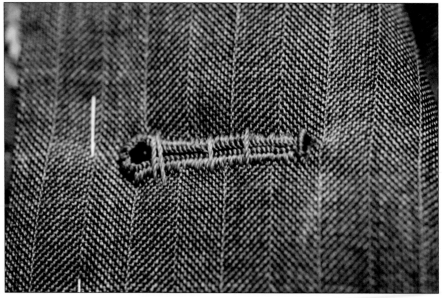

18

19

20

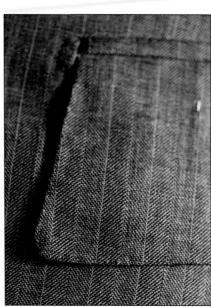

21

PART 3 – JOINING THE ARMS TO FINISH THE JACKET

The final part of the suit jacket to be assembled is the sleeves. Each one is assembled, marked out and cut to length (**image 22**). The location where the sleeves are to be fixed is marked with prick stitches (**image 23**); a modified type of backstitch.

They are loosely assembled and the cut and hang is checked (**images 24** and **25**). When fitted correctly the arms will hang to the correct point so that when the wearer is standing, the suit will fit perfectly without any tension in the fabric. The tailor works around the join with sturdy backstitches (**images 26** and **27**), after which the basting

stitches are removed with a stitch picker (or seam ripper) (**image 28**).

The shoulder pads consist of wadding secured to ermazine (soft and durable woven viscose) with a series of pad stitches to stop any slippage (**image 29**). They are assembled into the body, concealed beneath the colourful lining, stitched in place (**image 30**) and trimmed to fit (**image 31**).

The basting stitches are left in place, holding the padding and lining until the final fitting (**image 32**), after which the seams will be finished with colour-matched thread and the basting stitches removed.

22

23

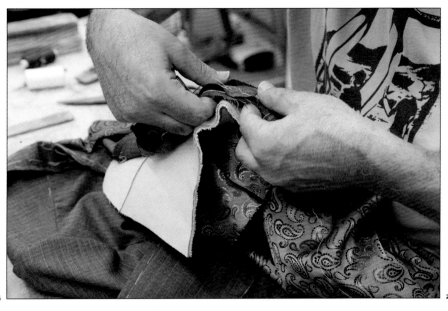

24

25

26

27

28

29

30

31

32

Featured Company

Anderson & Sheppard
www.anderson-sheppard.co.uk

→ Hand Finishing Leather Gloves

Hand stitching a pair of luxury Dents leather gloves takes a hand sewer three to four hours. In preparation, the leather is die cut (page 340) to the correct shape. This part, which makes up the hand of the glove, is known as the trank (**image 1**).

Before they are hand stitched, the holes for the points (linear-stitched detail that runs down the back of the hand) are pre-punched. The start of the lines is marked by hand (**image 2**). The perforator produces pairs of holes in straight lines (**images 3** and **4**). Pre-punching the holes ensures the highest-quality finish.

The gloves are sewn around the outside with running stitches to produce what is called a prixseam. The points are stitched and the thread is pulled tight, forming the leather into a rippled ridge profile (**image 5**).

1

2

3

4

5

Featured Company

Dents
www.dents.co.uk

→ Stitching on Wood

The Climatology collection was created by Elaine Ng Yan Ling and produced in collaboration with Wu Hao in Beijing. Each piece consists of a mix of material – wood, fabric, thermochromatic pigment, reflective surfaces and shape memory polymer – built up in layers with hand stitching and adhesive bonding. The mix of material creates objects that elegantly change shape and appearance with fluctuations in humidity, temperature and exposure to the elements.

Each of the ingredients is laminated (page 188) into a sheet material and laser cut (page 328) into the correct profile with the holes needed for stitching (**image 1**). The surface is covered with masking tape to prevent laser burns.

The designer uses gold-wrapped novelty yarn, folded three times to make eight-ply. A needle is used to sew through the layers of veneer with a running stitch. After making a complete line of stitches the designer doubles back to produce a solid line of thread, which appears the same on the front and back (**image 2**).

Several smaller pieces are sewn with metallic thread in preparation (**image 3**). They are joined onto the main structure in pairs (**image 4**). When all the layers have been joined together, which may total 12 or more pieces, they are steamed to soften the veneer. This allows the designer to manipulate the veneer into three-dimensional shapes. Once they have cooled, the laminates form a self-supporting structure responsive to the elements (**image 5**).

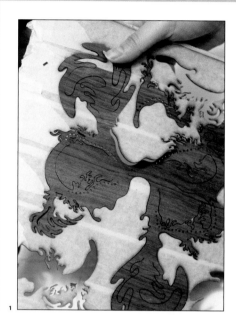

1

2

3

4

5

Featured Company

The Fabrick Lab
www.thefabricklab.com

Construction Technology

Machine Stitching

Stitching is a high-speed mechanical process used to seam, embroider, hem, overlock and flatlock woven and knitted textiles, nonwovens, paper and other sheet materials. Also known as sewing, it is combined with adhesives for high-performance applications, such as waterproof garments or lightweight racing sails.

Formats	Materials	Costs
• Lock stitch, chain stitch and fancy stitch • Overlocking and cover stitching • Taped seam (waterproof)	• Fabric, paper, card, plastic and thin metal sheet materials • Natural or synthetic thread	• Low to moderate, depending on complexity
Applications	**Quality**	**Related Technologies**
• Apparel • Packaging • Technical textiles	• High strength • Semi-permanent • Range of patterns and finishes	• Adhesive bonding • Hand stitching • Welding

INTRODUCTION

Machine stitching processes are versatile, flexible and suitable for joining and finishing a wide range of materials, including combining dissimilar materials. They are the dominant processes used to join textiles in garment construction. They are semi-permanent, which is an advantage for many applications, because it means it is possible to correct, adjust and rework stitched materials. The other joining methods, including welding (page 376) and adhesive bonding (page 370), cannot be so easily undone.

The two principal types of machine-made stitch are lock stitch (opposite), and chain stitch (page 357). Machine stitching is a form of mechanical joining that requires a complex series of operations to execute. Lock stitching is the process of interlocking two threads from either side of a sheet of material to form a strong mechanical joint. A wide variety of stitch patterns can be achieved, including straight and zigzag.

Chain stitch is constructed with one or more continuous lengths of thread, which form an interlooping chain-like structure. The downfall of this method is that if the thread is broken at any point it readily comes apart. But this is used to advantage to make joins that are easily opened by hand, such as basting operations in tailoring (page 314) and closures in heavy-goods bags.

Multi-thread chain stitch overcomes the vulnerability of chain stitching with a single thread by interlacing and interlooping the threads on the underside of the join. This creates a stronger stitch than equivalent lock stitching. It does not require a bobbin and so is used to produce high volumes of long-length stitching, such as jeans trouser legs (twin-needled folded seam).

The Lock Stitching Process

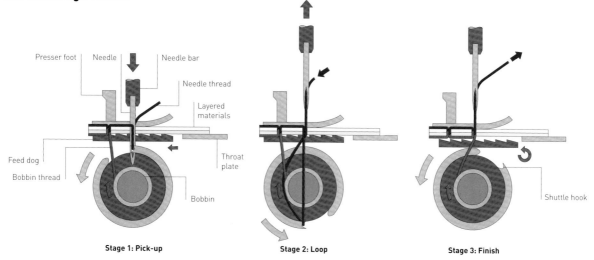

Stage 1: Pick-up   Stage 2: Loop   Stage 3: Finish

Labels: Presser foot | Needle | Needle bar | Needle thread | Layered materials | Feed dog | Bobbin thread | Throat plate | Bobbin | Shuttle hook

Overlocking is also referred to as overedging, merrowing or Serging (a trade name in North America). It is the process of finishing the edge of fabric with multiple loops of thread and can be combined with trimming and sewing. It protects the fabric from fraying by casting a net of interlacing stitches over the edge. By contrast, blanket stitching, or crochet stitching, is used to finish the edge of thick materials.

Covering chain stitches, also known as cover stitches, are the most complex and contain at least three groups of threads. A cover stitch can contain up to nine threads, including four needle threads. They are used to create large flat seams between two pieces of material. Known as flatlock, the needle threads are passed through the material, through loops of the bottom thread and between each other, to bridge the seam and form a strong joint.

There are other types of stitching used for niche applications, such as fancy stitching to give the appearance of hand stitching.

APPLICATIONS

Machine stitching remains the primary method used to join textiles and sheet materials for many applications including apparel, textile packaging (consumer and industrial), upholstery (furniture and automotive) and technical textiles (sailmaking, canopies, drapes and agro textiles).

TECHNICAL DESCRIPTION

Lock stitching is a mechanized process that interlaces two or more groups of threads. The needle and shuttle hook are synchronized by a series of gears and shafts, powered by an electric motor. In stage 1, the needle thread is carried through the fabric by the needle, and the bobbin thread is wound on a bobbin.

The needle pierces the layers of material and stops momentarily. In stage 2, a spinning shuttle hook picks up the needle thread. The shuttle hook loops the thread behind the bobbin thread, which is held under tension on the bobbin. In stage 3, as the shuttle continues to rotate, tension is applied to the top thread, which pulls it tight, forming the next stitch. The threads are interlaced tightly, making the stitch very secure. Meanwhile, the feed dog progresses forward, and the fabric is caught on its toothed surface and pulled into place for the next drop of the needle. With this feed system, known as drop feed, the layered materials are supported between the presser foot and the feed dog. The feed dog is surrounded by a throat plate, which the materials are pressed onto, preventing them from

rising and falling with the needle. The drawback of this system is that the friction between the feed dog and the bottom layer is generally greater than between the materials themselves. Therefore slippage may occur, leading to misalignment of the layers, or producing a puckered or bowed seam. Alternative systems include a walking presser foot (whereby the presser foot is split in two to be able to grip and move the top ply while the feed dog grips and moves the bottom ply), a moving needle (to hold the layers of material together as they are moved forward) and a roller (to grip and pull through the layers of material at the same rate from above).

A straight stitching is the most rapid and efficient to apply. Industrial machines can repeat this sequence 4,000 times per minute. The main disadvantage of lock stitching is that a bobbin is required, which can only hold a certain length of thread. This means production has to be paused to replenish the bobbin. It also restricts lock stitching with more than two closely placed needles, owing to the space taken up by the bobbin.

VISUAL GLOSSARY: MACHINE-STITCHED SEAMS AND JOINS

Superimposed Seam

Material: Nylon thread and fabric
Application: Waterproof jacket
Notes: Two panels are combined with a single line of stitches and turned inside out to hide the thread.

French Seam

Material: Nylon thread and fabric
Application: Waterproof jacket
Notes: Known as a French or piped seam, the two or more lines of stitching add strength and smoothness.

Taped Seam

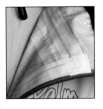

Material: TPU–polyester laminate tape on laminated polyester fabric
Application: Waterproof jacket
Notes: Sewn seams are sealed and reinforced with tape to create a waterproof join.

Clear Taped Seam

Material: TPU film tape on coated nylon fabric
Application: Waterproof jacket
Notes: Tapes are available in gloss, matt, clear and a range of colours, or they can be printed.

Stretch Seam

Material: Coated polyester and elastane thread
Application: Sleeve cuff
Notes: Elastic thread is chain stitched (for maximum elasticity) under tension, so it pulls the material in.

Bar Tack

Material: Nylon thread and fabric
Application: Safety harness
Notes: Tens, or even thousands, of stitches are applied over a designated area to reinforce joints.

Bar Tack Embroidery

Material: Nylon thread and fabric
Application: Safety harness
Notes: As well as reinforcing joins, bar tacking can be used to apply patterns and graphics similar to embroidery.

Bound Seam

Material: Nylon thread and fabric
Application: Safety harness
Notes: Raw edges are finished and protected with a folded strip of material, stitched from both sides.

Overlock

Material: Cotton fabric and nylon thread
Application: Packaging
Notes: Overlocking is used for functional and decorative purposes. The edges are cut, sewn and finished in a single operation.

Flatlock

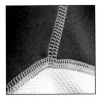

Material: Nylon thread and fabric
Application: Jacket lining
Notes: This high-speed chain-stitching process cuts, sews and finishes edges and hems in a single operation.

Chain Stitch

Material: Cotton thread and fabric
Application: Packaging
Notes: Two threads are interlaced and interlooped to create a secure join. Single-thread chain stitch will ravel.

RELATED TECHNOLOGIES

Stitching is the most versatile method for joining textiles and things onto textiles. In some cases, other joining processes may be more suitable or cost-effective. A great deal depends on the materials.

Welding (page 376) techniques, such as ultrasonic and hot bar, are used to join to thermoplastic materials such as nylon and polypropylene (PP), or materials laminated (page 188) with thermoplastic films (page 180). Joins are permanent, smooth and high-strength. Film and sheet materials, such as polyvinyl chloride (PVC), can be joined with a hermetic seal (impermeable to air and water). These processes are utilized for inflatables, tents, tarpaulins and packaging, for example.

Laser welding is also limited to thermoplastic materials. Joints are formed by heating and melting the joint interface, which solidifies to form a high-integrity weld. Since the 1970s, it has been utilized for joining films, sheets and textiles, mainly for specialized applications, in the automotive, packaging and medical industries.

There is a range of adhesive bonding (page 370) techniques. Depending on the materials and requirements there are many reasons why adhesives may be required, such as for dissimilar materials, fragile materials, hermetic joins, and combining sheet materials with non-sheet materials.

Welding and adhesive bonding affect the feel of fabrics more than stitching, because they create localized areas of stiffness, and they may alter the appearance too. These qualities can be used to the advantage of the design. For high-end and niche applications, and certain materials, machine stitching may compete with hand sewing (page 342). Stitching equipment has been developed to replicate hand-sewing patterns, such as fancy stitching a prixseam (see page 360) or whipstitch. Machine stitching is more economical. For example, hand stitching a pair of luxury Dents leather gloves takes a hand sewer three to four hours. Equivalent machine-stitched gloves take a skilled glove maker approximately 30 minutes to put together (see page 360).

The Chain Stitching Process

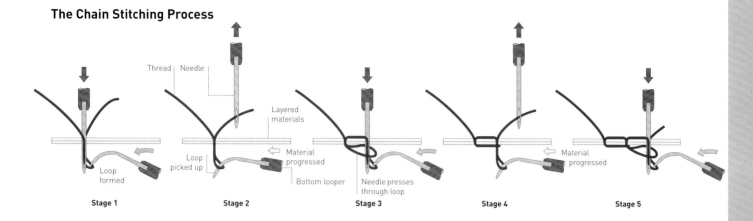

Thread | Needle

Stage 1 — Loop formed

Loop picked up — Stage 2 — Layered materials — Material progressed — Bottom looper

Stage 3 — Needle presses through loop — Material progressed

Stage 4

Stage 5 — Material progressed

QUALITY

A smooth seam that is even with no missed stitches must remain intact for the lifetime of the garment. Owing to the variety of materials and configurations, a range of techniques is employed to achieve this. Performance is measured according to strength, elasticity, durability, comfort and specific properties (such as water resistance or flame resistance). Stretch is critical for applications that require high elongation, such as providing a good fit in swimwear.

Lock stitching is used to create straight, zigzag and shaped lines of stitching. It is the only technique capable of reliably forming seams that travel around corners (even pivoting on the needle is possible). The stitch is secure, because breaking one stitch does not lead to the whole row unravelling. The end of the line of stitching is secured by back tacking (reversing to lock in the thread).

Many mechanical and decorative qualities can be achieved with this versatile process, depending on the mechanical requirements. For example, a stretch stitch is a zigzag pattern with overlap, so that as the fabric is stretched the stitch opens up without breaking.

The stitch pattern impacts on the mechanical properties of the construction. A straight stitch can weaken certain materials more than other patterns, because it creates a continuous, and often tightly packed, row of holes, whereas zigzag increases the distance between the punched holes and so has less impact on tear strength.

The materials being joined determine

TECHNICAL DESCRIPTION

Chain stitching interloops one or more threads. In single-thread chain stitch, each stitch is dependent on the previous one. Therefore, if the end of the row is not properly tied off or a loop is broken, the stitch can unravel relatively easily. To overcome this, two or more threads are interlaced and interlooped to create a more secure join (sometimes referred to as double-locked). It is quicker than lock stitch: up to 8,000 stitches per minute can be achieved.

In stage 1, to form the first stitch, the bottom looper catches the thread, lightly touching the needle as it passes. This requires very precise set-up and a skilled technician to ensure that the machines run smoothly, because needle maintenance is critical for ensuring high-quality seams. In stage 2, the material passes underneath the needle

and the bottom looper pulls the loop into alignment with the needle. In stage 3, the needle passes through the materials again, through the first loop and forming a second loop, which is caught by the bottom looper. The process is repeated in stages 4 and 5 to form the second stitch and loop.

Chain stitching multiple threads (such as double-locked) involves interlacing at least one additional thread through the bottom loops (known as a looper thread). Multiple needles and bottom loopers may be used and the bottom loops interlaced with an additional thread, or threads. This technique is used to overlock (loops passed over the edge of the material, page 368) and flatlock (loops passed over and through butt join between two materials, page 369).

the visual impact of the needle hole. For example, woven yarns give way to the needle and are not cut, while sheet materials such as leather are punched through, leaving a permanent and visible hole. The visibility of the punched hole depends on the size of the needle and the elasticity of the materials: elastic materials recover, whereas stiffer materials, such as paper and card, will not.

Quality of thread varies according to the material type and preparation technique (see Staple Yarn Spinning, page 56, and Filament Spinning, page 50).

DESIGN

Machine stitching is used to join parts with minimal visual impact (such as on patterned fabric), or as a feature (such as on leather, or bunching up the seam). Both approaches have the same mechanical requirements and the considerations are seam, stitch, sewing equipment, needle and thread.

Seams are described as superimposed, lapped, bound, flat, decorative or edge-neatening. There are many variations of each technique, depending on the visual and mechanical properties required. And each type of stitching, such as

TECHNICAL DESCRIPTION

Many specialist techniques have been developed to accommodate specific materials, geometries (such as sewing fingertips in gloves, see case study on page 361) or appearance (such as to mimic hand sewing). For example, this fancy stitching process creates a join by passing the needle, and single thread, right through the layered materials. Therefore, the thread is only visible on one side at any point, mimicking a handmade running stitch (page 344).

In stage 1, the needle is prepared with thread through a centrally located eye and held in the upper needle bar. In stage 2, the needle and thread are passed through the layered materials and the lower needle bar picks up the needle. In stage 3, the layered materials are moved through a specific distance (this forms the length of stitch). In stage 4, the needle and thread are passed back through the layered materials and the upper needle bar picks up the needle. In stage 5, the needle is pulled through. A loop, which has been formed in the thread, is picked up (usually by a spinning disk) and pulled through. The length of thread is limited to what can be pulled through. Therefore, this technique is only suitable for short seams. The material is moved forward, so that in stage 6, the process can be repeated to form the next stitch.

The Fancy Stitching Process

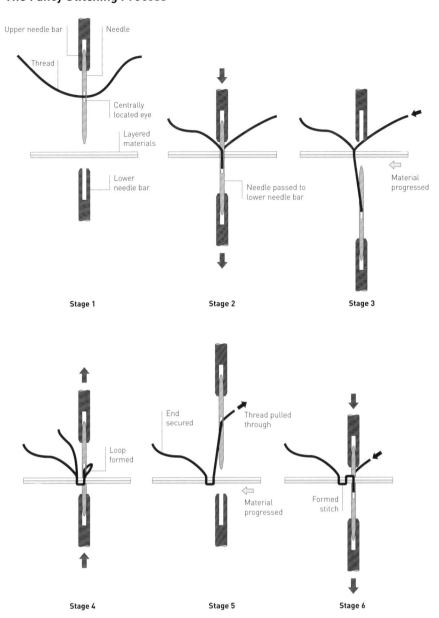

lock stitch and chain stitch, may be used to create more than one type of stitch simultaneously. This can reduce operations and so reduce cost.

Applying a fold, overlap or row of stitches puts additional stiffness into the construction (stitching can, however, be used to create flexibility, such as a hinge in rigid materials). Therefore, the placement of these join lines is critical to the final shape of the design. One, two, three or more lines of stitching may be used. At least two lines are required to produce a French seam and lock in both sides of folded joins in garments, for example, to provide a high quality

and smooth finish. The colour, pattern and number of rows of stitches can be made subtle or contrasting, depending on the visual quality desired.

A stitch pattern can follow straight, curved, convex or concave profiles. Very tight corners are also feasible with lock stitching, but this will increase sewing time, because the machinist must slow down or even stop the machine to maintain accuracy. For improved comfort, or aerodynamics, seams can be made flat, without any overlap (page 369). The number of stitches (measured per inch or per metre) is determined by the weight of material being joined. Lighter

materials typically require more stitches for the same length, although very fine fabrics may require fewer stitches to prevent the fabric from distorting.

Care must be taken to ensure the parts being joined can fit onto, or within, the sewing machine. Typical considerations are size of parts, size of openings, length of seam or hem, and thickness of material. Much of the decision making in garment production is concerned with achieving high-quality seams.

The stitching technique, thread and materials being joined determine the choice of needle. Dressmakers use very fine needles, whereas needles with a

The Taped Seam Process

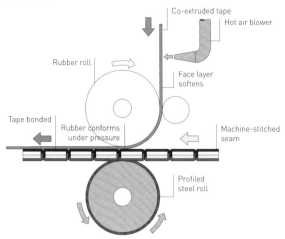

Co-extruded tape
Hot air blower
Rubber roll
Face layer softens
Tape bonded
Rubber conforms under pressure
Machine-stitched seam
Profiled steel roll

narrow wedge point are required to penetrate leather and create a suitably large hole for the thread to pass through.

Thread is often coated with lubricant or wax to reduce friction and aid its running through the needle and fabric. This is especially important for thermoplastic threads, such as polyester and nylon, which will soften and melt if they get too hot.

Tapes are applied to waterproof seams to increase durability and improve resistance to abrasion and fraying. The backing layer may be clear, coloured or printed (see Screen Printing, page 260, and Digital Printing, page 276) to match or contrast with the base material.

MATERIALS

Natural and synthetic yarns are used as sewing thread. Cotton, polyester and nylon are the most common, because they are strong and resistant to abrasion. In most cases synthetic threads outperform natural fibres mechanically and visually. Even so, natural fibres are not sensitive to temperature (thermoplastics will melt if they get hot) and may be dyed along with the fabric in some cases.

Synthetic fibres are processed to ensure low shrinkage (such as in washing and heating treatments), they are long-lasting (not affected by mildew and bacteria), and they have high abrasion resistance and high tenacity. Polyester and nylon have good resistance to chemicals; they are available in a range of bright colours, but will degrade with exposure to

ultraviolet light. Polyester is low-cost and versatile, with low shrinkage. Nylon is stronger and more expensive, but has a tendency to stretch during sewing and so shrinks back afterwards.

Silk is sometimes used as a sewing thread, but is expensive. High-performance synthetic threads include para-aramid (also known under the trade name Kevlar) and ultra-high-molecular-weight polyethylene (UHMWPE) (also known under the trade name Dyneema).

Thread is available in a range of formats including staple, filament and core-spun yarn (including ply, cord and braid). Core-spun is made up of a core surrounded by staple fibres to combine the strength and elongation of the core with the characteristics of the staple fibres.

Any material that can be punched through with a sewing needle can be joined or finished by machine stitching. However, each material has different sewing characteristics that must be taken into consideration, including resistance to puncture, resistance to tearing, elasticity and thickness.

COSTS

There are no set-up or tooling costs. Labour costs are significant, because unlike many modern manufacturing processes, a person is required to operate each machine. Unit costs depend on the length and number of stitches – in other words, how long it takes for a person to complete the task. The cost of the thread

TECHNICAL DESCRIPTION

Stitched seams may be sealed with plastic tape to make them waterproof. Taping increases the strength of the join by bonding the two panels of textile together.

The tape is typically a co-extruded plastic film (page 180), made up of a thermoplastic face layer and a protective backing layer. Membrane layers are added for improved performance, such as incorporating polytetrafluoroethylene (PTFE) between the adhesive and the outer layer.

Two layers of fabric are first joined with a lock stitch. They are fed between two rolls: a steel roll on the bottom and a hard rubber roll on the top. The steel roll is shaped with a groove to allow space for the stitch. Under pressure, the rubber roll conforms to the shape of the stitched seam, so that even pressure is applied across the width of the tape.

Just before the tape is fed between the rolls it is heated with a jet of hot air. The rise in temperature causes the thermoplastic face layer, which is typically polyurethane (TPU), to soften and become tacky. It is pressed onto the seam, forming a strong bond as it cools.

represents only a very small proportion of the material costs.

ENVIRONMENTAL IMPACTS

Well-made and sturdy garments may be reused many times and passed on, reducing the need for new clothes or old clothes being disposed.

Joining two or more materials presents a challenge at the end of their useful life. Stitching can be unpicked (undone), but this is not yet practical on a large scale. Using a single type of thermoplastic material means the item can be recycled more efficiently, but this may not be suitable or practical.

Recycling involves cutting the panels out of garments, which can be reused to make similar but not identical items. Ideally, materials are remade into products with equal or higher value, not downcycled into lower-quality products.

→ Glove Making

These are some popular types of stitching used in glove making. A prixseam is a superimposed seam with the cut edges exposed (**image 1**). It gives a sturdy look to products. Inseam is the same but made inside out, so the raw edges are concealed within the glove to give a smoother appearance (**image 2**). This is especially used for unlined or silk-lined gloves. A piqué seam is overlapped, so one edge is visible on each side (**image 3**). An overlock stitch on the outside is known as a whipstitch, brosser or roundseam (**image 4**). The most expensive are hand-stitched gloves (**image 5**). The same appearance can be achieved with a machine-made fancy stitch.

1

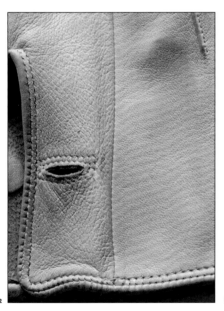

2

3

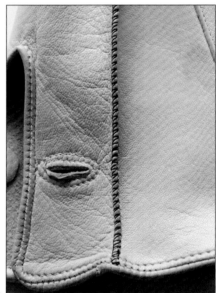

4

5

INSEAMING LEATHER GLOVES

The parts that make up the leather outer are first dampened, conditioned and cut out (see Die Cutting, page 336). The lock-stitching machine is fitted with two rollers, instead of a feed dog and presser foot. This is to ensure that the two layers of leather are fed through precisely and at the same rate. It is set up horizontally (as opposed to vertically), so the machinist can see the two edges being joined (**image 1**).

Making an inseam requires that the glove is sewn inside out. First the thumb and gusset are sewn into the hand, which is known as the trank (**image 2**). Then, the fourchettes (insides of the fingers) are joined to the trank. They are trimmed to length as they are joined together (**image 3**). Glove making is a very delicate and skilful operation. Apprenticeships take seven years. The fingertips are especially tricky, requiring tweezers to complete: as the machinist works around the tip where the two fourchettes meet, the seam is carefully arranged (edges pointing outwards and not all in one direction) to ensure a smooth and even finished glove (**image 4**).

Once the parts are joined together (**image 5**), the glove is closed with a continuous line of stitching around the outside edge (**image 6**). The sewn glove is turned inside out, revealing the finished inseams (**images 7** and **8**), and is ready for lining and welting (page 394).

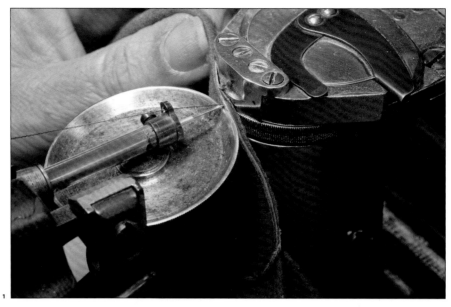

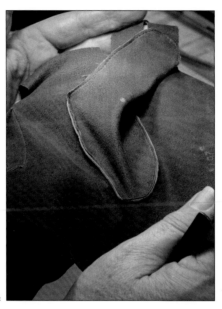

Featured Company

Dents
www.dents.co.uk

→ Fabricating Lightweight Racing Sails

While there have been many developments in the production techniques used to manufacture high-performance mainsails and genoas, downwind sails continue to be constructed with stitched seams. They need to be very light with large camber (**image 1**). Ripstop nylon provides just the right balance of strength and shape retention. The sturdier warp and weft fibres at measured intervals ensure the lightweight fabric can withstand the high loads applied during sailing (see Loom Weaving, page 76).

The two-ply polyester thread – coated to prevent the yarns slipping and improve resistance to ultraviolet light – is inserted into the grooved and air-cooled needle (**image 2**). The groove accommodates the thickness of the thread when punching holes through the material, and the cooling enables faster sewing speeds.

The seams are lapped and the stitches zigzagged (**image 3**) to ensure the load is spread sufficiently across the join. The parts are sewn together and once all the panels have been joined the sail is finished with mechanical fasteners and bound edges (**image 4**). Large areas of textile are joined by a machinist operating in a dugout with a sewing machine at the same level as the work surface. This reduces the stress applied to the fabric during construction (**image 5**).

The finished sail includes a range of stitches (**image 6**). Each one has been chosen to provide specific strength characteristics at the seam.

1

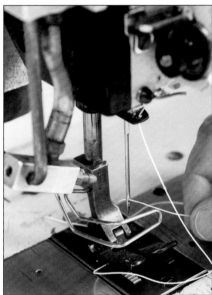

2

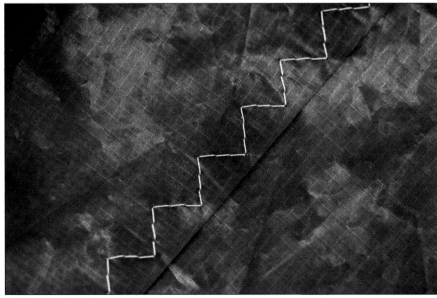

3

4

5

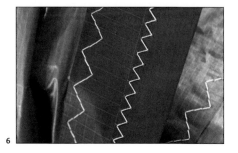

6

Featured Company

OneSails GBR (East)
www.onesails.co.uk

→ # Constructing a Waterproof Taped Seam

High-performance paddling gear needs to be lightweight, breathable and waterproof (**image 1**). Many different types of seams are used in the construction of outdoor garments and accessories. Several types are explored in the following pages.

The three-layer fabric is made up of a waterproof and breathable membrane, laminated between a durable outer layer and an abrasion-resistant lining, such as nylon or polyester (**image 2**). The laminates depend on the requirements of the application, such as resistance to abrasion, lightness or texture.

In this case, the edge of the garment is stitched with a superimposed French seam (**images 3** and **4**). A tape made up of an abrasion-resistant nylon outer, a PTFE membrane and a TPU layer (adhesive) is used to waterproof the seams (**image 5**). It is pressed onto the inside of the join by two rotating rollers (**image 6**). A jet of hot air warms up the adhesive just before it is applied, causing it to become tacky.

Finished seams are pressure tested at 0.2 to 0.5 bar (depending on the material) at designated intervals for quality control (**images 7** and **8**). Not all the products require testing, but checking the finished immersion suit with compressed air and water ensures that the product will be completely watertight (**images 9** and **10**).

1

2

3

4

5

Featured Company

Palm Equipment International Ltd
www.palmequipmenteurope.com

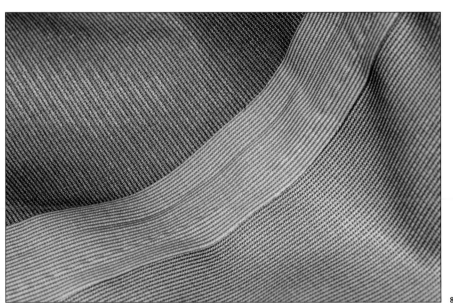

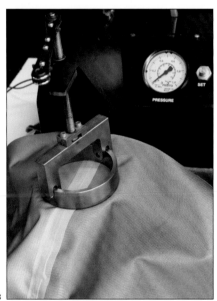

→ Bar Tacking a Safety Harness

Bar tacking is used to produce high-strength joins on lifesaving equipment, such as this search and rescue personal flotation device (**images 1** and **2**). It is also used to reinforce belt loops and pockets on garments. As a decorative technique, it may be used to embroider small areas with patterns of information (**image 3**).

It is a lock-stitching process, used to join and reinforce materials. Instead of the material moving, it is clamped in place and the needle moves. This means that up to

30,000 stitches can be applied very quickly to an area up to 100 x 100 mm (4 x 4 in.) without stressing and buckling the materials being joined.

In this example, amber nylon webbing is being joined to a buckle. The webbing is loaded into the bar tacker (**image 4**). The stitch pattern is pre-programmed, so when the pedal is depressed, identical rows of stitching are reproduced (**images 5** and **6**). Combining zigzag with straight stitches back and forth produces a strong interlocked seam.

Two lines of stitching are used to ensure a reliable join that is less prone to twisting and tearing (**image 7**).

1

2

3

4

5

6

7

Featured Company

Palm Equipment International Ltd
www.palmequipmenteurope.com

→ Stitching a Bound Seam

Raw edges may be bound with a tape sewn from both sides. On this lightweight outdoor jacket, the bound seam provides a smooth edge (**image 1**). It is used to finish constructions made up of more than one material, or if the materials resist folding (such as bulky or stiff fabrics). Additional material can be integrated into the seam too, such as for added stiffness or elasticity around a waistband. It is commonly used to finish the neck of t-shirts.

The case study demonstrates applying a twin-thread chain stitch (unlike the jacket cuff, which uses a single needle thread). The machine is prepared with two needle threads and one looper thread. The tape is fed from the side, folded as it meets the needles (**image 2**).

The fabric is folded to form a hem and loaded into the sewing machine (**image 3**). As the material is fed into the machine it is simultaneously combined with the edging and stitched (**image 4**). Bottom loopers pass the looper thread between the needle threads to create an interlacing stitch on the back side (**image 5**).

1

2

3

4

5

Featured Company

Palm Equipment International Ltd
www.palmequipmenteurope.com

→ Edge Finishing with Overlocking

An overedge stitch runs along the seam and over the edge (**image 1**). With this technique, a narrow band of stitching typically 3 to 5 mm (0.12–0.2 in.) from the edge is interlaced with loops passed over the edge. A suitably bulky thread is used, which spreads out to cover more surface area. In this case textured filament nylon is used (**image 2**). In garment construction, this process is often combined with trimming, to reduce the number of operations.

The materials to be joined are cut to size and aligned on the machine (**image 3**). The fabrics are seamed and edge-finished in a single process (**images 4** and **5**). Using five threads eliminates the need for two steps, by combining joining with overlocking. Contrasting coloured thread creates a bold finish. As such, overedge seams are used for decoration, reinforcement and construction. The threads are fed from cones (unlike lock stitching bobbin thread, which is restricted to a bobbin), thus it is suitable for high volumes of long seams.

1

2

3

4

5

Featured Company

Palm Equipment International Ltd
www.palmequipmenteurope.com

→ Joining with Flatlocking

Covering chain stitches are used to form flat seams (no overlap) between two pieces of material. These techniques are similar to overlocking, except that the interlooping threads are passed over and through a flat seam, making it the most complex stitching process. It requires at least three threads and can use as many as nine.

Flat seams reduce bulk and increase comfort. These qualities are utilized in undergarments (**image 1**). This blue neoprene insulation layer is flat-seamed with three needle threads interlooped on the outside surface for maximum comfort and performance (**image 2**).

In another example, sheets of neoprene are cut to size for constructing a wetsuit (**image 3**). The flatlocker is prepared with four needle threads and two looper threads to join the 3 mm (0.12 in.) thick material and ensure a very strong join (**image 4**). Two parts are aligned and loaded into the sewing machine (**image 5**). The edges are trimmed and stitched in a single operation (**image 6**). The underside shows the interlooping threads bridging the seam (**image 7**).

1

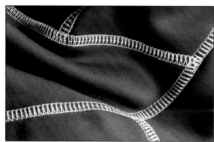

2

3

4

6

5

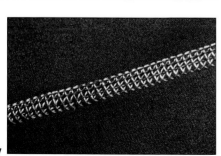

7

Featured Company

Palm Equipment International Ltd
www.palmequipmenteurope.com

Construction Technology

Adhesive Bonding

In this process, seams between two pieces of fabric are sealed with adhesive, eliminating the need for machine stitching and taping. Adhesives are engineered to match the performance of the base fabric, are less bulky than sewn seams and can be applied as a design feature or virtually invisible. Seams made in this way are also referred to as welded, laminated and fused.

Techniques	Materials	Costs
• Hot melt • Tape and film • Spray	• All types of sheet material, depending on the adhesive system	• Moderate to high

Applications	Quality	Related Technologies
• Apparel • Upholstery • Footwear	• Smooth and bulk-free • Waterproof • High strength	• Machine stitching • Welding • Laminating

INTRODUCTION

Adhesive bonding creates high-performance and streamlined seams between similar and dissimilar materials. It eliminates the need for machine stitching (page 354).

Adhesives are engineered with the same performance as the base materials they are being used to join, such as elasticity, waterproofing and tear strength. This unique ability is used in a range of applications from outdoor gear and footwear to upholstery.

The three principal systems are spraying, hot melt and continuous tape. The same techniques are used in laminating (page 188). The difference is

The Adhesive Bonding Process

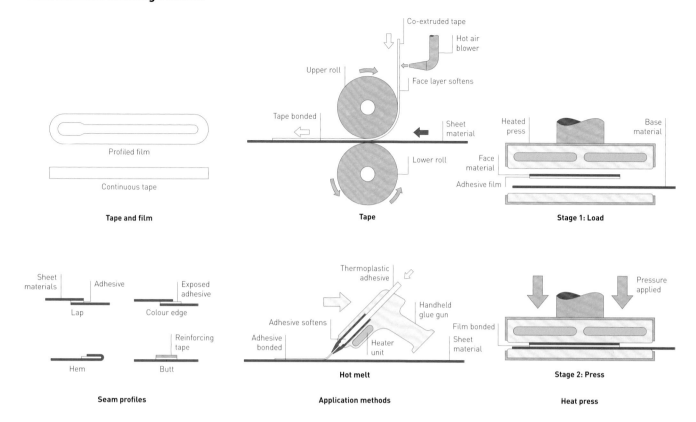

Tape and film

Profiled film

Continuous tape

Tape

Co-extruded tape

Hot air blower

Upper roll

Face layer softens

Tape bonded

Sheet material

Lower roll

Stage 1: Load

Heated press

Base material

Face material

Adhesive film

Seam profiles

Sheet materials

Adhesive

Lap

Colour edge

Exposed adhesive

Hem

Butt

Reinforcing tape

Application methods

Hot melt

Thermoplastic adhesive

Adhesive softens

Adhesive bonded

Heater unit

Handheld glue gun

Heat press

Stage 2: Press

Pressure applied

Film bonded

Sheet material

TECHNICAL DESCRIPTION

Adhesive bonding is used to form a range of seam profiles including lap (and colour edge), hem and butt. A lap seam is made up of two panels of material, joined along their edges. The width of overlap is determined by the requirements of the application. Exposing a set width of adhesive on the face of the base material will produce a coloured edge along the seam.

A hem is used to protect the edge of the material from ravelling. It may be combined with all of the other techniques. Doubling up material increases bulk and weight.

Cutting and joining the panels to form a butt joint produces a seam with a smooth face on one side. The cutting and bonding happen simultaneously, similar to the way in which a flatlock stitch is formed (page 369). The back side is reinforced with tape (similar to a taped seam, page 364) to strengthen the join and ensure the edges cannot pull apart.

Thermoplastic tape and film are supplied as continuous strip, or as sheets that are cut out to the exact shape of the seam.

Various techniques are used to apply adhesive. These are some of the more common. Tape is applied by hand or with rollers. The co-extruded tape is made up of thermoplastic, which softens as it is heated. The tacky surface is pressed down onto the sheet material. If fabric is used then the adhesive is forced into the structure. Once applied, the panel with adhesive is aligned with the second sheet and either rolled again, or heat pressed.

Hot-melt adhesive is applied with a heated gun. Thermoplastic is fed into the back and is heated as it is forced through the nozzle. As it melts it is deposited onto the surface of the sheet material. The second sheet of material is placed on top and pressed downwards, either in a press or between rolls. The thermoplastic cools to form a strong bond.

The heat press is used to produce one-offs and high volumes. In stage 1, the parts that are preloaded with adhesive film are placed onto the lower platen. Profiled (shaped) films may be used, and multiple

layers of adhesive and material may be built up in a single pressing cycle. Specific applications, such as constructing trainers, may require a three-dimensional press. In this case, an individual press will be required for left, right and each size.

In stage 2, the heated upper platen is lowered and pressure is applied for 20 to 30 seconds, or until the adhesive has heated sufficiently to migrate into the two materials. The part is removed and allowed to cool, by when full strength is achieved.

that for seams the application area is limited to the edges.

Spray techniques are good for covering large areas very rapidly, although they are not very precise. Hot-melt adhesive is limited to textiles that can withstand the high temperature. A thermoplastic adhesive is melted onto the join line and the panels are pressed together to form a strong bond.

Tape is utilized for seams that are both continuous and straight. Film (large area of tape) is used for seam profiles that can be pre-cut from film and applied to the fabric before or after pattern cutting. Both are clean and used on a wide range of materials. They are typically applied to one side of the join, activated with heat and pressure (or ultrasonic or radio frequency) and then the two panels are brought together and the heating and pressing cycle is repeated for a strong bond between the fabrics.

APPLICATIONS

Bonding, as opposed to stitching, avoids punching holes and so does not compromise the performance of the fabric. It is commonly used in combination with cutting and sewing techniques to apply pockets, zips and trim to garments.

In some cases, the majority of the product or garment may be joined with adhesive, such as racing sails, tents, sleeping bags, shoes, lingerie, technical outerwear and sportswear. In these products the extra cost is justified by the performance benefits.

Adhesive bonding may be the only option for joining textiles if one or more of the materials cannot be stitched, welded or mechanically joined to in any other way, such as very stiff or delicate materials.

RELATED TECHNOLOGIES

Particularly demanding applications that require very strong seams may be stitched and bonded (see page 370).

Taped seams (page 364) may be replaced by adhesive bonding. This approach eliminates the stitching and associated bulk, weight and cost. However, it is more complex and requires a higher level of skill, so only a few specialized factories are capable.

Welding is used to join thermoplastic or laminated materials. It competes directly with adhesive bonding in certain applications, with the advantage of cutting out the additional material that is required for adhesive bonding. For aesthetic parts adhesive bonding is preferable, because it does not affect the surface quality as welding does.

Laminating and adhesive bonding are the same, except that in laminating, the entire surface of the two or more materials are bonded as opposed to just the edges. The boundary between these two processes becomes blurred, such as in the production of laminated trainer uppers. The thermoplastic panels are bonded where they overlap, forming non-linear and unconventional seams. Compared to conventional cutting and sewing techniques, this approach yields lighter and less bulky trainers.

QUALITY

Seams formed in this way are very strong and resilient. The mechanical properties of the adhesive are tailored to the application, to ensure compatibility. For example, thermoplastic polyurethane (TPU) adhesive is elastic with very good stretch and recovery. When bonded into the seam it will closely match the performance of the base material, providing a soft hand, even when high stretch is required.

It is essential that the adhesive penetrate into both fabrics. Improperly bonded panels may peel apart in application.

Adhesive can cover a large area without affecting the process. This helps to spread the load on the seam in application. After gluing the exposed fabric edge is stuck down, producing a smoother edge less prone to ravel.

Adhesives are either clear or coloured. Clear types do not require colour matching.

DESIGN

Adhesive bonding surpasses other joining techniques when multiple parts with complex seams can be joined in a single pressing operation. This technique is used to make parts that are formed into three-dimensional profiles and so require a large number of seams, such as shoe uppers and bras.

As well as forming the seams, the laminating cycle may be utilized to apply overlay films without sewing, such as logos, knee and elbow patches, and reflective panels. This opens up opportunities for designers to incorporate material joins where they would otherwise be avoided, because stitching multiple layers together can lead to a bulky and poorly fitting garment.

Tape and film may be pre-cut to fit the profile of the seam. Combined with pattern-cutting techniques, such as laser cutting (page 328) and die cutting (page 336), complex shapes may be cut out in a single operation.

Adhesives are used to seal all types of geometry, including lap, hem and trim. Coloured films may be left exposed around a seam to create a precise coloured edge detail. This approach helps to reduce edge fraying too.

When designing for adhesive bonding, it is important to consider the compatibility of the adhesive and the textile. The performance requirements that make adhesive bonding a viable option are likely to determine material selection. Systems have been developed to fit specific requirements, such as lightweight, waterproof and windproof.

MATERIALS

All types of material can be joined, including woven, knitted, nonwoven, leather and plastic sheet materials. Openwork fabrics (see lace, pages 106–19, and knitting, pages 126–51) are bonded without delamination or wrinkles.

The most commonly used thermoplastic adhesives include TPU, polyamide (PA), polyethylene terephthalate (PET) and polyolefin. TPU has high tensile strength and elasticity, with good abrasion resistance. It is used as a film or hot-melt adhesive and is available transparent or coloured. PA film is used to bond a range of natural and synthetic fibre textiles, as well as foams. PET is a hot-melt adhesive. It is quite stiff

Waterproof
Material: Coated nylon
Application: Waterproof jacket
Notes: Zips and pockets are adhesive bonded, because stitching would compromise the jacket's performance.

Boot
Material: Molded elastomer sole with laminated fabric upper
Application: Footwear
Notes: Bonding is capable of producing lightweight and high-strength seams between flexible and molded parts.

Trainer
Material: TPU film, mesh and synthetic backing layer
Application: Footwear
Notes: TPU film is bonded directly onto the spacer textile below to produce a high-strength, smooth and stitch-free outside surface.

Unidirectional
Material: Carbon fibre
Application: Composite laminating
Notes: Glass filaments are adhesive bonded across the back of unidirectional carbon-fibre filament tows (untwisted yarns) to hold them in place in readiness for composite laminating.

Front Back

and so not suitable for applications that require flexibility. Polyolefin films include polypropylene (PP) and polyethylene (PE). They are used to join PP and PE fabrics that have low surface energy and so are very difficult to bond any other way.

It is important to consider the finishes applied to the fabric before joining (see fluid coating, page 226, and laminating). Treatments that include silicone or polytetrafluoroethylene (PTFE) will reduce the strength of the adhesive bond.

COSTS
Typically, the process is slower than machine stitching, a high level of skill is required and adhesive is more expensive than thread, making this an expensive

alternative. So for now it is limited to high-performance and high-specification applications, and only a limited number of factories have the capability.

In certain applications, however, adhesive bonding may be more cost-effective. For example, if multiple seams can be bonded simultaneously, then labour costs are significantly reduced for mass-production applications. In addition, bonding may reduce the total number of components required to make a garment.

ENVIRONMENTAL IMPACTS
Adhesive bonding is an efficient process, but it is additive. Thermoplastic adhesives are recyclable if they are combined with chemically compatible materials, for

example PET film adhesive with polyester fabric, TPU film adhesive with elastane fabric, and polyolefin adhesive film with PP- or PE-based fabric.

Dissimilar materials may be separated at the end of their life, because thermoplastics can be reheated and taken apart, although this is not usually practical or efficient.

→ Adhesive Bonding Racing Sails

There has been a great deal of development in recent years to reduce the weight burden of sails on racing boats. Computer software, such as finite element analysis (FEA), is a powerful tool used to assess and improve the potential performance of the sail prior to manufacture. In this case strain is measured across the surface to check the flying-shape performance (**image 1**).

The panels are made up of aramid (gold-coloured) and carbon (black) fibres laminated between sheets of polyethylene terephthalate (PET) film (**image 2**). Compared to stitched panel sails, adhesive bonding the seams improves strength to weight, while creating a smoother surface for the air to flow over. The panels are delivered to the sail loft pre-cut, although the final trimming is left to the skilled sailmaker (**image 3**).

The thermoplastic adhesive tape is elastic and sticky at room temperature. It is applied to the base material by hand with the backing paper intact (**image 4**). The backing paper is peeled off when the two panels are ready to be bonded together (**image 5**).

The panels are carefully aligned and pressed together (**image 6**). The tape is applied along the middle of the lap join, leaving space on either side for hot-melt adhesive. Masking tape is applied to the outside of the join to ensure the waste adhesive can be easily removed (**image 7**). Then, the hot-melt adhesive is run into the seam and pressure is applied to ensure full contact across the width of the overlap

1

2

3

4

5

6

(**image 8**). The masking tape is removed and PET tape is applied to the bonded seam (**image 9**). This ensures a smooth and durable surface finish.

The seam will be put under considerable load in use, so it is compacted with a roller to squeeze out any remaining air and ensure maximum surface contact for the adhesive (**image 10**). The finished sails are installed on a Ker 39 boat (**image 11**).

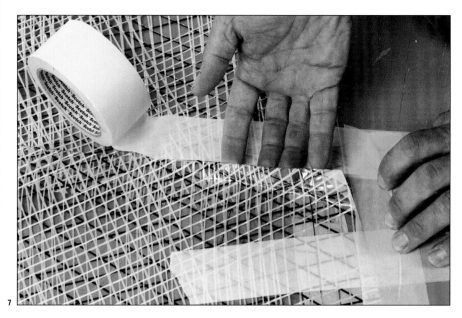
7

8

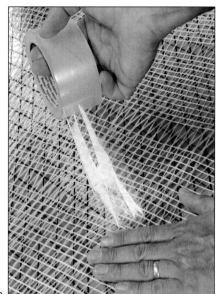
9

10

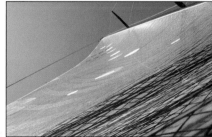
11

Featured Company

OneSails GBR (East)
www.onesails.co.uk

Construction Technology

Welding

Welding is used to join thermoplastics by heating and pressing the surfaces together. With these techniques strong, homogenous bonds are formed, which may be embossed, straight, shaped or three-dimensional. Hermetic welded seams are utilized in the production of items ranging from tension structures to high-performance outerwear.

Techniques		Materials		Costs
• Ultrasonic • Hot air and hot wedge	• Impulse • Radio frequency (RF)	• Thermoplastic woven, knitted and nonwoven	• Thermoplastic coated and laminated fabrics	• Rapid cycle time • Low to moderate cost depending on complexity

Applications	Quality	Related Technologies
• Apparel and packaging • Tension structures • Medical and cosmetics	• High strength, lightweight and hermetic • Visible on the surface • Overlapping join	• Adhesive bonding • Machine stitching • Stitch bonding

INTRODUCTION

Also called bonding and heat sealing, these processes eliminate the need for stitching (pages 342–69) and adhesive bonding (page 370) when using thermoplastic fabrics and films. High-strength seams are formed in straight lines or around three-dimensional joins, and may be embossed or textured. They are used in place of taped seams (page 364) on high-performance outerwear, reducing the construction to a single operation.

There are several different techniques, including ultrasonic, radio frequency (RF), impulse, hot air and hot wedge. RF is widely used and so known under

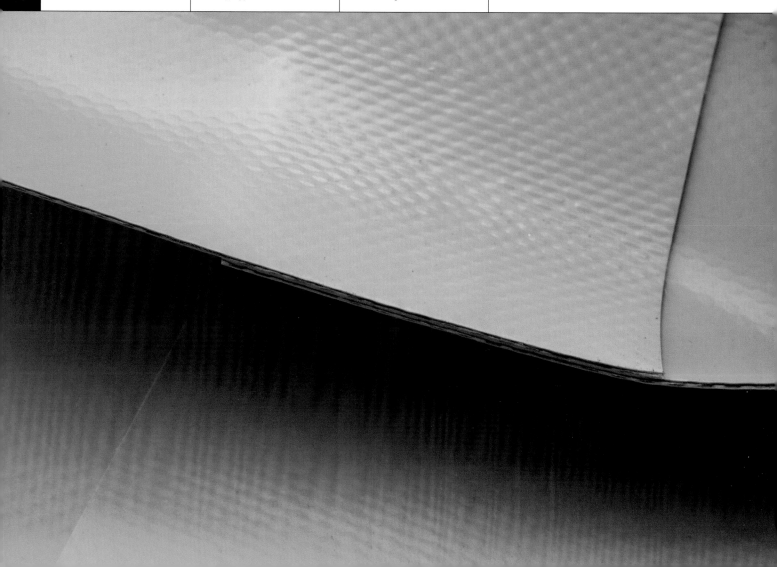

various names, such as high frequency (HF), dielectric and electronic welding; ultrasonic welding is often simply referred to as sonic welding; and hot air is also known as hot gas welding.

Rotary techniques, which include ultrasonic, hot air and hot wedge, are very rapid. For example, continuous ultrasonic welding is capable of speeds up to 500 m (1,640 ft) per minute. And flat-bed ultrasonic welding is capable of producing a single weld in less than a second.

Welding thin sheets means that an overlap is required at the seam. This is typically in the region of 5 to 70 mm (0.2–2.7 in.). Seamless joins are possible, but only on one side, because the back side will need to be reinforced.

APPLICATIONS

A growing number of products are being constructed by welding. Examples include acrylic covers, blankets and garments; polyester tents, hot air balloons, packaging, garments, laminates (page 188), quilts (page 196) and mattresses; nylon carpet, sports apparel, packaging, emergency evacuation slides, seat belts and hook and loop fasteners (such as Velcro); polyethylene (PE) packaging; polypropylene (PP) packaging, upholstery (page 404), outdoor furniture, carpet and tents; ethylene vinyl acetate (EVA) blood transfusion bags, medical products and packaging; polyvinyl chloride (PVC) book covers, stationery, tarpaulins and outdoor furniture; and elastane outdoor clothing.

Welding is used to make straight seams as well as joining mechanical fasteners (page 396) such as eyelets, hooks and zips onto thermoplastic sheets. In a process known as appliqué welding, it is used to join materials onto the surface of products for decorative purposes (see Embroidery, page 298).

RELATED TECHNOLOGIES

Welding has replaced adhesive bonding for joining like materials in many applications. The advantages of welding compared to adhesives include eliminating the use of messy compounds; the join cannot chemically break down over time or with cleaning agents; no bleed-through; joining multiple layers simultaneously; intermittent patterns; embossing; and immediate bond strength (no drying or curing time).

Stitching is a versatile process capable of making all types of seams from handmade (page 342) to mass-produced (page 354). The advantage of welding is that waterproof and hermetic seals are possible, because no holes are required for the thread to pass through. Eliminating needle holes also promotes a stronger join. However, sewing is still less expensive for most applications.

The drawback of welding is that it is limited to materials with sufficient thermoplastic content to form a weld. By contrast, adhesives and stitching may be used to join all types of material, including dissimilar materials.

Laser welding (see Laser Cutting, page 328) is emerging as an alternative to sewing and bonding for certain applications. It is currently quite limited – applications include inflatables and protective garments – owing to the cost and complexity of setting up. The upper material must be transparent and a light-absorbing material is used (or applied as a coating) for the thermoplastic base material. The laser beam passes through the transparent top layer and is converted into heat when it comes into contact with the light-absorbing base layer. This causes the material to melt and form a strong weld.

QUALITY

Joint strength is very good and can be equivalent to the strength of the base material. Common factors that influence this are uniformity of material thickness, type of thermoplastic and content, yarn orientation, elasticity and density of the materials being joined.

In addition, the type and style of construction of woven and knitted fabrics affects bond strength in different directions, owing to the surface flatness and yarn orientation. The random orientation of fibres in nonwovens promotes welds with very high strength.

The quality of ultrasonic welding is determined by the amplitude of vibrations and force. The optimum amplitude ensures maximum efficiency and the force applied determines the strength. Increasing the force applied improves strength.

RF is influenced by the uniformity of radio waves passing through the die onto the materials. Shorter dies tend to produce a higher-strength seam with uniform properties along the length.

The quality of joins formed with hot air, hot wedge and impulse welding are determined by temperature, pressure and speed. Seams formed with these welding techniques are generally not as strong as with RF and ultrasonic.

DESIGN

With these techniques a range of joint geometries is possible, including lapped, hemmed (flat or with pocket or rope inside), flat with tape on one or both sides, and reinforced. Lengths of rigid profile, known as keder (see case study on page 385), are sealed into fabric seams and used to fix the edge to a matching profile, which is typically an extruded aluminium rail. Keders are used to make assembling and disassembling easier, such as for tents, billboards, banners, truck sidings and yacht sails (in this case they are known as luff tapes).

Embossed patterns are applied during welding by utilizing the heat and pressure. They are functional as well as decorative, helping to direct the thermal energy and promote a stronger join.

Thermal processes may be used to produce very long joins in a single stroke or even continuous welds. For example, impulse welding is capable of producing welds 6 m (20 ft) long, and hot air welding is used to make continuous welds as long as the roll of material.

Continuous ultrasonic welding is done by repeatedly welding short strips side by side, or using a specially shaped drum (an embossed pattern on the surface efficiently directs the welding energy to the join as it rotates; see technical description overleaf). Long lengths of material 3 m (9.8 ft) or more wide, such as in mass production, are welded in a continuous process with multiple drums stacked side by side.

The Welding Process

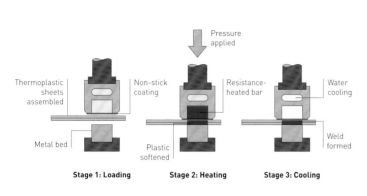

Stage 1: Loading   Stage 2: Heating   Stage 3: Cooling

Impulse welding

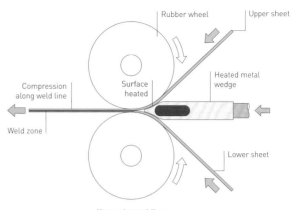

Hot wedge welding

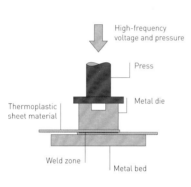

Radio frequency welding

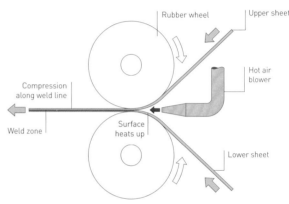

Hot air welding

TECHNICAL DESCRIPTION

These are thermal joining processes: under pressure the materials are heated and cooled to form a strong join. The amount of time for each stage of the process depends on the techniques used and materials being joined.

Hot air and hot wedge welding operate in the range of 200 to 750°C (400–1,380°F). The advantage of hot air welding is that there is no contact between the heating element and the fabrics being joined. Even so, hot wedge is preferred for some applications, because it is not noisy like hot air and does not produce as many fumes. Hot wedge is also better suited to joining very thin films that require support during welding, and the equipment is portable, so can be used on-site and outdoors.

Impulse welding uses one or two bars heated with the use of high-resistance wire, typically nickel-chromium. The bar is coated with a non-stick material to stop the thermoplastic adhering to the heated bar when it melts. With heat and pressure, the material at the joint interface is encouraged to melt and mix. After welding the bars are flushed with water to cool them down and re-solidify the plastic, thus increasing cycle time and helping to eliminate wrinkles.

RF welding joins thermoplastics (and some metals) using dielectric heating (electromagnetic field). The sheet materials to be joined are clamped together between a die (also called an electrode) and a metal plate. High-frequency voltage is applied and the resulting alternating electric field causes the molecules in polar thermoplastics to vibrate. The material heats up and as the die is pressed down on the surface the joint interface mixes.

A strong join is formed as the materials cool. Laser-cut patterns are not compatible with RF welding, because burned material along the cut edge (produced by the very high temperature of the laser) contains carbon, which will have the same effect on welding as metal in a microwave.

Ultrasonic welding works on the principle that electrical energy can be converted into high-energy vibration by means of piezoelectric discs. Electricity is converted from mains supply (50 Hz in Europe or 60 Hz in North America) into 15 kHz, 20 kHz, 30 kHz or 40 kHz operating frequency. The frequency is determined by the application; 20 kHz is the most commonly used frequency because it has a wide range of application.

The converter consists of a series of piezoelectric discs, which have resistance

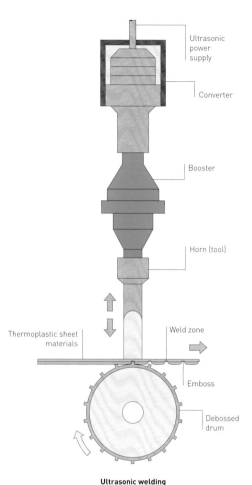

Ultrasonic welding

to 15, 20, 30 or 40 kHz frequencies. The crystals that make up the discs expand and contract when electrically charged, so converting electrical energy into mechanical energy with 95% efficiency.

The mechanical energy is transferred to the booster, which modifies the amplitude into vibrations suitable for welding. The horn transfers the vibrations to the workpiece. The size and length of the horn are limited to around 254 x 305 mm (10 x 12 in.), because it has to resonate correctly. The ultrasonic vibrations are transferred to the joint, generating frictional heat at the interface, which causes the material to plasticize. Pressure is applied, which encourages material to mix. When the vibrations stop, the material solidifies to form a strong, homogenous bond.

Continuous welding may be used to join three-dimensional seams. By guiding the layered materials between two rotating wheels (hot air and hot wedge), similar to machine stitching, a continuous seam is formed.

RF welding is used to create three-dimensional seams and shaped profiles with dies. Another advantage of RF is that it may be used to join and cut simultaneously: the cutting edge typically runs parallel to the weld line. This is why it is often referred to as tear-seal welding. This technique is used to make packaging, such as liquid-filled pouches, in a single operation.

RF and ultrasonic welding transfer the energy to the joint interface by vibration, rather than conduction. Therefore, when joining materials using the outside surfaces (as opposed to continuous roller processes), RF and ultrasonic will be more efficient than, for example, impulse welding – which is limited to materials under 0.5 mm (0.02 in.) thick – and suitable for sealing thicker materials.

Welding presents unique opportunities. Developments include sealing optical or conductive fibres in the join to create 'smart' seams capable of storing and transmitting data about the performance of a garment, the wearer or a structure.

MATERIALS
Suitable materials include thermoplastic fabrics – all types of construction including woven (pages 76–105), knitted (pages 126–51), nonwoven (page 152) and film (page 180) – as well as other types of fabric that are coated (page 226) or laminated (page 188) with thermoplastic. Materials that have been printed (pages 256–79) or dyed (page 240) may also be used.

Thermoplastic fabrics and laminates are available with a range of properties, from hard-wearing and weather-resistant to lightweight and breathable. Commonly used textiles include nylon, polyester, acrylic, PP and PE; laminates and coatings include polyester, PVC, thermoplastic polyurethane (TPU) and EVA.

RF is limited to certain materials – TPU, PVC, EVA, polyester, nylon and a modified PE – because the molecules in the plastic need to be excitable by the rapidly alternating electric field.

In addition to joining flexible materials, welding is used to join rigid extruded parts as well as injection moldings.

COSTS
Hot air and hot wedge are very rapid and as a result are the most cost-effective. Hot wedge requires the least energy of the two to heat the materials. RF requires dies and ultrasonic uses horns, which increases set-up cost. Ultrasonic is the most expensive.

Welding is more expensive than sewing for many conventional applications because it is compatible with fewer materials and applications and so is less common and more difficult to set up. By contrast, sewing is versatile and widely available. However, for other applications, such as when a taped seam is required for waterproofing, the price of welding will be comparable to, if not less than, sewing.

ENVIRONMENTAL IMPACTS
Welding eliminates the use of additional materials, which makes thermoplastic items much more efficient to recycle. It reduces weight by eliminating the need for mechanical fastening or adhesives.

The process itself is very efficient: thermal energy is directed to the seam by convection, conduction or friction. Ultrasonic and RF are the most efficient; almost all of the electrical energy is converted into vibrations at the joint interface, meaning that there is very little heat radiation.

With welding techniques there is no risk of contamination, which makes these process suitable for food packaging, toys and medical products.

→ Constructing a Marquee with RF Welding

Heavy-duty woven polyester fabric coated with PVC is here used in the production of a large bespoke marquee. The panels are engineered and calculated and a plan is made for cutting and welding (**image 1**).

In preparation, the individual panels are CNC knife cut and stacked ready for assembly (**image 2**). RF welding may be done on a large bed with a moving head, or using a small bed whereby the fabric is moved by hand (**image 3**). In this case, a tape-reinforced hem is being RF welded along the edge

of a panel. The tape is located (**image 4**), the die is placed on top (**image 5**) and the RF welding takes place (**image 6**). Typically, welding takes only around three seconds and the pressure is maintained for a further six seconds to allow a strong bond to form (**image 7**).

Next, the edge of the fabric is folded over (**image 8**) and the welding procedure is repeated to form the hem (**images 9** and **10**).

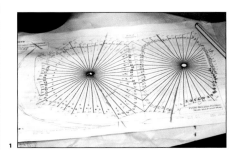

1

2

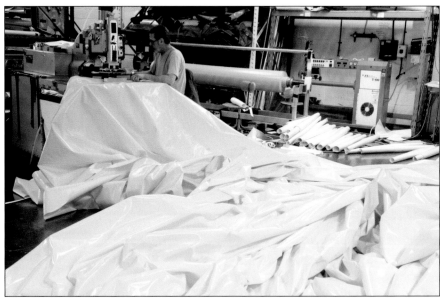

3

4

5

6

7

8

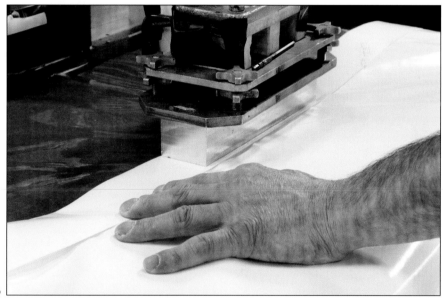

9

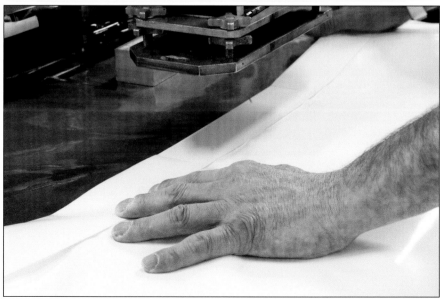

10

Featured Company

Flexitec Structures
www.flexitec.com

→ Forming a Watertight Butt Weld

It is possible to form a butt weld so that one side of the assembly is smooth. In this case, PVC-coated polyester is RF welded to make a water tank. This requires that a strong and watertight seal be formed.

The join is reinforced on the back side with a width of the same fabric (**image 1**). The die has a relief surface, which is transferred to the fabric as an embossed pattern (**image 2**).

The join is completed: the back side is impressed with the texture from the die and the front surface is flush (**image 3**).

1

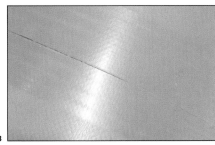

3

2

Featured Company

Flexitec Structures
www.flexitec.com

→ Hot Air Welding a Reinforced Seam

This is a high-speed process. Set-up is critical to ensure a high-quality weld. Therefore, it is particularly useful for producing continuous lengths of long straight welds.

The two layers of thermoplastic-coated fabric are pressed together between two rubber wheels (**image 1**). Just before they come into contact, a hot air blower heats the surface to above the softening point of the material. This is different for each type of plastic.

When the surfaces come into a contact the materials mix to form a strong join (**image 2**). Tape is used to reinforce the join when additional strength is required (**image 3**). The clear tape – which does not need to be colour matched for each application – bridges the seam on the surface (**image 4**). The side of the tape that comes into contact with the fabric is textured to improve bond strength.

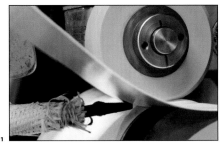

1

2

Featured Company

Flexitec Structures
www.flexitec.com

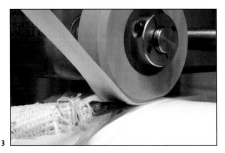

3

4

→ RF Welding Mechanical Fasteners onto Fabric

RF welding uses dies, shaped to accommodate a two-dimensional profile (**image 1**). It is used to apply mechanical fasteners onto thermoplastic fabrics, as well as welding shaped flat parts (such as packaging).

An insulated sheet window is assembled into PVC-coated fabric in a single operation (**image 2**). Molded plastic fixings are joined to fabric and the weld is as strong as the base materials. These fasteners reinforce eyelets around the base of a large watertight structure (**image 3**).

A D-ring is assembled into a length of fabric by folding it over and applying a weld along the overlap (**image 4**). Webbing constructed from thermoplastic is joined to PVC-coated fabric (**image 5**), and a metal eyelet is riveted (page 396) through the layers to provide a sturdy fixing point.

Heavy-duty webbing is used to join a D-ring onto the face of PVC and provide a sturdy fixing point without compromising the strength of the base material (**image 6**).

1

3

4

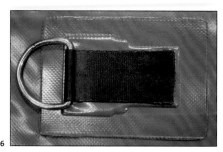

6

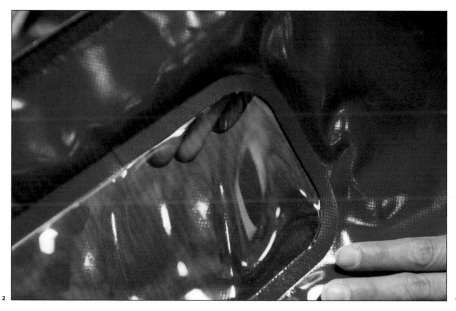

2

5

WELDING

Featured Company

Flexitec Structures
www.flexitec.com

→ Hot Wedge Welding a Hermetic Seam

Hot wedge welding is a straightforward process and the equipment is compact, which makes it very portable. The fabric is drawn over a heated metal wedge between two rollers (**images 1** and **2**).

This example demonstrates forming a hermetic seal in PVC-coated fabric (**image 3**). A channel is left along the middle of the join, which is used to test the join and make sure the weld is secure by applying internal pressure. Pulling the weld apart shows the structure of the join (**image 4**). The bond is very strong and when pulled apart the fabric delaminates before the weld zone fails.

1

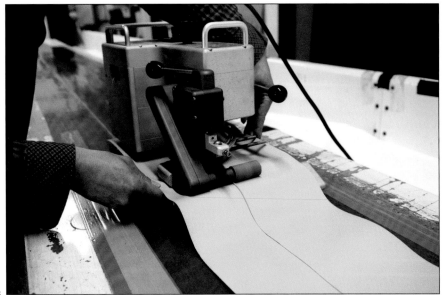

2

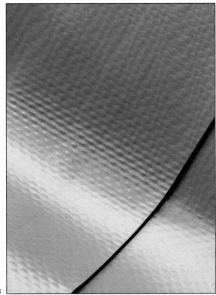

3

Featured Company

Flexitec Structures
www.flexitec.com

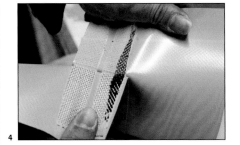

4

→ Impulse Welding Long Straight Seams

Impulse welding applies even heat across a long bar, making it particularly useful for long straight seams (**image 1**). Two layers of fabric are placed together and fed into the machine (**image 2**). This version automatically aligns the edges and may be used to form a hem or pocket.

The textured impulse-heated bar has a non-stick coating (**image 3**). The bar is heated and brought into contact with the lap join configuration. After a short heating, cooling and clamping cycle the join is formed and the welded fabrics are removed (**image 4**). With this technique, thin thermoplastic fabrics produce a quick and clean weld line (**image 5**).

Keders are rigid plastic profiles welded into the perimeter of products such as banners, sails and tents (**image 6**). They provide the means to temporarily fix fabric into a rail or track, such as an extruded aluminium frame. The long straight profile is ideally suited to impulse welding.

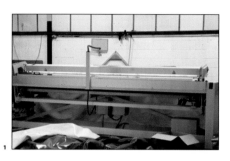
1

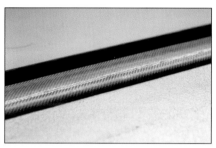
3

2

WELDING

385

Featured Company

Flexitec Structures
www.flexitec.com

5

4

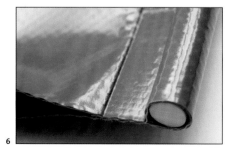
6

Construction Technology

Linking, Looping and Closing

Linking is the process of making up the various parts of a knitted garment, such as joining the sleeves to the body or finishing the collar. A single thread is knitted through the open edges to form a coherent structure and virtually invisible join. It is known as closing when employed to finish the toe of circular-knitted socks and tights (pantyhose).

Techniques	Materials	Costs
• Hand • Machine (automatic, lin-toe and rosso)	• Weft-knitted fabrics	• Low

Applications	Quality	Related Technologies
• Apparel • Hosiery	• Coherent structure • Invisible seam • Ridge on the technical face	• Stitching • Warp knitting • V-bed weft knitting

INTRODUCTION

Linking, also called looping, evolved from hand techniques, which are still used in the production of bespoke and very high-quality items. It is a fundamental step in the production of complex weft-knitted (page 126) garments. Items knitted as separate panels, such as a jumper comprising of arms, a body and a neck, must be joined together. Linking the end loops with thread produces a seamless and high-quality finish.

Tubular weft-knitted (page 138) items, such as socks and tights, are made as a pipe, and so have an open end. In a process known as closing, the end loops are joined together with thread, sealing the end and finishing the item.

APPLICATIONS

Linking is used to join the panels of garments, such as the body, arms and neck of knitted jumpers and cardigans. Closing is used to finish the toes of socks and tights (pantyhose).

RELATED TECHNOLOGIES

Construction methods such as warp knitting (page 144) and seamless weft knitting (see page 135) have been developed to produce complex three-dimensional garments and hosiery without the need for linking. With these processes, pattern and fit are created in a single knitting cycle, eliminating the need for additional production steps.

Machine stitching (page 354) is widely used to join panels together, including parts typically joined by linking. However, stitching requires a hem and so adds bulk. It is typically done while the garment, or sock, is turned inside out so that the outside surface remains smooth. The advantage of stitching is that additional layers of material may be introduced, such as padding, beading,

The Machine Linking Process

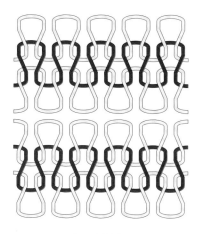

Stage 1: Knitting

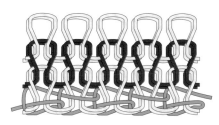

Stage 2: Linking

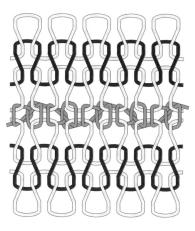

Stage 3: Finish

elastic and trim; and non-knitted parts may be joined to knitted parts.

QUALITY

Linking and closing are carried out manually, by an operator using a machine, or automatically. Manual processes are very high quality, but are more time-consuming and expensive than machine processes. Therefore the majority of garments and hosiery are finished by machine.

There are five principal techniques: stitching (manual, page 342, and machine, page 354); hand linking; machine linking (or automatic linking); lin-toe and rosso.

Hand linking is considered superior, because it yields a high-quality and seamless join, but it is the most time-consuming and so is the most expensive. The knitted loops are individually loaded onto a machine, which then links the two halves of the open toe or garment with a single thread.

Machine linking is the automated version of loop-by-loop closing. The join is seamless and similar quality to hand linking, but it is much more cost-effective, because the knitted toe is loaded directly onto the linking head without needing to be transferred by hand.

Lin-toe is the older mechanized method of linking the knitted loops of tubular socks. It is less commonly used nowadays, because it creates a slightly bulkier seam (virtually seamless), compared to machine linking.

The rosso seam, named after the Italian manufacturer who invented the process in 1946, produces a seamless join on the outside of a toe, but generates bulk and waste on the inside.

DESIGN

Depending on the technique used, linking produces a smooth join. A single thread is used, so the linked line will be one colour. Therefore, to maintain visual integrity, the linked area of knitting is often kept to one colour. If a pattern runs to the edge or if multiple colours are used in the linked area, then the line of thread will be more clearly visible.

When knitting fully fashioned garments, the edges of the panels are finished with courses of plain stitches. This ensures a coherent loop structure for linking.

Linking individual loops helps to maintain the fit and hang better than machine stitching or adhesive bonding (page 370). The join may be invisible, or emphasized as a design feature by changing the colour and type of thread, depending on the requirements of the application.

There are several different types of weft-knitting machine used in the production of garments and hosiery, including bar frame, V-bed and circular. Depending on the choice of machine, parts of the garment may be knitted as an integral part of the panel, such as the arms, welts and cuffs. This leaves

TECHNICAL DESCRIPTION

Machine techniques link and close with a single thread. In stage 1, the two sides of the knitted structure – two sides of a tubular knitted sock, or two parts of a garment – are completed with an exposed row of loops on either side. When using knits other than plain, a row of plain knitted loops is added to allow for linking.

In stage 2, the two layers of fabric are brought together, one on top of the other and the loops in each wale are aligned. In machine linking this may be automated, or the loops may be hand loaded onto the linking machine. Automatic processes are much more cost-effective.

A single thread is passed through each loop to form a chain of interlooped stitches. In stage 3, the fabric is opened up to reveal the finished seam. With the move, the legs of each loop go from horizontal to vertical. This produces a slight ridge on the face (two lines of thread join each loop to form a linear pattern) and a smooth finish on the back (criss-cross pattern provides flat cover for the join area).

only the collar to be added during garment make-up.

Single-bar machines, which produce each panel as a flat piece of fabric, require the most assembly processes to complete the garment. In this case, the panels are constructed into tubes (arms and body) by cup seaming (stitching), and the parts are then joined together by linking.

MATERIALS

Linking and closing utilize the same materials as the knitted item. This may include natural and synthetic staple (page 56) and filament (page 50) yarns. Ideally, yarns will be smooth, fine and strong. Filament yarns are the most consistent and have lower shrinkage, and so are the most straightforward.

The compatibility of the linking thread and knitted yarn need to be considered in case of finishing treatments, such as dyeing (page 240) and ironing. The thread must be compatible with what kind of use the item will be subjected to, such as outdoors, wet weather, washing, dry cleaning and ironing.

COSTS

In some cases linking will require a seamster to load the loops onto the needles and finish the linked join with hand stitches. This increases cost, but not significantly.

Toe closing is typically fully automated and carried out directly after circular knitting, eliminating any additional unloading and loading steps, making this a very efficient part of the process.

ENVIRONMENTAL IMPACTS

Linking is a relatively efficient process and generates little waste. The total environmental impacts will be determined by yarn selection and processing, knitting and finishing (see Machine Knitting, page 126).

VISUAL GLOSSARY: LINKING AND CLOSING

Machine Toe Closing
Material: Wool, nylon and elastane
Application: Technical sock
Notes: The nylon used to close the Bridgedale Trekker circular-knitted sock (see page 138) is colour matched to make the join less visible.

Technical face   Technical back

Linking Structure
Material: Wool and nylon
Application: Technical sock
Notes: Yellow nylon is used to demonstrate the structure of the linked thread through the black knitted loops.

Technical face   Technical back

Machine Linking
Material: Cashmere
Application: Apparel
Notes: The face of the join is seamless and virtually invisible. Bulk is formed on the back side owing to fashioning (shaped).

Technical face   Technical back

Machine-Stitched Seam
Material: Cashmere
Application: Apparel
Notes: Seams that are not practical for linking are joined with a line of stitches.

Case Study

→ Body Linking

The panels of this grey V-neck sweater are individually knitted on a bar-frame machine (page 132). Linking the edges together creates the complete garment. The panels are joined in stages. The shoulders are already joined (**image 1**). Next, the seamster will link the seams that run down the inside of each of the arms and the sides of the body (**image 2**).

The loops from each edge are aligned and loaded onto the linking machine needle bed, one on top of the other (**image 3**). The additional courses of knitting accommodate the fashioning (**image 4**). They fall outside the join and so form bulk on the back side.

The loops are threaded onto the grooved needles by hand (**image 5**). A needle passes along the groove, picking up each set of loops (**image 6**). A yarn is wound around the needle once. As it retracts through the fabric, the hook catches the yarn to form a loop (**image 7**). The loop is drawn through the knitted loops, as well as the loop on the shank of the needle, to form a stitch.

The process is repeated along the length of the arm and down the side of the body to form a continuous seam. Once complete, the linked seam is removed from the row of needles (**image 8**). The inside of the join shows the extra ridge of material, used to ensure a coherent loop structure for linking, which is necessary to join the different-shaped parts (**image 9**).

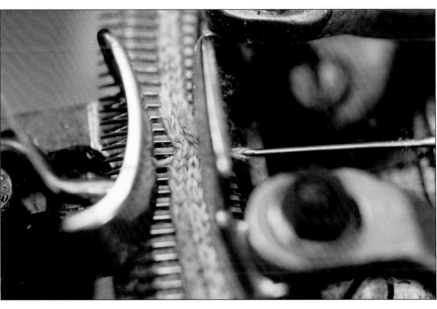

Featured Company

Johnstons of Elgin

www.johnstonscashmere.com

→ # Hand Finishing

Once all the parts have been assembled the collar is added around the V-neck. It is linked up to the point where it meets the shoulder and the two ends come together (**image 1**).

At this point, the only practical way to produce a high-quality loop-by-loop join is to stitch the loops together by hand. Each loop is picked up with the needle and thread (**image 2**). The seamster works along the join, from front to back (**image 3**), creating an invisible join (**image 4**).

The finished seam completes the collar (**image 5**) and the sweater is ready for final quality control, steaming and pressing.

Featured Company

Johnstons of Elgin
www.johnstonscashmere.com

1

2

3

4

5

→ Cup Seaming

Joins that are not practical for linking because of the loop structure, pattern or placement of the join are seamed with a row of machine stitches. Stitching is very rapid and does not require that each loop be individually joined. Instead, a row of tightly packed chain stitches (or other type of stitch) is made to catch each loop at least once.

The two edges are carefully aligned by hand (**image 1**). They are placed into the sewing machine, which makes a row of chain stitches (**image 2**). A lever on the opposite side to the needle catches the thread with each stitch to form a row of interconnected loops.

The blue part is finished and the thread is trimmed (**image 3**). The pink section is then sewed with colour-matched thread (**image 4**). Changing threads ensure the joins are virtually invisible on the face when complete (**image 5**).

Overlap in the seam, which is essential to ensure a strong join that will not unravel, forms bulk on the back side of the join (**image 6**).

1

2

3

4

5

6

Featured Company

Johnstons of Elgin
www.johnstonscashmere.com

Construction Technology

Lining and Labelling

A lining provides a smooth and clean surface on the inside. It can be decorative and functional. As an optional extra, it defines the value of a well-made item. For example, in jackets and gloves silk provides a smooth, insulating and luxurious surface, and in shoes leather insoles provide comfort and support while concealing the construction.

Techniques	Materials	Costs
• Hand and machine stitching • Adhesive bonding	• Woven, knitted and stitch-bonded fabrics • Leather, plastic, rubber • Interlining and insulating layers	• Low to high, depending on materials and construction

Applications	Quality	Related Technologies
• Apparel • Headwear • Footwear	• Smooth, ergonomic and comfortable • Insulating	• Weaving • Knitting • Laminating

INTRODUCTION

Linings and labels are fitted into a cover, an outer or an upper to provide a smooth and clean surface. Depending on the application, they may be constructed separately and installed at the end, or joined to the outer and inherent to the structure. In the construction of a suit jacket (see Hand Stitching, page 342), the body lining is attached to the jacket and the sleeve lining to the sleeve; the final seam is made at the end around the shoulder. By contrast, a the lining of a glove is made separately and fixed at the cuff (see page 395).

Linings may be a subtle part of the construction and purely for functional benefit, such as improving the cut and hang. Alternatively, they may be contrasting, branded or decorated in some way as to create a design feature.

APPLICATIONS

Linings may be used for all types of garment, headwear and footwear, but they are not always required. It depends on the application and construction. By providing an internal layer whose quality can be tailored to fulfill a specific function, they are particularly useful for formal outfits, ceremonial clothing, luxury items and insulated clothing.

RELATED TECHNOLOGIES

The functions and appearance of a liner may be incorporated directly into the outer material. Fancy weaving (page 84), knitting (pages 126–51), laminating (page 188) and stitch bonding (page 196) are all capable of producing double-sided and multi-layered fabrics. Thus, the face and back of an item constructed with these techniques can have different properties, eliminating the need for a separate lining.

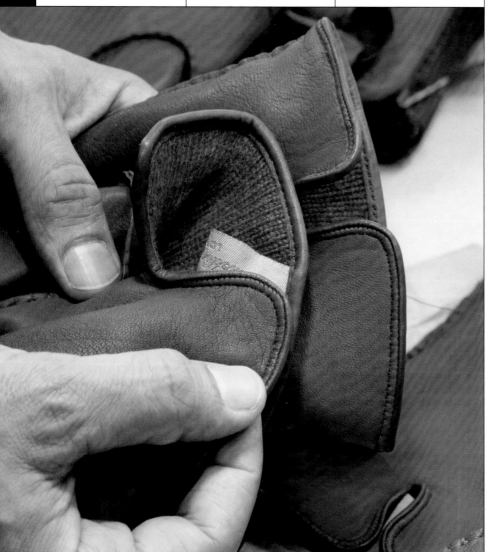

Anatomy of a Glove

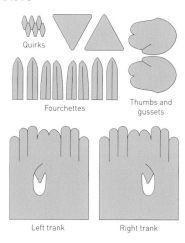
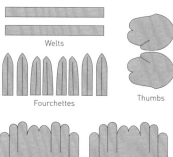
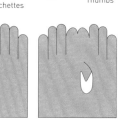
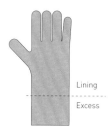

Quirks

Fourchettes

Thumbs and gussets

Welts

Fourchettes

Thumbs

Left trank

Right trank

Left trank

Right trank

Lining

Excess

Parts of a sewn glove

Parts of a sewn lining

Seamless knitted lining

QUALITY

Linings provide a smooth finish that conceals the construction. If they are used to make an item easier to put on, such as lining a jacket, then the quality of material and attachment is important, because it should provide a crease-free and seamless layer without affecting the fit. This same quality improves the hang of the jacket, because it helps to stop snagging and interference with other items of clothing.

They add structure, weight and body to items. Several layers of interlining may be built up between lining and outer to provide additional volume. For example, padding is used on the chest and shoulders of jackets to smooth the shape.

Linings are useful for improving insulation. Air is trapped between layers of material, such as quilt (page 196), fur (page 498) or foam (page 172). As a poor conductor, air helps to retain the wearer's body heat.

DESIGN

The method of attaching the lining will depend upon the item and application. Hand stitching, machine stitching (page 354) and adhesive bonding (page 370) may all be used.

Hand stitching is the most expensive and is generally reserved for bespoke and high-value items, although in some cases it may be the only practical way to finish a seam. Machine stitching is high quality and repeatable. The type of stitch will depend on the materials being combined and the location of the seam.

Linings are generally attached at the seams. Therefore the method of attachment may be completely concealed in the hem, or revealed as a design feature. In some cases, the lining may be joined across its surface, either by stitching or bonding. This reduces the likelihood of slippage and misalignment, but may increase stiffness in those areas.

Bonding is considered inferior for items of clothing. However, it is very useful for parts of the construction that are difficult to reach, such as the inside of a glove's fingertips. It can affect the handle more than stitching and so is less commonly used in garments.

For maximum efficiency in production, plain linings are preferable. Print (see Screen Printing, page 260, and Digital Printing, page 276) and other graphics make linings more complicated and time-consuming to assemble. Therefore, branding and care information (which is typically country-specific) are usually supplied on a separate label.

Linings are made from leather (page 158), woven (pages 76–105), knitted and stitch-bonded fabrics. Each material or construction has its own advantages. Selection depends on the requirements of the application.

MATERIALS

Materials used to provide insulation include silk, cotton, fur, wool, foam,

TECHNICAL DESCRIPTION

A fitted glove is made up of many parts (see Machine Stitching, page 361). The glove, which is typically leather, consists of a trank (hand), thumb, gusset (triangles for the thumbs), fourchettes (that go in between the fingers) and quirks (join the fourchettes to the trank at the base of the finger). The parts are profiled by pattern cutting (page 314), die cutting (page 336), laser cutting (page 328) or other suitable process. They are assembled with hand or machine stitching.

A lining made up of sheet materials, such as woven fabric or chamois leather, is produced in the same way, with the addition of welts to join the lining to the glove. The parts are assembled separately and then fitted into the glove around the cuff.

A seamless lining is produced by weft knitting on a V-bed. It is joined into the glove around the cuff, using a welt, and may be joined at the tips of the fingers with a dab of adhesive to stop the lining pulling out of the glove with the wearer's hand.

The excess is trimmed from the lining during fitting. The amount of excess depends on the length of glove.

feather, down and chamois leather. Each of these materials has different qualities and so is used to make different items. Fur, feathers and down are bulky and warm and so used for heavyweight winter items. By contrast, silk is lightweight and thin, and yet provides a reasonable amount of insulation. As a result, it is used to make much more delicate items.

Silk is smooth and this makes it easier to take a lined jacket on and off. Synthetics are also quite slippery, including nylon, polyester and acetate. Plastic fibres have varying levels of durability and tend not to breathe as well as natural-fibre fabrics.

It is important to consider the compatibility of liners and labels with the qualities of the item and materials and how it will be used, such as resistance to ultraviolet light or repeated washing.

COSTS

The cost of a lining may be similar to the price of the outer, such as a woven lining in a woven outer. It may be higher, such as when using high-value materials to create a luxurious feeling or improve the performance of the product. Or it may be lower, such as integrating a seamless knitted lining into a hand-stitched outer.

Hand processes are labour-intensive and require highly skilled craftspeople. Machine processes are much more cost-effective. Cycle time depends on the materials and construction techniques.

VISUAL GLOSSARY: LINING MATERIALS

Fur
Material: Lambswool
Application: Gloves
Notes: Wool from stillborn lambs is extremely soft and luxurious, and used to make bulky heavyweight gloves.

Knitted
Material: Cashmere
Application: Gloves
Notes: Weft knitting eliminates the seams in a complex three-dimensional lining to provide a smoother surface.

Chamois
Material: Leather
Application: Gloves
Notes: Chamois is napped to give it a smooth and soft finish. It provides gloves with a very comfortable fit.

Woven
Material: Silk
Application: Apparel
Notes: Jacquard-woven silk, in a range of bright and decorative patterns, is often used to line wool suits and ties.

Leather
Material: Leather
Application: Footwear
Notes: Leather insoles are die cut to size, embossed and foil blocked with the company logo.

ENVIRONMENTAL IMPACTS

Like all joining processes, lining adds one or more materials to the construction. The compatibility of the materials and assembly should be considered alongside the rest of the materials in terms of their end of life.

Case Study

→ Lining a Handmade Leather Glove

The purple linings for the right and left hand are weft knitted (page 126) in cashmere to provide a comfortable and insulating inner layer (**image 1**). Knitted on a V-bed in a single operation, the linings are seamless.

The leather gloves are handmade at Dents, in Wiltshire, England. Each lining is inserted and pushed to the fingertips to measure the correct length (**image 2**). Excess is trimmed from the cuff (**image 3**). The liner is pulled onto a hand-shaped mandrel. A small dot of glue is put on the fingertips (**image 4**). This

helps to keep them in place. The leather glove is pulled tight over the top.

Next, a strip of leather, called a welt, is used to join the lining into the glove. The three layers of material are joined and the edges are sealed with a whipstitch (see page 344). The small lever catches the thread from the needle after each stitch and passes it back over the edge and into the path of the needle in time for the next stitch, forming a series of interlooped stitches (**image 5**). The welt is trimmed to the correct length

and the gloves set aside (**images 6** and **7**). The welt is rolled over and lock stitched in place with the label attached (**image 8**). Excess is trimmed (**image 9**). The whipstitch is concealed within the hem and the only visible stitching is the lock stitch parallel to the edge (**image 10**). The welt forms a durable cuff that can withstand being taken on and off.

The gloves are pressed flat on a hand-shaped iron for a minute or so (**image 11**). Finally, the cuff is finished with a set of brass buttons (see page 399).

1

2

3

4

5

6

Featured Company

Dents

www.dents.co.uk

7

8

9

10

11

Construction Technology

Mechanical Fastening

Fasteners are utilized when textiles on their own cannot perform a specific function. They interlock, fold, thread or snap together to form strong joins. They may be permanent or temporary, and depending on the application, they may be emphasized as a design feature or concealed. There are many variations possible including size, material, colour and customization.

Techniques	Materials	Costs
• Snap fit • Press forming • Threaded	• Metal and high-strength (engineering) plastics • Medium- to heavy-weight fabrics	• Low to high cost depending on the performance and complexity of application
Applications	**Quality**	**Related Technologies**
• Buttons and fasteners • Apparel • Marine	• Temporary or permanent • Strong join	• Welding • Machine stitching • Adhesive bonding

INTRODUCTION

Rigid connectors are required when machine stitching (page 354) or other joining techniques are insufficient. They can be engineered to perform a specific function that fabrics on their own cannot deliver. For example, a snap button is designed to interlock when the front and back are pressed together. It must hold fast until the user chooses to pull the two sides apart. The amount of force required to join and pull apart the button is very carefully engineered to be convenient and comfortable in use.

By contrast, an eyelet riveted on the foot of a sail must not fail even under extreme load. It is designed and integrated to spread force around the sail, thus reducing stress concentration and the likelihood of failure.

APPLICATIONS

Fasteners are a versatile means of connection, so applications range from everyday to high-performance products.

Buttons are used in conjunction with buttonholes, or purely for decoration. They are used to join multiple parts of the same item, two sides together or two items together, or to help fashioning. They are commonly used in garments, footwear, bags, accessories, jewelry and interior textiles. Buttons focus load onto a relatively small area of fabric.

Snap buttons (also called press studs, poppers and snap fastenings) are low-profile and avoid holes in fabric, which makes them useful for applications from children's to outdoor clothing and inflatables to industrial products.

There are many forms of zip fastener, including exposed types with metal or plastic teeth, low-profile types to be concealed (such as in skirts), open-ended, made to length and waterproof seams. Zip fasteners are utilized in the construction of footwear, apparel, bags and removable upholstery.

Eyelets (also called grommets) are used for decorative and functional purposes. Functional applications include holes for laces, drawstrings and increasing airflow underarm. Similar types, but on a different scale, are used on sails (also referred to as cringles), curtains, pool covers and even book binding.

Rivets are used to provide additional strength at points of high stress, such as denim jean pockets and on workwear and carryalls.

RELATED TECHNOLOGIES

Stitching, whether by hand (page 342) or machine, is often used to join mechanical fasteners onto fabric. It is a very rapid means of assembly. Techniques such as bar tacking are used to provide strong joins between materials.

Sewn joins are flexible and retain many of the inherent qualities of the fabric, such as softness, handle and affinity for dyeing (page 240). However, the thread is prone to abrasion, both from the fabric it is joining and other surfaces. While broken threads do not usually cause items to come apart – unless chain stitches are used, which are prone to ravel if broken at any point along the seam – this can cause problems in areas of high stress, especially because each stitch applies load to just a few yarns in the fabric construction. Mechanical fasteners, such as rivets on denim jeans (patented by Levi Strauss and Jacob Davis in 1872), overcome this problem, because they are very durable while spreading the load across a larger area.

Welding (page 376) forms a strong and permanent join in plastic films and fabrics. It is only compatible with thermoplastics and some blends that include a high percentage of thermoplastic fibre. Suitable materials include nylon, polyester and laminates, including thermoplastic polyurethane (TPU), for example.

Adhesive bonding (page 370) is more versatile than welding and suitable for a wider range of materials. Designers and clothing manufacturers are starting to explore the potential of this technique for

Mechanical Fasteners

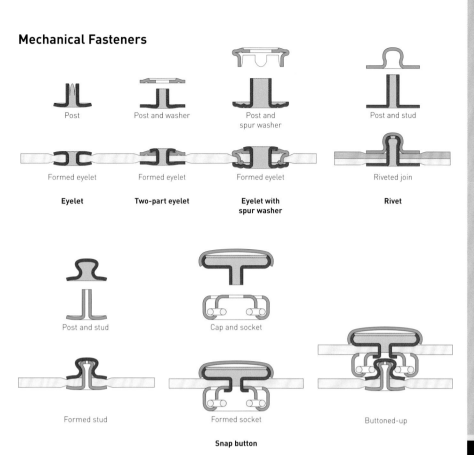

Eyelet • Two-part eyelet • Eyelet with spur washer • Rivet

Snap button

TECHNICAL DESCRIPTION

Eyelets are assembled from one or two parts. The simplest one-part eyelet, such as used to reinforced lace holes on shoes, are formed from a single post. They are inserted into a pre-cut hole and pressed from above and below with a metal tool shaped to match. The top of the post is flattened onto the back side of the fabric. The top of the cylindrical post has spurs to allow the metal to fold (avoiding stretching, which would cause it to split) when pressed to grip the fabric from both sides.

Two-part eyelets consist of a post and washer. Using a washer reduces the amount the post needs to spread out on the back side. Therefore, there is no need for the spurred profile: the top of the post is rolled over to capture the washer and form a strong join. They are more durable than one-part rivets.

Spur washers are used to produce the most durable eyelets. This type of fixing is used to reinforce a sail's clew, for example (page 402). The spur eyelet has

a row of teeth that pierce the fabric as the eyelet is formed. Thus, both the washer and post grip the fabric from both sides.

A rivet is formed from a post and stud. As they are pressed together, the post hits the top of the inside of the stud and so bends outwards to fill the cavity. They are used to join two layers of material, such as reinforced workwear.

Snap buttons are formed from two halves, each of which is made up of two parts. The post and stud are pressed together and form a join similar to the rivet. The cap and socket are also riveted together: the join is formed with a metal tool that has a blunt end to force the end of the post flat (gripping the socket).

The socket is loaded with a bent spring, which engages the button as they are pressed together. It holds the button in place until enough force is applied to open the spring and release the stud.

There are many variations on these configurations, including size, material, colour and potential for customization.

applications that currently use machine stitching and mechanical fasteners. To make a strong join force needs to be applied while the adhesive hardens. Therefore it is mainly restricted to flat seams (although these may be shaped afterwards) or handmade pieces.

QUALITY
The quality of finish is largely dependent on the materials and the method of integration. Snap buttons, rivets and eyes have been designed so the method of joining is invisibly concealed within. The outside surfaces may be exposed as a design feature, or hidden inside a hem.

Hard materials tend to protrude from the surface of the fabrics and so are exposed to additional abrasion. Metals have good resistance to abrasion. Even so, surface patina in the form of scratches, dents and deformation is inevitable. In some cases this may be desirable. Items made from steel are prone to rusting. For items where this could be a problem, brass, stainless steel or plastic are used.

Coatings help to produce a high-quality finish and colour match with the base material, and reduce the visual impact of wear and tear in the short term; if the coating is a different colour from the base material it will show through eventually. Using base metals that do not need a coating, such as stainless steel, is the only way to ensure long-lasting high-quality appearance.

DESIGN
Mechanical fasteners are utilized for applications where fabric on its own will not provide adequate mechanical performance. They have become a design feature in many applications, such as riveted jeans and branded buttons.

Graphics, colour and durable finishes are created on the surface of certain metals and plastics with coating technologies, such as painting and metal plating. In addition to these processes, aluminium is anodized to produce a bright and long-lasting finish that can be dyed a range of colours.

On solid mechanical fasteners, these finishes look quite different than on fabric. They can be produced with a range of finishes, from the natural appearance of metal to a bright mirror polish. Plastic fastenings may be transparent or coloured.

Within reason, any type of mechanical fastener that has the means may be joined to a textile construction. There are many thousand standard, or manufacturer's own mechanical fasteners that can be used and modified, including injection-molded plastic, rubber and pressed metal.

Fasteners and connectors may be connected to textiles with a range of joining processes. Common techniques are stitching or riveting; others such as welding and adhesive bonding may be used, depending on the materials and design. Fasteners that are sewn on can be moved and replaced more easily.

It is important to consider the compatibility of the materials and the application. For example, fasteners on items that will be washed regularly must be able to withstand the cleaning action while not causing shrinkage or puckering in the fabrics; certain thermoplastics soften at relatively low heat and so should not be included in items that are ironed at elevated temperature; and metals will get hot in the sunshine, so should be avoided next to the skin on beachwear, for example.

MATERIALS
All types of construction may be joined mechanically. The effectiveness will depend on the combination of material, interlining and fastening.

Interlinings are used to reinforce delicate face fabrics, such as around buttonholes. In other words, they are not always required. They are generally sewn into apparel and in some cases may be laminated (page 188). They are available in a range of weights and constructions, including woven, nonwoven and stitch-bonded.

The most commonly used metals are brass and steel. Stainless steel is used for applications that require added strength and resistance to corrosion. Thermoplastic fasteners are typically produced in polycarbonate (PC), polyamide (PA) or polyethylene terephthalate (PET).

The choice may be partly down to compatibility. For example, PA and nylon fabric are the same material, as are PET and polyester fabric.

Certain plastics are compatible with metal coating technologies (as are metals), such as vacuum metallizing and electroplating. Metal coatings improve surface quality by improving reflectivity, wear and corrosion resistance, as well as colouring capability. Expensive metals, such as gold or silver, may be applied to a less expensive base material.

Gold is a unique precious metal with a vibrant colour that will not oxidize and tarnish. The purity of gold is measured in carat (ct), or karat (kt) in North America: 24 ct is pure gold; 18 ct is 70% gold by weight; 14 ct is 58.3%; 10 ct is 41.1% and 9 ct is 37.5%. Silver is less expensive. It is bright and highly reflective, but the surface will oxidize more readily. Silver's tendency to oxidize is used to emphasize details by blackening it in a chemical solution and then polishing exposed areas back to bright silver. Many other types of metal may be applied as a thin coating, such as aluminium and nickel.

COSTS
Cost for standard types is low, though of course they will be more expensive than not using a fastener at all. Integration is a rapid process, but there may be several preparatory steps, such as interlining for strength and measuring the location of each fixing. The application of buttonholes, snap buttons and rivets may be a matter of seconds.

ENVIRONMENTAL IMPACTS
Mechanical fasteners introduce extra material, although this is normally very small. It is often impractical to remove them for recycling, even though valuable metals are used, and so they contribute to landfill. Thermoplastic fabrics joined with thermoplastic fasteners may be fully recyclable if compatible materials are used. This approach is used to make, for example, recyclable advertising banners.

Mechanical fasteners allow items of fabric – such as upholstery and mattress covers – to be removed and cleaned or repaired, extending the life of the item.

→ Applying a Snap Button

The button used to decorate Dents' handmade gloves (see also page 360) is made up of two halves. Each half is made up of two parts, which grip the leather from either side. The bottom half includes a stud and post (**image 1**). While the stud is loaded into the upper press, the post is placed onto the lower press (**image 2**).

The leather is marked with the correct location and the two sides are brought together. As pressure is applied the upright post punches through the leather into the stud and folds outwards to fill the inside of the domed head, thereby holding the two parts firmly together (**images 3** and **4**).

The upper part, which includes the cap and socket, is riveted in place to finish the glove (**image 5**).

399

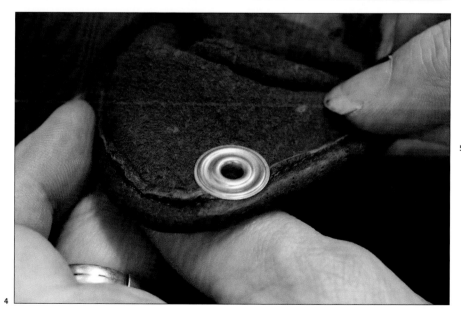

Featured Company

Dents
www.dents.co.uk

→ Sewing Buttonholes

Buttons and buttonholes are applied to garments last of all. In this case, Johnstons of Elgin is applying buttons to its knitted garments. The bodies of the garments are knitted in a single piece (page 132) and the arms joined by linking (page 386). It is more straightforward to knit the body whole and cut openings afterwards.

The neck and front opening are marked out with a chalk-like marker (**image 1**). The location of the buttons is also marked (**image 2**) and the seamster cuts along the lines with scissors (**images 3** and **4**).

This is a different garment, but the techniques used are the same. A strip of woven tape is prepared as an interlining to reinforce the buttonholes (**image 5**). The soft knitted cashmere would stretch and deform quickly without support along the edges.

It is sewn in position along the back side, securing the cut edge (**image 6**). As it is joined, the alignment along its length is checked (**image 7**).

Buttonholes are stitched to reinforce the edges and stop the yarn from ravelling. Tests are made to check that the colour and dimensions are set up properly for any particular garment (**image 8**).

Each buttonhole takes a couple of seconds to sew (**image 9**). Hundreds of stitches are made in a small area, similar to bar tacking (see page 366). As the edges are stitched the opening is cut to form the finished buttonhole in a single operation (**image 10**).

1

2

3

4

5

7

6

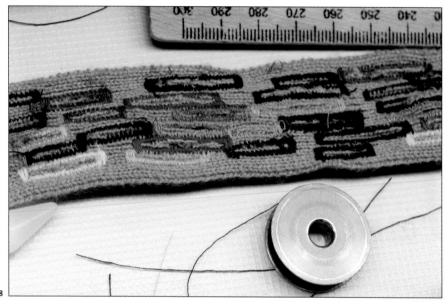

8

9

10

Featured Company

Johnstons of Elgin

www.johnstonscashmere.com

→ Reinforcing a Racing Sail with Eyelets

Eyelets are riveted or sewn into the lower corners of sails, known as the clew and tack, or reef points, to create durable openings to tie the sails down to the boom or deck. When used with spur washers, an eyelet offers superior gripping strength.

The sail is constructed from panels, and adhesive bonded or machine stitched together. The corners are substantially reinforced with layers of fabric, tape and zigzag stitching to help spread the high loads applied in application.

In preparation, the opening is marked and die cut to form a neat hole through the thickness of the sail (**image 1**). The two halves of the rivet are loaded into a press with the spacer in between (**images 2** and **3**).

The press forces the two halves together, forming the post and washer around each side of the sail to create a strong eyelet (**image 4**). The rivet is permanent and provides a secure high-strength fixing point between the flexible sail and the boat's rigging.

1

2

3

4

Featured Company

OneSails GBR (East)
www.onesails.co.uk

JOINING

402

→ Finishing Wire Rope with a Swageless Terminal

Wedged sockets grip the wire rope with mechanical force. They are used instead of swaged terminals – cold forged to create permanent terminations in wire rope – and so are often referred to as swageless.

The steel rope, which in this case is made up of 19 laid (twisted) filaments (known as 1 x 19), is cut to length and the outer strands are unwound (**images 1** and **2**). The four parts of the terminal are prepared (**image 3**). The socket is fed onto the wire and the long wedge is fed over the core strands (**image 4**).

It is pulled tight in a clamp and the cone is placed over the wedge (**images 5** and **6**). The threaded cap is screwed onto the socket (**image 7**), forcing the wire strands to bend around the wedge and compress the join tightly inside the terminal.

The assembled terminal retains around 80% of the breaking force of the rope (**image 8**). It can be moved, removed and replaced on-site without having to cut the rope.

Featured Company

Rig Magic
www.rigmagic.co.uk

Construction Technology

Upholstery

Upholstery is a traditional and highly skilled process that utilizes a range of hand and machine techniques. The final shape and appearance of a piece of furniture is largely dependent on the skill of the upholsterer and his or her ability to combine hard and soft materials into textile-covered structures. A range of materials is used and the choice depends on the application.

Techniques	Materials	Costs
• Joinery • Machine stitching • Hand sewing	• Structural frame, padding and cover materials	• Moderate to high depending on complexity and materials

Applications	Quality	Related Technologies
• Interiors • Domestic, public and commercial	• Dependent on skill and experience of upholsterer • Durability dependent on cover material	• Loom weaving • Warp knitting • Weft knitting

INTRODUCTION

Upholstery is the process of covering a structural frame with foam padding and a textile cover. Within this seemingly straightforward manufacturing technique, there are many opportunities for material and process innovation. And elements of upholstery may be used in a broad range of other applications.

The structural framework is generally fabricated in wood or metal. The majority of padding is polyurethane (PU) foam. It is either molded (see Foam Molding, page 410) or cut from block (see Foam, page 172). Modern foams have reduced the need for sprung seat decks, and many current upholstered sofas and chairs no

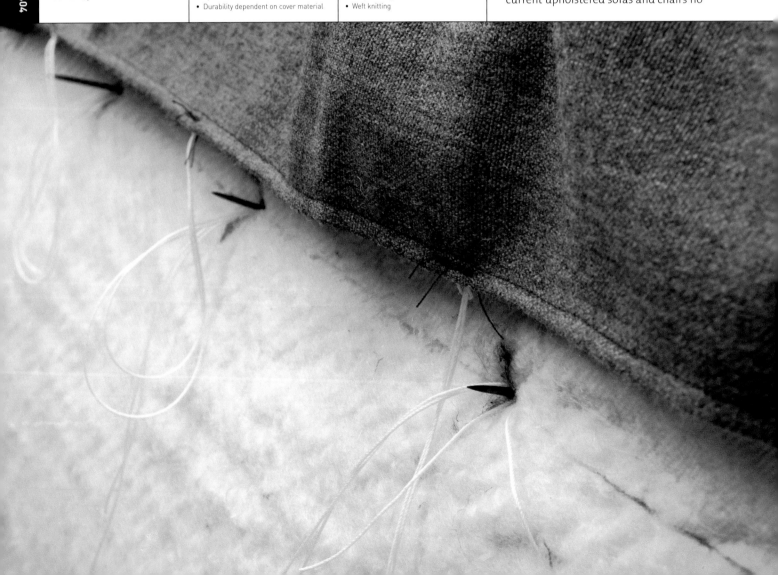

longer have springs. Instead, foam density is calculated to give maximum comfort.

There are many alternatives to conventional padded upholstery. This includes suspending three-dimensional knitted textile (see Warp Knitting, page 144, and Weft Knitting, page 126), or woven fabric (pages 76–105), in a frame to provide comfort and withstand heavy use. This technique is used for outdoor furniture (where padding is impractical), as well as modern desk chairs (for improved air circulation). Alternatively, foam is molded over a framework and covered, such as in car seats and some modern furniture.

APPLICATIONS

The majority of upholstery is used for sofas and chairs for the home and office. Although the techniques may differ, upholstery is utilized in the covering and decoration of furniture and interiors in the automotive, aerospace and marine industries.

RELATED TECHNOLOGIES

The majority of upholstery involves a combination of manual skills, making it a versatile production technique.

Constructing furniture from molded high-strength composites (page 446) and laminated wood veneer enables slimmer and lighter structures to be realized. By integrating flexibility into the structure, matching the shape to the body, or combining layers of padding, comfortable furniture is possible with these techniques.

Alternatively, entire pieces of furniture may be injection molded in a single piece. This approach is commonly used for outdoor furniture. Rotation molding and die casting metal can similarly be used in this way. These items of furniture will not have the comfort and luxury associated with upholstery.

QUALITY

The selected materials for the frame, springs, webbing, padding and cover will affect the overall appearance and durability. Furniture is often produced with a range of material options, so a suitable level of durability, cost

The Timber Frame Upholstery Process

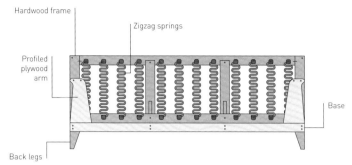

Hardwood frame
Zigzag springs
Profiled plywood arm
Base
Back legs

Stage 1: Timber frame

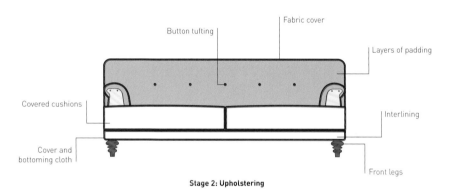

Button tufting
Fabric cover
Layers of padding
Covered cushions
Interlining
Cover and bottoming cloth
Front legs

Stage 2: Upholstering

TECHNICAL DESCRIPTION

Conventional upholstered furniture consists of a timber frame, springs, padding and covering fabric. In stage 1, the wooden frame is constructed. Because of the level of durability required, high-strength timber is used, typically beech. Profiled wooden parts are cut from plywood.

Springs are mounted onto the frame or suspended within it. There are three main spring systems: eight-way hand-tied, drop-in and sinuous springs.

Hand-tied springs are the most expensive and considered superior. Using them increases cycle time. A wooden box is fabricated, webbing is stretched across it and the springs are hand-tied to the webbing up to eight times.

Drop-in and sinuous springs are machine-made. Drop-in springs are a prefabricated unit fixed into the framework, and sinuous springs are continuous lengths of steel wire bent into zigzag shapes and fixed at either end.

Elasticated webbing is often used as an alternative to springs. It is lighter, thinner and comfortable, but its suitability will depend on the application.

In stage 2, the padding is attached to the structural framework. Several layers are typically used and built up according to the location on the product. For example, only a thin interlining of padding is required on the base, because thick cushions will be placed on top. Where required, fire-barrier padding (such as wool) is also stapled onto the structural frame. The fabric or leather cover is stretched tightly over the layers of padding. They are often adhesive bonded to ensure there is no slippage when in use. The covers are applied in two or more stages. The edges must be concealed, such as by folding the seams inwards.

The distinctive relief pattern of button tufting is created by pinning the cover to the support structure, through the padding, thus creating a three-dimensional effect similar to quilting.

→ Fabricating and Upholstering the Oscar Sofa

PART 1 – PRODUCING THE HARDWOOD FAME

Matthew Hilton designed the Oscar sofa in 2010 (**image 1**). It is built to last and is reasonably lightweight. The combination of a European-sourced hardwood frame, jute webbing, hessian straps, animal hair, natural fibres and wool cover means this product has low environmental impact.

The frame and internal woodwork are made from a combination of plywood and sawn beech timber. The sheets of birch plywood are used in the arms and cut out around a profile (**image 2**). A variety of joinery techniques are used to assemble the solid wood frame. The glued butt joints are reinforced with dowels and crimped metal plates (**images 3** and **4**). The metal plate ensures the joint remains tight while the glue hardens.

The back legs are stained with walnut-coloured stain (**image 5**) and the frame is ready to be upholstered (**image 6**).

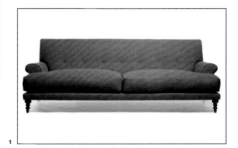

1

effectiveness or luxury can be achieved depending on the application.

Upholstery is a complex process and no two pieces of handmade furniture are exactly the same, so the final quality is largely dependent on the skill and experience of the upholsterer.

DESIGN

Foam is a versatile material: the density and hardness of foam is adjusted to suit the application. The textile is adhesive bonded to the foam, so overhanging shapes and recesses are feasible.

Shape limitation is partly dependent on the textile. Virtually any shape can be produced by cutting (pages 312–41) and sewing (pages 342–69). However, it may not be feasible or practical to do so. Soft forms are possible with technical weft knitting, which is capable of producing seamless three-dimensional covers.

Elastic fabrics may be stretched over some shapes, but they will bridge recesses and so cannot be used in all applications. Concave shapes can be upholstered, but the cover will need to be secured in place with a panel, pins, ties or adhesive to maintain the cover in place.

Printed fabrics (pages 256–79) are used to cover furniture. Aligning the graphics can lead to additional waste being produced in the pattern-cutting stage. Other finishing processes that may be used include flocking (page 280),

embroidery (page 298), button tufting and fancy machine stitching (page 354).

There are various techniques used to conceal the open end of the fabric cover. Conventionally, the fabric cover is pulled on and the edges stapled onto a single face, typically the bottom or back of the product. This is then covered with a separately upholstered panel.

MATERIALS

Any fabric can be used for upholstery. The location of use, such as home or office, determines the most suitable types. Synthetic fabrics such as nylon, polyester, elastane, polyvinyl chloride (PVC) and polypropylene (PP) are the most commonly used for high-wear applications. Of the natural materials, certain types of wool and leather are sufficiently durable to be utilized in furniture that will have to sustain heavy use.

For items used outdoors, materials are selected according to their resistance to exposure to ultraviolet (UV) light, salt mist (such as in marine applications) and weathering. Thermoplastic polyurethane (TPU), PVC and material coated (page 226) or laminated (page 188) with these are typically used in such applications.

Materials suitable for less abrasive applications include flocked fabric, raffia, mohair and cotton.

Fabrics are supplied on rolls that are typically 1.37 m (4.5 ft) wide, although

it is possible to get some fabrics much wider. Leather comes in various sizes: for example, cow hides range from 4 to 5 m² (43–54 ft²) and sheepskins can range from 0.75 to 1 m² (8–10 ft²).

COSTS

Overall cycle time is reasonably long owing to the number – and complexity – of the operations. Labour costs are relatively high.

The choice of covering material is an important factor in determining the unit price. High-value materials, such as high-quality leather from a top supplier, can double the price of the product.

ENVIRONMENTAL IMPACTS

Upholstery is the culmination of many different processes and as a consequence a typical sofa will include different materials that have varying environmental implications. Ensuring that each material is from a renewable and local source will ensure minimal environmental impact. Durability is determined by the least strong part in the construction, whether it is the frame, the padding or the cover.

Much more waste is produced when upholstering with leather, which is partly why it is more expensive. On fabric, the net shapes can be nested very efficiently, producing only 5% waste, whereas leather may have imperfections that cause up to 20% waste.

2

3

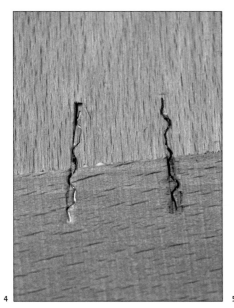

4

5

6

PART 2 – PREPARING THE FRAME FOR UPHOLSTERY

The base of the frame, where the seat will be, is crisscrossed with elasticated webbing (**image 7**) and the back is covered with steel zigzag springs. The webbing, layers of hessian and needle-punched wool (see page 157) padding are stapled to the timber frame.

The cover to be used for the seat area is sewn to black bottoming cotton cloth. Upholstery is a highly skilled process: the craftsman marks and cuts the padding and cover to fit each part individually (**images 8** and **9**).

The arms have jute webbing applied and are then covered with a latex-coated nonwoven coconut fibre and needle-punched wool to produce sturdy armrests with a padded outer layer (**images 10** and **11**).

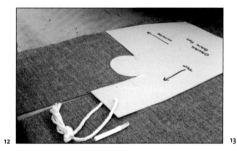

PART 3 – UPHOLSTERING THE SOFA

The wool cover is marked up using stencils and is then cut out (**image 12**). Each of the panels is machine stitched (page 354) inside out (**image 13**), so that when it is reversed on the sofa the stitching and excess material are concealed inside (**image 14**).

Where the arms meet the base a curved needle is used to make concealed slip stitches (page 344) (**images 15** and **16**). The seams around the back panel are concealed by folding the wool into bent metal clasps (**image 17**), which are closed with a plastic-headed hammer once the fabric is uniformly stretched out along the base (**images 18** and **19**).

The backrest is supported by a line of zigzag springs, which are secured to the wooden frame (**image 20**) and covered with nonwoven coconut fibre padding (**image 21**). The back has a line of sewn-in pulls (**image 22**) where the wool is pulled tight against the padding. This gives the appearance of a buttoned back (**image 23**).

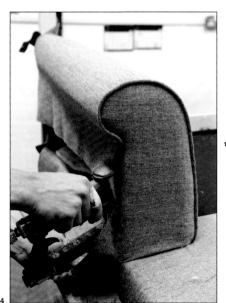

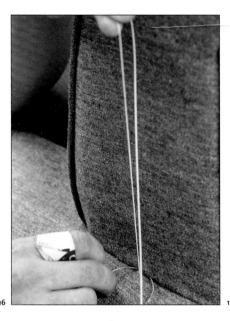

16

17

18

19

20

21

22

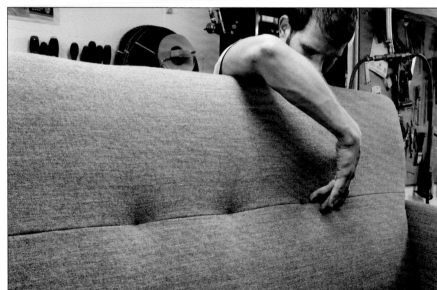

23

Featured Companies

Coakley & Cox
www.coakleyandcox.co.uk
SCP
www.scp.co.uk

Foam Molding

Molded foam is utilized in upholstery, car interiors, footwear and fitness products. Multiple parts, colours, electronics and structural components are combined into a single piece by over-molding preformed inserts, which may be foam or other material. Density, elasticity, rigidity and cell structure are all tailored to the requirements of the application.

Techniques	Materials	Costs
• Cold-cure foam molding	• Polyurethane (PU)	• Moderate mold costs • Low to moderate unit costs

Applications	Quality	Related Technologies
• Upholstery • Footwear • Fitness products	• High-quality surface • Uniform density • Accurate dimensions	• Knitted and woven spacer fabrics • Injection-molded foam

INTRODUCTION

Polyurethane (PU) is formed into three-dimensional parts by cold-cure foam molding. PU is extremely versatile and is available in a wide range of densities, colours and tactility (such as memory foam). The same technique is known as reaction injection molding (RIM) when used to form rigid parts. PU is available as sheet material used in laminating (page 188) and stitch bonding (page 196). Foam molding is preferred for shaped parts and produces a superior surface finish.

APPLICATIONS

Cold-cure foam-molded products include upholstery (chairs and sofas), car interiors,

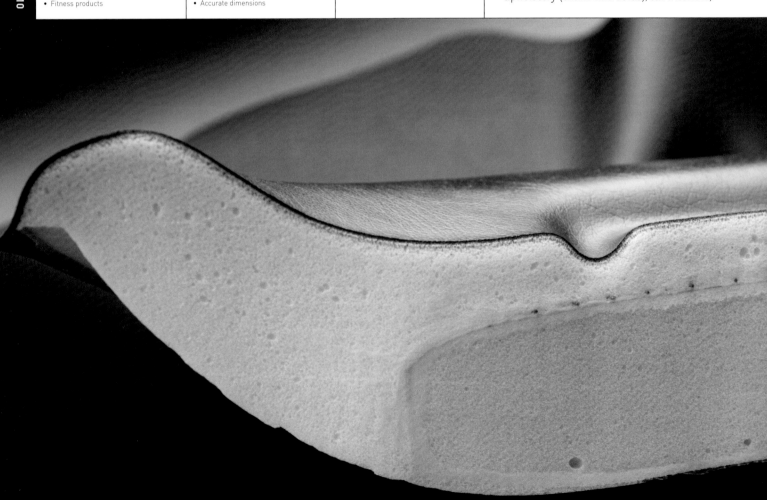

The Cold-Cure Foam Molding Process

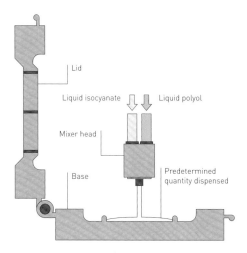

Lid

Liquid isocyanate Liquid polyol

Mixer head

Base

Predetermined
quantity dispensed

Stage 1: Filling

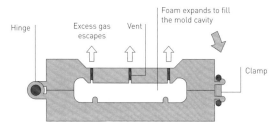

Hinge

Excess gas
escapes

Vent

Foam expands to fill
the mold cavity

Clamp

Stage 2: Molding

TECHNICAL DESCRIPTION
**The two ingredients – polyol and
isocyanate – that react to form
polyurethane (PU) are fed into
the mixing head, where they are
combined at high pressure. In stage 1,
the predetermined quantities of liquid
chemicals are dispensed into the mold
at low pressure (1–2 bar). As they
are mixed they begin to go through
a chemical exothermic reaction.**

**During stage 2 the blowing agent
causes the polymerizing mix to
expand to fill the mold. The only
pressure on the mold is from the
expanding liquid, so molds have to
be designed and filled to ensure even
spread of the polymer while it is still
in its liquid state. Vents in the upper
part of the mold allow gas to escape
to reduce the build-up of pressure.
The part is fully cured in around six
minutes in a one-way reaction.**

aircraft seating, padding (backpacks and protective clothing), fitness products (weights and trainers) and shoe soles.

RELATED TECHNOLOGIES

While cold-cure foam molding is limited to PU, injection molding is capable of processing a wide range of thermoplastics, among other materials. It is a widely used technology in other industries, especially consumer electronics, furniture, lighting and automotive applications. Foam-based products made in this way include shoes, medical products and sports items.

Knitting (pages 126–51) and weaving (pages 76–105) are capable of producing three-dimensional materials called spacer fabrics. These materials provide padding without compromising the flow of air, because they have an open structure. However, they are only available as sheet materials. Foam is molded with a profile and can have varying wall thickness, which helps to produce more ergonomic parts. Spacer textiles are also relatively expensive compared to foam, which limits their application somewhat.

QUALITY

PU is a durable and long-lasting material. The foam replicates the surface of the mold as it expands; higher-density moldings will reproduce textures with higher definition of detail.

The density of the foam is determined by the quantity of blowing agent. Seat padding is typically around 50 to 60 kg/m³ (3–3.75 lb/ft³), sound insulation is around twice that and self-skinning foam is up to 400 kg/m³ (25 lb/ft³). Additives are incorporated to improve flame retardancy or provide antimicrobial functionality, for example.

DESIGN

Over-molding provides the opportunity to incorporate a metal structure (for self-supporting parts, such as sofa armrests), electronics (for sensors and under-seat heating) or fabric cover (eliminating the need for a separate upholstering step). Complex multifunctional parts are produced by molding an insert, which is assembled with all the necessary components, followed by over-molding with a seamless foam skin.

Increasing the density of the foam produces a stiff and tear-resistant skin. Combined with texture, the surface eliminates the need for upholstering. This technique is utilized for car and aircraft interiors.

One of the advantages of reaction-polymerization plastics is that the cavity can be much larger than conventional injection molding, because the polymer is very liquid prior to being catalysed, so will easily flow around large and complex molds. As a result cold-cure foam-molded parts can be up to 3 m (10 ft) long.

A range of pigments is available to produce metallic, pearlescent, thermo-chromatic and photochromatic effects, as well as vibrant fluorescent and regular colour ranges. Multiple colours and parts may be combined in a single product by over-molding. This is used to combine materials with different functional qualities, or for decorative reasons.

Texture

Material: PU
Application: Aviation
Notes: The PU replicates the surface of the mold.

Self-Skinning

Material: PU
Application: Armrest
Notes: High-density molded foam forms a layer of skin around the surface, precisely reproducing the texture finish of the mold.

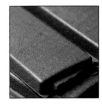

Over-Molding

Material: PU
Application: Automotive
Notes: Over-molding is used to incorporate electronics and fabric covers in seating, for example.

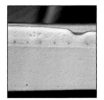

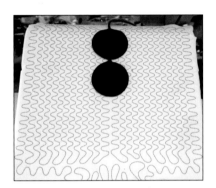

Metal Structure

Material: PU and aluminium
Application: Aerospace
Notes: High-strength aluminium panels are over-molded with foam to produce stiff and lightweight padded armrests.

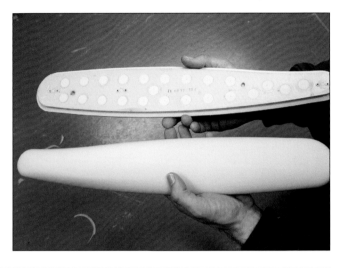

Embedded Electronics

Material: PU
Application: Automotive
Notes: Electronic components are laminated onto pre-molded pads, which are subsequently over-molded.

Chip Foam

Material: PU
Application: Automotive
Notes: Recycled foam is used to reinforce the sides of seats that have to withstand repeated sideways loads.

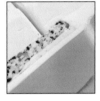

Colour

Material: PU
Application: Automotive
Notes: Naturally off-white, PU foam is produced in a range of vivid colours. Lighter colours will gradually yellow.

MATERIALS

PU is available as a thermosetting and a thermoplastic material. The principal difference is that thermosetting plastics form permanent cross-links in the polymer structure during polymerization. This means that the reaction is one-way and the material cannot be heated and melted in the way that thermoplastic can. Both types are available in a range of densities, colours and hardnesses. They can be very soft and flexible (Shore A range 25–90) or rigid (Shore D range).

Blowing agents are either chemical or physical. Chemical types are mixed into the raw ingredients and are triggered by reaction with isocyanate or thermal decomposition. They react to give off gas and thus cause the material to foam. For example, water reacts with isocyanate to generate carbon dioxide (CO_2).

COSTS

A mold is required, which limits this technique to production runs of at least a few hundred. For high-value applications it may be feasible to set up cold-cure foam molding for just a few units.

Mold costs are low to moderate, because there is no pressure required during molding, so they do not need to be as strong as those used for molding solid materials. The least expensive molds consist of two halves, known as the male tool and female tool. Undercuts are shapes not in the line of draw (linear travel of the mold) and so require molds made up of more than two parts (to allow the part to be demolded). For example, one-piece molded shoes are manufactured using complex molds made up of multiple parts that come apart in sequence.

Cycle time is around six minutes. The part is demolded and the foam is sufficiently cured after 24 hours to be trimmed and upholstered. Incorporating functional parts through over-molding will increase costs, but will reduce assembly and finishing later on.

ENVIRONMENTAL IMPACTS

Thermosetting scrap cannot be directly recycled. Instead, it is mechanically recycled by grinding up into small particles, which are subsequently added to virgin material used in molding, or bonded into sheets used as carpet underlay or upholstery padding.

Plastics contain relatively high embodied energy compared to other waste materials and so may be incinerated to recover this.

→ Producing an Office Seat

Molded foam offers many advantages to designers, such as the ability to create shapes that would be complex and costly to produce from foam slab, as well as producing parts with a superior surface finish. In addition, functional and decorative elements may be seamlessly incorporated into the part during molding. In this case, a molded plastic base is over-molded into an office seat to reduce assembly operations and ensure the highest-quality join between the structure and the padding.

The molded plastic (**image 1**) is prepared beforehand and loaded into the upper part of the mold (**image 2**). During molding, the air rises to the top. This reduces the surface quality of the top side of the foam molding. Therefore, incorporating the plastic panel from above means any slight imperfections will be concealed beneath.

The mold is sprayed with wax release agent followed by a predetermined measure of polyol and isocyanate (**image 3**). The mold is clamped shut (**image 4**) and the two liquids react and expand to form lightweight and flexible foam. The density is controlled by the amount of foam relative to the size of the mold.

Within six minutes or so the foam has expanded to fill the mold and cured in a one-way reaction (**image 5**). A physical bond has formed between the rigid plastic base and the soft foam. The parts are stacked up (**image 6**) and the flash is trimmed (**image 7**). The seat is ready for assembly and upholstery (page 404).

1

2

3

4

5

6

7

Featured Company

Interfoam
www.interfoam.co.uk

→ Foam Seat with Integral Fabric Cover

Interfoam has developed a process for incorporating a fabric cover directly in the molding process, thus eliminating the need for a separate upholstery step. Polyvinyl chloride- (PVC-) coated fabric (page 226) is heated to increase flexibility (**image 1**). This allows it to conform to the profile of the mold without tearing.

It is draped over the mold and clamped down (**image 2**). The operator aligns the fabric (**image 3**). Once in place a strong vacuum is applied, which draws it onto the surface of

the mold (**image 4**). This is essential, because the foam itself does not apply adequate pressure to force the fabric to the deep contours of the seat during molding.

The mold is filled with an exact measure of liquid polymer (**image 5**) and clamped shut for six minutes (**image 6**). The foam expands to fill the mold and form a strong bond with the back of the fabric (**image 7**). Excess fabric is trimmed and the molded seat is ready for assembly (**image 8**).

1

2

3

4

5

6

Featured Company

Interfoam
www.interfoam.co.uk

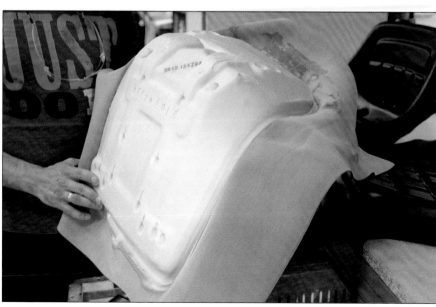

7

8

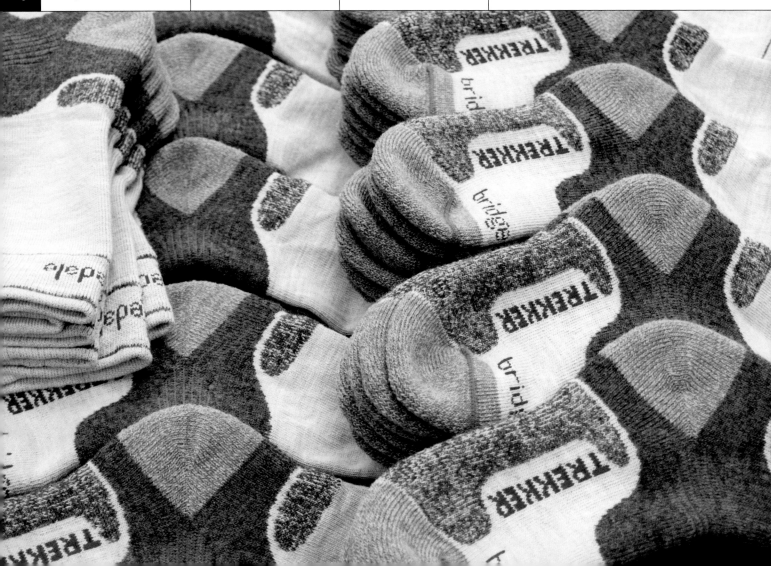

become soft when they are heated and so can be shaped, then solidify as they cool.

APPLICATIONS

Heat setting is used to relieve the stress in filaments (page 50) and textiles (see Shrinking and Stretching, page 210), as well as for pleating. Boarding is used to shape and finish hosiery, such as socks, tights, leggings and thigh-highs.

RELATED TECHNOLOGIES

Hosiery is woven (pages 76–105), fully fashioned (weft knitted and seamed), or circular knitted. Knitting is generally preferred for tight-fitting items, thanks to its higher stretch and recovery and more open structure than woven fabrics. However, until the development of thermoplastics and boarding, circular knitting was not capable of producing hosiery that fitted as well as woven and fully fashioned items. The development of nylon meant that by the 1940s tights were less expensive and more widely available than before. Improved quality and fit, combined with higher productivity and seamless construction, helped circular knitting to become the dominant means of production for such items.

It is possible to eliminate boarding in low-cost items with elastomeric yarns, such as elastane, and produce a 'one size fits all' garment.

QUALITY

The process of raising the temperature of thermoplastic yarn to above its glass transition temperature (Tg) and applying oriented pressure performs many important functions. By relieving the tension that has built up in the yarns during plying (page 60) and knitting, it improves dimensional stability and resistance to washing and reduces the likelihood of creasing and shrinking when in use.

Heat setting produces the finished shape, known as a durable press finish, which cannot be altered unless the thermoplastic yarn is heated up to the same temperature. Heat-set items do not require ironing, especially if they are tumble-dried after washing.

The Boarding Process

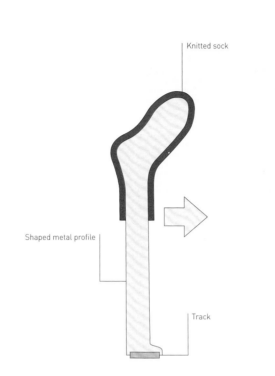

Hot steam chamber

Knitted sock

Shaped metal profile

Track

TECHNICAL DESCRIPTION

This is a straightforward process. Circular-knitted items, such as socks, tights or leggings, are pulled onto the shaped metal profile. Stretching over the form applies pressure. There are typically several metal profiles mounted onto a track and run in groups.

Together they are moved into the heating chamber, where the temperature is raised to above the 'glass transition temperature' (Tg) of the thermoplastic yarn. At this point the bonds between the polymer chains weaken, allowing them to move against one another. As a result the plastic begins to soften.

The type of material and additives determine the heat-setting temperature. Each polymer will behave differently, depending on whether it is amorphous, semi-crystalline or fully crystalline.

The elevated temperature is maintained. The knitted item gradually conforms to the shape of the metal profile. After sufficient time has passed, it is cooled, causing the polymer to solidify.

This determines the final shape of the part. With heat treatment creases, pleats, folds and other features are permanently set into garments. This technique is used to finish woven as well as knitted items.

Thermoplastics can be reheated and reshaped without significant loss of quality. Therefore, the shape is permanent, unless it is reheated to above its Tg and pressure reapplied.

→ Boarding Technical Socks

Technical socks are circular knitted in a wide range of sizes and styles (**image 1**). Each one of Bridgedale's socks is carefully boarded to create a durable press finish.

Boarding is essential for producing items that fit comfortably on the leg and foot. The designer uses a leg shape form with key measurements to test ideas (**image 2**). After boarding, the sock will fit the shape better than if it was taken directly from the circular-knitting process. After knitting the socks are washed and dried (**image 3**). The tension that has built up during production causes the socks to bunch up in their relaxed state (**image 4**).

They are stretched over metal profiles (**image 5**) and subjected to steam treatment for around 60 seconds. After this time the knitted nylon yarn has conformed to the shape of the metal foot (**image 6**). The socks are cooled to room temperature, causing the shape to set. Finally, they are automatically removed from the boards (**images 7** and **8**).

1

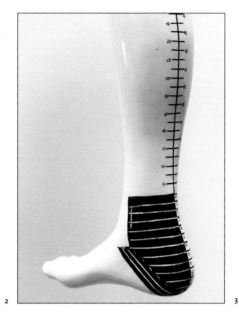

2

3

DESIGN

The final shape of the item is determined by knitting design, yarn selection, temperature and pressure. While designers do not usually have much say about temperature and pressure, they do affect the knitting process and mix of yarn that is used.

Circular knitting is capable of producing a range of diameters, from small socks up to wide tubes that are slit into flat sheets used to make a wide range of products. Modern computer-guided knitting machines can reproduce complex and intricate patterns involving multiple types and colours of yarn.

Boarding is standardized through size. In other words, the shaped metal profiles are small, medium or large. The shape can be adjusted for a specific project, but this is not common. Normally, all styles of the same product in the same size category are boarded on the same equipment. Right and left socks and tights are usually identical, rather than being made in pairs, to reduce complications in production.

MATERIALS

Thermoplastics and thermoplastic blends are suitable for boarding. Common types include nylon, polyester, acrylic, polypropylene (PP) and elastane. Semi-synthetics such as acetate are also suitable.

Blended and plied yarns are suitable if there is sufficient thermoplastic fibre content. Care must be taken to ensure that the high temperatures used do not affect other fibres in the mix, such as cotton or wool.

COSTS

Standard shaped metal or wooden boards are a fundamental part of hosiery production with circular knitting. Specifying custom shapes increases cost.

Cycle time is around 60 seconds and multiple items are heat set in a single operation. Automated processes are able to transfer hosiery directly from the toe-closing operation to boarding, reducing cycle time and labour costs.

ENVIRONMENTAL IMPACTS

While the process is not highly energy-intensive, it does require considerable heat. Temperature range is 120–200°C (248–392°F), depending on the type of thermoplastic. As a result, several tonnes of steam may be consumed every hour.

Boarding improves the dimensional stability of hosiery, providing a comfortable fit for longer.

4

5

6

7

8

Featured Company

Bridgedale Outdoor Ltd
www.bridgedale.com

Construction Technology

Hat Blocking

This is the process used to give the final shape to felt, fabric and straw hats such as fedoras, boaters, sombreros and bowlers. The prefabricated materials – in the form of flat sheets or preformed hoods (cone or capeline shape) – are dampened, heated and pulled over a block mold. The three-dimensional shape is permanently set as the material dries and cools.

Techniques	Materials	Costs
• Hand blocking • Machine blocking (French press and hydraulic press)	• Felt and leather • Straw and fabric	• Highly skilled labour • Cycle time ranges from one day to more than a week

Applications	Quality	Related Technologies
• Hats • Interiors	• Permanent shape • Slight variation in dimensions	• Machine stitching • Weft knitting • Basket weaving

INTRODUCTION

The finest hats continue to be made by hand over a wooden block and can take several weeks to finish. Machine-made hats are formed on over an aluminium mold using a mechanical press. There is still a great deal of handwork, even in machine processes, and as a result high-quality hats are expensive to make.

Blocking is capable of producing a finished shape from a flat sheet of material, although there are limits to how far materials can be stretched when damp. More commonly, blocking is used to press the final shape from a preformed sheet of material.

The preforms, known as hoods, are prepared using a variety of processes, which depend on the material. For example, straw hoods are handmade by basket weaving (page 120), whereas felt cones may be handmade or press-molded. They are prefabricated as a cone or capeline (wide brim), depending on the final form of the hat to be blocked.

APPLICATIONS

Blocking is used to make a wide range of hats. Some well-known styles include fedora, cowboy, bowler and sombrero. They are traditionally made from both felt and straw. Other types of straw hat include boater and conical.

Fedora hats have a brim and a crown that is typically creased lengthways with pinched-in sides. Some styles have their own name, such as trilby and panama. Cowboy hats have a higher crown and wider brim. They have become associated with the American West. Stetson is a famous brand of cowboy hat, partly because the Texas Rangers and several other law enforcement agencies in the United States adopted them as part of their uniform. The bowler hat, also known as a coke hat or (in the US) derby,

The Blocking Process

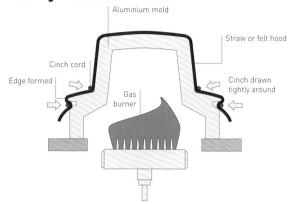

Single-sided mold

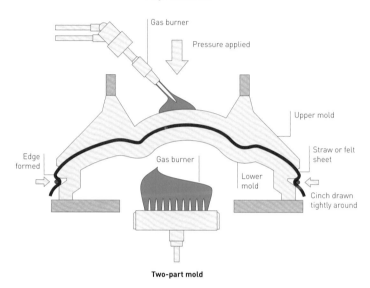

Two-part mold

Machine blocking

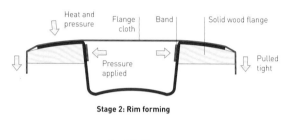

Stage 1: Crown forming

Stage 2: Rim forming

Hand blocking

TECHNICAL DESCRIPTION

The molds (or pans) used for machine block processing are typically cast aluminium. Single-sided molds are used to block shapes that do not have undulating profiles that require pressure from above. The hood of straw or felt is pulled over the heated mold and drawn tight with lengths of cord, known as cinches. Tying the brim first, followed by the base of the crown, increases the tension that can be applied.

Heat is supplied to the mold by a gas burner. As the sheet of dampened material is pulled down tightly steam is produced and it dries rapidly. Within three or four minutes the formed sheet is ready to be demolded.

Two-part molds are used to block hats with a profile that cannot be made over a single-sided mold, such as a concave crown, pinched sides or rippled profile.

The presteamed and dampened hood of felt or straw is stretched over the heated lower mold (also known as male mold). The heated upper (or female) mold is brought down onto it, compressing the material in between.

Hand blocking is used to shape straw, felt and leather. Although the processes are quite different the principles remain the same: steaming and softening; forming; and drying.

In stage 1, the crown is formed. The generic shaped hood is given its finished shape by pulling it tight onto the mold, tying with cord and drying. The placement of the cinch determines where the crown ends and the brim starts. This point is known as the break line.

The hat is pulled tight onto the block and a wooden plug, known as a tipper, is pushed

in firmly to shape the top of the crown. The hat is left on the mold to dry for 24 to 48 hours, depending on the climate.

Once the hat has fully dried out, it is removed from the mold. If the crown is not accurate or symmetrical at this stage then it may be reblocked.

In stage 2, the brim is shaped over an oval-shaped wood mold, called a flange. Like the crown, the shape of brim is determined by the style of hat being made. The hat is placed crown-down into the mold and a brass band block is placed inside to hold the hat securely in place.

A dampened flange cloth is pulled tightly over the brim and heat and pressure are applied, usually with a hot iron. Once smooth, a heavy weight is applied and the brim is left to dry for another 48 hours or so.

has a round crown that was originally developed for gamekeepers in England (more practical than a top hat when riding). A sombrero is a wide-brimmed hat from Mexico worn by horsemen and musicians. The boater is a formal style of hat with a flat top and brim. Conical hats are worn by farmers and outdoor workers in China and Vietnam.

Similar techniques are used for a range of other applications, including lightshades, trim and footwear. Outside hat making the process is referred to as pressing and molding.

RELATED TECHNOLOGIES

Felt hats are preferable for colder seasons, whereas straw hats are better suited to warmer climates. In this way, felt and straw may be used for the same shape and style of hat, but destined for use in cold and warm weather respectively.

Many techniques are used to make hats for warmth or to provide shade from the sun. Weft knitting (page 126), used to make fabric hats such as beanies, produces seamless one-piece hats in numerous colours, patterns and textures. A wide range of styles is possible on a single machine, without the use of molds, making this a cost-effective process. Alternatively, hats may be constructed from panels of material that are stitched together (pages 342–69). This is versatile, as all types of material and multiple layers can be joined, including fleece, fur and fabric. However, hats made with these techniques, which utilize pliable materials, tend to lack structure. To create a rigid hat, fabrics are laminated (page 188) onto a rigid backing fabrics, called blocking net, such as buckram (woven cloth coated with stiffening agent), which is available in a number of colours and from fine to coarse. This is less pliable than straw or felt so the shape is normally achieved by blocking the crown and brim separately on specific blocks. They are then sewn together by machine.

It is possible to skip the blocking phase by basket weaving the hat directly in the finished shape. This technique is limited because weaving cannot produce the curvature, well-defined edges and details that are possible with blocking.

QUALITY

The choice of material and design of the hat will affect the performance. The crown covers the head and provides protection from sun, wind and rain. Felt is waterproof and insulating. By contrast, the open structure of woven straw means that hats made from this material tend to be lighter and more airy. However, very fine woven materials, such as used to make panama hats, provide dense head coverage. Linings (page 392) are used to improve cover and warmth.

The durability, dimensional stability and weather resistance of hats is improved by soaking the hood in a stiffening agent, such as shellac (derived from the lac bug, which is indigenous to India and Thailand), polyvinyl acetate-based (PVA-based) adhesive, diluted starch or gelatin sizing prior to blocking. The stiffening agent softens when heated, allowing the fibres to move against one another and form into a three-dimensional shape. As it cools it solidifies into a stiff, resilient material.

The concentration of stiffening agent determines the feel and drape of the hat. Depending in the type of material and desired effect, it may not be necessary to use a stiffening agent at all.

Hats produced from handmade hoods will have some dimensional variation. As they are wetted and heated the material will expand in all directions. Then, as it cools and dries it will shrink onto the block. Slight variation in material quality or weave density, for example, mean the expansion and shrinking are not consistent. The potential variation will depend on the type of material. The highest-quality hats may be blocked several times.

Hats produced from consistent materials, such as felt, will have less dimensional variation. As a result, they can be produced at higher volumes and lower cost.

DESIGN

The extent to which a material can be formed depends on its malleability and ductility. Most materials cannot be stretched far without breaking. Therefore, hoods are produced with a three-

dimensional profile that is close to the final molding shape.

Certain sheet materials, such as sinamay (see page 428), are converted directly from flat sheet into the finished shape by blocking. In other words, they do not require preforming into a hood. With this technique a hat is formed from sheet in just a few minutes.

Hoods are pressed to make the final shape either in a single step or with progressive molds. In other words, if the required shape cannot be produced in a single mold then several steps may be undertaken. This technique is also used to make more accurate hats and undercut profiles (recessed shapes) that would otherwise be impractical.

Blocking is usually the final process, other than finishing with trim. Therefore hoods are normally dyed (page 240), printed (pages 256–79) and embossed (page 220) before molding. After forming, it is possible to coat the materials using spray and dip techniques.

Edges may be finished with metal wire, which ensures a crisp edge. Alternatively, they are cut and left bare, overlocked, bound (see Hand Stitching, page 342, and Machine Stitching, page 354) or woven (if the hood was preformed by basket weaving, for example). The required finish depends on the vulnerability of the material to fraying and unravelling, combined with the desired effect.

Colour comes from the material, or from dyeing or printing prior to blocking. The design opportunities depend on the suitability of the raw material, which must be able to stretch during forming. Therefore, while dyed colour works well, printed geometric and repeating patterns tend to be avoided.

Loom-woven sheet materials, such as sinamay, may be produced with different colours in the warp and weft, as well as integrating various types of yarn.

Straw hoods utilize many different basket-weaving techniques, including plain, twill, fancy, patterns with different coloured strands and openwork structure.

MATERIALS

Many different types of fibre are woven into straw hoods, including abacá

Felt
Material: Sheep's wool
Application: Hat
Notes: Wool fibres are converted into nonwoven textiles by felting (matted together with pressure and heat) or needle punching (mechanical entanglement).

Fur
Material: Rabbit fur
Application: Hat
Notes: Fur felt hats are superior quality and typically take several weeks to make by hand. Once made with beaver, they are now produced using rabbit and hare fur.

Straw
Material: Straw
Application: Hat
Notes: Many different types of straw may be woven and dyed, including panama, sisal, wheat, jute and raffia.

Twill
Material: Straw
Application: Hat
Notes: A range of weave structures is used, including plain (basket) and twill, depending on the style and application.

Woven Edge
Material: Straw
Application: Hat
Notes: The edge of hand-woven hoods is finished by interlacing the strands back into the weave structure.

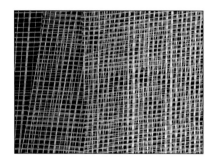

Sinamay
Material: Abacá
Application: Hat
Notes: Sheets of sinamay are layered together to make a sturdy construction. Density can range from solid to open.

(banana), panama, sisal (and the finer and more expensive parasisal), rush, wheat, jute and raffia. Sinamay is woven from abacá, and sennit may consist of palm, straw or grass.

Abacá straw hoods are woven in China and the Philippines, and panama and sisal straw come mainly from South America. They may take days or even weeks to make by hand, depending on the quality and raw materials.

Felt hoods are manufactured with a wide range of fibres. The most commonly used are wool and rabbit fur. The fibres are combined using heat and pressure to form a durable and resilient material, which is molded into a cone-shaped hood.

Leather is also suitable for blocking, but is less commonly used. It is more often formed into three-dimensional shapes for shoes, saddles and holsters, for example (page 430).

COSTS

The cost of the hood depends on the quality of the raw materials. Both felt and straw range from low- to high-cost. For example, raffia is inexpensive and relatively coarse, so it does not take long to weave. By contrast, panama takes a great deal of preparation, handwork and finishing. Producers of the finest panama hoods make very few each year, which means they can be very expensive.

Blocking ranges from the production of bespoke hats to high-volume production with metal molds. By contrast to the several weeks that handmade hats can take to make, mass-produced items may take just a few minutes. As a result, cost ranges from low to very high.

ENVIRONMENTAL IMPACTS

Hats made from natural and renewable materials, using traditional manual techniques, have virtually no impact on people or the environment. These hats are an essential parts of many people's lives. Factory-based production, using steam and hydraulic presses, requires more energy and waste is inevitable. However, by sticking to natural and renewable materials, the environmental impacts remain low. In addition, well-made hats that have lost their shape can be reblocked to extend their life.

The source of material is an important consideration; for example, materials may need to be transported over large distances, or may be from countries that do not impose high standards of labour. Certification is key to ensure materials are sourced ethically and responsibly.

→ Felt Blocking in a Split Mold

A different mold is required for each size of each style (**image 1**), although many styles are made in only one size. This fedora hat requires a two-part mold, which is loaded into the two-sided blocking press (known as a French press) and preheated (**image 2**).

The black felt cone is prepared by dampening and steaming. It is evenly pulled over the lower mold (**image 3**). The top mold is brought down and pressure is applied. The outside edge is tied down to draw the brim tight and ensure a well-defined edge (**image 4**).

After three or four minutes the molds are separated and the shaped hat is removed (**image 5**). It is passed on to the finishing section, referred to as machining. Excess is trimmed from around the brim (**image 6**). Then a length of wire is placed inside the fold to make a crisp edge (**image 7**), which is locked in place by a line of stitching (**image 8**). The headband is sewn inside the crown and a band is sewn around the base of the crown to finish the hat (**image 9**).

1

2

3

4

5

6

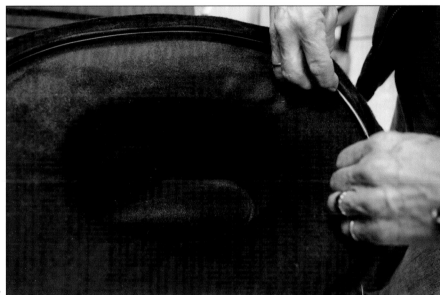

Featured Company

Whiteley Fischer Limited
www.whiteley-hat.co.uk

7

8

9

→ Straw Blocking with a Single-Sided Mold

Straw is dyed, basket woven into a capeline and coated with stiffening agent before blocking. There are many colours and styles to choose from (**image 1**). This wide-brimmed ladies' hat is made of green straw, which is wetted and steamed to soften the fibres and stiffening agent (**image 2**).

This design requires only one mold (**image 3**), because the shape does not have any indents or other design features that require pressure from above. The straw hood is placed onto the mold, pulled tight to the profile and held with string (**images 4** and **5**). The hat is still hot when it is removed from the mold (**image 6**). Once it has cooled, the shape is permanently set. The edge is carefully cut away and a length of wire is sewn into the brim to produce a crisp and sturdy edge (**images 7**, **8** and **9**). Floppy-brimmed hats are made without the wire.

The headband is sewn into the crown (**images 10** and **11**). Once the base is complete, ribbons and other decoration may be added (**image 12**).

1

2

3

4

5

6

7

8

9

10

11

12

Featured Company

Whiteley Fischer Limited

www.whiteley-hat.co.uk

→ Blocking from Flat Sheet in a Split Mold

Rolls of sinamay are available in many colours and combinations (**image 1**). Sheets are cut to size and layered in groups of three (**image 2**). Laminating multiple layers ensures a stiff structure across a wide brim.

Like the felt hat (page 424), this design requires a two-part mold, because the shape dips in around the crown (**image 3**). The straw is pulled tight over the lower mold and the two halves close together (**image 4**). Once clamped shut, the brim is pulled over the lower edge and secured with string (**image 5**).

Steam is pumped through the sinamay and after three to four minutes the molds separate to reveal the finished shape (**image 6**). The layers of sinamay are permanently bonded together (**image 7**).

The blocked hat is trimmed, the brim is secured and finishing touches are made (**image 8**).

MOLDING

428

1

2

3

4

5

6

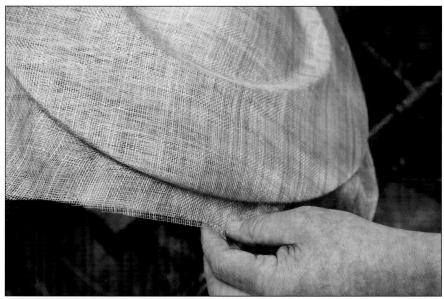

7

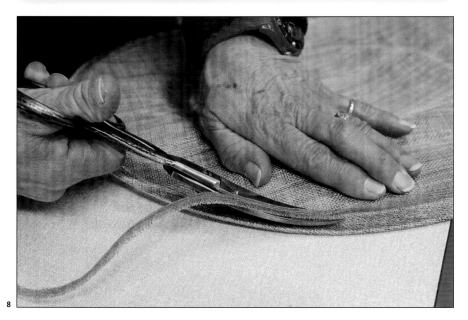

8

Featured Company

Whiteley Fischer Limited
www.whiteley-hat.co.uk

Construction Technology
Leather Molding

In a process also known as blocking, leather is softened to make it pliable, molded and set into a three-dimensional shape. Leather is inherently durable and breathable, while being comfortable to use. Combined with expert craftsmanship, this unique and natural material is formed into long-lasting and desirable products from formal shoes to phone covers.

Techniques	Materials	Costs
• Machine molding • Wet molding • Lasting	• Leather • Suede	• Highly skilled labour • Cycle time ranges from one day to more than a week

Applications	Quality	Related Technologies
• Footwear • Saddles • Accessories	• High quality and durable • Natural leather finish	• Injection molding • Hat blocking • Machine stitching

INTRODUCTION
Leather is a luxurious material used for its toughness and low stretch, even when under repeated bending and twisting loads. To make it suitably pliable for molding into three-dimensional shapes, the leather must be softened with water or steam. Hot and wet, it is stretched over the mold and held until dry. With time it sets to form a durable item.

Leather molding is carried out either by hand over a wooden form, or with hydraulic pressure over a plastic or metal mold. In the case of shoemaking, the foot-shaped plastic mold is called a last.

The Leather Molding Process

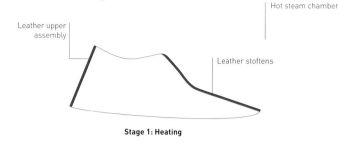

Stage 1: Heating

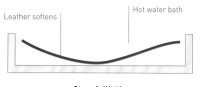

Stage 1: Wetting

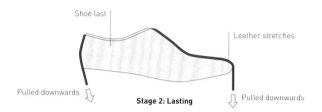

Stage 2: Lasting

Stage 2: Loading

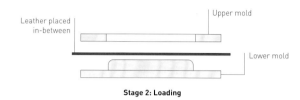

Stage 3: Nailing

Stage 3: Forming

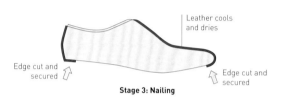

Lasting

Wet forming

TECHNICAL DESCRIPTION

LASTING

Leather molding is carried out either by hand or with the assistance of a hydraulic press. The processes are remarkably similar, mainly because machine molding has evolved directly from manual techniques. And the desire is that equally high-quality products can be achieved with hydraulic-assisted pressing. For example, high-quality factory-based production combines both, utilizing machines only where they bring advantage and do not compromise quality.

The majority of molded leather is used in lasting. This technique has been practised for generations to make high-quality leather uppers for shoes and boots. In stage 1, the leather upper, which has been stitched together into a three-dimensional form, is prepared by heating in a steam chamber. It is heated to 180°C (356°F) for a minute or so, at which point the leather becomes pliable enough to mold. Steaming softens the bonds between the fibres in the material,

allowing them to move against one another without affecting the strength.

In stage 2, the leather construction is placed onto the last and stretched over the form. Lasts are traditionally made from wood, and may also be referred to as a mold. Modern lasts are made from nylon.

In stage 3, the leather is pulled around the profile, stretching it in all directions to take the shape of the last. Once in place, the leather is nailed or mechanically held onto the last. This ensures it stays taut to the surface as it dries and so reproduces the three-dimensional shape exactly.

Although other products can be made in the same way, lasting is a complex process and requires a high degree of skill to execute well and therefore simpler molding methods are used on parts that do not have such a complex profile as a shoe.

WET FORMING

Wet forming is typically carried out as a manual process. High-volume production, such as for phone covers, is carried out on

a hydraulic press. The principal difference between manual and hydraulic pressing is that in stage 1 of molding by hand the leather is submerged in a bath of hot water as opposed to steaming in a chamber. The function is the same: to heat and soften the leather so it is pliable enough to be molded.

In stage 2, the leather is placed into a two-part mold. Profiles that do not require internal bends may be produced on a single-sided mold, similar to lasting. In stage 3, the molds are brought together and pressure is maintained until the leather has dried. This usually takes 12 hours or so, depending on the climate and drying conditions.

In both cases, thinner materials are easier to mold, but will ultimately be less strong than thicker leather. Thin leather is laminated or adhesive bonded onto a stiff backing layer, or liner, to help it maintain the molded shape throughout its life.

APPLICATIONS

The majority of molded leather goes into making uppers for shoes and boots. Other applications include saddles, seats, holsters, sheaths, satchels, carry cases (such as for phones and tablets), accessories and masks. Leather hats are made by blocking (page 420), in the same way as felt and straw.

RELATED TECHNOLOGIES

Molding creates a three-dimensional shape from a single piece of material. Similar-shaped shoe uppers are fabricated from other types of material, such as synthetic leather (produced by needle punching, page 152, or laminating, page 188); dip-molded rubber (page 442); weft-knitted polyester (page 126); or injection-molded thermoplastic polyurethane (TPU) or ethylene vinyl acetate (EVA). These materials have particular qualities that make them suitable for other types of shoes. For example, three-dimensional weft-knitted uppers make lightweight and flexible running shoes; rubber is ideal for waterproof boots; and injection molding produces very low-cost shoes in a single process. So, while they can be used to make the same-shaped part, the use is quite different.

Leather will comfortably bend in one direction. Molding is required only if the shape bends in more than two directions. Therefore, pouches and similar items may be fabricated by simply stitching (pages 342–69) or bonding (page 370) flat parts together.

QUALITY

Leather is a luxurious material, desired for its unique and natural properties. It is durable and breathable, which makes it ideal for footwear. Even so, there are many factors that determine the final quality of a molded leather product including the raw material, skill and experience of the craftsperson, and the finishing processes.

Leather is a natural material and each hide or skin has unique properties. It is graded according to quality, because there are many factors that can affect the look and feel, such as the type of animal, breed, age, lifestyle and where on the animal it was taken from. In addition,

the visual quality is affected by the life of the animal, such as scars, veins and wrinkles. Imperfections are removed during pattern cutting (page 314).

The type of leather contributes to the tactility and appearance: the skin side is tough and smooth or has visible grain; the flesh side is softer and may be napped (page 216) to form suede; or the fur may be left on for a hairy finish.

Leather is a forgiving material, but because each piece is unique a high degree of skill and experience is required to make the highest-quality molded leather goods. Incorporating seams adds complexity, because they will highlight alignment inaccuracies during molding.

Leather is molded on its own, or with adhesive to increase stiffness, reduce stretch and improve shape retention. The type of leather (and where it was cut from the animal) is determined by the application. For example, riding boots need to be very stiff and so require thick leather. Using leather with the flesh side facing outwards means scuffs and scratches can be easily removed by rubbing, whereas scratches on the skin side will be permanent.

The finishing processes determine the shade of colour (page 240), gloss (polishing and buffing) and durability of molded leather (waterproofing and so on)."

DESIGN

Molding geometry is limited to the amount leather will stretch. Certain types are able to stretch more than others. For example, leather from the belly region typically has more stretch and can be molded into deeper profiles. Stiffer leathers will mold less well, causing wrinkles to occur on tight bends.

Leather stretches more easily than it compresses. In other words, leather will form more easily over outside curves than it will compress into channels and recesses. Leather may be removed to allow the material to be drawn into a deep profile without wrinkling.

The size of a skin is determined by the type and age of animal. This will impact on the maximum size of the molding that can be achieved without having to join

two or more parts together. For example, cow hides range from 4 to 5 m² (43–54 ft²) and sheepskins can range from 0.75 to 1 m² (8–10 ft²).

Single-sided molds are the simplest. Leather is stretched over the mold, such as a shoe last, and held until the shape is set. Forming undulating shapes that require a two-part mold is more complex and expensive (see also Hat Blocking, page 420).

Certain fabrics may be incorporated with molded leather before or after forming. Fabrics joined in before molding will not be formed like leather, but will be held in shape by the surrounding material. It is important that they are not damaged by the heat and pressure required. A wider range of material may be incorporated after molding: stitching (pages 342–69) and bonding (page 370) techniques determine the limits of what can be achieved.

Leather is available in a wide range of forms, colours and finishes suitable for molding. It can be punched, embossed (page 220), printed (pages 256–79), foil blocked and embroidered (page 298).

MATERIALS

Leather is a high-performance material and because it is natural, each product made with it will have its own unique qualities. It comes from the skins and hides of mammals, reptiles, fish and birds. Fur is an animal skin that has the hair, wool or fur fibres still attached.

All types of leather are suitable for molding, although cow is the most common. It is smooth, tough and durable (more so than sheep, but not as durable as goat or kangaroo, for example).

COSTS

The cost of the raw material has a great impact on the cost of molded leather goods, because the cost varies significantly depending on the animal and quality. Pigmented leather is used for lower-cost shoes. The surface is sprayed to correct the colour, reducing the visual flaws.

Molding adds considerable cost to the price of the raw material, because many processes are involved and high-quality products take several days to complete.

→ Developing the Concept

John Lobb has been making the finest men's shoes and boots since 1866. It continues to provide a bespoke shoemaking service from its Paris atelier, while expanding its ready-to-wear collection.

Designing new products remains a hands-on process. Concepts are developed through an iterative process of making and refinement until the designer's vision is realized. Starting from a sketch, the shoe engineer makes templates for pattern cutting the first prototype (**images 1** and **2**).

Each stage in the development process is reviewed and changes are drawn onto the prototype (**image 3**). The finished design is made by hand in preparation for production development (**image 4**). It takes the highly skilled shoemakers at John Lobb around 190 steps to make a pair of leather shoes, so it is critical that every detail is worked through before production for the next season can commence.

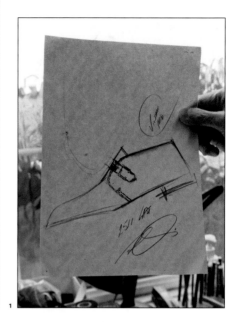

1

2

3

4

The molds are not too expensive and in principle any form can be molded over as long as it is strong enough to hold up to the pressure. For example, sheaths and holsters are often handmade directly over the object they will conceal. And bespoke shoes are produced on lasts that are made to meet each individual customer's requirements exactly.

ENVIRONMENTAL IMPACTS

Cow leather is a by-product of meat production and the animal must be slaughtered to remove the skin. A lot of leather comes from countries where animals are not properly looked after. And several species of animal that have been used for leather in the past are endangered and illegal to use for leather in many countries. Therefore, it is very important to consider where the material comes from.

The tanning process uses chemicals that can cause damage to people and the environment if they are not properly controlled. There is a great deal of work going on in many factories to reduce the environmental impacts and water consumption (see pages 162–65).

The molding process does not add significantly to this. The waste generated during pattern cutting – higher-quality leather goods typically result in more waste, because the natural finish of the

leather is maintained and therefore imperfections have to be avoided during pattern cutting – may be reconstituted and reused. Heat and energy are required and the efficiency of this depends on the workshop set-up.

Featured Company

John Lobb
www.johnlobb.com

→ Closing the Upper

The process of assembling the leather upper is called closing. The leather parts that make up the assembly are pattern cut from a hide in a process known as clicking (see Die Cutting, page 336). John Lobb uses French calfskins, tanned, dyed and finished into aniline leather in Italy. The highest-quality raw material of this type comes from France and the best finishing comes from Italy.

The patterns are skived (pared) with a sharp rotating blade to produce a chamfered edge (**images 1** and **2**). This reduces the edge thickness and so reduces the step in overlapping joins. The stiff leather that goes into the toe is preformed by blocking (**image 3**). Pressure is applied between two molds, which are heated to 62°C (143.6°F), to gently stretch the leather down the middle. This helps with the lasting process.

The upper is assembled from several parts (**images 4**, **5** and **6**). The total number of parts depends on the style and design. High-quality shoes are made with a lining. The seams are bonded and stitched (**images 7** and **8**).

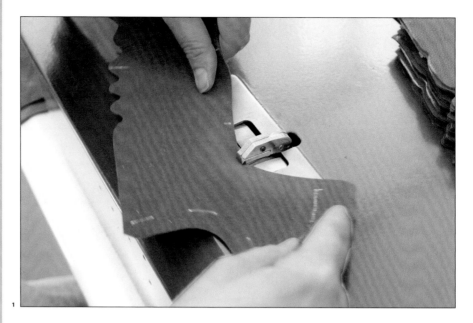

1

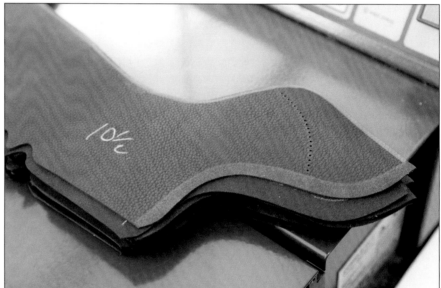

2

3

4

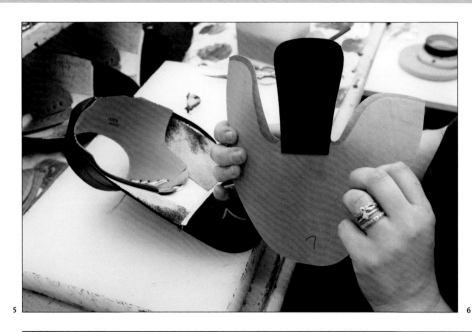

5

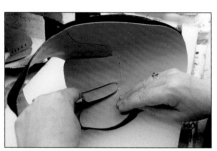

6

7

8

Featured Company

John Lobb
www.johnlobb.com

→ Lasting

Lasting may be carried out by hand, or with hydraulic assistance. Using hydraulics increases productivity, but only to a point. High-quality shoes still involve a considerable amount of handwork.

The yellow lasts are prepared with an insole, which is secured to the underside (**image 1**). The upper is prepared with a toe stiffener (**image 2**), which is made up of a nonwoven textile impregnated with adhesive. This adds strength to the toe when the shape is set.

First of all, the upper is secured on the heel of the last (**image 3**). It is heated until the leather becomes soft, then placed in the jig and the leather is pulled over the mold (**image 4**). When the desired shape is achieved the clamps close on the toe and the leather is bonded onto a ledge on the underside of the insole (**image 5**). The dimensions are checked and the shoe inspected (**images 6** and **7**).

The shoes are molded in pairs and set aside (**image 8**), ready for closing the heel.

1

2

3

4

5

6

7

8

Featured Company

John Lobb

www.johnlobb.com

→ # Toe Lasting

As before, the last is prepared with the insole and the upper is aligned and fixed in place at the heel. The shoemaker then gradually works his way around the outside, fixing the lining and outer leather in place (**image 1**). He is helped by hydraulic-assisted clamps, which operate individually to gently stretch the leather over the form (**image 2**). The toe measurements are checked to ensure the uppers are exactly right (**image 3**). The shoes are constructed in pairs to ensure they are a perfect match (**image 4**).

The sides are stretched over at the jointing bench and secured against the ridge on the insole with staples (**images 5** and **6**).

The toe of a different shoe is ready to be lasted (**image 7**). It is placed in a clamp, which stretches the leather over the toe profile (**image 8**) and is held in place by a length of wire secured by two nails (**image 9**). The length of wire will be removed later on, once the leather has set in place.

1

2

3

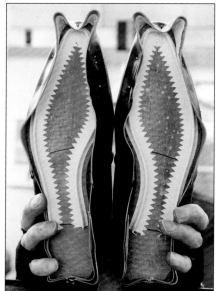

4

5

Featured Company

John Lobb
www.johnlobb.com

6

7

8

9

→ Heel Lasting

The heel is stiff and is stretched as little as possible as it is pulled over. The leather is gathered and secured in a process known as seat nailing (**images 1** and **2**).

At this stage the whole perimeter of the upper is secured onto the insole and held tight over the last. To finish it off, the leather is steamed and gently hammered into shape (**image 3**). This smoothens the profile and removes any imperfections.

1

2

3

Featured Company

John Lobb
www.johnlobb.com

→ Making

The finished upper is covered with plastic to protect it from contamination throughout the making process.

These shoes are Goodyear welted, so called after the inventor. This means that they can be repaired, because the welt attaches the upper to the insole and allows the parts to be separated without damage should the need arise.

The welt is stitched onto the insole (**image 1**). Because the whole shoe is not welted a seat lift is added to maintain the height of the heel (**image 2**). A wooden shank is inserted into the cavity to reduce bending (metal is also used but wood has better natural spring) and the underside of the insole is filled with cork filler (**image 3**). Cork is used because it is naturally breathable (cheaper shoes use foam). It takes around 24 hours to fully cure.

The leather sole is stitched onto the welt (**image 4**). It is sewn at an angle and into a channel cut into the bottom of the sole. The ridge is bonded back down with polyurethane (PU) adhesive using a brass wheel (**images 5** and **6**). The heel is bonded onto the sole in a heel press (**image 7**). This holds it in place until the last is removed and it can be secured with nails right through the insole.

The shoe has been on the last for around seven days by this point and the shape is fully set. The last is removed by steaming the leather to soften the ankle and allow it to stretch without damage (**image 8**). The last is designed to bend in the middle, like a foot, so that it can be removed (**image 9**).

With the plastic cover still intact, the sole is shaped by sanding (**image 10**). This is done by eye and requires a high degree of skill. Next, the sole edge is stained black and the sole is sanded and stained (**image 11**). The edge of the sole is ironed to produce the dark outline (**image 12**).

A pair of finished shoes is hand polished and ready to be packed (**image 13**). The leather is treated with cream and hot air to rehydrate the surface. Waxing, known as hand glazing, takes around 8 to 10 minutes for a regular finish and 15 to 20 minutes for a high polish.

1

2

3

4

5

6

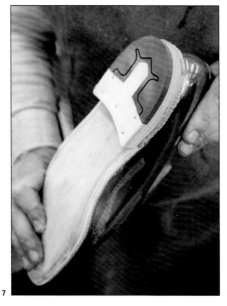

7

8

9

10

11

12

13

Featured Company

John Lobb

www.johnlobb.com

Construction Technology
Dip Molding

This low-cost molding process is used to manufacture seamless three-dimensional plastic and rubber products. Common examples include gloves, condoms and balloons. A few materials are suitable, including polyvinyl chloride (PVC), latex and neoprene. As a coating technique, it is used to produce colourful and durable finishes on metals and textiles.

Techniques	Materials	Costs
• Dip molding • Dip coating	• Latex • Neoprene • Polyvinyl chloride (PVC)	• Low tooling costs • Low unit costs • Fluid coating and laminating

Applications	Quality	Related Technologies
• Apparel • Accessories • Seals	• Gloss or matt finish • Range of colours • Seamless construction	• Machine stitching • Welding

INTRODUCTION
Dip molding is used to manufacture seamless three-dimensional items, such as rubber gloves and swimming hats. Removing the seams has many advantages for certain applications, such as increasing comfort and waterproofing.

There are two techniques. The first uses hot metal molds, which are coated with a release agent and dipped into liquid plastic (plastic suspended in plasticizer); the heat causes the plastic to gel onto the surface of the mold. The second uses a coagulant, which is applied to the surface of the mold. When dipped in liquid latex, the rubber particles bind together on the surface of the mold.

The plastic or rubber is cured on the former after dipping.

Flexible materials can be formed into hollow profiles and parts with severe undercuts, because once cured the material can be stretched over undulations in the mold. It can be easily converted into a coating process, such as for tool handgrips, by exchanging the release agent for a primer.

APPLICATIONS

It is relatively low-cost and so used for many industrial applications, such as electrical insulation covers and tool handgrips. Other applications include gloves, balloons, condoms, wrist and neck seals (used in drysuits, survival suits and immersion suits for example, see page 365), one-piece rubber outfits, and waterproof socks, hoods and hats. It is utilized in the production of medical products too, such as bags, surgical gloves, sheaths, tourniquet straps, bladders and waterproof plaster-cast covers.

RELATED TECHNOLOGIES

Dip molding is used to make many low-cost three-dimensional items that cannot be practically made in any other way, such as balloons and rubber gloves.

However, parts that do not need to be seamless or airtight can be constructed by machine stitching (page 354), welding (page 376) or adhesive bonding (page 370) panels together. These techniques are more versatile than molding, because the properties do not have to be uniform throughout. For example, different materials, thicknesses and colours may be combined in a single item.

QUALITY

Dip molding and coating produce parts with a smooth and seamless finish. The outside surface (that does not come into contact with the single-sided mold) tends to be glossy, but can be matt or foam-like.

The side that comes into contact with the mold is precise; embossed details and textures will be precisely reproduced on the mold-side of the molding. It is possible to turn parts inside out after molding so this finish is on the outside. Because the material is liquid until it gels

The Dip Molding Process

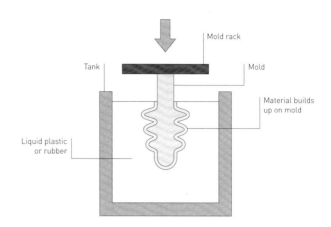

TECHNICAL DESCRIPTION

There are two main dip-molding techniques. The first uses pre-heated metal tools. For PVC the oven is set between 300 and 400°C (572–752°F). The length of time taken to pre-heat the mold depends on the size, but it typically between five and 20 minutes. The molds are commonly made from cast or machined aluminium, but steel and brass are also used.

The hot metal mold is coated with a dilute silicone solution for dip molding and with a primer for dip coating. The tool is positioned above the tank of liquid plastic or rubber. Plastic is suspended in plasticizer, making it liquid at room temperature. In the case of PVC it is known as plastisol. The same material is used as ink to screen print textiles (page 269).

The mold is submerged in the liquid. On contact with the surface of the mold, the liquid gels to form a polymer. The wall thickness rapidly builds up, reaching 2.5 mm (0.1 in.) within 60 seconds or so. PVC polymerizes at 60°C (140°F), so as the mold cools down and the wall thickness builds up, polymerization slows down. Once the PVC has polymerized, it cannot be returned to its liquid state and so cannot be recycled directly.

Dwell time in the liquid is usually between 20 and 60 seconds. To increase wall thickness the tool is heated up for longer, or steel is used to maintain a higher temperature for longer.

The second dip molding technique is used to form latex parts. Instead of heating up the mold, a coagulant is applied to the surface. When dipped in liquid latex, the rubber particles bind together on the surface of the mold.

In both techniques, parts are left on the mold to solidify. Once cured (polymerized) and vulcanized (rubber), they are removed from the mold to be washed and dried.

onto the mold, it can run and sag like paint. Therefore, the wall thickness at the bottom of the part is likely to be thicker than the top. To minimize this, molds are inverted after dipping, which also reduces the formation of a dip at the base of the molding. The speed of the dipping operation

(and the temperature of the mold) affects the quality of the molded part. 'Creep' lines are caused by slow dipping speeds (and high tool temperature) and air bubbles can form if the dipping speed is too fast. Therefore, prototyping is essential to determine the correct molding parameters.

VISUAL GLOSSARY: DIP-MOLDED PARTS

Colour
Material: Latex
Application: Medical
Notes: Latex is naturally milky white and so can be coloured, but the colour will not be as vivid as PVC or neoprene.

Coating
Material: PU-adhesive-coated latex
Application: Apparel
Notes: Adhesive is applied to the surface of latex to provide a strong bond in construction.

Texture
Material: Latex
Application: Apparel
Notes: The surface finish of dip-molded parts will be the reverse of the mold.

DESIGN

A major advantage is that molds for this process tend to be relatively low-cost and the same molds used for prototyping can be used for production. They are made from aluminium or ceramic, depending on the dipping process and materials.

Using a single-sided mold means that the accuracy of the outside surface cannot be precisely controlled and the material will smooth over sharp features. This results in variable wall thickness, which can be a problem with protruding details, because sufficient material may not build over them.

Mold design is affected by the nature of the liquid material, which gels on contact with the hot surface or coagulant coating. Therefore, flat surfaces, undercuts and holes must be designed so that they do not become air traps. A 5 to 15° draft angle is usually sufficient on surfaces that run parallel to the surface of the liquid. Air traps stop the material coming into contact with the mold, resulting in depressions and holes where the material has not gelled sufficiently. In some cases, air traps may be deliberately designed into the mold to produce holes that would otherwise need to be punched afterwards.

For hot-tool dip molding the wall thickness is determined by tool temperature and dip time (dwell). Within reason, high temperatures and long dips produce thicker wall sections. Wall thickness is generally between 0.2 mm and 5 mm (0.007–0.2 in.).

Wall thickness is built up with multiple dips. In such cases, it is possible to change the colour and hardness of each layer. As well as the obvious aesthetic advantages, multiple dips can provide functional benefits, such as electrical insulation that wears to indicate material thinning.

Several materials can be molded, and are available in a range of colours from transparent to opaque (including effect pigments, such as metallic). Surface finish can range from gloss to matt.

Dip-molded parts can be finished with a range of processes. For example, household gloves are lined with flock (page 280) for comfort and insulation.

MATERIALS

Polyvinyl chloride (PVC) is the most commonly used dip-molded material. Additives are used to improve the material's flame-retardant properties, chemical resistance, ultraviolet (UV) resistance and temperature resistance, and to reduce its toxicity for food-approved grades. It is available in a wide range of Shore hardness from A 30 to A100 (standardized measure of material hardness, where 30 is very soft and 100 is semi-rigid). It is available in the widest range of colours and effects, and can be colour matched.

Latex is also widely used. For applications that require contact with people or food, a modified grade is used, which has lower levels of protein and so is less likely to cause an allergic reaction. Latex is relatively low-cost and has excellent mechanical and elastic properties compared to synthetic polymers. It is naturally milky white and so can be coloured, but the colour will not be as vivid as PVC or neoprene.

Neoprene is a synthetic rubber that was developed by DuPont in the 1930s as an alternative to latex. It is also known by its chemical name, polychloroprene (CR). It has good resistance to alcohols, oils and acids, good elasticity at a wide range of temperatures and good resistance to physical failures, such as cutting.

COSTS

Tooling costs are low. Cycle time is rapid and multiple tools are usually dipped simultaneously to further reduce cycle time and cost.

ENVIRONMENTAL IMPACTS

PVC is often criticized for containing harmful ingredients. For example, phthalates and bisphenol A are subject to restrictions in many countries. Non-phthalate plasticizers are available for sensitive applications, such as toys and medical devices, made from PVC.

Natural rubber and latex can be fair trade and certified by the Forest Stewardship Council (FSC). This means the rubber contains no added chemicals, PVC or toxins, and was tapped from a tree in a sustainable manner, which prevents deforestation or displacement of indigenous people or wildlife.

Some people are allergic to the proteins in latex and in the most extreme cases the reaction can be life-threatening. Modified grades with lower levels of protein are used for sensitive applications.

→ Dip Molding Rubber Gloves

The gloves are formed over ceramic molds, which may be 30 years old or more (**image 1**). They are cleaned, prepared and coated with a coagulant (**image 2**). Low volumes may be carried out by hand, but mass-produced items, such as gloves and condoms, are continuously produced on fully automated production lines.

The coated molds are submerged steadily to ensure the viscous liquid does not fold over on itself and trap air (**image 3**). Stimulated by the coagulant, the rubber particles bind together to form a thin layer of latex.

The molds are removed from the bath of liquid latex and inverted so that the drips run back in (**images 4** and **5**), but this is not always necessary. The latex is vulcanized in an oven at 75°C (167°F) for approximately two hours. If they have a neoprene coating they will be further cured for 45 minutes at 120°C (248°F). Vulcanization is an irreversible process, whereby the latex is subjected to high pressure and temperature, causing cross-links to form between the polymer chains. This increases the elasticity and durability of the material.

The dry gloves are removed from their molds (**image 6**). They are washed and dried (**image 7**) and then trimmed to the correct length. The inside surface mirrors the finish of the mold (**image 8**).

1

2

3

4

5

6

7

8

Featured Company

Precision Dippings Marketing
www.precisiondippings.co.uk

Construction Technology
Composite Press Forming

Continuous fibre-reinforced thermoplastics are converted into three-dimensional parts by press forming. The combination of high-performance polymer and fibre creates a composite that is stronger than the sum of its parts. The unique properties make it useful in diverse applications ranging from aerospace and automotive through to sports shoes.

Techniques	Materials	Costs
• Press forming • CNC machining	• High-performance fibres and thermoplastics	• Moderate to high depending on molds and materials

Applications	Quality	Related Technologies
• Automotive, aerospace and marine • Footwear • Sports products	• High strength to weight and impact resistance • Highly repeatable	• Autoclaving • Filament winding • Compression molding

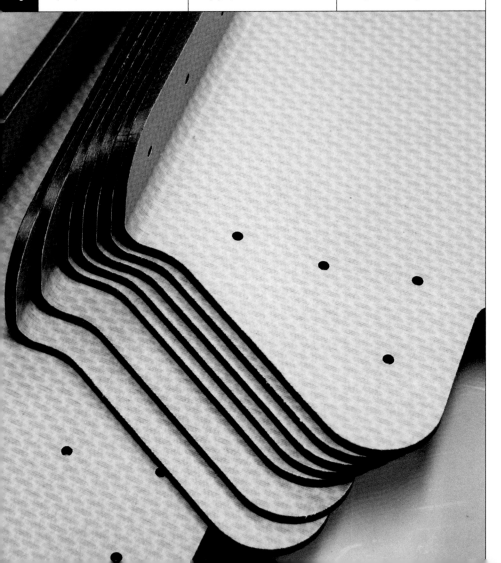

INTRODUCTION

Lightness is of growing importance in many industries. Reducing weight in the automotive, aerospace and marine industries helps to save a lot of fuel, which is economically beneficial and reduces impacts on the environment.

This demand has led to the development of thermoplastics reinforced with high-performance fibres. Compared to conventional composite laminates, which are produced with thermoset resin, they are less expensive to manufacture, more easily recycled and can have higher impact strength and fatigue resistance.

Application in industries with very high safety standards, such as aerospace, puts high demands on the accuracy and quality of parts. Destructive and non-destructive (see case study on page 453) tests are used to determine exact porosity, thickness and dimensions.

Parts are trimmed after molding by CNC machining. This is a precision-cutting technology used to profile and shape a wide range of materials. The number of axes that the CNC machine operates on determines the geometries that can be cut. In other words, a five-axis machine has a wider range of motion than a three-axis one (see also CNC Cutting, page 316).

APPLICATIONS

Thermoplastic composites are used in some applications for purely technical reasons. For example, in the automotive industry they help reduce vehicle weight and thus fuel consumption; in aerospace they provide lightness alongside fire resistance and low toxicity; and they help to make lightweight and high-strength safety gear, such as helmets.

Other industries use these materials for both technical and visual purposes. Sports products utilize thermoplastic composites for improved strength to weight, impact resistance (even at low temperatures) and energy return (spring back). Examples include football boots, ski boots, skis, snowboards and helmets.

RELATED TECHNOLOGIES

Molded thermoplastic composites have replaced metals and conventional composites in many applications, as well as creating unique application opportunities of their own.

They have replaced metals in applications that demand higher strength to weight (thermoplastic composites may be 50% lighter for the same strength), fatigue resistance, temperature performance (they perform effectively even when continually exposed to freezing or boiling temperatures), wear resistance and insulating properties. The advantages of using a thermoplastic matrix – as opposed to thermosetting – are ease of manufacturing and recycling, and lower toxicity.

Press forming is limited to sheet parts with a uniform wall thickness. In other words, the sheet material is pressed into a shape and cannot be easily drawn into tubes or parts with variation in wall thickness. Tubular parts are manufactured by filament winding (page 460) and bulk parts are fabricated by compression molding.

QUALITY

Thermoplastic composites are made up of high-performance fibres, which are encapsulated within a thermoplastic matrix. Factors that affect the ultimate strength that can be achieved include fibre type, method of construction, orientation and ratio to resin matrix.

High-performance fibres include filaments of glass, aramid and carbon. They are manufactured as unidirectional (see Spread Tow and Fabric Prepreg, page 234), biaxial (see Loom Weaving, page 76) or multiaxial (such as tri- or quadri-) fabrics. The method of fabrication will affect performance, because it will determine the thickness of fabric, how many fibres are laid in each direction and the number of overlaps.

Fibres are stronger along their length than across their width. Therefore,

composites are anisotropic and the mechanical properties in each direction depend on fibre orientation. For example, a composite made up of unidirectional fibres will be much more difficult to bend in the direction the fibres are laid. Ultimately, composites made up of multiple layers of unidirectional fabrics, with the fibres oriented along the lines of predetermined stress, demonstrate superior strength to weight.

Strength and stiffness generally increase as the ratio of fibre to resin is increased up to around 70%. After this point, there is insufficient resin to maintain bond strength across the surface area of the fibres.

These are technical products. The resins are expensive and manufactured in lower volumes than commodity materials. This means they are less likely to be available in a range of colours, or to have properties that are satisfying for cosmetic applications. In such cases, thermoplastic polyurethane (TPU), nylon, polypropylene (PP), or other suitable matrix may be used instead.

DESIGN

Fibres are prepared as unidirectional, biaxial or multiaxial fabrics. Unidirectional fabrics are adhesive bonded (page 370) or stitch bonded (page 196) to hold the fibres in place. The fibres are straight (they do not pass over or under any other yarns) and therefore provide the highest strength possible. They also enable the fibre to be placed exactly where it is needed in the composite construction.

Biaxial fabrics are produced with perpendicular warp and weft (by loom weaving). Several different types of weave can be used including plain, twill and satin. Plain fabric (one over, one under) is very stable, but difficult to drape around sharp profile changes. Using a heavy balance of fibres in the warp direction produces a near unidirectional format. Twill fabrics (two over, two under to create a diagonal pattern) have an open weave, readily draping and conforming to complex profiles. Satin weave (four over, one under, for example) is a much flatter fabric that can be easily draped to

a complex surface profile. However, owing to this construction, such weaves are unbalanced.

Multiaxial fabrics, such as triaxial (three directions) or quadriaxial (four directions), are stable in more directions and provide higher strength in more directions, corresponding to the additional yarns. They may also reduce the number of plies required, because the yarns cover more directions.

These fabrics are made from bundles of parallel filaments, known as tow. The number of filaments they contain determines their size. A 6K tow contains 6,000 ends and a 12K tow contains 12,000.

To make thinner fabric, the tow is spread out into a tape – known as spread tow – and interlaced by tape weaving. There are several advantages to this, including reducing the crimp (each filament is relatively straighter), which increases strength to weight; improving surface smoothness; reducing the gaps between the horizontal and vertical tows; and producing thinner composites.

Press forming shapes three-dimensional parts from sheet materials. The amount of forming is limited by the fibre reinforcement and depends on the construction of the material. Thin and pliable sheets may be formed over complex shapes with curvature in more than one direction.

The fibre reinforcement maintains sheet thickness during forming. Therefore, shapes with undulating profiles will draw material into the mold and may cause wrinkling and distortion. It is possible to design the mold so that any deformation occurs outside the edge of the finished part, which is trimmed after molding.

MATERIALS

The most common types of fibre reinforcement are glass, aramid (also known under the trade names Kevlar and Twaron) and carbon. Glass is the least expensive of the three and used for general-purpose applications. There are several types of filament, including A (alkali), C (chemical) and E (electrical). A is the least strong and most cost-effective; C has the best chemical resistance; E has the highest tensile strength and stiffness.

The Press Forming Process

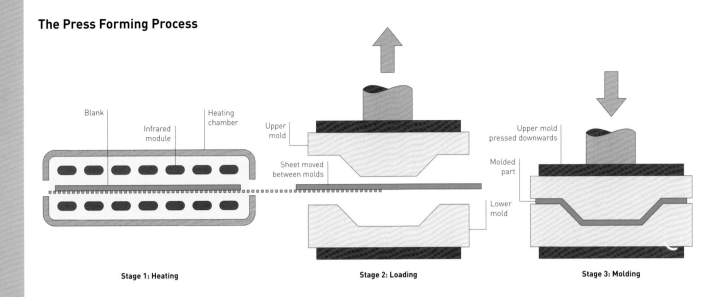

Stage 1: Heating   Stage 2: Loading   Stage 3: Molding

TECHNICAL DESCRIPTION

Press forming thermoplastic composites is a relatively straightforward process. Innovation comes from individual company's proprietary developments.

In stage 1, the blank, which is pre-cut to size to reduce waste, is loaded into the heating chamber. Infrared modules raise the temperature of the thermoplastic resin matrix to its softening point. For engineering plastics such as PC and nylon this is typically around 140°C (284°F), and for high-temperature plastics such as PEEK and PEI it is between 300 and 400°C (572–752°F) or more. The plastic is heated to its softening point so that it is suitable for pressing. To reduce the mechanical properties of the thermoplastic sufficiently so that forming can take place semi-crystalline types are melted and amorphous types are softened.

In stage 2, the sheet material is placed in the two-part mold. Press molds typically consist of an upper and lower mold, often referred to as tools. This allows the forming of simple three-dimensional sheet profiles. For undercut parts the mold is more complex, or a second pressing cycle is required.

Molds are typically made of metal. Low volumes and prototypes may be produced by pressing a single-sided metal mold against a rubber cushion (stamp). As well as reducing cost, using a rubber cushion distributes surface pressure more effectively than metal molds.

In stage 3, the mold halves come together, squeezing the softened thermoplastic against the mold surfaces. Deep profiles draw the sheet material inwards as it conforms to the shape. In this way, material thickness remains constant.

The whole cycle, from heating to molding, takes five minutes or so, depending on the size and complexity of part. After pressing, the mold halves separate and the finished part is removed.

Aramid has higher tensile strength than glass, and exceptional impact resistance, but it is significantly more expensive. Carbon is the most expensive of all, owing to the high cost of production. It has superior tensile strength, compressive strength and bending stiffness, but is quite brittle.

Natural fibres, such as hemp, flax and jute, are being used to reinforce composites for the automotive industry to reduce weight, cost and environmental impact. They are replacing conventional composites such as glass-filled plastics and sheet metal for both structural and decorative applications.

Biocomposites combine the properties of natural fibres with bioplastics to produce high-performance parts that are lightweight and low-cost and can be biodegradable.

Other types of fibre reinforcement, such as boron, quartz and silicon carbide, are used only for high-heat applications.

High-performance resin matrices include polyetheretherketone (PEEK), polyetherketoneketone (PEKK), polyphenylene sulfide (PPS) and polyetherimide (PEI). These materials are used for aerospace applications, because they have high operating temperatures (up to 260°C/500°F), high toughness, excellent chemical resistance and excellent fire, smoke, toxicity characteristics. These aerospace-grade materials can also be useful for demanding industrial applications.

For industrial and product applications, composites are based on nylon, acrylonitrile butadiene styrene (ABS), polycarbonate (PC), polyethylene terephthalate (PET), TPU and PP. Each has its own strengths and weaknesses. Nylon is a tough engineering polymer with good mechanical properties; ABS has good impact properties; PC has superior impact resistance; PET has relatively low processing temperature; TPU ranges from flexible to rigid; and PP has good impact strength even at temperatures well below freezing.

TECHNICAL DESCRIPTION

CNC machining is used to produce three-dimensional forms directly from CAD data. In conjunction with molding, it is used to trim parts after forming. The head operates on x- and y-axis tracks (horizontal) and a z-axis track (vertical). Machines with a rotating bed are known as four-axis, and adding head rotation creates a fifth axis. Each additional axis allows for holes and recesses to be made at another angle, for more complex parts. A seven-axis machine can cut from any direction.

Many different tools are used in the cutting process, including cutters (side or face), slot drills (cutting action along the shaft as well as the tip for slotting and profiling), and dovetail (wider at the bottom), conical (wider at the top), profile (for making a constant profile), flute and ball-nose cutters (with a dome head, which is ideal for 3D curved surfaces and hollowing out).

Some CNC machining facilities are large enough to accommodate a full-scale car (and larger, up to 5 x 10 x 5 m/16.5 x 33 x 16.5 ft). Many different materials can be machined in this way, including plastic, metal, wood and foam.

Once up and running, the CNC machine will repeat a sequence of operations precisely. Changing the set-up is the major cost for these processes.

Self-reinforced plastic (SRP) composites are a unique group of materials that are made up of a single type of material. In other words, the same material is used for the fibre reinforcement and resin matrix. Only certain materials are suitable: so far systems have been developed to make PP and PET variants. They are low-cost and can be much more easily recycled at the end of their useful life.

SRPs have so far been used to make luggage (the most well-known example of this is Samsonite's use of Curv self-reinforced PP), automotive body panels, safety helmets and ballistic protection for armoured vehicles.

The CNC Machining Process

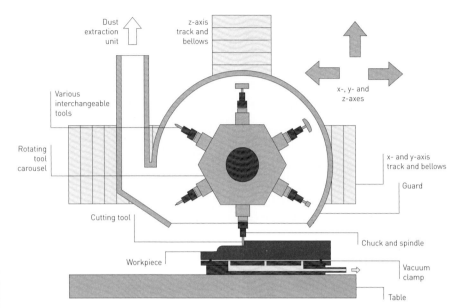

Three-axis CNC with tool carousel

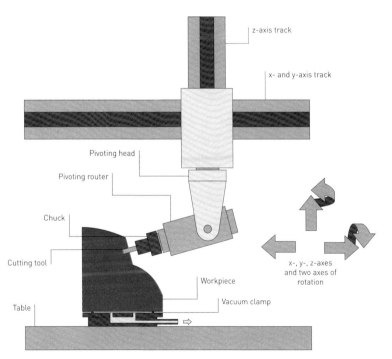

Five-axis CNC

Flax   Flax Molded

Biotex Flax and Thermoplastic Composite
Material: Flax and PP
Application: Automotive
Notes: Flax and PP fibres are woven together into twill
fabric. Heat and pressure cause the PP to melt and fill
the space between the fibres.

Red   Green

Biotex Flax with Coloured Resin
Material: Flax and PP
Application: Automotive
Notes: The PP fibres melt and flow around the flax to
produce an opaque coloured composite.

High-Performance Composite
Material: Carbon fibre and PEEK
Application: Aerospace
Notes: Materials used in aerospace have superior
mechanical properties and fire and toxicity
characteristics.

Tepex Thermoplastic Composite
Material: Carbon fibre and TPU
Application: Footwear
Notes: These boots are very lightweight (200 g/7 oz) and
the player does not feel pressure from the studs.

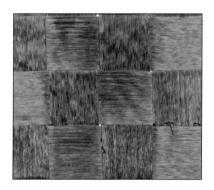

Carbon-Fibre Spread Tow
Material: Carbon and TPU
Application: Aerospace
Notes: Carbon-fibre tow is spread into thin tapes (down
to the thickness of just a few filaments) and woven into
very thin composite sheets.

Mixed-Filament Spread Tow
Material: Carbon, aramid and TPU
Application: Aerospace
Notes: Carbon is very strong but brittle, so it is
combined with aramid to provide more flexibility.

COSTS

Mold costs are moderate to high
depending on the complexity of the
design. Cycle time is rapid and typically
no more than five minutes. Labour
costs are relatively low, although there
may be a lot of finishing required,
increasing costs.

The price of carbon-fibre-reinforced
composite is roughly a tenth of what is
was ten years ago thanks to increased
global production and more efficient
manufacturing methods. Even so, it is
several times more expensive than glass,
and PEEK is many times more expensive
than nylon, which means that high-
performance materials are significantly
more expensive to produce.

ENVIRONMENTAL IMPACTS

The most important factors affecting
the total environmental impacts of
composites are the ingredients and end
of life. Conventional composites, such as
carbon-, glass- and aramid-reinforced
plastic, are very light, stiff and strong.
They are changing the way high-
performance products are designed and
built, improving efficiency and enabling
structures that would otherwise be
impractical. However, carbon and aramid
are energy-intensive to produce, and at
the end of their life composites are very
complex to recycle (although there has
been significant development in this area).

Thermoplastics are highly recyclable.
Parts may be heated and pressed several
times. This reduces waste, while material
that does become scrap can be recycled
and reused.

Biocomposites are made up of natural-
fibre reinforcement and bioplastic (also
called bioresin). Alternatively, they are
biobased and consist of a thermoplastic
matrix or non-bio fibre reinforcement.
(See page 238 for the recyclability of
biocomposites.)

→ Press Forming a Carbon-Fibre-Reinforced Thermoplastic

In preparation for molding, the thermoplastic composite is cut to exact dimensions. It is suspended in a frame, so that it can be placed accurately into the mold (**image 1**). The heating chamber and mold are in line, so the material can be transferred directly from one to the other and minimize heat loss prior to forming (**image 2**).

Once up to temperature, the sheet is held between the mold halves (**image 3**). The mold is promptly closed, forcing the sheet to bend

(**image 4**). Once it has cooled sufficiently to maintain the shape it is removed from the mold.

The part is carefully checked to make sure the mold is pressing accurately. Sheet uniformity is measured with a micrometer (**image 5**) and the operator passes over the surface with an ultrasound to check for porosity (**image 6**). For aerospace applications it is critical that the part is 100% accurate and repeatable.

Even more thorough checks may be carried out in the quality control (QC) department (page 453).

1

2

3

Featured Company

DTC Dutch Thermoplastic Components
www.composites.nl

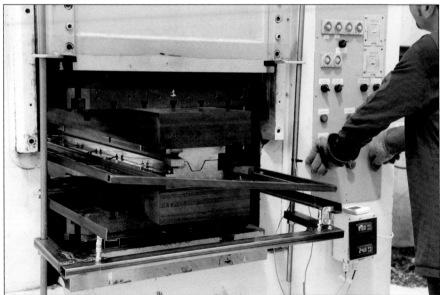

4

5

6

→ Trimming a Seat with Five-Axis CNC Machining

This is a backrest for an aeroplane seat. The unidirectional carbon-fibre-reinforced plastic is press formed in batches and stacked, ready to be trimmed (**image 1**). Trimming removes excess material and keeps weight to the absolute minimum.

The part is loaded onto a vacuum bed (**image 2**). The CNC machine follows a predetermined path, calculated from the original CAD file (**image 3**). The part is completed (**image 4**) and the waste is removed from around the perimeter.

It is checked and the edges are softened with abrasive (**image 5**). The shaped part is assembled onto a frame, which forms the structure of the seat. The molded thermoplastic composite helps to create a very light, strong and reliable seat structure.

1

2

3

4

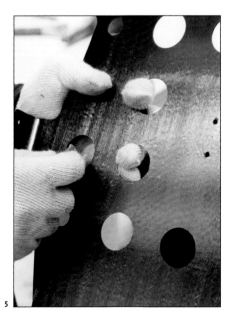

5

MOLDING

452

Featured Company

DTC Dutch Thermoplastic
Components
www.composites.nl

→ Non-Destructive Testing for Critical Parts

The process of molding thermoplastic composites starts with a 3D CAD file (**image 1**). The data is used for simulations, to check the performance of the part in application prior to manufacture. Once everything has been agreed, the data is extracted and used to set up manufacture, machine the jigs and molds and drive the CNC trimming operation.

Once the part is molded, the dimensional accuracy is checked against the original 3D data using a coordinate-measuring machine (CMM) (**image 2**). The CMM measures predetermined points to check for dimensional accuracy. Material thickness is checked using a micrometer (**image 3**).

The material is scanned for porosity and internal invisible defects using ultrasound C-scanning (**image 4**). The screen (**image 5**) shows the measured ultrasound reflection. The operator is trained to interpret the reflection image. All these tests are non-destructive. In other words, the structure of the part is unaffected, so these techniques may be used to check every part if necessary.

Destructive testing may also be used to check the material properties at the point of failure. This gives the most accurate data about how the part will perform in extreme conditions.

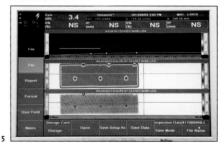

Featured Company

DTC Dutch Thermoplastic Components
www.composites.nl

MOLDING

454

Construction Technology
3D Thermal Laminating

This technology was developed by North Sails to make ultra-lightweight and seamless sails. Continuous lengths of high-performance fibres are placed by a computer-guided robot and laminated between thin films over a three-dimensional mold. This surpasses traditional methods of cutting, stitching and glueing to produce sails with the optimum flying shape.

Techniques	Materials	Costs
• Three-dimensional laminating (3DL) • Three-dimensional isotropic (3Di)	• High-performance filaments and thermosetting plastic • PET film, polyester fabric or nonwoven	• High unit costs due to materials and process

Applications	Quality	Related Technologies
• Sails • Potential applications include inflatables, high-altitude balloons, tension structures	• Very high strength to weight • Seamless	• Machine stitching • Welding • Adhesive bonding

INTRODUCTION

Sails have a three-dimensional 'flying shape'. In conventional sailmaking, patterns are cut out and stitched (page 354) or adhesive bonded (page 370) together to produce the optimal shape. In three-dimensional laminating (3DL), sails are molded in the optimum flying shape and so minimize cutting and joining. This produces sails that are up to 20% lighter than conventional equivalents.

Luc Dubois and J. P. Baudet constructed the first 3DL sail at North Sails in 1990. Since its was commercialized, 3DL has dominated Grand Prix sailboat racing. More recently, North Sails has developed an entirely new approach to sailmaking called 3Di (see case study on page 458). Based on Gérard Gautier's 2001 invention, twisted yarn (page 60) is replaced with spread tow (page 234). The metallic-looking sails (first tested by the Swiss Alinghi team in 2007) are much stiffer and have a balanced resistance to distortion in all directions. These properties are more akin to a rigid airfoil than to conventional fabric sails.

The laminated sails are formed over computer-guided molds. In the case of 3DL, the fibre reinforcement is laid down along predetermined lines of stress (wind load) and sandwiched between thin sheets of polyethylene terephthalate (PET) films coated with adhesive. 3Di does not need the additional outer layers of film, because the entire surface of the mold is covered with the composite membrane. The adhesive in the pre-impregnated tape (page 236) bonds the layers to form a superstrong laminate.

Laminating 3Di uses a similar technique to conventional thermoset composite molding, whereby a vacuum is pulled on the entire assembly while heat is applied. Indeed, North Sails has taken the same approach to rigid composites in a development called Thin Ply Technology (TPT). Using the unique tape-laying system originally developed for 3Di, ultrathin composite membranes are produced from multiple layers of spread tow (see page 458). The fibres are laid down according to the requirements of the application to maximize strength while minimizing waste. The final part will be lighter and stiffer than conventional composite laminates.

APPLICATIONS

Although these laminating technologies are used almost exclusively for sailmaking, there is potential for them to be developed for applications in a variety of products. If the 3D design fits within the limits of the mold – similar shape and materials – then a wide range of products could be produced. Examples include high-altitude balloons, dirigibles, tension structures, inflatable structures, installations and furniture.

RELATED TECHNOLOGIES

Conventionally, large fabric structures, such as tents, balloons, canopies and sails, are constructed with adhesives, welding (page 376), stitching or a combination of these. Welding and bonding eliminate the use of needle and thread, so help to maintain the full integrity of the fabric while reducing bulk at the join. The only processes capable of eliminating the seams in very large and seamless composite fabrics are 3DL and 3Di.

QUALITY

The sails are built to perform a very specific function. They are required to hold their shape even in strong winds, be durable and lightweight, and produce minimum resistance to the flow of air over the surface.

Aerospace vacuum and pressure techniques are used to cure the laminates on the mold. This ensures that the textile will not delaminate even when extreme loads are applied. The thermally formed PET film in 3DL maintains the position of fibres, which are laid down in alignment with the direction of stress over the sail.

3Di is made up of multiple tapes in all different orientations. The result is a

The 3D Thermal Laminating Process

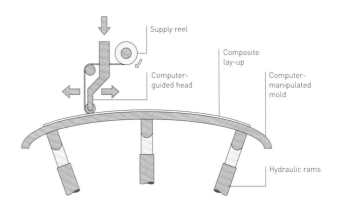

Supply reel

Composite lay-up

Computer-guided head

Computer-manipulated mold

Hydraulic rams

TECHNICAL DESCRIPTION

Each sail has a 3D shape. The designer produces a computer-aided design (CAD) file, which is transferred onto the surface of the mold by a series of pneumatic rams. The shape of the mold will determine the flying shape of the sail when it is made.

In the production of 3DL, a polyethylene terephthalate (PET) film reinforced with high-strength fibres is draped over the mold and put under tension. It is constructed from panels laser cut (page 328) and joined together with clear pressure-sensitive adhesive (PSA) tape. It is coated with an ultra-high-molecular-weight polyester adhesive, specifically designed for North Sails by its industrial partners for the 3DL process. Strands of fibre reinforcement are applied by a fibre placement head guided by a computer-controlled overhead gantry. It operates on six axes, so can follow the profile of the mold exactly. The fibres are coated with adhesive, which helps to keep them in place.

A second layer of PET film is laid on top of the fibres and forced to laminate under pressure at approximately 8.6 N/cm² (14.7 psi) from a vacuum bag. Heat is then applied with a blanket to cure the adhesive. After heating, the fibre adhesive is roughly 85% cured and is removed from the mold. It is then brought to the curing floor for a further five days

of post-cure to ensure that the adhesive is fully cured and will not delaminate.

3Di does not require PET films or any other type of substrate. Instead, it is a membrane composed of reinforcing fibre and resin matrix only, like any other composite structure. In the initial process fibre tows are spread out down to the filament level and impregnated with high-molecular-weight polyester thermoset resin in a tape format. The pre-impregnated tapes are then laid onto a plotting bed in multiple orientations with an automated tape-laying system (ATL) to produce sections of the sail, termed 'preforming'. The outer layers of the sail membrane, which are laid down first and last, have an 8 g (0.28 oz) nonwoven polyester embedded in a stiffer resin system, on top of the prepreg fibre tape. Once sufficient internal structural layers have been built up, a second layer of nonwoven material is laid on top and the whole assembly is placed inside a vacuum bag. Subsequently, heat and pressure are applied to consolidate and cure the resin to form a one-piece laminate. The tougher outer layers provide abrasion resistance, while the resin used to combine the internal filament tapes has higher elasticity. This combination ensures a durable fabric that will flex and fold without breaking.

600 Series
Material: Aramid fibre and PET film
Application: Racing sails
Notes: 100% aramid fibres produce a very durable and lightweight fabric.

860 Series
Material: Aramid and carbon fibres and PET film
Application: Racing sails
Notes: The carbon fibres increase tensile strength and reduce stretch.

900 Series
Material: Aramid and carbon fibres and PET film
Application: Racing sails
Notes: Increasing the carbon content produces a stiffer and stronger sail.

Marathon Series
Material: Aramid fibre with polyester fabric
Application: Sails
Notes: Incorporating plain-woven polyester fabric results in a more durable sail.

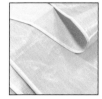

Spread Tow
Material: Carbon fibre and thermosetting polyester
Application: Various
Notes: Individual filaments are laid out side by side in tapes 100 mm (3.9 in.) wide and 35 microns thick.

fabric with quasi-isotropic properties, which means it is more balanced, with improved shape retention. This is critical for racing sails. The less a sail shape-distorts while the boat moves over waves, the greater the efficiency of converting wind energy into forward motion.

DESIGN
3DL molds are computer-guided and can be very large, up to 500 m² (5,380 ft²).

Specific fibre layouts and combinations provide varying levels of performance. 3DL is based on laying fibre along predetermined lines of stress. The flowing shape produced in the catenary curves join areas of stress with smooth, flowing lines of fibre reinforcement.

The 600 series is made of high-modulus aramid fibres to produce a durable sail that will maintain its shape. The 800 and 900 series combine carbon fibre with aramid to reduce weight and stretch. Different films can be used on the outside surface to protect against tearing and chaffing and provide ultraviolet (UV) stability. Plain-weave polyester, known as

taffeta, is used on the outside surface of the Marathon series to increase durability.

The multidirectional orientation of fibres in a 3Di construction mean the design is approached more like a composite boat hull or aeroplane wing. Combining multiple layers gives ultimate control to the designer. But as with all major innovations, it will take time for designers to learn how best to apply the impressive strength and durability to achieve the highest performance possible.

The outside layers on a 3Di sail can be pigmented and have UV inhibitors added. This helps to protect the underlying fibres from exposure to abrasion and sunlight. Most of the reinforcing required for high-load areas is accomplished with added internal layers.

MATERIALS
High-performance fibres including carbon, aramid and ultra-high-molecular-weight polyethylene (UHMWPE) are most commonly used. Carbon is stiff, with exceptional tensile modulus. However, its high stiffness is its downfall for flexible applications, because it is prone to breaking when bent or impacted. Aramid, commonly referred to by trademark names such as Kevlar and Twaron, has very high strength to weight (up to five times greater than steel), resistance to abrasion and energy absorption. UHMWPE, more commonly known

as Dyneema or Spectra, has exceptional strength to weight (up to twice that of standard aramid). It is more resilient too and so less prone to breaking when flexed under high load. These fibres are used in isolation, or combined for ultimate strength, lightness and durability.

The two outer films or woven fabrics used to sandwich the fibre lay-up in 3DL are made up of polyester. The high-strength PET film is also commonly referred to by the trademark name Mylar.

Both the 3DL and 3Di sail membranes are manufactured with high-molecular-weight polyester resin systems. These systems are highly elastic and capable of 200–300% elongation. Sails need to be flexible, so these high-elongation systems allow for some micro-adjustability inside the laminate, enabling the filaments to displace during bending and fold cycles, which helps prevent damage to the fibres themselves. This is very different from how the resin systems in rigid composite structures perform, where the resin matrix is an integral part of the strength of the finished part.

COSTS
The materials and manufacturing processes used in 3DL are expensive and therefore are limited to high-performance applications, largely because of the high level of customization required, and the significant investment made in large-scale reconfigurable molds and robotics.

Cycle time is up to five days, owing to the length of the time required for the bespoke fibre placement process. Labour costs are not high, because of the sophisticated fabrication strategies and the level of automation employed.

ENVIRONMENTAL IMPACTS
These sails minimize material consumption and reduce weight. However, once the composites have been bonded together they cannot be separated and are difficult, if not impossible, to recycle. There are emerging polymer technologies that include both recyclable resins and the equipment to perform the processing, so it is likely the environmental aspects will improve.

→ 3DL Sails on a TP 52 Type Racing Boat

The TP 52 type racing boat (**image 1**) has here been fitted with North Sails 3DL sails. The process begins with a bespoke sail design, according to the flying shape, because each boat requires a slightly different profile for the best performance. There is a range of sizes possible from 10 m² to 500 m² (108–5,380 ft²). The shape of the full-size mold is computer-guided, so it can be adjusted for many different sail shapes (**image 2**).

The aramid-reinforced PET film is draped across the mold (**image 3**) and pulled tight with tensioning straps. The six-axis computer-controlled gantry lays down strands of adhesive-coated carbon fibre following the anticipated load lines (**image 4**).

An airtight bag is laid over the films and fibres and a vacuum is drawn over the entire surface. The operator moves over the sail applying heat to thermally form the laminate in the shape of the mold. A conductive heating system is used in contact with the surface (**image 5**), or an infrared heating system that travels above the surface (**image 6**) applies the heat required. The choice of heating system depends on the fibres in the sail.

When the sail has fully cured, corner reinforcements, batten pockets, eyelets and other details are applied by sailmakers using traditional cutting and sewing techniques.

1

2

3

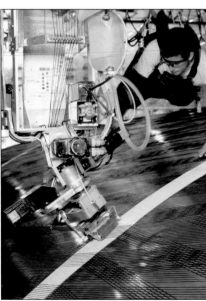
4

5

6

Featured Company

North Sails
www.northsails.com

→ 3Di Sails on a Farr 400 Racing Boat

The hull and rigging of the Farr 400 One-Design (**image 1**) are made entirely from carbon fibre and epoxy composite, and the sails are carbon-fibre 3Di, making this one of the highest-performing racing boats in its class.

The process of sailmaking begins with design and simulation, using a custom software suite consisting of linked finite element analysis (FEA) and computational fluid dynamics platforms, to determine the optimum structure and aerodynamics for any given sail. Once the results are satisfactory with the modelling then the composite construction is worked out in layers based on the FEA results (**image 2**). Each of the layers of tape is made up of individual filaments running in parallel formation, which can be laid in any orientation. Fibre density is increased in more highly loaded sections of the membrane by stacking tapes to create density clusters. Each fibre tape is positioned exactly where it is needed. Combining multiple layers in this way gives ultimate control to the sail designer to prescribe the sail membrane precisely in terms of strength and stiffness.

Spread tow is prepared by pre-impregnating with high-molecular-weight polyester thermoset resin, and a spool of carbon and UHMWPE is loaded onto the tape-laying head (**image 3**). Controlled on x- and y-axes, the computer-guided head applies the tape to a flat surface and takes up the paper backing (**image 4**). The sails are constructed in panels called preforms for optimum

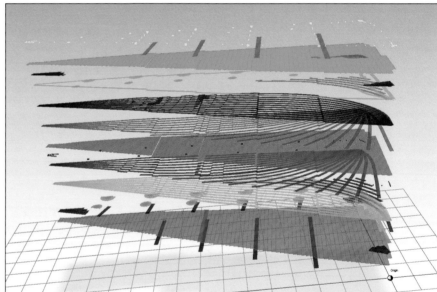

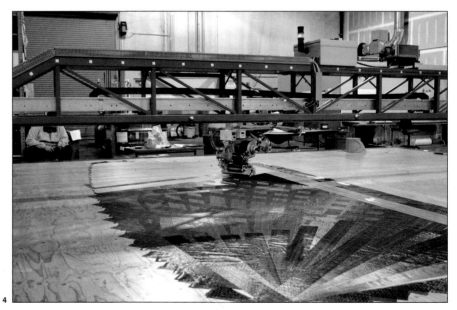

utilization of space and equipment, which are transferred to the reconfigurable molds.

The mold is set up according to the flying shape of the sail (**image 5**). The panels are placed together on the mold and the edges are overlapped to form scarf joins. This ensures a very strong construction with seamless joins.

An impermeable film is placed on top and a strong vacuum is drawn on the composite membrane, pressing the layers together. It is heated up (**image 6**), causing the resin that binds the fibres together to coalesce and cure to form a strong and coherent matrix (**image 7**). When the sail is removed from the mold it is not possible to pull the individual tapes apart.

5

6

7

Featured Company

North Sails
www.northsails.com

Construction Technology
Filament Winding

Often simply referred to as winding, this process is used to produce hollow parts with ultimate strength-to weight-characteristics. Layers of high-performance fibres mixed with resin are wound onto a shaped mandrel, which is either removed and reused, or permanently encapsulated by the fibres as a functional part of the construction.

Techniques	Materials	Costs
• Mandrel winding • Bottle winding (liner)	• Carbon, aramid and glass fibres • Thermosetting or thermoplastic resin matrix	• High owing to cost of set-up and raw materials

Applications	Quality	Related Technologies
• Aerospace, automotive and pressure vessels • Furniture and sculpture	• Very high strength to weight • Patterned finish	• 3D thermal laminating • Composite laminating • Composite press forming

INTRODUCTION

Winding creates hollow parts, which are typically cylindrical in profile, by wrapping continuous lengths of fibre around a mandrel. There are two principal systems, defined by the resin: thermoset and thermoplastic.

The two types of thermoset winding are 'wet' and 'prepreg'. Wet winding draws the filament tow through an epoxy bath prior to application. Prepreg winding uses high-performance filament tow pre-impregnated with epoxy resin, which can be applied directly to the mandrel without any other preparation.

Thermoplastic winding is a more recent development. Fibres are

The Filament Winding Process

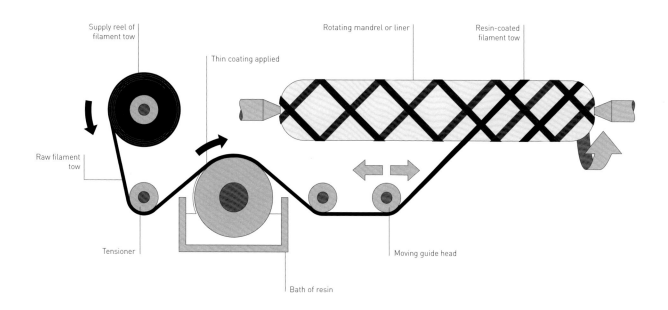

Supply reel of
filament tow

Thin coating applied

Rotating mandrel or liner

Resin-coated
filament tow

Raw filament
tow

Tensioner

Moving guide head

Bath of resin

impregnated with a thermoplastic, which is heated just before application, causing the surface to melt. It is wound onto the mandrel and bonds to the surface as it cools and re-solidifies. The advantages of thermoplastics include reduced volatile organic compound (VOC) emissions and increased recyclability.

APPLICATIONS
Filament-wound products can be found in high-performance applications in the aerospace, deep sea and automotive industries. Examples include portable gas tanks, oxygen tanks (for breathing apparatus), wind turbine blades, helicopter blades, deep sea submersibles and automotive suspension systems. Designers and engineers have begun to explore the unique technical opportunities of this process for small volumes of furniture and sculpture.

RELATED TECHNOLOGIES
Composite laminating and thermoplastic composite press forming (page 446) are used to produce similar parts. The benefit of filament winding is that the direction

TECHNICAL DESCRIPTION
The carbon-fibre tow is applied to the rotating mandrel by a guide head. The head moves up and down along the mandrel as it rotates, and guides the filament into the overlapping pattern.

The width of the tow is chosen according to the material being used and the layer thickness required. The fibre is continuous and broken only when a new supply reel of fibre reinforcement is loaded. This is the wet lay-up process; the fibre reinforcement is coated with a thermosetting resin (typically epoxy) resin by a wheel rotating in a bath of the resin.

A complete circuit is made when the guide head has travelled from one end of the mandrel to the other and back to the starting point. The speed of the head relative to the speed of mandrel rotation will determine the angle of the fibre. A single circuit may have several different angles of tow, depending on the requirements.

Bulges can be made by concentrating the tow in a small area. These are used either to create localized areas of strength or to build up larger diameters that can be machined for improved accuracy.

Thermoplastic winding uses the same set-up, except that the resin bath is replaced with an infrared heating unit. The commingled plastic and fibre reinforcement are heated to just above the melting temperature of the matrix, which is around 200 °C (392 °F) for polypropylene (PP). Pressure may also be applied to ensure consolidation.

Colour Winding
Material: Glass with epoxy
Application: Pressure vessel
Notes: Pigment is added to the resin to create a uniform inherent colour. Alternatively, the surface is coated.

Taped Finish
Material: Carbon with epoxy
Application: Automotive
Notes: Tape is wound over the last layer of carbon to produce a smooth finish and compressed laminate.

CNC-Machined Finish
Material: Carbon with epoxy
Application: Automotive
Notes: CNC machining (page 449) is used to produce a very accurate surface as well as channels and recesses.

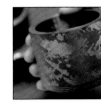

Multifibre
Material: Carbon and aramid with epoxy
Application: Automotive
Notes: Depending on the requirements of the application, multiple different high-performance fibres may be used.

Self-supporting Structure
Material: Carbon with epoxy
Application: Sample
Notes: High-performance composites are stiff and used to create rigid three-dimensional openwork structures.

of the strand can be adjusted precisely throughout production from 0° to 90°. The same freedom of design does not exist in composite laminating and press forming, because it uses pre-woven or unidirectional fabrics.

The 3D thermal laminating (3DL) (page 454) developed by North Sails is similar to filament winding. The difference is that in 3DL filaments are laid onto a static mold in a process known as tape laying. In this case the molds can be very large and up to 500 m² (4,305 ft²). 3DL is best suited to sheet parts, whereas filament winding is more suited to hollow parts.

QUALITY
There are four main types of finish. These are 'as wound', taped, machined and epoxy gel coat. 'As wound' finishes have no special treatment. Applying a clear tape to the surface results in a smooth surface finish. The principle is the same as vacuum bagging in composite laminating, but the surface finish is more controllable. Surfaces are machined where precise tolerances are required.

Smooth, glossy finishes can be achieved with an epoxy gel coat.

Stiffness is determined by the thickness of lay-up and tube diameter. Winding is computer guided, making it precise to 100 microns. The angle of application will determine whether the layer is providing longitudinal, torsional (twist) or circumferential (hoop) strength. Layers are built up to provide the required mechanical properties.

DESIGN
Opportunities for designers are limited to cylindrical and hollow parts. But they need not be rotationally symmetrical; oval, elliptical, sharp-edged and flat-sided profiles can be filament wound.

Parallel-sided and conical mandrels can be removed and reused, while three–dimensional hollow products that are closed at both ends can be made by winding the filament tow over a hollow liner, which remains as part of the final product (see case study, page 464). This technique is known as bottle winding and is used to produce pressure vessels,

housing and suspension systems. Other than shape, the benefits of winding over a liner include forming a water-, air- and gas-tight skin.

Parallel-sided parts can be made in long lengths and cut to size. Conventional filament winding is typically limited to 3 m (10 ft) long and up to 1 m (3.3 ft) in diameter. However, much larger forms are produced by filament winding, such as space rockets, which may take several weeks to make.

Winding over a hollow liner produces parts that have inward- and outward-facing corners such as a bottleneck profile. Even though it is possible, it is generally not recommended to wind over outward-facing corners with a radius of less than 20 mm (0.8 in.) because product performance will be affected. There is no lower limit on the radius for inward-facing corners.

MATERIALS
The fibre (typically glass, carbon or aramid) is applied as a tow that contains a set number of fibre monofilaments. For example, 12k tow has 12,000 strands, while 24k has 24,000 strands. An outline of their particular qualities is given under composite press forming.

Thermosetting plastics have cross-links in their molecular structure, which means they have high resistance to heat and chemicals. Combined with carbon fibres these materials have very high fatigue strength, impact resistance and rigidity. Epoxy is the most commonly used.

Thermoplastics may be heated and reformed many times over. If they can be separated from the fibre reinforcement at the end of their useful life then they may be recycled. The most commonly used types are polypropylene (PP), polyethylene terephthalate (PET) and polyetheretherketone (PEEK). However, owing to the increased complexity of production this technique is not as widespread as thermoset winding. PP is used for its high impact strength, especially at low temperatures; PET and nylon are engineering polymers that have high stiffness and chemical resistance; and PEEK is a high-performance plastic used in aerospace applications.

→ Design and Testing

ComposicaD software is used to generate the patterns needed by CNC equipment to produce filament wound parts. It uses parametric modelling and the results, seen in the computer-generated images, indicate the required angle of tow and number of layers (**image 1**).

Finite element analysis (FEA) is used to determine the stress and strain on the product prior to manufacture. Prototypes are then fabricated to test the computer theories (**image 2**).

Pressure is applied to test the rupture strength (**images 3** and **4**). The plastic liner on its own is gas-tight, puncture-resistant and strong, but it lacks circumferential strength. Therefore, it will rupture when very high pressure is applied. The composite skin provides the necessary strength and helps to make this a very lightweight and compact container. Composite pressure vessels are stronger and lighter than equivalent metal pressure vessels.

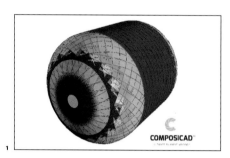

1

2

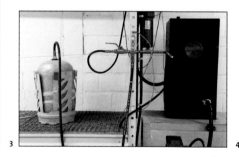

3

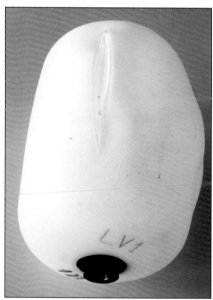

4

Featured Company

Seifert and Skinner & Associates
www.seifert-skinner.com

COSTS

This is a high-cost process for low volumes. Increasing the volume reduces unit cost. Even so, the materials are expensive and so every effort is made to reduce material consumption while maximizing strength. Winding over a liner will increase the cost of the process because a new liner is manufactured for each cycle.

Filament winding cycle time is 20 to 120 minutes for small parts, but can take several weeks for very large parts. Curing time for thermoset winding is typically four to eight hours. The winding process is computer guided, but there is a high level of manual input, so labour costs are moderate to high.

Currently thermoset winding is the more cost-effective, but that may change as thermoplastic winding continues to be improved.

ENVIRONMENTAL IMPACTS

Laminated composites reduce the weight of products and so reduce fuel consumption. Thermosetting materials cannot be recycled, so scrap and offcuts have to be disposed of. However, the new thermoplastic systems being developed have the potential to reduce the environmental impacts of the process.

Operators have to wear protective clothing to avoid the potential health hazards of contact with the materials.

→ Filament Winding a Pressure Vessel

Seifert and Skinner & Associates is an innovative company that specializes in the design and development of winding solutions for a range of applications. This case study demonstrates encapsulating a blown polyethylene (PE) liner with carbon-fibre-reinforced epoxy. This type of composite pressure vessel is used for liquefied petroleum gas (LPG) and compressed natural gas (CNG) storage and transport.

The moulded white liner is loaded onto a rotating jig and the surface is carefully cleaned with acetone (**image 1**). Carbon-fibre tow is supplied from a spool (**image 2**). Many different spools will be running simultaneously. The epoxy resin is mixed in preparation (**image 3**) and poured into the bath. The fibres are coated as they pass over the wheel (**image 4**), which is partly submerged in the resin, and emerge fully impregnated (**image 5**).

The first layer of carbon is applied at almost 0° (**image 6**). The rounded ends allow adequate tension to be pulled on the tow without slipping on the mandrel at shallow angles (**image 7**). Multiple layers are built up with the tow placed at predetermined angles to provide maximum strength to weight (**images 8** and **9**).

After winding is complete (**image 10**), the composite is placed in an oven, to cure the resin.

1

2

3

4

5

Featured Company

Seifert and Skinner & Associates
www.seifert-skinner.com

6

7

8

9

10

Construction Technology
Additive Manufacturing

Also referred to as 3D printing, rapid prototyping and digital manufacturing, these layer-building processes are revolutionizing the way designers are creating and manufacturing textiles and fashion. Data is transferred directly from CAD to make physical parts from sheet, liquid or powder-based materials, thus bypassing conventional forming techniques.

Techniques		Materials	Costs
• Lamination	• Stereolithography	• Any material that can be melted, fused, cured or sintered including bioplastics, thermoplastics, epoxy, sand, plaster, wax	• Low to moderate for prototyping and one-offs
• Fused deposition	• Laser sintering		• Expensive for high volume
• PolyJet	• Powder printing		

Applications	Quality	Related Technologies
• Prototyping	• Visible layered structure	• CNC machining
• Jewelry and footwear	• Strength depends on materials and method of manufacture	• Laser cutting
• Industrial, automotive and aerospace		• Lost-wax casting

INTRODUCTION

These processes are used to construct simple and complex geometries by combining together sequential layers of material. The process starts with a computer-aided design (CAD) model sliced into horizontal cross-sections. Each slice is then cut, fused, bonded, sintered (coalesced powder) or cured to the preceding one to gradually build up the complete 3D form. This approach to creating products is termed additive manufacturing to differentiate it from reductive processes, such as cutting, that create shapes by removing material.

The type of material, strength of part and quality of finish depends

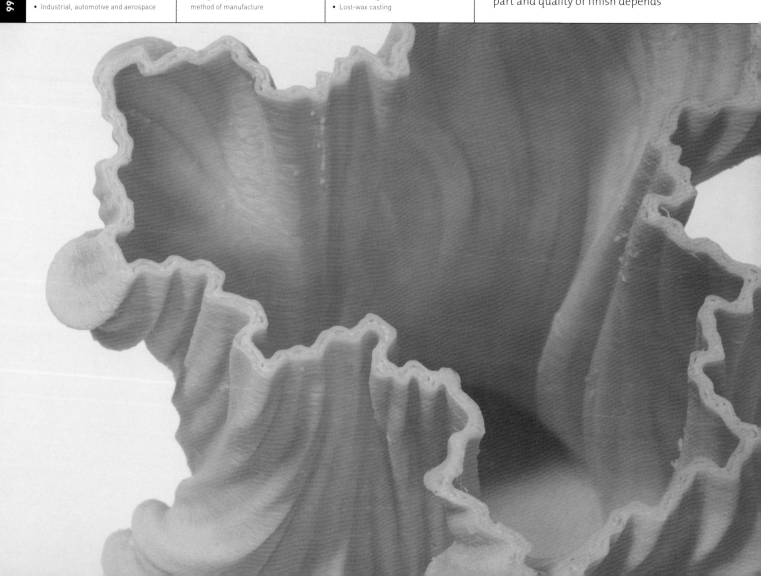

on the process. There are many different techniques used today. This section illustrates laminated object manufacturing (LOM), fused deposition modelling (FDM), PolyJet, stereolithography (SLA), selective laser sintering (SLS) and direct metal laser sintering (DMLS).

SLA is a widely used additive manufacturing technique and produces the finest surface finish and dimensional accuracy. Parts are built up from liquid epoxy cured by a computer-guided laser. Parts are finished to a very high standard by painting, polishing or metal coating. The process has been adapted to produce very small parts with ultra-high resolution (micro) as well as very large parts in a single build (mammoth).

LOM builds up 3D parts by cutting (knife or laser) and laminating sheet materials. It is the least expensive additive manufacturing technique, but lacks the precision and versatility of equivalent processes and so has not been as commercially successful.

FDM was invented by Scott Crump, founder of Stratasys, in 1988. It is also referred to as fused filament fabrication (FFF). Layered plastic models are built up using a continuous extruded filament. The plastic is heated and fused to the proceeding layer through a heated and computer-guided nozzle. A similar technique is used to fabricate 3D parts in wax. These form patterns used to produce high-quality metal items by lost-wax casting (also called investment casting).

PolyJet is a relatively new process – patented by Stratasys – that is essentially digital printing in 3D. Coloured photopolymer is applied in successive layers and cured under ultraviolet (UV) light. The process is rapid and versatile; multiple different thermoplastics can be used from rigid to elastic and transparent to opaque. With this process it is possible to combine multiple colours and materials in the same part.

Powder-based processes include SLS and DMLS. With these techniques, each cross-section is mapped onto the surface of the additive manufacturing material by a laser, which fuses the particles together. A variety of materials such as polymers, composites, ceramics, metals and even sand can be formed in this way. Powdered plaster is a cost-effective material to print. It is laid down with a binder (such as water) and infiltrated with cyanoacrylate (superglue) to create a strong part suitable for sanding and spraying.

APPLICATIONS

These processes are utilized mainly for development and prototyping in order to reduce the time it takes to get a product to market. They are also being used to direct-manufacture products that have to be precisely reproduced with very accurate tolerances.

Fashion, footwear and jewelry designers are utilizing a range of these techniques to produce low volumes of identical parts as well as one-offs and bespoke items. Metal items, such as jewelry and footwear, are produced using DMLS or a combination of wax prototyping and lost-wax casting. Plastic parts, such as shoe soles, chainmail and accessories, are manufactured using FDM, SLS and SLA.

Additive manufacturing techniques are capable of producing 3D objects in any material that can be fused together in layers. They are therefore being explored for less conventional applications, such as bespoke confectionery, made-to-measure medical implants and even replacement parts for NASA equipment in space.

RELATED TECHNOLOGIES

Additive manufacturing encompasses several unique processes. They have evolved to fulfill specific design and material requirements. In most cases, the choice of process is determined by the desired quality and material requirements. However, in some cases, it is possible to prototype the same material using two different processes. For example, whereas nylon filament is utilized in FDM, nylon powder is used in the SLS process. The different layer-building techniques will affect the material properties.

Similar to additive manufacturing, CNC cutting (page 316) and laser cutting (page 328) are computer guided. Therefore, they may be used to produce parts directly from CAD data and to precise tolerances. CNC cutting, when using three or more axes of movement (x, y, z and so on), is suitable for producing 3D parts. This technique has been used for many years to prototype as well as mass-produce high volumes of identical parts.

QUALITY

As a result of manufacturing 3D forms in layers, contours are visible on angled surfaces. The amount depends on the technique: LOM is compatible with sheet from 0.01 mm (0.0005 in.) thick and is accurate to around 0.1 mm (0.005 in.); FDM is capable of producing parts with a resolution of 0.2 mm (0.008 in.) and minimum wall thickness of 0.5 mm (0.2 in.); PolyJet builds in layers 0.016 mm (0.006) thick, accurate to 0.1 mm (0.004 in.) and with feasible details down to 0.6 mm (0.025 in); SLS builds in 0.1 mm (0.004 in.) layers and is accurate to 0.15 mm (0.006 in.), with a minimum wall thickness of 1 mm (0.04 in.) and live hinges down to 0.3 mm (0.01 in.) thick; SLA forms in 0.05 to 0.1 mm (0.002–0.004 in.) layers and is accurate to 0.15 mm (0.006 in.); and DMLS builds in 0.02 to 0.06 mm (0.008–0.0024 in.) layers and is accurate to 0.05 mm (0.002 in.).

Micro-modelling SLA can be used to produce intricate and precise parts up to 77 x 61 x 230 mm (3 x 2.4 x 9 in.). This process builds in 0.025 mm (0.001 in.) layers, which are almost invisible to the naked eye, and so eliminates surface finishing operations.

To achieve a high-quality surface the parts may require finishing. The techniques used and practicality of finishing depends on the materials and process. For example, it is possible to acquire a very high 'glass-like' surface finish on the water-clear SLA epoxy resin by polishing.

DESIGN

The ability to directly link the outputs of CAD and 3D scanning with additive manufacturing – thus cutting out physical restrictions associated with reductive or manual processes – means that designers are able to produce complex, intricate and previously

The Additive Manufacturing Process

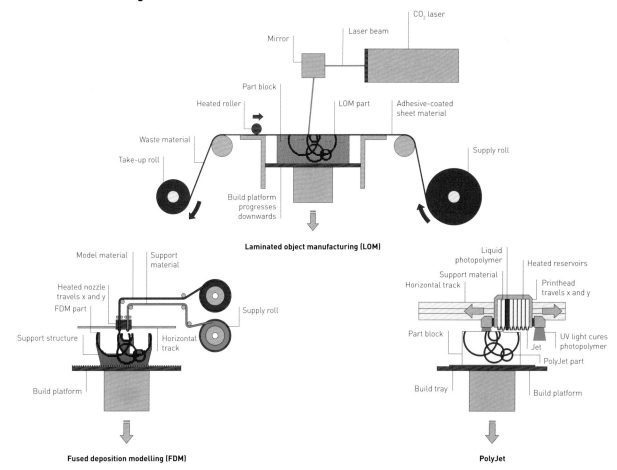

Laminated object manufacturing (LOM)

Fused deposition modelling (FDM)

PolyJet

TECHNICAL DESCRIPTION

LAMINATED OBJECT MANUFACTURING (LOM)

All of these additive manufacturing processes start with a CAD drawing sliced into cross-sections. Each cross-section represents a layer in the build, which in the case of LOM is equivalent to the thickness of the sheet material.

The raw material is supplied on a roll coated (page 226) on one side, or impregnated (page 234) with resin. First, the take-up roll draws through the correct length of material. Next, the heated roll passes back and forth across the surface under pressure to firmly bond the paper onto the part block below. The laser cuts the outline of the part to separate waste from the build. Once complete, the roll of paper is moved on, leaving behind the part block and the LOM part. The waste material is carefully removed. This is aided by cutting

a gridded pattern into the sheets that make up the part block.

FUSED DEPOSITION MODELLING (FDM)

FDM utilizes thermoplastic material. These materials are softened with heat and resolidify as they cool. The material is fed through a heated nozzle, which causes it to become semi-liquid. Guided along computer-controlled paths, the nozzle applies a 0.2 mm (0.008 in.) thick layer. As the plastic cools a strong bond is formed with the layer below.

The support structure is formed with a second material. It is placed only where it is needed, such as underneath overhangs or curving surfaces. The two heated nozzles work in tandem: the model material is extruded, followed by the support structure. Once the layer is complete, the build platform moves down a specific distance

and the next layer is applied on top. For simple geometries it is possible to use exactly the same material for the support as for the model. It is built as a loosely packed structure to aid removal. For complex parts, water-soluble PVA or PLA are utilized. These materials are removed by washing the model in warm water and detergent.

POLYJET

With this process model material is applied in layers 0.016 mm (0.0006 in.) thick by a precisely controlled printer head. It is similar to digital printing, except that the ink is replaced with liquid photopolymer. The polymer is jetted onto the preceding layer and cured with ultraviolet (UV) light as the printer head passes overhead. It is fully solidified after exposure to UV and does not require post-curing treatment.

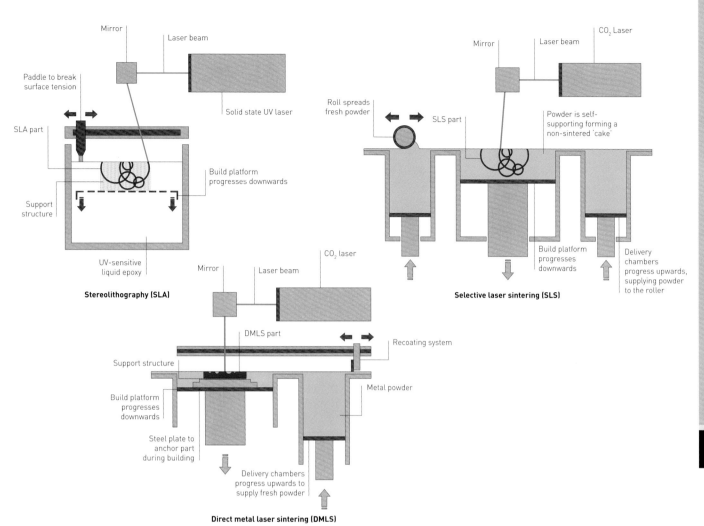

Stereolithography (SLA)

Mirror

Laser beam

Paddle to break
surface tension

SLA part

Solid state UV laser

Support
structure

Build platform
progresses downwards

UV-sensitive
liquid epoxy

Selective laser sintering (SLS)

Mirror

Laser beam

CO₂ Laser

Roll spreads
fresh powder

SLS part

Powder is self-
supporting forming a
non-sintered 'cake'

Build platform
progresses
downwards

Delivery
chambers
progress upwards,
supplying powder
to the roller

Direct metal laser sintering (DMLS)

Mirror

Laser beam

CO₂ laser

DMLS part

Recoating system

Support structure

Metal powder

Build platform
progresses
downwards

Steel plate to
anchor part
during building

Delivery chambers
progress upwards to
supply fresh powder

Up to eight materials or colours may be applied in a single pass. Along with photopolymer, the printer applies support material, which is required when building complex geometries and overhanging shapes. This is easily washed away afterwards.

STEREOLITHOGRAPHY (SLA)

With a layer thickness of between 0.05 mm and 0.1 mm (0.002–0.004 in.), a UV laser beam, directed by a computer-guided mirror, cures the surface layer of the UV-sensitive liquid epoxy resin. The UV light precisely solidifies the resin it touches.

Each layer is applied by submersion of the build platform into the resin. The paddle sweeps across the surface of the resin with each step downwards, to break the surface tension of the liquid and control layer thickness. The part gradually develops below the surface of the liquid and is kept off the build platform by a support structure. This is made in the same incremental way, prior to building the first layer of the part.

SELECTIVE LASER SINTERING (SLS)

A CO₂ laser fuses fine nylon powder in 0.1 mm (0.004 in.) layers, directed by a computer-guided mirror. The build platform progresses downwards in layer thickness steps. The delivery chambers alternately rise to provide the roller with a fresh charge of powder to spread accurately over the surface of the build area.

Non-sintered powder forms a 'cake', which encapsulates and supports the model as the build progresses. The whole process takes place in an inert nitrogen atmosphere at less than 1% oxygen to prevent the nylon from oxidizing when heated by the laser beam.

DIRECT METAL LASER SINTERING (DMLS)

A considerable amount of heat is generated during this process because a 250 watt CO₂ laser is used to sinter the metal alloy powders. An expendable first layer of the part is anchored to the steel plate to stop distortion caused by differing rates of contraction. Such a layer also means that the part is easier to remove from the steel plate when the build is complete.

During the sintering process, the delivery chamber rises to dispense powder in the path of the paddle, which spreads a precise layer over the build area. The build platform is incrementally lowered as each layer of metal alloy is sintered onto the surface of the part. The whole process takes place in an inert nitrogen atmosphere at less than 1% oxygen to prevent oxidization of the metal powder during the build.

impossible geometries to fine tolerances and precise dimensions.

Additive manufacturing allows designers to iterate 3D CAD much more intimately and quickly than before through a process of printing, scanning, updating and reprinting. The cycle may be repeated many times to perfect the 3D both virtually and physically.

These processes are restricted by the size of machine. However, equipment capable of producing much larger parts is continually emerging, such as the Mammoth SLA machine by Materialise, Belgium, which can print parts up to 2100 x 680 x 800 mm (83 x 27 x 31.5 in.). LOM typically produces parts up to 508 x 762 x 610 mm (20 x 30 x 24 in.); FDM up to 915 x 610 x 915 mm (36 x 24 x 36 in.); PolyJet up to 490 x 390 x 200 mm (19.3 x 15.4 x 7.9 in.); SLS up to 580 x 380 x 700 mm (22.8 x 15 x 27.6 in.); SLA up to 500 x 500 x 500 mm (19.7 x 19.7 x 19.7 in.); and the DMLS process is confined to parts 250 x 250 x 185 mm (9.8 x 9.8 x 7.3 in.).

PolyJet is the most versatile additive manufacturing technology. Many types of material can be incorporated, ranging from structural to elastic. Based on PolyJet technology, Objet Connex 3D printing is capable of combining multiple materials in the same part. Up to eight different materials may be printed to form seamless composite parts. The surface is left as built, or improved with polishing, painting and metal coating.

The Thermojet process is used to produce 3D parts from wax in preparation for lost-wax casting metal. The wax part forms the pattern, around which a ceramic shell is formed. The 3D wax pattern is melted from the ceramic shell during firing and the cavity left in its place is filled with metal.

The SLA technique produces water-clear, translucent and opaque parts. The surface finish can be improved with polishing, painting or metallizing. SLA materials mimic conventional plastics, which means when used for prototyping, SLA is good for making parts with the visual and physical characteristics of the final product. However, the material degrades in UV light and so is not suitable for long-term use.

The SLS technique forms parts with physical characteristics similar to that of injection-molded plastics. Because parts can be made with live hinges (flexibility integrated into single material) and snap fits, this process is ideal for functional prototypes, one-offs and low volumes of identical parts. A live hinge is always constructed so that the hinge is on the horizontal plane, for strength; if it were built vertically the layers would be too short to withstand the stress of opening and closing, and it would fail.

In the SLS system, multiple parts can be built simultaneously and on different planes because the non-sintered powder supports the sintered parts. By contrast, parts made by PolyJet, FDM, SLA and DMLS processes need to be supported and undercuts must be tied into the build platform. This means that fewer parts can be formed at the same time and fine undercuts are more difficult to achieve.

The orientation of the part can affect its mechanical properties – strand-filled materials having better strength in certain geometries. The orientation of SLS parts in regular nylon powders has to be considered to maintain accuracy. For example, a tube is built vertically to keep it round; if built horizontally the tube would be very slightly oval. A large flat plane should be manufactured at an incline because if it is built flat there will be too much stress in the part, which will result in warpage.

DMLS provides an alternative to machining or casting metal. It is very accurate and produces parts with fine surface definition. As a result, DMLS parts are used as prototypes as well as production pieces. However, it is a tricky process to work with. Design for DMLS should take into account that the part is divided into the outer skin, the inner skin and the core. In a typical build, for every three layers of metal powder that are spread, the outer skin is sintered three times, the inner skin is sintered twice and the core is sintered only once.

DMLS parts with surfaces that face downwards at less than 30° to the build platform require a support structure to be included. This will need to be removed by cutting. Also, the first layer of the DMLS build will be anchored to the platform and must be cut free afterwards.

LOM is the most straightforward of these processes and is used to form a range of sheet materials. It does not require a support structure, because the waste from each sheet is left in place during the build and so supports undercuts and overhangs. Therefore, multiple parts may be produced simultaneously. LOM parts produced from paper have wood-like characteristics and the surface may be improved by sanding.

FDM is based on materials commonly used in mass-produced consumer products. While the properties of a layered part will be different from that of a molded one, there are enough similarities to make this a desirable process for prototyping. No special finishing is required, although the surface may be improved with gentle sanding and polishing.

Soluble materials used for the support structure mean that cavities and other complex geometries may be created and the structure removed with relative ease. However, these materials are more expensive and so wherever possible manufacturers will use commodity material for the support structure and remove it by hand.

MATERIALS

LOM is most commonly used to fabricate 3D forms from sheets of paper. Other sheet materials have been tested and used for niche applications, including plastic, metal and prepreg composite.

FDM is compatible with polycarbonate (PC), acrylonitrile butadiene styrene (ABS), PC/ABS, polylactic acid (PLA), polyvinyl alcohol (PLA), polyphenylsulfone (PPSU) and polyetherimide (PEI). ABS is the least expensive of these materials and is available in a range of colours (including white, grey, red, yellow, green, blue and black). PLA and PVA are water-soluble and so used to produce an easily removed support structure. PC, PPSU and PEI are tough and relatively expensive plastics used for high-end applications. PC is available in white and PEI is used natural (beige). Even though FDM uses production-grade materials, the

Layered Structure
Material: ABS
Application: Interior
Notes: FDM utilizes plastic filaments to create 3D parts. The layered structure is clearly visible on the surface.

Chrome Finish
Material: Metal-coated epoxy
Application: Prototype
Notes: The surface of SLA parts may be polished and coated with metal to produce a mirror-like finish.

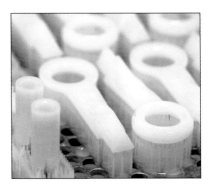

Support Structure
Material: Epoxy resin
Application: Prototype
Notes: Support structures are required when building overhangs and undercuts with SLA.

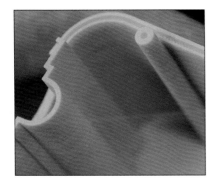

Sintered Finish
Material: Nylon powder
Application: Prototype
Notes: SLS produces high-quality parts, but the surface finish is not as good as parts produced by SLA.

Complex Geometries
Material: Nylon powder
Application: Model
Notes: With SLS, the non-sintered powder supports the build, allowing complex undercut geometries.

Carbon-Fibre Composite
Material: Carbon-fibre-reinforced nylon
Application: Prototype
Notes: Minute carbon fibres are incorporated into nylon powder and prototyped using SLS technology.

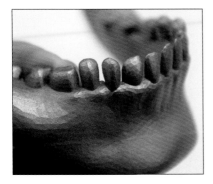

Metal Prototyping
Material: Stainless steel
Application: Model
Notes: Fine metal powder is built up by DMLS to produce dense, high-quality parts.

Interlinked Metal Parts
Material: Steel alloy
Application: Watch strap
Notes: DMLS is capable of producing multiple interlinked metal parts, permanently joined during the build.

Micro Metal Parts
Material: Nickel-bronze alloy
Application: Model
Notes: DMLS uses very fine metal powder – 0.02 mm (0.0008 in.) in this case – to build precise, intricate parts.

properties will not be equivalent to that of injection-molded parts, owing to extruding and fusing a continuous filament as opposed to filling a mold under high pressure.

PolyJet is compatible with thermoplastics including polypropylene (PP), high-density polyethylene (HDPE), polystyrene (PS), high-impact PS (HIPS), polymethylmethacrylate (PMMA), PC and ABS. Several of these materials are available transparent and coloured, such as PP, PS, HIPS, PMMA, PC and ABS. Elastomeric materials used to produce flexible parts include styrene-based elastomers and ethylene propylene diene monomer rubber (EPDM).

The SLS process is compatible with a variety of nylon-based powders, including a range of colours and powders with rubber-like qualities. The tough nylon 11 is heat-resistant up to 150°C (302°F) and so can be used to produce functional prototypes for engineering applications. Composite nylons, such as carbon- and glass-filled materials, are used to build parts for high-load applications.

The SLA process uses liquid epoxy resin polymers that are categorized by the thermoplastics that they are designed to mimic. Some typical materials include ABS mimic, PP/PE mimic and water-clear polyester (PBT)/ABS mimic. Materials that can withstand temperatures of up to 200°C (392°F) have been developed for the SLA process.

The DMLS process is compatible with specially developed metal alloys. Examples include aluminium alloy for lightweight and high-strength parts; nickel–bronze, which is harder-wearing than aluminium; stainless steel; steel alloy, which has similar characteristics to mild steel; and titanium alloy.

COSTS

The cost of additive manufacturing is largely dependent on build time. Cycle time is slow, but these processes reduce the need for any preparation or further processing and so turnaround can be rapid. Individual part cost is reduced if multiple products are manufactured simultaneously. The SLS powders are self-supporting and so large numbers of components can be built around and inside one another to reduce cost.

LOM is the quickest and typically the most cost-effective, because the laser only has to track around the outside edge of the profile. Using thicker sheet materials will reduce cycle time to a certain extent. Whereas LOM, PolyJet and FDM can be handled immediately, the cycle time for SLS and DMLS includes cooling time and SLA requires curing with UV light.

SLS parts are generally less expensive than SLA ones because they need less post-building processing. DMLS parts are typically cut from the steel plate and polished. This can be a lengthy process, but it depends on the product and application.

ENVIRONMENTAL IMPACTS

All the scrap material created during additive manufacturing can be recycled, except fibre-reinforced powders. These processes are a fairly efficient use of energy and material, as they direct thermal energy to the precise point it is required.

Case Study

→ 3D Scanning Pleated Structures

The ability to create 3D data from real world objects within minutes has transformed design and development. Architects and designers are continually exploring new opportunities associated with the ability to scan, print, tweak and remake in an iterative design process that is yielding exciting and never previously seen results.

In this case study, textile artist Elaine Ng Yan Ling (The Fabrick Lab) creates a series of pleated forms with sheets of paper (**image 1**). In a digital manufacturing lab set up by 360Fashion (www.360fashion.net) in China, the object is scanned and converted into 3D data so that it can be rapid prototyped using FDM technology.

The 3D scanner (**image 2**) builds up a point cloud by measuring the distance of each surface using light and triangulation (**image 3**). With this data it is possible to reconstruct a 3D model compatible with CAD software.

Scanners can only collect information from within their field of view: hidden surfaces and undercuts that are not within the line of sight will be missed. Therefore the object is rotated (**image 4**) to capture each side, and reference points on the table allow the scanner to pinpoint the surfaces relative to one another.

The point cloud is instantaneously reconstructed in 3D CAD data and the operator carefully adds back in parts missed by the scanner (**image 5**). Once completed, the data is manipulated to create a complex series of interlocked pleated objects ready for manufacture (**image 6**).

1

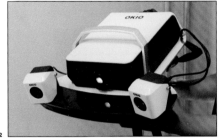

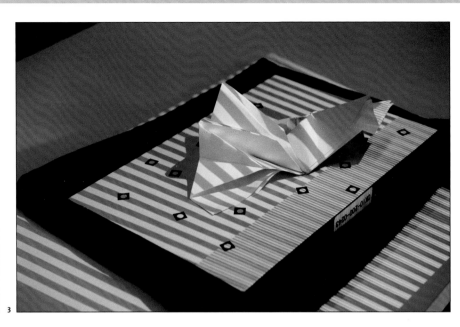

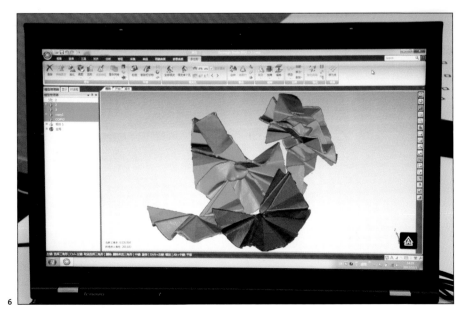

Featured Companies

The Fabrick Lab
www.thefabricklab.com
Tianyuan
www.3dscan.com.cn

→ Fused Deposition Modelling

The pleated design is 3D scanned and remodelled using CAD (see previous page). The data is loaded into the additive manufacturing software and oriented on the build platform to reduce the amount of support structure required (**image 1**).

Layer-building processes work on the principle that the build platform progresses downwards (z-axis) in increments as each layer of the model is built on top. The computer-guided head sits above the platform, guided by a horizontal track that provides x and y movement (**image 2**). Continuous filaments of plastic, in this case ABS (**image 3**), are loaded onto the machine and supplied to the nozzles. There are two nozzles, because one filament is used as the support structure, while the other guides the material for the actual part (**image 4**).

The process begins by building the support structure directly onto the build platform (**image 5**). The structure is less dense than the actual part to make it easier to remove afterwards.

The support structure maps the outline of the 3D shape. With each new layer, support structure (grey) is constructed simultaneously with the part (white). Support structure is built on top of the part if the shape doubles back on itself (**image 6**).

With each new layer the build platform progresses downwards (**image 7**). The chamber is heated to reduce the amount of energy required to melt the plastic filament, as well as reduce warpage caused by non-uniform cooling.

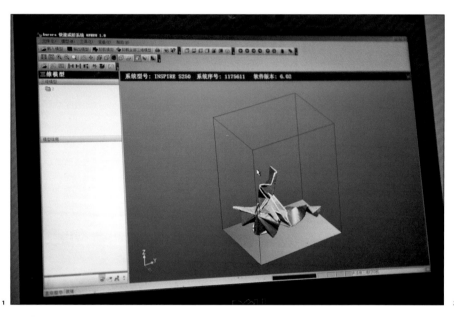

1

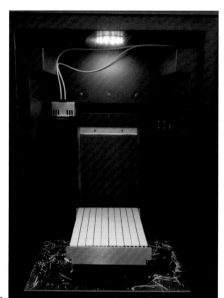

2

3

4

5

The completed model, which is a replica of the CAD drawing, is ready to be removed from the build platform (**image 8**). In this case, the support structure on the underside is removed by hand. It is built with an open structure to reduce its density and strength (**image 9**). This makes it more economical as well as aiding separation from the model (**image 10**). The finished part is completed (**image 11**).

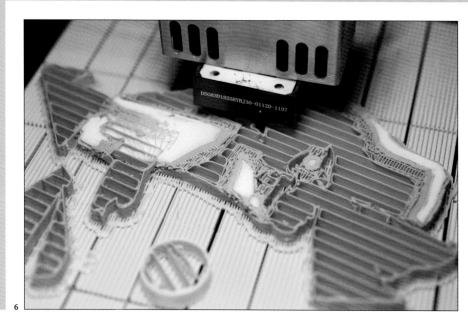

6

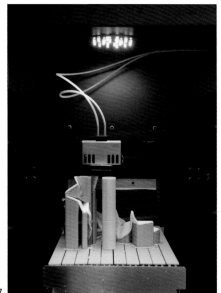

7

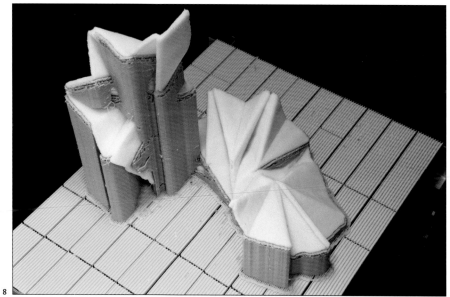

8

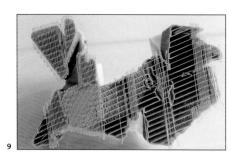

9

10

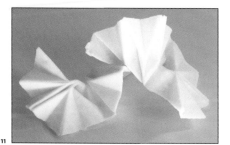

11

Featured Companies

The Fabrick Lab
www.thefabricklab.com
PP3DP
www.pp3dp.com

Part

4

Materials

Plant Fibres

These fibres are derived from the stems, leaves, seed pods, fruits and other parts of plants. The individual fibres are characterized by a hollow tubular construction with thick cell walls made up of cellulose, hemicellulose, lignin and other waxy substances. They are extracted from the plant as fibre bundles, which can be as long as the plant itself. Plants used in the production of textile are cultivated all over the world, but each species is limited to certain climates and soil types. The yield and the techniques developed to extract the fibres, which have varying levels of effectiveness, mean that some are much more cost-effective than others to convert into yarn. They are used for all manner of applications, from fancy fabrics to hard-wearing denim workwear and the finest linens to industrial packaging.

PROPERTIES OF PLANT FIBRES

Plant fibres, whether harvested from the seed pod, stem or leaf, are made up of the same ingredients. Cellulose is a strong linear polymer made up of repeating glucose units. It forms very fine strands, called microfibrils, in the plant cell walls. These are bonded together with a matrix of hemicellulose, lignin and pectin. As the plant grows, new layers of cell wall are formed around the outside.

The proportion of diverse polymeric substances and structure of the fibre is determined by its role in nature. Whereas flax (structural bast fibre) has a high proportion of hemicellulose, cotton (seed hair) is virtually pure cellulose. Their contrasting properties – strength, handle and colourability – are the result of these differences.

Thanks to the fibre structure, they have a high level of absorbency and so are often utilized in pile fabrics to make towels, and woven into warm-weather clothing that remains dry. Water absorption causes the fibres to swell and

so closes up the gaps in tight weaves. This means that some cotton fabrics are naturally water-repellent (without the need for coating, page 226, or laminating, page 188).

One of the downsides of plant fibres is that they tend to have low resiliency, resulting in wrinkles and fibre breakage. A good example of this is linen woven with flax and used to make summer suits. Coatings help to reduce wrinkling, but they may also affect the natural quality of the material and are considered a health hazard if formaldehyde-based resins are used.

As a natural material, the fibres are susceptible to ultraviolet light, mold and insect attack. However, certain types exhibit very high resistance to one or more of these factors. For example, hemp is naturally antibacterial, which means it is resistant to mold and mildew. Combined with the fact that hemp is strong and resistant to heat, ultraviolet light and salt water, this is a very desirable material for a wide range of decorative and technical applications.

Many of these fibres are suitable for dyeing – cotton is produced in the widest range of bright and pastel colours – although the necessary bleaching and chemical treatments can affect the strength and resilience of the fibre. Therefore, many plant fibres are used primarily in their natural colours. Omitting bleaching and dyeing greatly reduces the environmental impacts of producing plant fibre textiles.

TYPES OF PLANT FIBRES

Plant fibres are divided according to where the fibre was extracted from, which are bast (stem), seed hair (seed pod), fruit hair, leaves and other (such as roots and bark). Each of the plant fibres has unique properties, but they also share many of the same qualities.

Bast fibres are obtained from the stem of plants and include flax, hemp, jute, ramie, nettle, kenaf and bamboo. They are typically long, stiff and strong as a result of the need to keep the plant upright during its lifetime. They may be extracted by hand, as they were in ancient times, or by biological and mechanical means, which makes them much more cost-effective but can affect their strength and other properties.

Bamboo is utilized to make textile in several different formats. Bast fibre from the stem of the grass is produced in relatively small quantities because of lengthy and complicated processing steps. It is much more commonly converted into bamboo viscose (page 508) through a series of mechanical and chemical processes. The resulting fibre has some very desirable properties, but the characteristics and environmental impacts are very different from natural bamboo plant fibre, and the two should not be confused (bamboo viscose is often marketed as being like natural bamboo, but this is not the case). It is also an important building material, split into veneers and utilized in basket making (page 120).

Cotton, consumed in the largest quantities of all textile fibres, is derived from the seed pod of the cotton plant. Single-celled strands of cellulose grow from the seeds, and having not been put under any load during their lifetime, tend to be soft and supple. The fibres are relatively short and straightforward to convert into yarn (page 56). Other types of seed pod fibre, such as coir, have varying characteristics, governed by the life of the plant.

The leaves of several types of plant, including the abacá, sisal and pineapple, are cultivated to make rope and textile. Extraction is a laborious process (page 40) and as a result they are much less common and often quite expensive once exported from the country of origin. The quality ranges from coarse and wiry sisal and abacá, to lustrous and luxurious piña (pineapple).

NOTES ON MANUFACTURING

The production techniques are different for each type. Cotton stands out as the most successful of the plant fibres. With the invention of the cotton gin and several other processing innovations at the end of the 18th and early 19th century, cotton production became a very lucrative business. The balance of properties –

softness, drape, strength and affinity for bright coloured dyes – combined with the relative ease with which it can be extracted from the plant and processed into fibre, means it has remained the most widely produced textile fibre. It is a versatile material used to make all sorts of fabric, from hard-wearing workwear to delicate, decorative lace (see image right).

Bast fibres are much more difficult to extract and process, in part because they are long, coarse and tough. They are traditionally converted from plant stems into bundles of fibre by hand, and mechanization has not been as successful as cotton production. Cultivation is relatively straightforward and bast fibre plants grow throughout the world. As a result, they have been used since ancient times to make all manner of rope, net and textile. Industrial production is not so straightforward. Once the plants have been harvested, the cellulosic fibres must be separated from the stem and each other. This is typically achieved by retting: a natural process whereby the plants are left outdoors or in water to allow microorganisms and enzymes to break down the pectin that binds the fibres together. This can take a few days to several months, depending on the climate, fibre type and technique.

Once retted, the fibres are put through a series of mechanical processes to fully separate and align the fibres. The length varies according to the species and dealing with such long fibres is complicated for machinery. Therefore, the strands are often shortened to make smaller staples, similar in length to cotton, so that similar equipment may be used to convert them into yarn.

Leaf fibre production is the most laborious and difficult of all. It is mainly done by hand and the expertise exists in small communities in relatively few countries. Once mature, the leaves are removed from the plant and scraped with a blunt implement to remove the fleshy matter. This reveals the fibres, which then need to be soaked, dried and combed to make strands that can be converted into yarn.

Natural materials have inevitable variation in quality and consistency

Lace Woven by Cluny Lace (see pages 108–11), this delicate cotton fabric is produced on a Leavers loom, guided by jacquard-punched cards.

(see image overleaf top left). This presents a challenge for production and means that these materials will never be as efficient to produce as synthetic equivalents. However, it is that same variability that gives natural textiles much of their unique quality and charm. These properties cannot be underestimated, because we purchase items as much for their emotional appeal as for functionality. This fact is illustrated by the successful development and use of production techniques to manufacture synthetic fibres that mimic the appearance of natural materials.

Cotton and other cellulosic fibres are mercerized (page 209) to improve their strength and surface lustre. It is nearly always carried out before printing (pages 256–79) and dyeing (page 240), because the colour will appear brighter and more saturated.

The type of fibre and its natural colour will determine the potential effectiveness of dyeing. Certain plant fibres, such as cotton and ramie, have a strong affinity for dyes. Others are more difficult to

Natural variation
There is inherent variation in the structure and appearance of natural materials. Flax fibre is characterized by small nodes occurring along its length.

Vegetable dyes Leaves, seeds, bark and berries are converted into dyes using techniques that have been practised for thousands of years.

colour, either because they cannot be bleached effectively or because they break down when exposed to ultraviolet light and so are not colour fast.

Natural or vegetable dyes are derived from locally available materials, such as plants, animals or minerals. They have been practised for thousands of years and are frequently combined with resist dyeing (page 240). Since the development of synthetic dyes – in particular vat dyeing with its range of consistent, rich colours that have excellent fastness on plant fibres – natural dyes have become restricted to handicrafts and small communities practising traditional crafts (see image above right).

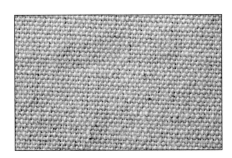

**Left top
Organic cotton** The environmental impacts of cotton production are greatly reduced by eliminating the use of harmful chemicals.

**Left
Hemp pulp packaging** Hemp fibres are combined with conventional wood pulp to make lighter and stronger packaging for consumer electronics goods.

PLANT FIBRE MATERIAL DEVELOPMENTS

Since the widespread adoption of synthetic fibres (page 510), which are typically less expensive to produce and less reliant on climate and soil conditions, plant fibre textiles have been in decline. They are, however, potentially much more sustainable to produce than their synthetic equivalents. Certain types, in particular bast fibres such as hemp and bamboo, are fast-growing and produce a high yield without the need for environmentally damaging pesticides, herbicides and fertilizers. These very beneficial properties mean that plant fibres are being re-explored as sustainable and renewable materials.

A notable exception to the sustainable nature of plant fibres is cotton, which accounts for more than 10% of global agrochemical consumption. Combined with the vast quantities of water and chemicals used in the production of yarns and textiles, this material leads to harm for people and the environment.

Organic cotton (see image below) is grown without the use of harmful chemicals, which greatly reduces the environmental impacts. The natural colour variation of cotton, which ranges from creamy white to red, green and brown, is being exploited to eliminate the need for bleaching and dyeing. Such materials can have a positive impact on the environment, which is in stark contrast to conventional industrial cotton production.

In recent years, several pulp-based developments have emerged that use these fibres, such as molded pulp packaging made up of hemp with wood fibre (see image below). Many types of plant fibre – in particular bast fibres from hemp, flax and kenaf – are used in the production of paper and board. They are used to increase strength and reduce weight. Another advantage of these fibres is that thanks to their relatively low lignin content, bleaching may be chlorine-free. However, production is more expensive than for wood pulp and so they are mainly used for specialized applications, such as cigarette papers and banknotes.

Another significant area of development, which is also closely linked to the potential sustainability of plant fibres, is the use of these materials as alternatives to high-performance fibres used in composite laminates (page 239). A range of different weaves is produced: they are tailored to the mechanical requirements of the application. Composites Evolution developed a 'twistless' spinning technique (see image, opposite top left), which ensures higher strength to weight by maintaining straighter fibres.

Composites Evolution uses flax to manufacture Biotex, a range of high-performance natural composites (see image, opposite top right). This technology is being explored and developed for application in automotive and marine applications. It has even been used to make parts of racing cars, such as the Lola Drayson Le Mans prototype electric racing car and the WorldFirst racing car sponsored by Warwick Innovative Manufacturing Research Centre (WIMRC) at Warwick

Manufacturing Group. The ultra-high-performance electric vehicle, based on an existing Lola chassis, was launched in 2011.

High-strength plant fibres, such as flax, hemp and jute, are used to reinforce synthetic or natural resins. When combined with a natural resin, such as polylactic acid (PLA), they are known as a biocomposite. These materials are helping to reduce weight, cost and environmental impact in the automotive industry by replacing conventional composites such as glass-filled plastics and sheet metal.

Twistless spinning Straight yarns are woven into a range of fibre reinforcements. This technique was developed by Composites Evolution to produce composites with maximum strength to weight.

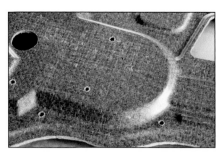

Plant-fibre-reinforced composite Produced by Composites Evolution, this prototype interior panel for Land Rover by is made from Biotex flax/polypropylene (PP). It is 60% lighter than the current steel part at the same stiffness.

COTTON

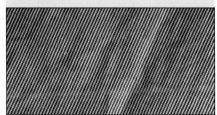

Description
A single-celled fibre harvested from the seed pod of the cotton plant (genus *Gossypium*). The staple fibre, which is typically between 13 and 38 mm (0.5–1.5 in.) long, is separated from the plant by ginning (page 32). It is carded, combed and twisted to make yarn (page 56).

Strength ● ● ● ○ ○
After scouring and bleaching it is virtually pure cellulose. Heavy-duty woven cotton has a good balance of lightness, durability and pliability, a characteristic utilized in denim. Its strength improves by 20 to 30% when wet.

Resilience ● ● ○ ○ ○
It has low elasticity and does not recover well when stretched (this is why the knees and elbows of cotton garments become baggy over time). It is generally not used for high-wear applications, upholstery or carpet. Wrinkle resistance is improved by finishing (see Fluid Coating, page 226).

Quality ● ● ● ● ○
Cotton feels soft to the touch, has good drape, prints well and is highly absorbent when fabricated from low-twist yarn. Cotton fibres swell in the wet and so can be used to produce weatherproof fabrics, such as Ventile, which is used in performance garments for military, medical and workwear applications.

Colour ● ● ● ● ○
It is naturally off-white, red, green or brown. Bleaching out the natural colour produces white cotton. It has very good affinity to dyeing, in particular vat dyes (page 240), which can produce a wide range of colours with excellent fastness. Cotton will yellow when exposed to the sun for prolonged periods.

Versatility ● ● ● ● ●
As a result of its widespread and historical use cotton is considered one of the most versatile materials. It is used to make a wide variety of garments, hosiery, bedding, towels, curtains, lace, embroidery, packaging and nonwovens (page 152). It is also blended (page 242) with many other types of fibre to take advantage of their properties.

Availability ● ● ● ● ●
It is the most commonly used textile fibre, accounting for about half of global textile production. Cultivation is limited to warm climates. The largest producers are China, southern USA, South America, India and Pakistan. Egyptian cotton is considered the highest quality, along with American pima, but there are many types, so quality varies.

Cost ● ● ● ● ○
Low to high depending on quality and source. Very fine quality Egyptian cotton and those grown using organic methods are the most expensive owing to the relatively low volumes produced each year.

Environment ● ○ ○ ○ ○
Cotton is one of the most heavily sprayed field crops, accounting for a significant proportion of global consumption. Converting cotton fibre into textile by scouring, bleaching and mercerizing requires even more water and chemicals to be used, making this one of the most environmentally harmful textile fibres.

FLAX

Description
The woody stalk of the flax plant (*Linum usitatissimum*) yields long bast fibres ranging from 25 to 900 mm (1–35 in.). The longest are converted into high-quality yarn, called line, and the shorter fibres are converted into tow. Linen has been produced from flax for thousands of years and is believed to be one of the oldest textiles.

Strength ● ● ● ● ○
The fibre is cellulose-rich (70–75%) and strong, second only to silk among the natural fibres. The strength increases by up to 20% when wet. It has good resistance to abrasion.

Resilience ● ● ○ ○ ○
It has relatively low resilience and is stiffer than cotton, quite brittle and susceptible to creasing. The fibre consists of nodes, or slubs (perpendicular dislocations), along its length, which adds some flexibility, but decreases strength.

Quality ● ● ● ● ●
The fibre ranges from coarse to fine, depending on the type, source and method of conversion from plant to fibre. Very fine linen will have long straight fibres with a consistent diameter (free from visible nodes). High-quality linen is prized for its strength, soft hand and high lustre. It is highly absorbent and can gain up to 20% water without feeling damp.

Colour ● ● ● ○ ○
Flax is naturally light brown to dark grey. The tough fibre is whitened in the sun (grass bleaching) or with chemical bleaching. Using chemicals weakens the fibre and so is avoided where possible, or it is dyed muted colours to avoid heavy bleaching. It is less susceptible than cotton to ultraviolet light.

Versatility ● ● ● ○ ○
Flax is less widely available and more expensive than cotton. It is used for more specialized applications, such as high-quality woven garments, towels, bedding and handkerchiefs. Damask tablecloths are traditionally made from flax and distinguished by the repeating jacquard pattern (reversible). Its high strength is utilized to make biocomposites (page 239). In the past it was used to make rope (page 60) and paper, and it continues to be used to make banknotes.

Availability ● ● ● ○ ○
Flax is grown mainly throughout Europe and China. The strongest and highest-quality flax comes from Ireland and Belgium.

Cost ● ● ● ○ ○
Linen is a relatively expensive fabric, because the cultivation and preparation of flax are time-consuming and labour-intensive. The highest-quality linen continues to be processed with manual techniques.

Environment ● ● ● ● ○
Retting involves laying the flax stalks out in the field where they were cut and allowing naturally occurring microorganisms to breakdown the pectin that bonds the fibres to the stem. In the past, flax was retted in ponds and streams, which polluted the surroundings waterways. More recently developed chemical retting techniques are quicker but also potentially harmful to the fibre and environment.

JUTE AND KENAF

Description

Fibres from the stems of jute plants (genus *Corchorus*) are converted into hessian (burlap). One of the least expensive plant fibres, it is obtained from two species of *Corchorus*: white jute (*C. capsularis*) and tossa jute (*C. olitorius*). The fibres are up to 2 m (7 ft) long. Kenaf (*Hibiscus*) yields a similar bast fibre ranging from 1 to 3.6 m (3–12 ft) in length.

Strength ● ● ○ ○ ○

Jute is coarse and tough, becoming weak and brittle with time and exposure. Kenaf has similar strength, but is degraded by processing into yarn. Both lose around 10% of their strength when wet.

Resilience ● ○ ○ ○ ○

They have relatively poor resilience: their brittleness makes them prone to creasing and shedding fibres, which in the case of jute gets worse with age.

Quality ● ● ○ ○ ○

Jute is breathable and made flame-retardant with a suitable coating (page 226). It is stiff and does not drape well. It has a pronounced z-twist due to the biased spiral of the fibre structure. Kenaf fibre has varying levels of quality. It is typically coarser and harsher than jute with a brittle hand. Unlike jute, it has a balanced spiralling structure and may be constructed into z- and s-twist yarns with equal properties.

Colour ● ● ○ ○ ○

Naturally off-white, grey, gold, red, brown or black, jute is dyed with vat and other techniques (page 240). As a low-cost industrial fibre it is mainly used in its natural state. Kenaf is naturally light coloured, from off-white to yellow.

Versatility ● ● ● ○ ○

In the past, the cost-effectiveness of jute meant it was used in a wide range of applications including interior fabrics and shipping sacks. Nowadays, it has been replaced by durable and low-cost synthetics. Even so, it remains an important textile fibre and is used to make sacks, upholstery (page 404), shoes and geotextiles. It is also used in nonwovens (page 152) and biocomposites (page 239). It is often blended with other natural fibres, as well as synthetics. Kenaf is used for rope, twine and coarse fabric similar to hessian. Like hemp, jute and flax, it is being explored as an alternative to glass fibre for lightweight and high-strength composites. It is used as an alternative to wood pulp in paper production.

Availability ● ● ● ● ●

The main producers of jute are Bangladesh and India, although it is cultivated in several other countries including China, Indonesia, Myanmar, Nepal, Pakistan, Thailand and Vietnam. It is one of the most widely consumed textile fibres, second only to cotton. China, India and Bangladesh are the main producers of kenaf.

Cost ● ● ● ● ●

Jute is fast-growing, widely available and relatively easy to process, making this a low-cost fibre. Kenaf is even cheaper than jute, because cultivation is simpler and it grows in a wider range of climates.

Environment ● ● ● ● ●

They grow with little need for pesticides and herbicides, which makes them a low-impact crop. However, like flax they require retting, which must be carefully controlled to avoid pollution. Kenaf has high yield and its relatively light colouring means it requires less bleaching than other fibres.

HEMP

Description

Industrial hemp (genus *Cannabis*) is fast-growing and yields soft and strong bast fibres from its stem. The fibres can be very long and up to 4.5 m (15 ft) in length. The use of this fibre dates back many thousand years in the production of rope and textile.

Strength ● ● ● ● ○

The fibre is strong and has natural antimicrobial properties (making it resistant to mold). It also has natural resistance to degradation by heat, ultraviolet light and salt water. As a replacement for glass-fibre reinforcement in modern composite development, it has been demonstrated to contribute to significantly lighter and stronger parts.

Resilience ● ● ● ○ ○

Processed hemp has high cellulose content (up to 80%), but a little more lignin than flax, which makes it slightly stiffer and coarser.

Quality ● ● ● ○ ○

It has a soft hand similar to flax, but can feel harder, and the surface has a lustrous appearance. As a result it is easily mistaken for linen.

Colour ● ○ ○ ○ ○

Naturally off-white, brown or green, it is typically used in its natural state, because bleaching weakens the fibre.

Versatility ● ● ● ○ ○

Historically, hemp was a very versatile material and used in rope, textiles, sails and paper. It was also traditionally the primary material used to make canvas. Today, applications are relatively limited and include garments, interior textiles, carpets and some papermaking. It is frequently blended with cotton and other fibres to impart some of its strength and resilience, for example to make more durable denim and upholstery fabrics.

Availability ● ● ○ ○ ○

Hemp thrives in most climates. However, cultivation is limited to only a handful of countries owing to its close association with marijuana (a psychoactive drug). China and Europe are the largest producers.

Cost ● ● ○ ○ ○

Cultivation and production are labour-intensive and it is grown in relatively small quantities, which make it an expensive fibre to produce.

Environment ● ● ● ● ●

Hemp is cultivated and processed with similar techniques to flax. It can have a more positive impact on the environment, because it yields much more fibre for the same area and it can be grown without pesticides, herbicides or fertilizer. It yields many important by-products too, such as oil, mulch and animal bedding. Textiles finished without chemical bleaching or dyeing are sustainable and can be safely composted at the end of their useful life, returning the nutrients back to the soil.

RAMIE

Description

Ramie (*Boehmeria nivea*) yields a high-strength bast fibre, which can be bleached to a bright white with a high surface lustre. Length ranges up to around 0.55 m (21.6 in.). It is in the same family of plants (*Urticaceae*) as the common stinging nettle (*Urtica dioica*), which is re-emerging as a sustainable textile fibre.

Strength ● ● ● ○ ○

Ramie has a similar structure to flax and is one of the stronger plant fibres.

Resilience ● ○ ○ ○ ○

After processing it has high cellulose content (over 90%). It is stiff and brittle; the fibres break relatively easily and the surface is susceptible to wear. Therefore, like cotton, it is treated to enhance wrinkle resistance.

Quality ● ● ● ○ ○

These fibres share many similarities with hemp and flax, being closely related, including a natural resistance to mold and bacteria. Depending on the source they may be lighter than flax, and for the same weight they are stronger and more absorbent.

Colour ● ○ ○ ○ ○

Ramie is naturally off-white to brown and can be bleached to bright white. It is suitable for dyeing a range of bright colours, with a silk-like lustre.

Versatility ● ● ○ ○ ○

Due to the high cost of ramie, and its low resilience, it is often blended with cotton and other natural fibres to impart some of its strength.

Availability ● ● ○ ○ ○

Ramie is produced mainly in China (it is also known as China grass) as well as in South America and India, among other countries. Nettle is a successful perennial that grows throughout temperate regions of North America, Europe and Asia, but as a textile fibre it is cultivated in very low quantities.

Cost ● ● ○ ○ ○

Ramie production is time-consuming and labour-intensive, and nettle fibre is not widely available, making these relatively expensive fibres, although in some cases they may be less expensive than flax.

Environment ● ● ● ○ ○

The common nettle grows without the need for herbicide and pesticide. It is a perennial and so does not need to be grown from seed. Ramie is more susceptible to disease and so is more likely to be sprayed with chemicals, but may still be grown in a sustainable manner.

NATURAL BAMBOO FIBRE

Description
Bamboo plant (*Bambuseae*) bast fibre is different from regenerated bamboo (page 508). With recently developed mechanical techniques, fibres up to 150 mm (6 in.) are extracted from the stem of the plant and converted into yarns to make fabrics.

Strength ● ● ● ● ●
Bamboo fibre is made up of bundles of parallel fibrils held together with layers of lignin and pectin. This makes it reasonably strong: it is finer and stronger than viscose (regenerated) bamboo. It has superior resistance to bacteria and ultraviolet light.

Resilience ● ● ○ ○ ○
It has similar resilience to hemp and flax.

Quality ● ● ● ○ ○
It has a reasonably soft hand, comparable with flax, depending on how it is processed and constructed. The surface of the fibre is uneven, with inherent variations in thickness, which promotes rapid water absorption and evaporation, enhancing the breathability of fabrics (several times more than cotton).

Colour ● ● ● ● ○
It is naturally white to light brown with a high affinity for dyestuffs. It can be produced in a wide range of colours.

Versatility ● ● ● ● ○
It is used to produce fabric and pile for garments, hosiery and towels. It is woven (see Loom Weaving, page 76) as pure bamboo, but is typically blended when knitted (see Weft Knitting, page 126) to reduce the hardness while imparting some of its strength, softness and breathability to the finished fabric. Similar to other plant fibres, it is being explored as fibre reinforcement in lightweight composites.

Availability ● ○ ○ ○ ○
Bamboo is mainly cultivated in China, but is not commonly converted into fibre without chemical processing (viscose).

Natural bamboo is difficult to process by spinning and as a result there is a limited number of factories capable of producing the yarn.

Cost ● ○ ○ ○ ○
Processing bamboo into fibre is complex and time-consuming, requiring many steps, and it is produced in very small quantities, all of which makes it a relatively expensive fibre.

Environment ● ● ● ● ●
Bamboo is fast-growing and high-yield and does not require pesticides, herbicides, fertilizer or irrigation for cultivation. It is considered a renewable and sustainable crop.

COIR

Description
Coir is obtained from the husk of coconuts harvested from the coconut palm (*Cocos nucifera*). The two main types are brown and white. Brown is harvested from mature coconuts and white is obtained from coconuts before they have ripened (known as green). The stiff and wiry fibres are typically more than 150 mm (6 in.) long and can be up to 350 mm (14 in.).

Strength ● ● ○ ○ ○
Coir fibres are relatively large in diameter. They are hollow, with thick walls made up of one third cellulose and an equal proportion of lignin. The fibre is stiff and breaks easily when bent. It is resistant to abrasion, weathering and salt water.

Resilience ● ○ ○ ○ ○
Coir is brittle and has low resilience. It is combined with latex to produce a long-lasting biocomposite material, known as rubberized coir, which demonstrates much higher resilience. In this format it is used to pad upholstery (page 404).

Quality ● ○ ○ ○ ○
The fibres are stiff and hairy, resulting in a coarse and abrasive surface.

Colour ● ○ ○ ○ ○
Naturally brown or off-white, it may be bleached and dyed. However, the high concentration of lignin makes colouring very difficult and so it is mostly used in its natural state.

Versatility ● ● ○ ○ ○
Coir is tough and so is used to make flooring, mats, batting and twisted rope (page 60). It is used for geotextiles, such as for erosion prevention, thanks to its low decomposition rate, although, like all natural fibres, it will eventually break down.

Availability ● ● ○ ○ ○
Even though coconuts grow in many countries, the majority of coir fibre comes from India and Sri Lanka. It is typically produced by small-scale farmers who utilize local mills to obtain the fibre, and is exported by large distributors.

Cost ● ● ● ● ○
It is relatively low-cost, especially in the countries of origin, where it is an abundant material.

Environment ● ● ● ○ ○
Coir is a renewable crop, but it is produced mainly in developing countries and is labour-intensive. Therefore, the source is critical to ensure fair and safe working conditions and that materials are harvested in an environmentally sensitive manner. The retting, bleaching and dyeing of coir are all potential sources of pollution. Fibre production provides a valuable income for many small-scale producers, but they often do not have the investment to bring in new, more sustainable practices and so may be forced to stick with polluting processes.

LEAF FIBRES

Description
Leaf fibres extracted from sisal (*Agave sisalana*) and abacá (*Musa textilis*) are known under the same name as the plant. Fibres obtained from the leaves of the Red Spanish variety of pineapple plant (*Ananas comosus*) are known as piña. And fibres from the young leaves (*cogollos*) of the panama plant (*Carludovica palmate*) are used to make the famous hat of the same name (see Hat Blocking, page 420).

Strength ● ● ○ ○ ○
Sisal fibres are strong, coarse and typically up to 1.2 m (4 ft) long. Abacá produces strong fibres up to 3 m (10 ft) in length with a relatively high proportion of lignin (15%).

It has good resistance to salt water and good buoyancy. Piña produces strong fibres up to 600 mm (24 in.) long.

Resilience ● ● ● ● ○
Sisal and abacá are quite stiff and brittle. Piña is more elastic and reasonably resilient, which makes it suitable for a wider range of fabrication techniques and applications. Panama fibres have good resiliency as a result of the lengthy preparation that goes into making a high-quality hat.

Quality ● ● ● ● ○
Sisal and abacá are hard fibres similar to coir. By contrast, piña is very soft with a hand similar to flax. Piña has a lustrous surface and is utilized in the production of luxurious, sheer and ceremonial garments.

Colour ● ● ○ ○ ○
Sisal is cream-coloured; abacá is light beige; and piña and panama are off-white. They may be dyed (page 240) but are typically used in the natural colour or bleached in the sun.

Versatility ● ● ● ● ○
Prior to the widespread adoption of synthetic fibres, leaves were an important material used in the production of rope, textile and paper. Today, sisal and abacá are used to make a small amount of twisted rope (page 60) and certain types of paper. They have been explored to a lesser extent as fibre reinforcement for composite laminates). Abacá is converted into sinamay for hat blocking (page 420), and sisal's toughness is utilized in carpets (often blended with wool) and metal buffing cloths, as it is coarse but will not scratch. Piña is hand-woven (pages 44 and 90) into high-quality fabrics including sheer, lace (pages 106–19) and embroidery (page 298), but in very small quantities. Panama is used almost exclusively to make hats.

Availability ● ○ ○ ○ ○
Sisal is produced in parts of South America (mainly Brazil) and Africa. Abacá (also known as Manila hemp) and piña are grown in the Philippines. The leaves, stems and roots of other types of banana plants (*Musa* genus), to which the abacá is related, are converted into fibres of varying quality. Panama grows in Central America and tropical South America (mainly Ecuador).

Cost ● ○ ○ ○ ○
Turning leaves into fibres is a laborious process, involving scraping the pulp from the leaves by hand, which makes these relatively costly fibres to produce.

Environment ● ● ● ● ○
These plants may be grown without herbicides, pesticides and fertilizers and tend to be cultivated by smallholders in a reasonably sustainable manner. After harvesting they are often decorticated in large centralized facilities.

Fibrous Wood, Grass and Leaves

Sheets, strands and strips of naturally fibrous material are used in their raw state, or combined with other yarns and textiles in the production of garments, headwear, upholstery, baskets and interior fabrics. Wood is sliced into thin veneers down to 0.1 mm (0.004 in.), which are flexible and paper-like, and may be laminated together to make strong panels; leaves are woven and dried into rigid textiles and lightweight packaging; and bark is stripped from trees to make durable and weather-resistant fabrics. Many types of plant are utilized, ranging from the core and sheath of climbing palms to stems and leaves of flowering grasses. The types and techniques vary according to locally available materials and traditions that may have been practised for thousands of years.

PROPERTIES OF WOOD, GRASS AND LEAVES

These are cellulosic materials containing lignin, pectin and other resinous substances (see also Plant Fibres, page 478). The type of fibre, which is determined by where it was derived from on the plant, will define the characteristics of the material. Strips of material taken from the stem are made up of bast fibre. Their role is to keep the plant upright during its lifetime, making them particularly stiff and strong. They are extracted and processed as raw fibres in the production of many important fabrics, such as linen and hessian. Strips of material made up of these fibres, such as obtained from the stems of flowering plants, share many of the characteristics of these plant fibres. The difference is they have not been removed from the surrounding tissue, shortened and plied into yarn, so they retain more of their inherent strength and stiffness. The properties of the material may vary along its length, which is limited to how much of the stem is usable.

Softwood, hardwood, rattan and bamboo are made up of cellulose and lignin, the same as plant fibres, but the quality is quite different. Trees grow very large over many years. This results in growth rings, which vary according to the seasons. Early in the growing season tree growth is rapid and the wood is typically lighter in colour because the cells are larger. Darker rings indicate slower growth from later in the growing season. Rings are intersected with rays, which are structures radiating from the centre of the tree to transport food and waste laterally. The combination of rings and rays produces patterns and flecks of colour on the surface of veneer. The fibrous material is stiffer and more brittle than that derived from younger and faster-growing plants.

The properties of rattan and bamboo are comparable to those of hardwoods in many ways. The difference is that the solid core of rattan does not increase as it grows lengthways, while the core of bamboo is hollow. Bark ranges from lightweight cork to

fibrous and durable palm. Their properties can be attributed to the protective role each type plays on the tree trunk. For example, cork is insulating, virtually impervious to water and naturally resistant to microbes. Palm bark also has good water-shedding properties, and is utilized in China to make outdoor clothing such as hats and capes.

Leaves are a complex natural composite made up of strong fibres surrounded by spongy tissue. Layers of chlorophyll-containing tissue, called mesophyll, are sandwiched between tough skin. The leaf produces a waxy substance on its surface (cuticle), which is waterproof (to prevent water loss) and protects it from insects and bacteria. Like bast fibres that run up the stem of plants, the veins are made up of lignified cellulose tissue, which provide structural support as well as transport water and nutrients. This combination of properties creates a strong, flexible and waterproof material that can be twisted into rope, woven to make fabric, or simply used to wrap food parcels.

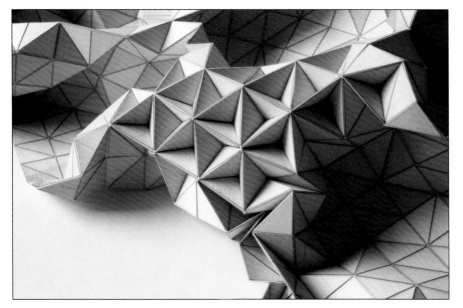

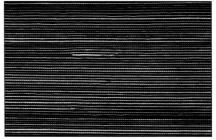

Wooden textile
Created by Elisa Strozyk, this unique fabric combines the properties of wood veneer with woven textile. The behaviour of the fabric – whether used as a flooring, curtains or upholstery – depends on the geometry and size of each wooden tile.

Top
Grass roots Even the roots of flowering plants may be turned into useful items, such as the hardy bristles of this Chinese scrubbing brush.

Above
Loom-woven grass Straight lengths of grass are inserted into synthetic filament warp with each pick to form a stiff cloth.

TYPES OF WOOD, GRASS AND LEAVES

There are two main categories of wood: softwood and hardwood. The terms are misleading, because, for example, balsa is very soft and light but is classified as a hardwood, while certain softwoods are dense and hard-wearing like some hardwoods. The principal difference is that softwood is derived from coniferous evergreens, whereas hardwood is from deciduous broadleaved trees. Softwoods tend to be faster growing and used to make commodity materials such as panels, paper and board.

Hardwoods range from commonly used types, such as oak, birch and maple, to more exotic and expensive types such as rosewood, teak and mahogany. In addition to the diversity of rich and beautiful colours, they may also grow with highly figured grain patterns. These can be rare and desirable, depending on the pattern and colouring, and greatly increase the cost of the veneer.

Tree bark is more difficult to harvest and process. As a result, fewer types and applications have been exploited. A notable exception to this is cork, which is produced on an industrial scale. Like all natural materials, the development of cost-effective synthetic alternatives has resulted in a steady decline in the demand for cork. Even so, it continues to be an important material for both industrial and domestic applications thanks to its outstanding properties.

Rattan and bamboo are fast-growing. They produce hardy fibrous materials and have been used for generations to make rigid textiles, such as baskets (page 120) and upholstery (page 404).

Fibre crop is harvested after only one or two growing seasons, unlike wood, which has a much longer growing period. Available throughout the temperate and tropical regions of the world, flowering plants such as grass, rush, bulrush and sedge, do not need much processing to make them usable, and many can be applied straight from harvest.

The stems and leaves of plants are used to make textiles for clothing and packaging; rope; floor covering; upholstery; brushes (see image above, top right) and brooms (for example, sorghum, or broom corn). Many different types of fibrous material are woven into straw hoods for hat making (page 420), including leaves, bark and grass. Even though they are natural materials and more commonly associated with hand techniques, grass is combined with synthetic fibres and converted into technical fabrics (see image above, bottom right, and page 92).

NOTES ON MANUFACTURING

The wood and bark from trees and heavy-stemmed plants must be processed into usable thin sheets or strips. This involves slicing veneers with a large knife, or splitting long strips. Producing veneers on an industrial scale requires heavy-lifting equipment and cutting machinery, which limits processing to centralized factories.

Veneers and other sheets derived from woody materials tend to be quite stiff when dry. When moistened with steam they can be formed with pressure into three-dimensional shapes. This technique is used to make furniture, packaging and even disposable wooden plates (see image overleaf). The minimum bend radius is determined by the thickness of the individual sheet, so many thin sheets may be stacked up and bent very tightly. However, this depends on the direction of bend, because fibrous materials are anisotropic and will bend easily across

Bamboo plate The growing awareness of bamboo's positive environmental impact has led to many new products and applications. Even though bamboo is a renewable and sustainable material, the total life cycle should be considered, from harvest and processing to transportation and disposal.

Embossing wood Used to make bespoke and high-end packaging, as well as to decorate wooden furniture, embossing with a heated metal tool reproduces patterns permanently and cleanly on the surface of wood veneer.

the grain, but less well along the grain (in the direction of growth).

Rigid textiles and sheets of veneer may be painted to produce a weather-resistant and coloured finish. Alternatively, veneer is laminated (see left-hand image on page 485) to combine the properties of high-performance fabrics with the unique and natural qualities of wood.

The surface may be engraved with laser cutting (page 328) before or after colouring. This produces darker tones in wood and is used to apply graphics as well as decoration. Embossing wood with a heated metal tool produces smooth surface graphics with a clean outline (see image above right).

Plants and leaves may be converted into usable material with little processing, and several types are the by-products of farming food crops, which means they can be very low-cost. However, this also means that the quality is variable, and such unpredictability presents a challenge for factory-based production. As a result, many of these materials are processed by hand using techniques that have been practised for generations. So while the raw material may be cheap and readily available, the high labour cost means that high-volume production is limited to countries with low labour costs. This presents the difficulty of maintaining sustainable high-quality living standards and working conditions, while continuing to sell natural products at a reasonable

price. Fair trade schemes promote better prices, decent working conditions, local sustainability and fair terms of trade for farmers and workers in the developing world, and can help to identify the best sources of material.

Fibrous materials are coloured with dyes, which are applied as wood stain, or using techniques similar to fabric dyeing (page 240). The colour will not be as vibrant, because the natural colour of the plant will affect the end result, making it more pastel-like (see image below). Darker colours tend to work best.

MATERIAL DEVELOPMENTS

While the application of these types of fibrous material has evolved very little, they are being processed in innovative new ways to create sustainable and competitive alternatives to plastics.

Packaging has become a significant part of a product's life cycle, because items are shipped between countries and may sometimes be shipped back and forth between production, assembly

and distribution as a result of very large and specialized factories. In the past, locally available and renewable materials, such as leaf and bark, were used to protect food and other items. Plastics have since become the dominant material used in packaging, along with many other applications, thanks to their relative lightness, low cost and ability to be molded into an infinite number of shapes and sizes.

Bioplastics combine the versatility of plastics with the sustainability of fibrous materials. Recently developed PaperFoam is a leading example. It is made up of wood fibres mixed with starch and a small percentage of other materials (see image opposite top). It can be molded and is fully biodegradable at the end of its useful life. It is used or developed for all types of packaging applications, from disposable food packaging to delicate electronics goods.

Fibrous woody materials have properties unlike any other sheet or fibre used in textile making. Inquisitive

Coloured raffia Wood, leaves and grass may be dyed like yarn and textile. Without bleaching, the colour will appear washed out.

designers are continually reinterpreting their rigidity and ability to be shaped using molds or weaving techniques. These properties have been used since ancient times, and with developments in textile construction and yarn technology, designers are finding new and unexpected combinations that use wood, plants and leaves to produce new forms of textile.

These materials are also being explored as potential fibre reinforcement in composites. Modern developments that combine bark fibre with natural or synthetic resin are reminiscent of traditional bark canoes. Development is not as widespread as natural-fibre reinforcement, mainly owing to limited availability of materials. With the correct mix of materials, bark has the potential to create lightweight and high-strength materials with interesting visual and tactile properties unlike any other composite.

PaperFoam PaperFoam is made up of 70% starch, 15% wood fibres and 15% premix (secret ingredient). It is molded into containers and packaging that are 100% biodegradable at the end of their useful life.

SOFTWOOD

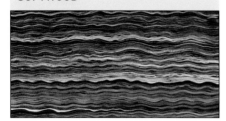

Description

Softwoods are predominantly evergreen and relatively fast growing. They include pine (genus *Pinus*), spruce (*Picea*), Douglas fir (*Pseudotsuga*), cedar (*Cedrus*) and larch (*Larix*). Softwood veneer thickness ranges from 1 to 5 mm (0.04–0.2 in.).

Strength ● ● ● ○ ○

Wood is anisotropic and stronger along its length than its width. It is also prone to shrinking or expanding as it dries or absorbs water from the atmosphere. Therefore, on its own, veneer is not very strong, but depends on the type of wood (higher density equals greater strength). Combining several layers with adhesive creates a material with significantly improved strength to weight.

Resilience ● ● ● ● ○

While the fibres have reasonably good elasticity on their own, veneer is prone to splitting and breaking where they join together. Very thin sheets will be reinforced with a backing fabric.

Quality ● ● ○ ○ ○

They have low resistance to abrasion and dent easily. Softwoods have a distinctive odour that is produced by the resin. Cedar produces oils that protect it against insects. These same oils protect it against decay, meaning that it can be used outdoors untreated (the same as larch). The rest of the softwoods must be impregnated with preservative if they are to be used outdoors.

Colour ● ● ● ○ ○

They have a coarse and straight grain, and are naturally off-white to dark brown in colour. Over time they will change colour. Cedar and larch turn a lovely silvery grey when left outdoors. They may be stained darker colours, such as brown, grey and black.

Versatility ● ● ○ ○ ○

All softwoods are used as veneers, but perhaps the most common is pine, which is used to make laminated furniture. They are used as large sheets, the size of which depends on the tree, or cut into strips and woven (page 90) or laminated. Spruce's light colour and long fibres make it ideal for producing pulp and paper.

Availability ● ● ● ● ○

Coniferous softwoods are grown all over the world, but some species are limited to certain latitudes and altitudes. The rate of growth determines the strength and economic value of the wood.

Cost ● ● ● ● ○

Slower-grown trees have fewer knots and more tightly packed grain. They are harder and more expensive.

Environment ● ● ● ● ●

Wood is non-polluting, biodegradable and can be recycled or used as biofuel at the end of its life. In most cases, the energy used to harvest, convert and transport wood is less than the energy it has stored by photosynthesis. Each 1 m³ (35 ft³) of tree growth absorbs around 0.9 tonnes of carbon dioxide, so replacing materials such as concrete or steel with wood can significantly reduce carbon dioxide emissions. Certification schemes, such as the Programme for the Endorsement of Forest Certification (PEFC) and the Forest Stewardship Council (FSC), verify the flow of wood from forest to factory and end use, which is essential to ensure that the timber comes from sustainable sources.

HARDWOOD

Description

Hardwood materials are derived from deciduous and broad-leaved trees. There are many types including birch (genus *Betula*), beech (*Fagus*), ash (*Fraxinus*), maple (*Acer*), oak (*Quercus*), walnut (*Juglans*), fruitwood, and exotic hardwoods (ziricote, rosewood, teak, mahogany and so on). Common hardwoods such as birch are available in veneers up to 5 mm (0.2 in.) thick. Exotic and rare hardwoods are typically produced much thinner, and range from 0.1 to 1.5 mm (0.004–0.06 in.) thick. Rods are typically from willow (and sometimes hickory). They are harvested in spring or winter when the rods have reached 1.2 m (4 ft) or more in length.

Strength ● ● ● ● ○

Decorative hardwood veneers are typically thin and brittle. Like softwood, it is anisotropic and stronger along its length. There are many different types, from soft and lightweight to dense and heavy. The interlocking grain of certain hardwoods, including types of maple, improves strength. Laminating several layers with adhesive yields the strongest material.

Resilience ● ● ○ ○ ○

Hardwoods have low elasticity and are brittle. Thin veneers are difficult to produce and are typically laminated with a nonwoven (page 152) cellulose backing. Otherwise the fibres are likely to come apart. Rods are more resilient, but will still crease easily unless they are softened with steam. Young thin shoots, known as switches, retain a high level of flexibility.

Quality ● ● ● ○ ○

Hardwoods range from soft balsa to very dense ebony. Quality depends on the type and where on the tree the veneer was taken from. Since hardwood veneers typically only form a thin decorative layer, the most important quality is appearance. For structural applications birch is often preferred owing to its balance of lightness, colour, durability and cost.

Colour ● ● ● ● ○

Natural colour ranges from off-white birch and light-brown oak to dark-brown mahogany, black ebony and even purple exotic hardwood. As a tree matures the heartwood centre darkens, which contrasts with the lighter-coloured sapwood. Figured grains, such as burl, bird's eye and quilted, result in interesting and often highly desirable visual patterns.

Versatility ● ● ● ● ○

Hardwoods are used for decorative and structural applications. The rods are used in baskets and furniture.

Availability ● ● ● ● ○

This depends on the type and location. Oak and walnut grow across Europe, North America and Asia, and the quality and properties vary accordingly. The exotic hardwoods are mainly derived from South America. Figured grain occurs in many types, but certain patterns are more common in specific trees. The phenomenon is natural and uncontrollable, which means availability is low.

Cost ● ● ○ ○ ○

Hardwoods are more expensive than softwoods, as they are grown more slowly and harvested in smaller quantities.

Environment ● ● ● ● ○

With hardwoods, especially exotic types, it is important to avoid using wood that originates from controversial sources, such as countries where deforestation and exploitation take place. If properly sourced, using hardwoods can have a positive impact on people and the environment.

BAMBOO

Description

Bamboo (*Bambuseae*) is a type of grass of which there are many different species and genera. Certain species grow up to 1 m (3.3 ft) per day and reach heights of 30 m (98 ft) or more. It is cut to length and used as cane, split into strips for basket weaving (page 120), or sliced and laminated into veneer. Its properties are comparable with softwood, hardwood and rattan, but it is faster growing and has a straighter grain.

Strength ● ● ● ● ○

Like wood, the grain structure of bamboo means that it is stronger along its length than across its width, and the higher the density the stronger it is. It is also cut and laminated to improve strength.

Resilience ● ● ○ ○ ○

Thin veneers are inherently brittle. Rods are more resilient, but still bend very little unless softened with steam. Once formed into a three-dimensional shape, such as a basket, they bend very little thanks to their high stiffness.

Quality ● ● ● ● ○

Bamboo has a straight and consistent grain. It is lighter and harder than many hardwoods. It has a high sugar content, which makes it prone to insect attack. As a precaution, it is harvested when the sugar content is at its lowest.

Colour ● ○ ○ ○ ○

It is light-coloured, ranging from light yellow to brown. Like wood, it can be dyed. As a natural material there will inevitably be colour variation.

Versatility ● ● ● ● ●

Bamboo, in all its forms (see also Natural Bamboo Fibre, page 483), is extremely versatile and is used to make baskets, furniture, packaging, utensils, and shelter.

Availability ● ● ● ● ○

Due to the growing awareness of the benefits of bamboo there has been an increase in production and export. China is the main producer, but it is also cultivated in North America and Africa.

Cost ● ● ● ● ○

The raw material is relatively inexpensive owing to the low cost of production and the high yield. It can be harvested every four years or so, compared with several decades for many hardwoods.

Environment ● ● ● ● ●

Bamboo is very fast-growing and high-yield and does not require pesticides, herbicides, fertilizer or irrigation for cultivation. Therefore, as veneer or lumber (solid wood), it is considered a renewable and sustainable material.

RATTAN

Description

Rattan is a diverse group of climbing palms that produce vines up to 200 m (656 ft) long and up to 70 mm (2.75 in.) in diameter. The outer bark is removed and the core is dried. Both are used in wicker baskets (page 120) and furniture (page 404) (see also Flowering Plants, opposite). Unlike bamboo, these materials have a solid core.

Strength ● ● ● ● ○

Used as rods and thin strips, rattan retains most of its strength along its length. It has high resistance to splintering, is lightweight and has a similar hardness to many types of softwood.

Resilience ● ● ○ ○ ○

Once dry, the inner core is quite pliable and can be formed around tight bends. It holds its shape well, because it has low elasticity and high stiffness.

Quality ● ● ○ ○ ○

Like hardwood rods and the stems of flowering grasses, the natural quality is variable. It is susceptible to mold, mildew and bacteria. The surface of rattan core is smooth. The core diameter does not vary along its length, because it does not get thicker as the vine gets older. When split into strips, the outside surface, distinguished by the rounded profile of the core, will be smoother and slightly harder.

Colour ● ● ○ ○ ○

Naturally light brown in colour it can be stained or dyed (page 240) a range of darker colours. Colour uptake will depend on the species and whether it is split or whole.

Versatility ● ● ● ● ○

Within the rattan-producing countries it is used for a broad range of applications, from baskets, furniture and musical instruments to temporary structures and staves for martial arts. Exported rattan is mainly used to produce baskets and furniture upholstery.

Availability ● ● ● ● ○

The largest producer is Indonesia, but it grows across Asia. The choice of material will depend on local availability.

Cost ● ● ● ● ○

Rattan is typically not very expensive, but this depends on the source.

Environment ● ● ● ○ ○

This is a potentially renewable and sustainable material. However, it is often sourced from regions where natural materials are exploited, leading to deforestation and negative impact on the local people and surrounding ecosystem. In addition, some companies use potentially harmful chemicals to process the raw material.

BARK

Description

Fibrous bark used to make textile has traditionally been obtained from a variety of trees including conifer (*Cupressaceae* family), palm (*Arecaceae*), birch (*Betulaceae*) and paper mulberry (*Broussonetia papyrifera* species). The type of tree and techniques employed to harvest and utilize the bark depend on the location (see also Cork, opposite). Thickness ranges up to around 3 mm (0.2 in.).

Strength ● ● ● ○ ○

Strength depends on the type, but all barks are stronger lengthwise than widthwise. Backing sheets with a nonwoven (page 152) or with stitch bonding (page 196) or weaving (pages 76 and 90) helps to keep the fibres in place and so make the most of the longitudinal strength of the bark.

Resilience ● ● ● ○ ○

Raw bark is brittle and has low elasticity, except for birch, which is elastic. Beating the bark into paper-thin textile, such as tapa and cedar bark, produces a much more resilient material that can be flexed without breaking.

Quality ● ● ○ ○ ○

This will vary according to the type of tree and method of extraction. The outside and inside will have different properties – the inside surface will be softer – unless it is the inner bark, in which case the two surfaces will be similar.

Colour ● ● ○ ○ ○

Bark is used in its natural colour, which varies according to the type of tree and time of year when it was harvested. Natural colour ranges from silvery-white birch to brown palm. If decorated then it is typically block printed (page 256) with natural dyes in keeping with tradition.

Versatility ● ● ● ● ○

The sheet material is used flat or sliced into strips. Over the years bark has proved to be an important textile used to make paper, clothing, packaging, transportation and shelter. Palm bark is used to make waterproof clothing, such as capes and hats. The supple properties of the mulberry tree are used by drying, soaking and beating the bark – a technique from the Pacific Islands that has almost entirely disappeared – to produce soft and pliable tapa textile. Fibres from the paper mulberry were used to make paper in China and Japan. Birch has been adopted for many different uses ranging from baskets to canoes.

Availability ● ○ ○ ○ ○

Even though the trees that yield suitable bark grow through most parts of the world, there are few people who have the knowhow to harvest it without damaging the tree. Coniferous trees used to produce cedar bark are native to the west coast of the USA. Palm trees are native to the tropics, but they can be found across Europe and North America. The paper mulberry is cultivated across Asia and the Pacific. Many types of birch grow across North America, Asia and Europe.

Cost ● ● ○ ○ ○

Because of the high level of skilled required to harvest bark and the few people capable of doing it, these are expensive materials. Birch and palm are typically the least costly.

Environment ● ● ● ● ●

Harvesting bark can be as sustainable and renewable as softwood if done mindfully. It can be taken from certain trees without causing irreparable damage, if only the outer bark is removed and the inner bark – which transports water and nutrients up the tree – remains intact.

CORK

Description
Sheets of cork are harvested from the bark of the cork oak (*Quercus suber*), which grows across the western Mediterranean. It is available as raw material, compressed blocks, or laminated sheets.

Strength   • • • • •
On its own, cork is not very strong under tension, but is highly compressible. It is combined with synthetics, such as neoprene (page 518) to improve tensile strength.

Resilience   • • • • •
Cork is highly resilient and elastic. The surface is hard-wearing and conforms well under load. Unlike most materials, it does not spread laterally when pressed. It has high impact-absorption characteristics.

Quality   • • • • •
It is lightweight (up to 85% atmosphere by weight) and highly insulating against sound, electricity and temperature. Its low density means it is very buoyant. The surface has high coefficient of friction (grip). Suberin, a waxy substance, is naturally present in the cork and protects it from water penetration and microbial attack. It is resistant to decay and difficult to burn. As a natural material there will be inherent variation in the properties.

Colour   • • • • •
It is light tan, but can be coloured like softwood.

Versatility   • • • • •
It is most widely used as a bottle stopper and an insulating material. It is an important material in the production of shoe soles (page 440), flooring, padding and packaging. It is also used to a lesser extent in composite laminates. Agglomerated sheets are made up of waste cork (typically from bottle stopper production) bound together with natural resins.

Availability   • • • • •
Portugal produces more than half of the world's commercial cork and it remains one of the most important exports for the country.

Cost   • • • • •
For its quality it is relatively cost-effective and comparable with commodity hardwoods.

Environment   • • • • •
Cork is renewable and sustainable. Once a tree is old enough, the bark is peeled every ten years or so. Harvesting the bark does not damage these trees permanently; most other trees will die if their bark is cut around the trunk. At the end of the product's useful life it can be recycled or composted.

FLOWERING PLANTS

Description
Several types of grass (*Poaceae* family, which includes reeds) (see also Bamboo, opposite), rush (genus *Juncus*), bulrush (genus *Typha*), and sedge (*Cyperaceae* family) are traditionally used to make wicker baskets, hats, shoes and furniture. Many grow wild across the temperate and tropical regions of the world, and some are a by-product of farming. Depending on the species, the stem or leaf may be used.

Strength   • • • • •
Stems and leaves kept intact retain their inherent strength, which will be greatest along its length, in line with the direction of fibres.

Resilience   • • • • •
The fibres are long and flexible, but bending around tight corners will result in creases, as they have low elasticity.

Quality   • • • • •
The quality depends on the species with wide variation. Items made from plant wicker can be used outdoors, but they will eventually start to be degraded and so are protected from the elements with paint or varnish.

Colour   • • • • •
Natural colour ranges from light grey, brown and green for living plants to shades of yellow, gold and brown for dried varieties. They may be dyed (page 240) with colour.

Versatility   • • • • •
The stems (culms) are long, up to 1.5 m (5 ft) or more, and typically the most useful part of the plant. Grass has a rigid and hollow stem, except at the nodes where leaves grow in alternating directions; grass stems are often sheathed in leaves; rushes have a solid round core often filled with pith; and sedges are distinguished by a triangular stem. In the case of bulrushes, the long leaves are hand twisted to make strands suitable for weaving. They are turned into baskets (woven very tightly it is possible to make baskets watertight), mats (Japanese tatami are constructed using common rush), ceremonial and decorative items. Hats are created in the

desired shape, or blocked (page 420) from sheets or pre-forms of woven material. In the past plant stems and leaves were commonly twisted or braided to make rope. They are used wet or dry, depending on the desired effect and end use.

Availability   • • • • •
Flowering plants grow throughout the temperate and tropical regions of the world. Certain species are more successful than others – common rush (*Juncus effuses*), common reed (*Phragmites australis*) and sweetgrass (*Muhlenbergia filipes*) – and have spread throughout Europe, North America and Asia. Some types are by-products of farming, such as wheat and barley straw. Others are harvested from the wild, such as along riverbanks. Many are not cultivated because they are used in such low volume, which limits availability.

Cost   • • • • •
These materials are readily available and take very little preparation, making them cost-effective (except for bulrush leaves, which require a high level of skill to twist into usable strands). But this depends on what is available locally.

Environment   • • • • •
These types of plant grow very well, even if heavily cut or grazed. They will continue to grow back and in some cases are considered a problematic weed. Care must be taken to ensure that harvesting them does not affect the local wildlife.

LEAVES

Description
Leaves are harvested from trees and plants to be used as sheet material for apparel and packaging, or rolled up into strands suitable for weaving. The most commonly used leaves come from the banana tree (genus *Musa*), chestnut tree (*Castanea*) or one of many in the palm (*Arecaceae*) family such as the areca (genus *Dypsis*) and raffia (*Raffia*).

Strength   • • • • •
Leaves are reasonably tough as a result of their fibrous structure, but are easily torn lengthways between the fibres. They are comparable to leaf fibres (page 483) and the exact strength will depend on the type of leaf and age of plant.

Resilience   • • • • •
Fresh leaves are flexible, but turn brittle as they dry out.

Quality   • • • • •
Banana and chestnut have a smooth and clean surface. Palm and nettle leaves have a slight texture. They are watertight and suitable for food contact. Some are edible (stinging nettles are used to make soup, for example), but not all.

Colour   • • • • •
Natural colour changes as the leaves dry out, going from green to silvery grey and shades of brown. Certain leaves may also be dyed (page 240) once they have dried out sufficiently.

Versatility   • • • • •
Banana and palm leaves are large and used in many countries to make packaging or wrap food parcels. In parts of Asia and South America banana leaves are popularly used in cooking, because the leaf imparts some sweetness to the food while protecting the ingredients from burning. Certain types of soft cheese are wrapped in chestnut or nettle leaves to provide protection and impart some 'woody' flavour. French Banon cheese is probably the most well-known example and comes wrapped in chestnut leaves tied with raffia. Other uses of raffia leaf include twisted rope (page 60) and woven hats, baskets (page 120) and floor coverings.

Availability   • • • • •
Trees and plants that produce suitable leaves for use in textiles grow in many countries. Availability will be affected by the growing seasons, especially in temperate climates.

Cost   • • • • •
Relatively cost-effective in the countries of origin. The cost depends on the amount of preparation required.

Environment   • • • • •
These materials are sustainable and renewable, especially when fallen leaves are used. At the end of their useful life, these materials can be composted. As with softwood and hardwood, it is important that the source is known. There is no certification scheme dedicated to leaf products, which makes it difficult to ensure there are no negative impacts on people or the environment associated with their production.

Natural Protein Fibres

Natural protein fibres used in the production of textiles are derived from animals. They are made up of amino acids formed into long polymer chains. The genes of the animal define the structure and chemical attachments, which determine the characteristics of the fibre. There are two main groups, which are keratin (wool, hair and fur) and fibroin (secreted). Keratin is derived from mammals, such as sheep, goats, camels and rabbits. Fibroin is obtained from insects, such as moth larvae. The most appropriate is silk derived from the cocoon of the domesticated mulberry silkworm. Created as one continuous strand, it is the only natural filament used in textiles. Several types of natural protein fibre are considered high-value and luxurious thanks to their soft and lustrous qualities.

PROPERTIES OF NATURAL PROTEIN FIBRES

The two different groups of protein fibre, keratin and fibroin, are made up of similar ingredients, but have some contrasting properties owing to the structure of the polypeptide chains and strength of the chemical attachments.

Wool, hair and fur consist of keratin. The fibres are made up of a cortex (core), which consists of a complex layered structure of spiralling microfibrils, wrapped in a protective cuticle. Wool's natural crimp is the result of the bilateral cortex. The two halves have slightly opposing properties and so cause the fibre to twist and bend. This contributes to wool's elasticity and resilience.

Hollow fibres, such as llama and alpaca hair, are often referred to as medullated, or partially medullated if the core does not run all the way along (the hollowness is formed by the deterioration of a specific group of cells that make up the core). The medullated structure does not seem to affect the strength, or affinity for dyeing, but does provide lightweight insulting

properties. These qualities are closely linked to the hair's function of having to keep the animal warm in the extreme climate of the Andes. When used to make textiles, hollow fibres are desirable for both warm and cold weather clothing, because they keep the heat in, or out, depending on the ambient temperature.

The coats of goats, rabbits and camels are made up of at least two types of hair. Guard hair protects the animal from the rain and prevents the fleece from matting. The role of the softer fibres that make up the undercoat is to keep the animal warm during the cold winter months. This is the highly prized part of the coat and the proportion differs according to the animal. For example, the cashmere goat has only a small proportion of very fine undercoat compared to the Angora goat, which has a thick coat consisting mainly of undercoat.

An advantage of keratin protein fibres is that they are inherently flame-resistant. This is in stark contrast to most plant fibres (page 478), which burn readily when exposed to an open flame. However,

unlike plant fibres, they become weaker when wet.

Silk monofilament is one of the strongest natural fibres used in the production of textiles. It is produced by insects – several species of moth larva produce a suitable silk cocoon – by mixing two liquids to form silk fibroin (complex insoluble protein made up of polypeptide chains), which is subsequently squeezed through a fibre-forming duct. As the fibre is formed and stretched, the polypeptide chains become highly organized in a linear fashion. This produces a strong fibre with little elasticity.

The role of the cocoon is to protect the larva during pupation. As a result, it has evolved some beneficially resistive properties, for example against alcohols, weak acids and water. The fibres are bonded together with sericin proteins excreted by the larva. These make up about one third of the cocoon once complete and are removed in the production of silk for textiles (page 46).

Silk's high affinity for dyeing (page 240) and its lustrous surface produce brightly coloured yarn. Combining silk with more durable fibres, such as wool, creates textiles that take advantage of the properties of both (see image right).

TYPES OF NATURAL PROTEIN FIBRE

There are many varieties of wool, hair and fur, the quality of which is defined by the age of the animal, breed, health, diet and local climate. Sheep's wool is produced in the highest volumes and with the widest range of properties. Depending on where the fibre was sourced and how it has been blended (page 242), the quality ranges from very fine to coarse.

Wool is considered high-quality and fibres from the Merino sheep are generally the finest and softest of all the breeds. The high cost of rearing animals compared to cultivating natural plant fibres, and the complex set of processes required to convert wool into usable fibres for making yarn (page 22), make this a relatively expensive material. Therefore, wool is often blended with synthetics and other natural fibres to reduce cost. As a result, high-quality materials are often labelled to indicate the proportion of different ingredients (see image overleaf, top left) and wools from particular breeds, such as Merino, are often distinguished from lesser-quality types.

Goat's hair includes cashmere and mohair. These fibres are generally superior to sheep's wool, though this will depend on the quality and age of the animal. Cashmere is famed for its soft and luxurious hand. These fibres are produced in lower quantities and are more time-consuming, which makes them more expensive.

Fibres obtained from the camelids – llama, alpaca, vicuña and guanaco – are produced in relatively small quantities and are more expensive than fibres from sheep and goat. The vicuña is the rarest of all, because the animal was hunted almost to extinction by the 1960s and so is now governed by strict South American laws.

Camel hair used to make yarns and textiles is produced in Mongolia and China, and is mainly used by local people to make clothes and shelter. It is difficult to obtain a consistent and steady supply.

Silk and wool
Designed by Elaine Ng Yan Ling, The Fabrick Lab, this rug incorporates wool (red) and silk (pink) to create a relief pattern that is enhanced by the different qualities of each material.

Fibres from camels and camelids are often blended with other high-quality fibres to make more from less.

Fibres from younger animals are superior and the first shearing or moulting produces the highest quality possible, because only one end is cut and the other is naturally rounded. The same is true for skins (page 498).

Angora is obtained from the highly prized Angora rabbit. This type of rabbit, of which there are four main breeds, grows a long coat much more quickly than other types. Often kept in isolation to prevent their hair becoming tangled (matted), they are shorn or combed several times a year and the hair is converted into luxurious yarn and textile.

The yield and quality of wool, hair and fur fibres depends a lot on the climate and farming conditions. Silk sericulture, by contrast, has been perfected in many countries and as a result high-quality silk can be produced with relatively consistent results. Of course, being a natural material there is variation, and certain producers are able to create far superior quality filament.

There are tens of thousands of silk-producing insects. As with fur, the characteristics of each have evolved to accommodate specific environmental and protective needs. Several species of moth larva produce a silk cocoon suitable for use in textile production. The most appropriate of these – in terms of both

domestication and filament quality – is the mulberry silkworm.

NOTES ON MANUFACTURING

Each of the fibre-producing animals yields a unique material, which presents different challenges for manufacturing textiles. The biggest difference among them is that wool, hair and fur are staples, ranging from just a few cm (1 in.) up to 0.5 m (20 in.), whereas silk is secreted as one long filament.

Staple fibres are converted into yarn by spinning (page 56). The amount of twist affects the bulkiness, the softness and thus the density of the yarn. Higher twist typically produces a stronger yarn, because there is more friction between the strands.

The outside layer of wool fibre is made up of overlapping cuticle cells. They act like hooks, gripping the scaled surface of other fibres. This quality helps to produce resilient yarn, because the fibres are prevented from slipping over one another. This makes it possible to produce openwork fabrics, such as knitted jumpers

Above
Pure wool label Pure wool items are often labelled to indicate the quality of the product. Wool from well-known breeds, such as Merino sheep, is often identified on the label, because it can be difficult to tell them apart from one another.

Left
Novelty yarn Various configurations of wool are held together with a fine thread and loosely weft knitted into fabric.

Above
Knitted wool jumper Wool has good elasticity, which is enhanced by the crimp of the fibres. This quality is utilized in the production of openwork fabrics, such as this jumper (see also page 135), which conform to the shape of the wearer's body as well as proving warmth and comfort.

(see image above top right), that are durable and hold their shape.

This same quality helps to make sturdy nonwovens (page 152). However, it also means that with use and washing and as the fibres move against one another, they gradually become matted. This causes wool fabrics to shrink and become stiffer.

Fibres with less pronounced scales on the surface, such as llama and alpaca, need to be twisted more tightly, because the surfaces do not grip as well.

The twisted yarns are then plied (page 60) together in twos, threes or more, to make more durable and bulky yarn. As well as combining two or more similar yarns to make a balanced plied yarn;

different types may be wound together to create novelty yarns with a unique look and feel (see image above).

Silk staples are converted into yarn in the same way. Filaments of silk, on the other hand, are reeled from the cocoons and plied simultaneously, cutting out all the preparatory steps as required for staple yarn.

Only about half the fleece sheared from a sheep is usable. The rest includes lanolin (fatty substance), dirt and other contamination. Separating the usable fibres involves several washing processes, called scouring (page 202). Fibre obtained from camelids, such as llama and alpaca, does not contain lanolin and so can be

turned into yarn more readily. This also means the fibres are hypoallergenic and less likely to cause an allergic reaction.

Natural protein fibres take dyes very well to produce long-lasting colour. The highest-quality colours are achieved with silk, thanks to the natural lustre of the material (see image opposite top). The softer and sometimes fuzzier surface on wool yarn means the colours appear more subdued (see image opposite bottom).

NATURAL PROTEIN FIBRE DEVELOPMENTS

There are continual innovations in manufacturing, in particular different types and methods of coating (page 226) to enhance specific fibre properties. For example, CSIRO, in a project funded by Australian Wool Innovation Ltd, developed a way to reduce the surface energy of wool and thus the ability of water to cling to fibres. As a result, the amount of water it can absorb is reduced while the speed with which it dries is increased. This coating technology, labelled QuickDry Merino (QDM), increases the efficiency of washing and drying, and reduces water absorption in wet weather.

There are many significant developments that help to reduce the environmental impacts of using natural protein fibres. Generally, converting wool into yarn is more sustainable than producing synthetic yarn (page 50). However, significant quantities of water and chemicals are required to sufficiently degrease, clean and prepare wool for use in textiles. Haworth Scouring (page 26) has demonstrated that it is possible to produce wool yarn using 100% organic methods and is certified as such by the Soil Association in the UK. Through investment in their production process and close integration with a modern effluent treatment plant, it has significantly reduced the environmental impacts of early-stage wool processing.

Dyeing silk uses chemicals and large quantities of water. In 2001, the Institute of Materials Research and Engineering (IMRE) in Singapore demonstrated a technique for producing brightly coloured silk yarn straight from the caterpillar. By feeding it brightly dyed leaves just before it spins the cocoon, the silkworm produces intrinsically coloured filament, which could potentially eliminate the energy, labour and waste associated with conventional dyeing.

Silk is used in medical applications, such as sutures. It is biologically compatible with human tissue so the immune system does not attack it. The dressing may be removed once the wound has healed, or left to deteriorate naturally. It is being explored for many other medical applications, including drug delivery systems and the creation of scaffolds for bone regeneration.

Silk cocoons consist of approximately one-third sericin, also known as silk gum

Dyed silk Mulberry silk filament has a triangular shape and dyes very well. This results in vivid and saturated colours that shift depending on the angle of view. This quality is emphasized with satin weaving techniques (page 84).

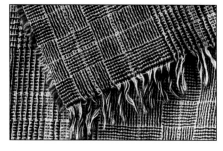

Dyed wool The individual yarns in this Merino wool scarf, produced by Johnstons of Elgin, were package dyed and then woven to create a decorative pattern. The napped finish (page 216) produces a softer colour.

(it holds the silk strands together). During reeling, when the silk is unwound from the cocoon, the sericin is heated so that the bonds become weak enough to pull the fibres apart. Recent developments include combining this by-product with cosmetics, highly effective moisturizers, mixing with polymer foam (page 172) to enhance moisture absorption properties, or simply using it as one part of a copolymer to make plastic film (page 180) or sheet.

The ability to recreate spider's silk has long been targeted as the holy grail of ultra-high-performance material development. It exhibits very high toughness as a result of high strength combined with high elasticity. It is being explored for medical, military and engineering applications. It is not practical to farm spiders, so researchers have been developing other ways to mass-produce or mimic the material. Nexia Biotechnologies made a significant step by genetically modifying a goat to produce the protein in its milk. The protein is extracted and processed in

a way that mimics the way in which spider's silk is formed, to create a superstrong and lightweight filament material they called Biosteel. However, the company went bankrupt and the next major breakthrough was not until 2013, when AMSilk demonstrated a prototype of a spider silk-like material manufactured from recombinant silk proteins (the result of a new combination of genetic material). The new material, also called Biosteel, is superstrong and similar to silk in colour and processing. It has been demonstrated to outperform even the highest-performance synthetic materials (page 510), such as nylon and aramid (also known as Kevlar).

WOOL

Description

Wool is obtained from the fleeces of domesticated sheep (*Ovis aries*). It grows from follicles on the skin and protects the animal from the elements throughout the cold months. The sheep is sheared in the spring, when it would naturally start to moult, to protect it from overheating and insect attach. Sheep are farmed all over the world and each breed yields a wool with its own unique characteristics, ranging from fine to coarse (10–50 microns) and straight to crimped. Depending on the quality and end application, wool is converted into either woollen or worsted yarn (page 56).

Strength   • • • • •

Fibre length ranges from 38 to 114 mm (1.5–4.5 in.). Woollen yarn is made up of shorter fibres and worsted consists of long combed fibres. Therefore, worsted is stronger and more compact. The low tenacity of the fibres is increased by twisting them into yarn or felting. Wool loses about one quarter of its strength when wet.

Resilience   • • • • •

Wool has a natural crimp. The amount depends on the breed of sheep. This contributes to the excellent elastic recovery of the fibre, making it very resilient and relatively wrinkle-free.

Quality   • • • • •

Worsted yarns are fine and compact. Woollen yarns have low twist and more loosely aligned fibres and so tend to be softer and bulkier. The quality is also influenced by the breed, upbringing, climate and shearing. The first shearing (lambswool) produces the finest and softest wool, because one end remains naturally rounded. Wool has very good drape, a quality emphasized by its elasticity. It is water-resistant and absorbs up to about 35% of its weight in water without feeling wet; fine wools effectively wick moisture away from the skin; and it is a good insulator (traps air). It is fire-resistant, in that it will burn in a flame but self-extinguish when no longer in it. It is vulnerable to insects and shrinking, and so is often coated (page 226) to improve durability.

Colour   • • • • •

The natural colour of wool ranges from creamy white to dark brown or grey. It is often woven (pages 76–105), knitted (pages 126–51) and felted into fabrics that utilize the broad spectrum of inherent colours. It is also possible to bleach and dye (page 240) wool a range of colours.

Versatility   • • • • •

Wool is available in many forms, including virgin (new), reclaimed or salvaged (wool remanufactured into new products, such as nonwovens used in upholstery, page 404), lamb (sheared from six- to eight-month-old lambs), hogget (first shearing from an older sheep) and wether (all subsequent shearings). The different types and qualities mean wool is utilized in applications that range from very fine suits and technical outdoor clothing to hard-wearing carpet and upholstery. Its natural affinity to felt is a quality used to make nonwovens, but the downside of this is that all wool fabrics have a tendency to become entangled and thus shrink when washed.

Availability   • • • • •

Wool is produced in many countries including North America, Australia, New Zealand, the UK and Argentina. Merino is considered the highest quality thanks to its balance of strength, elasticity and fineness.

Cost   • • • • •

Rearing animals is relatively expensive compared to cultivating plants. There are many different qualities of wool, including blends, ranging up to very expensive. Lower-cost synthetic, cotton and plant fibres have replaced wool in many applications. As a result, production has been in decline since around 1990.

Environment   • • • • •

Depending on the source, wool can be ethical, sustainable and renewable. Even so, the production of wool requires that sheep are reared on farms and protected with medicine and pesticides, harmful chemicals are used in the production of the yarn and in some countries animal welfare is an issue.

CASHMERE

Description

Cashmere is the fine, soft and straight hair obtained from the neck region of the undercoat of the cashmere goat (*Capra hircus laniger*). It is very fine, typically no more than 18 microns, which is comparable with superfine Merino. It can be finer, but these fibres are extremely rare.

Strength   • • • • •

Staples are typically around 50 mm (2 in.) long and very fine, which means they have relatively low tenacity.

Resilience   • • • • •

Cashmere is no more elastic than sheep's wool, but may be abraded more quickly as a result of its softness.

Quality   • • • • •

Cashmere is a lightweight luxury fibre graded according to its strength, colour and fineness. The finest-quality fibres are not contaminated with coarser hairs from the surrounding fleece. High-quality cashmere has excellent drape and a very soft hand. Like sheep's wool it is water-repellent and fire-retardant.

Colour   • • • • •

It is naturally white, grey or light brown. Similar to sheep's wool, it may be dyed a range of colours.

Versatility   • • • • •

Only the fine hair from the undercoat is classified as cashmere. It is used to make a limited number of fine-quality items such as suits, sweaters, scarves, socks and blankets. Cashmere shawls are also known as pashmina (traditionally from northern India). The lower-value coarse hairs – typically around 80 microns in diameter – from the rest of the fleece are used to make padding, among other things.

Availability   • • • • •

China is the largest producer (rearing the goats as well as buying in fleeces from other countries, sorting, dehairing and converting into high-quality fibre), followed by Mongolia, Turkey, Tibet and Iran. Each animal typically yields 150 g (5.3 oz) of cashmere per year.

Cost   • • • • •

The fine hairs of the undercoat make up only about one quarter of a typical 125 g (4.4 oz) fleece. They take time and care to extract – the highest-quality fibre is obtained from the goat by combing over one or two weeks during the moulting season – making this an expensive fibre.

Environment   • • • • •

Cashmere goats are reared like sheep. Depending on the source, the fibre can be ethical, sustainable and renewable. However, cashmere is produced in many countries where the animals' welfare may not be properly protected.

MOHAIR

Description

This fibre is obtained from the fleece of the domesticated Angora goat (*Capra hircus ancryrensis*). It has a similar diameter to sheep's wool, but is smoother, stronger and more resilient. The goats are typically shorn twice a year to produce staples 100 to 150 mm (4–6 in.) long and 25 to 40 microns in diameter. The fineness depends on the age of the goat.

Strength ● ● ● ○ ○

Mohair is significantly stronger than wool for the same weight.

Resilience ● ● ● ● ○

The natural wavy crimp adds to its durability and resilience.

Quality ● ● ● ● ○

The lack of scales on the shaft of the fibre, unlike sheep's wool, means the fibre is less likely to shrink and felt in use. This also means the surface of the fibre is less likely to attract dirt and other contamination. It is not as soft as cashmere and other luxury wools. Fibres around 25 microns in diameter are considered the best quality and obtained from goats within their first three shearings. After that, the fibres become coarser and thus less valuable. The whole fleece is used to make yarn. This means that some of the guard hairs (known as kemp) and hollow (medullated) fibres will be mixed with the finer-quality undercoat. These other fibres do not have such good affinity for dyeing and tend to stick out, because they are stiffer and shorter, and so will be visible in lower-quality fleece.

Colour ● ● ● ● ○

The high-quality undercoat is made up of uniform and solid fibres and so yields superior and even colour when dyed. Incorporating kemp and medullated fibres reduces the overall quality. Mohair has high affinity for dyes and can be produced in a range of colours with excellent fastness.

Versatility ● ● ○ ○ ○

Its high strength and highly insulating properties are used in garments, hats and upholstery. It is also used as a substitute for exotic hair. Mohair pile (see Pile Weaving, page 96, and Weft Knitting, page 126) is very durable with a lustrous finish. It is frequently blended with superfine synthetic yarn to produce high-quality and insulating garment fabrics.

Availability ● ● ○ ○ ○

The USA, South Africa and Turkey are the biggest producers of mohair. Each animal yields 3 to 5 kg (6.6–11 lb) of fleece per year.

Cost ● ● ● ● ○

Mohair is typically a little more expensive than sheep's wool, but this depends on the quality. Almost the entire fleece is used to make yarn, increasing the yield per goat compared to cashmere. Longer staple fleece and finer fibres are the most expensive.

Environment ● ● ● ● ○

Angora goats are farmed like sheep, but on a smaller scale. Those reared by small-scale farmers tend to be better looked after than the more industrially farmed animals. Goats raised in better conditions will produce higher-quality mohair.

LLAMA AND ALPACA

Description

Llamas (*Lama guanicoe glama*) and alpacas (*Lama pacos*), along with vicuñas and guanacos, are camelids. Originally from South America, they are related to the camel. Their long, soft and durable fleeces are sheared or combed every year or two and the fibres are used in the production of luxurious textiles. They are lightweight and highly insulating as a result of having to protect the animals from extreme temperatures high up in the Andes.

Strength ● ● ● ○ ○

The fibre is lighter and stronger than sheep's wool.

Resilience ● ● ● ○ ○

The fibres have less pronounced scales on their surface and so are twisted more tightly to make yarn. This produces stiffer fabrics with less drape. Therefore, it is often blended with fine-quality wool to improve the shape and produce a soft hand.

Quality ● ● ● ● ○

The fibres from these animals are partially hollow, making them lightweight with good insulation properties. They range from 15 to 40 microns in diameter. Finer hair is considered higher quality and hairs over 30 microns can be itchy against the skin. Alpaca is generally considered superior to llama, but this depends on the herd, fleece and proportion of guard hairs to undercoat. Suri alpaca is superior to huacaya alpaca fibre and is more suited to producing worsted (combed) yarn. Generally, the fibre is softer and finer than cashmere. Fibre sheared from young animals (crias) is the highest quality of all. The scales are not as pronounced as on wool, making it more comfortable against the skin and less likely to felt (although it can still be felted). It is naturally antimicrobial and unlike sheep's wool, it does not contain lanolin, which means it is lighter as well as being hypoallergenic. It is also fire-resistant.

Colour ● ● ● ● ○

Alpacas have a very wide range of natural colours – there are five main colours divided into 22 official shades – from creamy white to black and including shades of grey and brown. Lighter colours have a good affinity for dyes. Both natural and dyed colour has excellent fastness. Thanks to the less pronounced scales this fibre has better lustre than wool (suri is the most lustrous of the alpacas). Llamas were not traditionally reared for producing fibre and so have been bred in fewer colour varieties.

Versatility ● ● ● ● ●

As a result of the different properties, they are used for a range of textiles, from very fine garments to coarse and durable ponchos and rugs. High-quality fibre is used to make lightweight clothing, including suits, sweaters and jackets.

Availability ● ● ○ ○ ○

Each alpaca yields about 3.5 kg (8 lb) of fleece per year. Peru is the main producer, but herds are being established in the USA and other parts of the world. Suri is more difficult to come by, because there are far fewer of this breed.

Cost ● ● ○ ○ ○

These fibres are not widely available, are labour-intensive to produce and are desirable, which makes them quite expensive. Price ranges significantly according to the quality of the fibre.

Environment ● ● ● ● ○

The absence of lanolin means that much less cleaning and no chemicals are required to produce the finished fibre, which is in contrast to the preparation required for greasy wools (see Wool and Hair, page 22). They are produced in relatively few areas of the world and so the fibre is often shipped over great distances. Animals that are well kept will yield better quality fibre, because stress can have a impact on the quality of the hair.

VICUÑA AND GUANACO

Description

The vicuña (*Vicugna vicugna*) and guanaco (*Lama guanicoe*) are rare camelids that were exploited almost to extinction. Today, the fibre is obtained from farmed stocks as well as from wild herds in South America. Strict laws have been put in place to protect wild vicuña.

Strength ● ● ● ● ○

For soft and luxurious fibres they are surprisingly strong and very lightweight.

Resilience ● ● ● ● ○

They have good elasticity and resilience.

Quality ● ● ● ● ●

Vicuña fibre is exquisite, very fine and around 12 microns in diameter. It is highly insulating and soft – it protects the animals from the very cold weather high up in the Andes. The coarser guard hairs are typically removed from the shorn fleece by hand. Guanaco fibre is coarser – around 15 microns – and considered lower quality, but is still a luxury fibre comparable with cashmere.

Colour ● ● ● ● ●

Vicuña ranges from fawn to golden brown and guanaco is naturally off-white to brown.

Versatility ● ○ ○ ○ ○

It is so expensive and rare that it is generally reserved for only the finest-quality garments, such as suits and neckwear.

Availability ● ○ ○ ○ ○

There are a small number of producers and relatively few animals, each of which yields only around 225 g (8 oz) of fleece per year. Vicuña is probably the most difficult natural fibre to come by. Guanaco yields more than twice the amount of usable fibre per year and is not considered an endangered species like the vicuña.

Cost ● ○ ○ ○ ○

Vicuña is highly prized and produced in very small quantities, making it one of the most expensive fibres. Guanaco is cheaper, but still relatively high-cost.

Environment ● ● ● ● ●

The desirability of these fibres meant the animals were exploited almost to extinction in the 1960s. Since then, strict laws have been put in place to protect the animals from hunting and to ensure the fleeces are obtained in a renewable manner. However, even with laws in place, animal welfare may still be a concern.

CAMEL

Description

Camel hair is gathered by hand from the two-humped Bacterian variety (*Camelus bactrianus*) during the moulting months, which run from spring to summer. Hair from other types of camel may also be used, but is of inferior quality. The guard hairs can be up to 40 cm (15 in.) long, while the insulating undercoat is 19 to 24 microns thick and up to 125 mm (5 in.) long.

Strength ● ● ● ○ ○

The guard hairs are strong and the undercoat is softer and more delicate. As the animal gets older its hair becomes coarser and stronger.

Resilience ● ● ○ ○ ○

Like wool the fibres have crimp, although it is less pronounced on camel hair, which means it has some elasticity but not a great deal.

Quality ● ● ● ● ○

The guard hairs are coarse and waterproof, and the undercoat is very soft and highly insulating. The fibres are lightweight. Blending with wool will improve the hand and silk improves the drape.

Colour ● ● ● ○ ○

It is naturally tan and is not usually dyed, but will take colour as well as wool.

Versatility ● ● ● ● ●

From the guard hairs to the soft undercoat, it is used to make a wide range of textile products from tents (Mongolian yurts) and carpets to fine suits and hosiery. As an expensive fibre it is often blended to reduce the cost while maintaining many of its desirable properties, and dyed the natural tan colour of camel hair.

Availability ● ○ ○ ○ ○

The largest producers are Mongolia and China. The herders and local people use it to make clothing and shelter, and not much is exported, making this a rare fibre. Each animal yields up to 10 kg (22 lb) of textile fibre-making hair per year and about one third of this is used in apparel.

Cost ● ○ ○ ○ ○

It is exported in low quantities and is expensive.

Environment ● ● ● ● ○

The hair is collected by hand and is a by-product of keeping the camels that are used for transporting people and goods around the desert.

YAK

Description

Yaks (*Bos mutus*) live mainly in the Himalayan region of south-central Asia and in the wild they are endangered. The hair feels similar to cashmere and camel, but has not reached the same level of popularity, although local weavers make luxurious fabrics from it. Depending on the age of the animal, fibre from the soft undercoat ranges from 15 to 20 microns in diameter and is up to 50 mm (2 in.) long. The guard hairs are longer and coarser.

Strength ● ● ● ● ○

Yak hair is relatively strong for a keratin fibre, although it loses some of its strength when wet.

Resilience ● ● ● ● ○

The natural crimp means yak fibre is elastic and resilient, similar to sheep's wool.

Quality ● ● ● ● ○

The quality depends on the animal, how old it is and the local climate. The best-quality fibre comes from young calves. Yak undercoat is soft, insulating and lustrous. Fibres taken from the undercoat are solid (unmedullated). The guard hairs are partially medullated (hollow) and coarser, so they are removed prior to spinning. Like sheep's wool, yak hair contains lanolin, but in much smaller quantities. The fibres have less pronounced scales, making them smoother and less prone to felting and shrinking than sheep's wool.

Colour ● ● ● ● ○

Natural colour ranges from white to grey, brown and black. White is the most scarce and expensive. It is suitable for dyeing (page 240), but bleaching affects the hand.

Versatility ● ● ● ● ○

Coarse guard hairs are spun to make rope (see Plying and Twisting, page 60) and hardy fabrics (nomadic tents and blankets). The softer undercoat is converted into a luxurious yarn to make garments and accessories. Outside its native countries, Tibet and Mongolia, yak fibre is often blended with other luxury fibres, such as cashmere and silk, to reduce cost.

Availability ● ● ○ ○ ○

Farmed yak produces around 900 g (2 lb) of soft undercoat each year. Most of it is used locally and only a small amount exported as raw fibre. Most is exported as finished products.

Cost ● ● ○ ○ ○

The fibre is combed from the animal by hand during the natural moulting season, and relatively low quantities are available, making this an expensive material. It is possible to shear the fleece from the animal, but this will produce inferior-quality fibre due to the higher number of guard hairs.

Environment ● ● ● ● ○

Yak fibre is typically a by-product of farming yaks for milk and meat. Yaks are also commonly used as draught animals for carrying people and goods. Sustainability depends on how the animals are looked after.

ANGORA

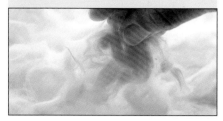

Description

Angora is a hollow and very lightweight fibre obtained from the domesticated Angora rabbit (*Oryctolagus cuniculus*). There are four main breeds, English, French, Giant and Satin. Angora is often confused with mohair from the Angora goat because of the name, but the two fibres produce quite different textiles. Angora is much softer and finer – typically only 10 to 15 microns thick.

Strength ● ● ○ ○ ○

Angora is fine and not very strong. Combining the soft fibres into a plied yarn (page 56) improves strength, but it is best to blend them with higher-strength and fine-quality wools.

Resilience ● ● ○ ○ ○

Pure angora has low elasticity, abrades easily and felts readily owing to its fineness and softness. Blending helps to overcome this.

Quality ● ● ● ● ●

Angora is a high-quality and luxurious fibre. It is characterized by its lightness, softness and highly insulating properties. It is used to make very warm clothes, or blended with other fibres to impart some of these properties. The quality depends on the breed and the proportion of topcoat (guard hairs) to undercoat. More guard hairs produce a silkier-looking coat, whereas a higher proportion of

undercoat produces a bouncier material with greater volume. The best-quality material comes from rabbits with a balanced mix of fibres. It also depends on where the fibres are taken from: the back and side of the rabbit produce the finest, softest and cleanest fibre.

Colour ● ● ● ● ○

There is large natural colour variation from white through various shades of grey, gold and brown to black. Mixed-colour coats produced mottled yarn. The fibre may also be dyed (page 240).

Versatility ● ○ ○ ○ ○

The soft fibre is relatively limited in application owing to its high cost and low resiliency. It is typically used to make woven and knitted garments and neckwear.

Availability ● ○ ○ ○ ○

It is produced throughout Europe, North America and Asia in relatively low volume.

Cost ● ○ ○ ○ ○

The fur is collected by plucking, combing or shearing. Plucking and combing are manual processes that produce the highest-quality fibre but are more time-consuming and so more expensive. Shearing produces higher yield but still only up to 400 g (14 oz) per year. The coats have to be combed regularly to prevent matting.

Environment ● ● ● ○ ○

Angora fibre may be renewable and sustainable, but this depends on the welfare of the animals. Care must be taken when sourcing Angora from countries that do not enforce adequate animal welfare laws. In these cases, the sustainability will depend on the individual farmers and the way in which the workers look after and shear the rabbits. They are often kept in separate cages to avoid their coats felting, and in near darkness.

SILK

Description

Silk obtained from the mulberry silkworm (*Bombyx mori*) is the only natural filament commonly used in textiles. There are other types of silk-producing insect, but this type is the most widely cultivated (sericulture) (see Silk, page 46). After around one month the mature caterpillar takes three to four days to produce a continuous filament around 1,000 m (3,280 ft) long, encapsulating itself in a protective cocoon.

Strength ● ● ● ● ●

It is a very fine solid filament, measuring around 10 microns in diameter with very high strength to weight, making it one

of the strongest natural fibres. However, it loses some of its strength when wet.

Resilience ● ● ○ ○ ○

Silk has low elasticity and does not recover well after stretching. This is not usually a problem, because silk is used in decorative applications that are not put under significant strain. The surface has moderate abrasion resistance.

Quality ● ● ● ● ●

Silk is lightweight with good insulating properties. As a natural material there is some variability. It is up to the operators unravelling the cocoons to combine filaments into a uniform plied yarn (page 60). Continuous filaments are used to make the highest-quality fabrics. Staple yarns harvested from broken cocoons – either from the wild or from caterpillars that have been allowed to metamorphose into moths – or from leftovers, produce lower-quality and less expensive yarn (see Staple Yarn Spinning, page 56). Silk staples are often combined with other fibres during spinning to impart some of the softness and strength of silk.

Colour ● ● ● ● ○

The natural colour of farmed silk ranges from white and yellow through brown and grey. Mulberry silk has a triangular, prism-like structure, which reflects light from all directions to give a bright and lustrous appearance. It has

very good affinity to dyeing (page 240) and is produced in a wide range of vivid colours. Wild silk is used in its natural colour, such as shades of brown, yellow and green, or is bleached and dyed. It has a flatter cross-section and so appears duller than farmed mulberry silk.

Versatility ● ● ● ● ○

The decorative qualities of this filament are utilized in satin fabrics (page 84), embroidery (page 298) and velvet (page 84) for fine garments and interior fabrics. Its high absorbency and insulating properties mean that it is comfortable and lightweight in garments designed for both cold and warm weather conditions.

Availability ● ● ● ● ○

The majority of silk is produced in China, and has been since ancient times. However, the highest-quality silk comes from Italy and France.

Cost ● ○ ○ ○ ○

It is a relatively expensive fibre as a result of the high labour costs and complexity of production. Filament yarn is the most expensive. It is less expensive than rare luxury fibres, such as vicuña.

Environment ● ● ● ○ ○

To make a continuous filament of silk the caterpillar must be stifled with heat before it can turn into a moth and break out of the cocoon. Using farmed or wild staple silk means the caterpillar has been allowed to grow into a moth (although some staple comes from the waste of cultivated filament). Used without dyeing, these yarns have very low environmental impact.

Leather and Fur

These natural materials have been used since ancient times to make clothing and shelter. Tanning is the process used to convert the fresh and biodegradable skin, or pelt, into leather and fur. It is a complex and lengthy process, which contributes to the relatively high cost of these materials. Leather ranges from tough and hard-wearing to soft and strong, but it has a tendency to dry out over time and fur sheds its hairs if not well looked after. Traditionally obtained from wild animals hunted for their meat, most leather and fur now comes from animals raised on farms. While desirable to some, others object to keeping animals for use in textiles, or any other commodity. Unlike all other textiles – except silk filament – leather and fur are obtained from animals that have been slaughtered.

PROPERTIES OF LEATHER AND FUR

The skin of mammals, birds, fish and reptiles is suitable for tanning (page 158), but the processes and properties will vary. During tanning, chemicals form cross-links between the fibrous bundles of protein to create a durable and long-lasting material.

Hides come from large mammals, such as cows, pigs, sheep and deer. They are either left with the hair on and used as fur, or converted into sheets of leather, in which case the skin is split into layers to make it more flexible. The proportion of fat, collagen and glycosaminoglycan (complex polysaccharides), and the compactness of the fibrous weave, determined by the type and age of animal, affect the ultimate quality of the leather. For example, cow is tough and harder-wearing than sheep, but not as durable as goat. Skin from younger and smaller animals tends to be softer and suppler. For example, horse leather is stiff and impractical, but pony is supple enough for apparel and bags. The quality also depends on lifestyle, such as diet,

health and damage (fighting or insect attack, for example).

Hides are split into layers. The first layer, which is the outside surface and has all the appearance of the original skin with grain and hair follicles, is called 'full grain'. 'Top grain' refers to leather from the top layer that has been shaved and buffed to produce a uniform finish free from natural imperfections.

The inner layers are known as 'split' leather. They are not as smooth as top grain and tend to wear less well, because the fibres have been cut through. They are usually embossed or buffed to create a more durable finish. Suede is produced from split leather, which is napped (page 216) to pull up the fibres.

Fur, hair and wool are made up of keratin (see also Natural Protein Fibres, page 490). Each fibre consists of a core encircled by macrofibrils and wrapped in a protective layer of cuticle. The fineness, crimp and elasticity affect the material's softness, resilience and insulating properties. Colour and lustre are affected by fibre smoothness, pigmentation and

hair density. Most farmed fur-bearers have been selectively bred to produce the colours and patterns we are familiar with.

Moulting is the natural shedding of old hair to make way for new growth. It is most noticeable in animals that grow different coats in the winter and summer. The warmest and most desirable furs come from animals that live in cold climates, such as rabbit and chinchilla, fox and mink. Their winter coats are dense and insulating.

Fur is made up of two main types of hair, which grow from different follicles: guard hair and ground hair (undercoat). The guard hairs are typically long, straight and hollow. They protrude from the undercoat to create a layer of protection against rain, dirt and sunlight, as well as providing much of the visible colour. The undercoat is soft, dense and water-shedding. Its role is to keep the animal dry and warm by trapping a layer of air next to the skin (see image opposite top). The difference between these two is not always clear, because some fibres will have the properties of both. The

Below
Snakeskin Reptile
skins, such as this viper
pelt, are prized for their
exotic-looking patterns
and surface texture.

Below
Snakeskin Reptile
skins, such as this viper
pelt, are prized for their
exotic-looking patterns
and surface texture.

Right
Fur The stiff upright
guard hairs are
easily distinguished
from a raccoon's soft
undercoat. They are
longer, straighter and

often coloured. The
undercoat is short and
dense and the hairs
are a little wavy, which
helps them to interlock
and keep the animal
warm during winter.

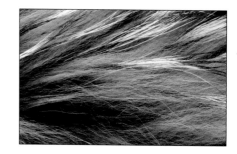

proportion, length and density affect the quality of pelt.

A densely packed and fine undercoat feels soft and luxurious. Chinchilla grows tens of hairs from each follicle to create a highly insulating coat, which in turn produces a particularly soft and desirable pelt. Long and brittle guard hairs can detract from the downy undercoat. In some cases, the guard hairs are plucked or sheared to produce a more uniform and velvety material. Some furs, such as sable, feel soft in all directions, while others have hair that lies flatter on the skin and so only feel soft in the direction of grain.

The fleeces of sheep (and Angora goats) are dominated by soft wool. They have proportionately fewer straight and brittle hairs, or kemp. The fibres range from fine to coarse, and straight to wavy, depending on the breed. Fine wool is more resilient and can withstand wear and tear without breaking. Merinos produce the finest wool and as a result their fleeces are the most expensive. Sheepskin is soft and not very strong compared to cow, pig and pony. It is traditionally used to make chamois leather. It is commonly found with the hair on (see Leather Tanning, case study on page 166), and is used for its highly insulating and comfortable properties.

By contrast, the hides of cows, horses and most goats are covered with mainly straight, short and brittle hair. Without the dense undercoat the hairs lie flat against the skin to produce a smooth surface with a distinctive grain. Breeds that have evolved in cold climates have developed a softer and more pronounced undercoat, such as buffalo, cashmere goat and Shetland pony.

Ostrich is the only bird commonly used to make leather. It is one of the strongest and most flexible leathers and so is desired as much for its durability as for the distinctive markings on its surface.

While it is practical to use ostrich leather for all types of applications, it is much more expensive than commodity leathers (cow and pig), because it is produced in significantly less quantity, and so its use is limited to only the highest-value items.

Fish and reptile skins are converted into leather using similar tanning processes to mammal hides, but they require a lot more work in preparation and this is often done by hand, making these types of materials much more expensive. For its thinness and lightness, fish leather is reasonably strong and flexible. Reptile skins are tough, with a hard-wearing outer layer of overlapping scales. They are mainly used for their decorative colour, pattern and texture (see image above).

TYPES OF LEATHER AND FUR

The skin of almost any animal may be turned into leather or fur, from moles to buffalo. This section covers the types most commonly used for fashion and interiors. Several are not included, because they are banned, endangered or

not commonly processed into leather of fur. Sealskins were popular in the early 20th century, both for traditional uses by people native to their territory and for fashion and textiles. Extensive hunting led to massive depletion in seal numbers and now it is illegal to kill seals in most countries. The vicuña and guanaco, two South American llamas, suffered a similar blow to their population and were virtually extinct by the 1960s (page 496). The desirability of beaver's pelt (genus *Castor*), which is one of the softest furs, led to overexploitation and massive depletion in numbers living in the wild. As a result, strict laws were enforced to help protect the animal and it has been observed that re-established beaver populations have helped to reinvigorate entire ecosystems. Nowadays the pelts are still traded, but they are a rarity.

Beavers and other rodents from cold climates have been so highly sought after in Europe that trading their fur, such as to make felt hats (page 416), fuelled the development in North America of cities including New York, Edmonton and

Winnipeg. Russia was also an important source of these pelts and trappers established many new settlements in the far north of the country.

Leather products made today are distinguished as either commodity (mammal) or exotic (ostrich, fish and reptile). Those produced in the largest quantities and that are by-products of meat production, such as cow and pig, are the least expensive. In some cases, the skin is taken from younger or smaller animals, such as calves and ponies. It is much softer and considerably more desirable. These skins are still a by-product of meat production, but come from stillborn or young animals slaughtered to make certain meat products. For example, the finest calfskin comes from animals that have fed on milk alone (see image below). As the cow gets older and its diet changes the hair becomes coarser and the leather tougher.

Fur is divided into fleece (sheep), hides or skins (for example cows, goats and horses) and furs (mink, chinchilla and so on). Fleeces and hides or skins are mainly the by-product of meat production and the quality ranges considerably depending on the source and animal. For example, lambswool is short and very soft, because the fibres have not been sheared (see page 502). However, as a by-product and thus a relatively low-value item, the quality of the product depends greatly on the abattoir and tannery. As they become

more valuable, then there is a possibility that meat becomes a by-product of fleece.

Furs are either a by-product of meat (rabbit) or taken from animals farmed solely for their fur (fox, chinchilla, mink, nutria and sable). Some furs come from the wild, but this is becoming increasingly rare and now accounts for only around 15%. Strict laws are in place to protect wild animal populations and their welfare.

NOTES ON MANUFACTURING

These unique materials are processed separately from other textiles until they are finally assembled into garments, upholstery, gloves and linings. Tanning is complex and potentially very polluting. Considering the source is therefore critical to ensure that the materials do not have negative environmental impacts.

After tanning leather is dyed through a process of retanning. All leather that goes through chrome tanning is dyed, otherwise it would have a light-blue tint. Leather and fur used in its natural colour has been treated without chrome, for example using synthetic or natural tannins extracted from vegetable matter.

There are several different finishes including aniline, napped (suede and nubuck), semi-aniline, pull-up, pigmented and patent. Aniline means the natural appearance of the surface has been retained – it is dyed but left uncoated (see image below). Napped finishes, which include suede and nubuck, are

also uncoated. The difference between these two is that whereas nubuck is produced by abrading the topside, suede is created on the inner surface. Therefore nubuck tends to be more durable. The other finishes include a coating, thereby concealing blemishes and other imperfections – semi-aniline employs a protective topcoat; pull-up leather is oiled or waxed to encourage the surface to develop a unique patina; pigmented leather has a durable plastic coating; and patent has a high-gloss finish (typically plastic-coated, see page 226). Full grain is typically finished as either aniline or semi-aniline.

Designers have explored many other finishing processes on the surface of leather, with varying results, such as laser cutting (page 328), screen printing (page 260) and digital printing (page 276). It is also possible to carve the surface of leather to produce three-dimensional effects, or create patterns by embossing or punching.

During the tanning process, after the flesh and fatty tissue has been removed, the decision is made whether the skin will be converted into leather or fur. Producing leather requires that all the hair be removed. If the fur, or fleece, is left on, then tanning follows a slightly different route. After tanning, the skin may be slimmed down to make the hide softer and more pliable. It is finished by tumbling in absorbent powder, such as sawdust, to remove any leftover grease, tossing in a mesh cylinder to remove the powder and lifting the hair into place. The best furs are used with their natural colouring and inherent properties, because dyeing reduces the quality.

The skins of fish and reptiles have hard and scaly surfaces. Techniques have evolved to accommodate the more difficult nature of the skins, such as making them more flexible or removing the scales. They may be used in their natural state or coloured with dye and spray, depending on the desired finish. After tanning, fish leather is often coated to ensure adequate durability of the surface.

As a result of the natural variation in size and shape, leather and fur items tend

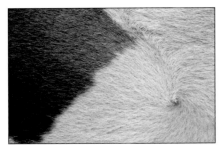

Calfskin Calfskin is not quite as tough and durable as cowhide, but it is very soft and pliable, making it a desirable material for boots and handbags.

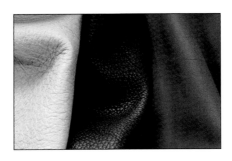

Aniline leather Leather is dyed during the tanning process. A range of bright colours can be achieved, such as on peccary (wild pig, left), cow (middle) and deer (right).

to have more seams than products made from woven or knitted fabrics. Stitching (pages 342–69) is used in the production of items larger than a single pelt, or for linings. The softest furs tend to come from small animals, such as chinchilla and mink. Therefore, stitching techniques have evolved to convert small pieces into larger garments without creating obvious seams on the fur side.

Owing to the natural irregularity of the materials, they are rarely joined by adhesive bonding or other processes that rely on two uniform surfaces. Having said that, leather shoes are adhesive bonded, as well as stitched, to provide optimal strength to weight (page 430).

Strips of leather or fur are combined into larger pieces by weaving (see image right top). This technique is used to produce lighter and less expensive items such as hats, scarves and jackets.

LEATHER AND FUR DEVELOPMENTS

Organizations such as the Coalition to Abolish the Fur Trade (CAFT), and People for the Ethical Treatment of Animals (PETA), have campaigned publicly and tirelessly against the use of fur in fashion and textiles. With the help of many other groups they have persuaded several large retailers to stop selling fur altogether. In Australia and the UK it is illegal to farm animals solely for their fur, while in some countries and communities fur and leather provide vital clothing and shelter, and may be an essential source of income. It is crucial that the source is well understood so as to ensure that when purchasing these types of product the animals have been humanely raised and killed.

In the quest for the production of increasingly cheaper pelts, there are countless examples of farming practices that would put off even the hardiest leather and fur lovers. Buying products that have been certified helps to ensure that animals have been treated properly. For example, the Humane Care Certification scheme by the Fur Commission USA was put in place to protect animals farmed for their fur; the Agreement on International Humane Trapping Standards (AIHTS)

Top
Woven mink fur
Woven strips of fur have a distinctive crisscross pattern. Less material is required, making the fabric lighter and less expensive than using whole pelts.

Above
Synthetic fleece Low-cost synthetic imitations, such as viscose, polyester and acrylic, are knitted and finished to resemble fleece or fur. They are lighter and easier to care for than the real thing.

Above
Dyed fur Fur (keratin) and skin (protein) are chemically different and so can be dyed separately (two colours), or together (one colour). Dual colour fur is achieved by dyeing and spraying to recreate natural-looking markings. High-quality furs are used in their natural state, as dyeing can affect the quality.

sets standards for wild animal trapping techniques and the International Trade in Endangered Species of Fauna and Flora (CITES) is set out to protect endangered and monitored species.

As a result of the relatively high cost of leather and fur, together with the animal rights issues, imitations have been developed using synthetic materials. Imitation leather is manufactured in several different ways. Synthetic leather, also sometimes referred to as leatherette, is the result of coating woven fabric with thermoplastic polyurethane (TPU) or polyvinyl chloride (PVC) and embossing (page 220) the surface with a grain pattern. One of the disadvantages of this material is that it is not breathable. A more recent development is nonwoven (page 152) leather made up of densely packed synthetic fibres, such as elastane or viscose (page 508). It is lightweight, breathable and long-lasting and can be hard to tell apart from split leather.

Fake fur is knitted (page 126), woven (page 96) or flocked (page 280). A wide range of visual and tactile properties is produced to mimic the natural variation found in nature (see image above). Very good imitations are possible, but the more realistic they are then the more complex and time-consuming they are to produce, which makes them more expensive. Although animals are not killed in the production of these alternatives, they have other environmental impacts that should be taken into account (see Filament Spinning, page 50).

Imitation leather and fur may be engineered to exact requirements, unlike the real thing that varies from animal to animal. This enhances the garment or textile, such as improving lightweight, antimicrobial or water-repelling properties (see image above right). With modern synthetic processes, several fibres may be combined to take advantage of more than one property. With all the technical benefits of synthetic materials it seems more exciting to explore their full potential outside the constraints of trying to imitate the natural, variable and unique qualities of leather and fur.

COWHIDE AND CALFSKIN

Description

The skins of cows (genus *Bos*) are by-products of meat and milk production. The size of the hide will depend on the animal and ranges from 4 to 5 m² (43–54 ft²). Calfskin is typically less than half this size. They are primarily used to make leather with the hair removed. Hair left on most meat-bearing breeds is straight and no longer than 10 mm (0.4 in.), while calf hair is much shorter and softer.

Strength   ● ● ● ● ○

Leather is relatively strong with good elongation before breaking. As a natural material its strength varies across the hide and under tension the weakest point will give way,

thus defining its ultimate strength. Where the skin was taken from also has an effect: leather from the belly region is the thickest and fullest. It may be laminated to increase strength, but this detracts from the inherent qualities. As it gets old, leather will dry out, becoming brittle and prone to splitting.

Resilience   ● ● ○ ○ ○

The surface is hard-wearing, but cowhide has little elasticity. Physical properties depend on a combination of the quality of the skin and tanning method. Vegetable tanning produces firm and hard-wearing leather; chrome tanning yields a more resilient material. Thinner leather, which has been split, is more pliable. Cowhide cannot stand repeated wetting and is generally protected with oils or waxes.

Quality   ● ● ● ● ○

This depends on the age of the animal and how it has been raised. Insect attack, barbed wire fences and fighting all leave marks and traces. Young calves that have drunk only milk yield the softest skins. Mature cows that have eaten grass and hay most of their lives produce a tougher skin with coarse hair. The finished appearance depends on the tanning, retannage, dyeing and finishing processes.

Colour   ● ● ● ● ○

The skin is typically light brown in colour, but can vary. It may be dyed during tanning or finished in a wide range of colours,

including metallic. Hair colour varies depending on the breed, ranging from off-white to shades of brown, grey and black.

Versatility   ● ● ● ○ ○

It is not the lightest, most elastic or comfortable, but as the most widely available leather it can be found in all forms and with all manner of finishes. Buffalo skin is also used to make leather goods, such as dress shoes and boots. It is tougher and heavier than regular cowhide.

Availability   ● ● ● ● ●

As a by-product of meat production, domesticated cow skin is widely available throughout the world. It is not available in certain places for religious reasons. It is used in all types of apparel. Interior applications include upholstery and rugs.

Cost   ● ● ● ● ○

It is one of the least expensive leathers. The price depends on the quality and ranges from moderate to high.

Environment   ● ● ● ○ ○

Leather production requires that the animal be slaughtered. The source of skin is critical, to ensure the animals were treated humanely. The environmental impacts of leather tanning depend on the tannery. In countries with strict environmental laws, tanneries are capable of dramatically reducing their ecological footprint, but this requires significant investment in modern equipment.

SHEEPSKIN AND LAMBSWOOL

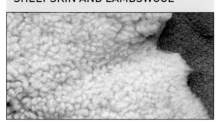

Description

Skins from domesticated sheep (*Ovis aries*) and lambs are commonly tanned with the wool intact, called fleeces. They are also used to make some very soft leather products. Like cowhide, the skin is a by-product of meat production. However, breeds that yield very high-quality fleece may be reared mainly for their skins. Adult sheepskins can range from 0.75 to 1 m² (8–10 ft²). Lambskins are smaller; the size range depends on the breed. The length of fleece depends on the age of animal and can be more than 70 mm (2.8 i.n), ranging from straight to highly crimped. Shearling refers to the pelt of a lamb up to a year old that has been shorn after

tanning to a uniform length. Lambswool refers to fleece from a young animal that has never been sheared.

Strength   ● ● ○ ○ ○

Sheepskin is thinner and much less durable than cowhide.

Resilience   ● ○ ○ ○ ○

It stretches, but is not very elastic, so will gradually lose its shape. Therefore, it is typically used in applications that do not stress the material, such as linings and rugs.

Quality   ● ● ● ● ○

Sheepskin is soft and supple and is traditionally used to make chamois leather. Wool provides excellent insulation and is antistatic, ideal for use in linings of jackets, hats, gloves and boots. Like wool yarn (page 494), it can absorb a lot of moisture before feeling wet, thus keeping the wearer dry. The softness of the wool and suppleness of the skin depends on the breed and age of the animal: stillborn lambs have the softest wool. The length of wool typically depends on age, but will also be affected by breed and time of year.

Colour   ● ● ● ○ ○

Skin colour is typically light brown to grey, depending on the tanning process. The colour of fleeces ranges from creamy white to dark brown or grey, depending on the breed. Wool may be bleached and dyed a range of colours, similar to wool

fibre, or left with the natural colour and patterns.

Versatility   ● ● ○ ○ ○

Fleeces are used to to make linings for boots, gloves and jackets, as well as throws, rugs and upholstery. Smaller pieces of fleece may be joined together with machine stitching (page 354), to make larger panels with a virtually seamless finish (on the front side) to cover larger areas. They may also be sheared and dyed to mimic the fur of wild animals, such as beaver.

Availability   ● ● ● ● ○

Sheep are farmed in many countries around the world. Fleeces will be different depending on the source. Spanish and Italian fleeces are considered high quality; the highest-quality Merino comes from New Zealand; African lamb fur is typically coarser; and colder climates, such as Russia and Austria, will produce woollier fleeces.

Cost   ● ● ● ○ ○

As a by-product of meat production, the fleeces are relatively low cost. Lambswool is more expensive.

Environment   ● ● ● ○ ○

Sheep are farmed for their meat much like cows. Converting the skin into fleece uses the same processes and has equivalent impacts on the environment.

GOATSKIN AND KIDSKIN

Description

The skins from domestic goats (*Capra aegagrus hircus*) and their young (kidskin) are comparable with sheep and lamb. They are converted into leather and fur products. Goat's hair ranges from short to long and wavy, depending on the breed. The soft and luxurious hair from the Angora (mohair) and cashmere breeds is considered too valuable to convert into fleece and so is typically only available as yarn (page 495).

Strength   ● ● ● ○ ○

It is strong, supple and comparable with deerskin.

Resilience   ● ● ● ○ ○

It is lightweight, slightly stretchy and durable. When used to make gloves or garments it is pre-stretched to avoid it stretching too much more when in use. In this way, the leather will mold to the shape of the wearer as opposed to becoming baggy and poorly fitted.

Quality   ● ● ● ● ○

It is water-repellent and hard-wearing, even though it can be quite soft. Kidskin is the softest and most supple. The quality depends on the breed, but also on where it was on the body. Fleeces tend to have pronounced patterns of different lengths of hair, ranging from coarse to soft. The hair ranges from straight to wavy and may lie flat or stick out from the pelt. In some cases, the guard hairs are removed, known as 'pulled goat', to produce a softer fleece predominantly made up of undercoat. The length and coarseness of hair will depend on the breed and source: goats from colder climates have thicker coats.

Colour   ● ● ● ○ ○

The skin is typically a shade of light brown. Goat hair tends to be more patterned than sheep's wool, something that is impossible to control, and comes in a range of colours.

Versatility   ● ● ● ● ○

Goatskin is used to make leather goods, such as gloves, because it has a good balance of softness and durability. As high-quality leather, it is also used for a variety of applications, from bags to bookbinding. Goat fleece is less common and is relatively cheap, so is used to mimic wild and less available animals, such as badger.

Availability   ● ● ○ ○ ○

The skin of several breeds of goat is converted into leather and fur. A range of colours, patterns and hair qualities is available. It is not as common as sheep or cow skin, mainly because of farming practices.

Cost   ● ● ● ○ ○

This depends very much on the breed and quality, ranging from moderate to high. Goat leather is highly prized for glovemaking, while the soft hair of the Angora goat produces insulating and comfortable fleeces.

Environment   ● ● ● ○ ○

As with sheep, the skins may be the most valuable part of the animal and so goats are frequently reared for the skins alone. Therefore the skin is not always a by-product of farming, although many goats are herded like sheep and kept for their milk and meat. The tanning process is the same as for skins taken from sheep and cows (see Leather Tanning, page 158).

DEERSKIN

Description

Deer (family *Cervidae*) are farmed for meat and fur, and wild animals are culled or hunted to manage their numbers. Deer leather is high-quality, comfortable and long-lasting. It is softer and more supple than most other types.

Strength ● ● ● ●

Deerskin is very soft and supple, and stronger and lighter than cowhide.

Resilience ● ● ● ● ●

It is not as resilient as cow, so will stretch more easily. Before fabricating into gloves or shoes it is pre-stretched to reduce the amount it will change when in use. Over time, items made from deerskin will mold to the shape of the wearer's body as a result of it continuing to stretch.

Quality ● ● ● ● ●

Deerskin is lightweight and breathable, which makes it comfortable to wear and is why it was so cherished by the Native Americans and continues to be used in high-quality garments, gloves and shoes. It remains soft even after repeated wetting, which means that it does not need to be so thoroughly protected as leather from cows.

Colour ● ● ● ● ●

Naturally off-white, light grey or brown, the skin can be dyed after tanning. The colour of hair depends on the breed, which is influenced by location. For example, reindeer have coats that range from white to grey, with a darker stripe down the centre of their back. Deer in warmer climates, such as red and roe, tend to be lighter shades of fawn and reddish brown, to blend in with their surroundings.

Versatility ● ● ● ● ●

Deer is considered very versatile thanks to its light and soft qualities. However, it is not produced in the same quantities as cowhide and so is not as widely used or adapted. Elk (moose) are large deer from North America, which yield strong and thick leather, but not as durable as regular deer.

Availability ● ● ● ● ●

As a raw material, deerskin is not as widely available as cow, sheep or pig. Therefore it is generally a little more expensive, but this depends on the quality and breed.

Cost ● ● ● ● ●

Deerskin is moderate to high cost, depending on the quality.

Environment ● ● ● ● ●

Wild and domesticated deer are killed for their meat. Their leather and fur is a by-product.

PIGSKIN

Description

One of the most versatile leathers, pigskin is soft and relatively easy to work. There are many different breeds of pig (genus *Sus*), including domesticated and wild types. The skins from domesticated pigs are as widely available as cowhide and available in as many forms and finishes, whereas the skins from wild animals (such as boar and peccary) are rare and expensive.

Strength ● ● ● ● ●

Pig leather is lighter than cowhide, soft and reasonably strong. Peccary yields very soft, strong and supple leather.

Resilience ● ● ● ● ●

It stretches, but is not very elastic, so will gradually lose its shape. In other words, it needs to be pre-stretched before use to make sure it does not stretch in application.

Quality ● ● ● ● ●

Pigskin is an all-rounder. It is used for a wide variety of applications and can be finished with patterns, colours and tactility that are similar to many more expensive leathers, such as ostrich.

Colour ● ● ● ● ●

Pigskin is naturally pinky-brown, but it depends on the breed and may even be patterned or mottled. It can be dyed a range of colours and finished with all the same pattern and colour effects as cow leather (see Leather Tanning, page 158).

Versatility ● ● ● ● ●

The leather is used for a wide range of applications, from shoes and fashion apparel to interiors and upholstery. It is not produced with the hair left on, because a pig's hair, if there is any, is coarse and brittle. Other than the more conventional applications, pigskin is used for medical application (such as the treatment of burns) and by tattoo artists to practise on.

Availability ● ● ● ● ●

It is widely available in pork-eating countries and regions.

Cost ● ● ● ● ●

One the whole, pigskin is cost-effective. However, this does depend a great deal on the breed: peccary is one of the most expensive types of leather available.

Environment ● ● ● ● ●

It has been said that the only part of a pig that cannot be used is its squeal. In many countries this is true and pigs have become an important farmed animal for their meat and skins. The tanning process is the same as for skins taken from sheep and cows.

HORSEHIDE AND PONY

Description

Horses (genus *Equus*), ponies (small horses), colts (male horse under four years old), donkeys (or asses) and mules (cross between a female horse and a male donkey) produce similar leather to cows, but it can be more durable. Leather from smaller and younger animals tends to be softer and suppler. The most desirable pelts have short and straight hair that lays flat against the hide.

Strength ● ● ● ● ●

It is tough and durable leather comparable with cowhide, but a little less supple.

Resilience ● ● ● ● ●

Like cowhide it has little elasticity and gets more brittle with age.

Quality ● ● ● ● ●

Horse skin is thick, stiff and close-grained. It is not suitable for use in apparel. However, the skins of ponies can be soft and suitable for working into items of clothing and accessories.

Colour ● ● ● ● ●

It comes in a wide range of natural colours and patterns and can be dyed. The most common colours are bay (dark chocolate brown) and chestnut (tan). Natural patterns and colours are typically more varied than found in calf.

Versatility ● ● ● ● ●

Like pigs and cows, horses have been domesticated a long time and as a result they have been widely used, including for textiles. Nowadays, horsehide is used for a limited range of apparel. Shell cordovan is leather made from the rump of a horsehide and is used to make shoe uppers (page 430).

Availability ● ● ● ● ●

Horses used to be more common than they are now, because they were important working animals on farms and in cities. The leather was used for general-purpose applications, much like cow and pig leather today. Since engines largely superseded horses there are fewer hides available.

Cost ● ● ● ● ●

The price depends on the quality and can be quite high for pony and foal. It used to be as cheap as cow but is now generally more expensive owing to reduced availability. This depends on the country, because in some regions they are still widespread and thus less expensive.

Environment ● ● ● ● ●

Horses have served humankind for many thousands of years. Predominately used as working animals; they have also provided meat, hide, hair and hoof. Converting the skin into leather or fur uses the same processes as cowhide.

RABBIT AND CHINCHILLA FUR

Description

Many species of rodent, which may be considered pests, have been used to make fur including semi-aquatic types (beaver, nutria and muskrat) and those that live on land (squirrel, rabbit, hare and chinchilla). Nowadays, the most important of these is the rabbit family (*Leporidae*), nutria (*Myocastor coypus*) and chinchilla (genus *Chinchilla*), which are farmed in Europe, North America and Asia for their fur. The hair on a mature animal is up to 20 mm (0.8 in.) long and the undercoat is about half that. Fur famed for its luxurious softness, such as from the Angora rabbit, is more typically converted into yarn (page 497) than pelts.

Strength ● ● ● ● ●

The skins of these rodents are relatively thin and not very strong. They are typically used for trim, lining (page 392) or insulation purposes. When made into large articles, such as coats, they are layered or woven with stronger fabrics.

Resilience ● ● ● ● ●

Like all skins the leather has some stretch, but is not very elastic and loses its shape easily and more so when wet. The fur is delicate, and softer furs, such as chinchilla, are particular vulnerable to abrasion.

Quality ● ● ● ● ●

The highest-quality fur comes from these animals' winter coats. They are prized for their softness, and because rabbit is relatively cheap it is often used to mimic the appearance of more exotic and expensive furs, such as mink. The most desirable coats are homogenous with a uniform distribution of silky guard hairs and undercoat. These are used for smaller items, such as collars, and the lesser-quality furs are turned into large items, such as coats. The rex rabbit has significantly less guard hair and so produces a denser and softer fur. Nutria, or coypu, is similar to the beaver and its fur is often sheared to produce a uniform and dense finish.

Colour ● ● ● ● ●

The natural colour of rabbit's fur ranges from white to brown.

It is often dyed. Chinchilla is a distinctive grey-blue to black and is most often used in its natural colour. Nutria is brown to yellowish brown in colour.

Versatility ● ● ● ● ●

The skins of these animals are used with the hair on and in their natural state, or sheared to a uniform length. They are used as trim and lining for their softness and insulating properties. Entire garments constructed from these pelts are expensive; chinchilla is worth many times more than rabbit.

Availability ● ● ● ● ●

Farmed across Europe, Asia and North America, rabbit is widely available. Chinchilla is less common and very costly.

Cost ● ● ● ● ●

Rabbit ranges from moderate to high and chinchilla is one of the most expensive furs.

Environment ● ● ● ● ●

Rabbit pelts are sometimes by-products of meat production, although not always, because obtaining the highest-quality fur (a fully developed coat takes two or more years) means meat is a by-product. Chinchilla and nutria were hunted to virtual extinction. The chinchilla is now protected in the wild, but the nutria is still considered a pest owing to its invasive nature and destructive habits along riverbanks. Like mink and fox they are farmed for their fur.

MINK AND SABLE FUR

Description

The mink family (*Mustelidae*) and other types of weasel – namely the marten (genus *Marten*), including Russian sable – are used to make soft and highly prized furs. Russian sable is considered to be one of the softest and finest furs, while mink has a silky and lustrous natural colour.

Strength ● ● ● ● ●

The skins are not very strong and are only used with the fur on. They are applied as trim or lining, or combined with strips or layers of fabric.

Resilience ● ● ● ● ●

The skin is soft and stretches, but it is not very elastic, so will gradually lose its shape. The very soft fur will have good spring-back, but its softness makes it prone to abrasion.

Quality ● ● ● ● ●

High-quality mink has a deep, soft and lustrous pile. The undercoat is the most desirable. It is dense and so helps to keep the guard hair erect. Shearing the fur to a uniform length (in particular removing the coarser guard hairs) produces a soft velvet-like fabric. North American and Canadian furs are typically denser, because they have evolved in a colder climate. Sable furs have the luxurious quality of smoothness in all directions.

Colour ● ● ● ● ●

As a result of selective breeding, the natural colour of mink fur ranges from light to dark grey and shades of brown and black. Sable ranges from light yellow to a deep reddish brown. The highest-quality fur is not dyed.

Versatility ● ● ● ● ●

This is a desirable and expensive fur, which limits applications to formal wear (coats and capes, for example) or small items such as neckwear, collars and trim.

Availability ● ● ● ● ●

Mink from American and Canadian farms is considered some of the best, thanks to the suitable climate and farming conditions. Other counties farming mink include Denmark, Finland and China. Sable comes mainly from Russia. A cheaper version of sable comes from Canadian farms, but is considered inferior.

Cost ● ● ● ● ●

Mink and sable are luxurious and expensive. The high cost is compounded by their small size: a typical mink coat is made up of between 40 and 80 pelts. It requires a high level of craftsmanship to put the pelts together to produce a uniform appearance.

Environment ● ● ● ● ●

Mink is farmed solely for its fur (it is the most commonly fur-farmed animal). Only taken into captivity in the last 100 years, it remains a wild animal by nature. Farming methods that confine animals to small wire cages stacked side by side have been heavily criticized, because this puts the animal through undue stress and conflicts with its naturally solitary and territorial nature. Sable fur comes from both farmed and wild animals. The highest-quality types come from wild animals that are protected in Russian nature reserves.

FOX, WOLF AND COYOTE FUR

Description

Animals from the dog family (*Canidae*) used to make fur include the wolf, coyote and fox, among others. They live wild and species from the northern, colder climates are hunted for their pelts. Foxes are farmed in several countries. The fur is lightweight and up to 70 mm (2.7 in.) or more in length with distinctive long and straight guard hairs supported by a wavy undercoat of soft insulating hair.

Strength ● ● ● ● ●

The pelts are only used with their fur on. The hair itself is reasonably strong.

Resilience ● ● ● ● ●

The skins are stretchy but not very elastic. The hair is lightweight and resilient.

Quality ● ● ● ● ●

The fur is thick and stands upright on the pelt. Winter coats are the most desirable, with a dense undercoat and smooth guard hairs. Pelts from animals from the northern climates, such as Siberia, have the highest insulating properties. Female animals tend to produce coats with the highest lustre, but they are also smaller. Coyote is not as high quality as fox and wolf, but still very soft and used to make luxury items.

Colour ● ● ● ● ●

Their natural colour varies according to the breed and climate. Foxes range from silver, white and red to black. Red is the most common and least expensive, silver and white are more valuable. Wolves are shades of white, brown, grey and black. Coyotes have less colour variation and are typically greyish-fawn with dark tips to the guard hairs. The pelts may be dyed, but the highest-quality fur is used in its natural colour.

Versatility ● ● ● ● ●

They are used to make coats, hats, scarves, collars and stoles. The highly insulating winter coats are used for

outdoor cold weather gear. For example, the hoods of cold weather parkas are lined with coyote fur to keep a warm pocket of air around the face and stop the skin freezing. The tails are used as accessories, such as to decorate bags.

Availability ● ● ● ● ●

Some of these animals are widespread. Foxes, in particular the red fox, are common in many countries. Most fox pelts now come from farmed animals. Coyotes and wolves are hunted wild and so used in much smaller quantities.

Cost ● ● ● ● ●

The fur is moderately expensive, but usually only applied as trim and so does not increase the price of a garment dramatically. They are relatively large compared to other luxury furs, such as mink and chinchilla, making them cost-effective by comparison. The cost also depends on the colour: silver fox costs significantly more than the more common red.

Environment ● ● ● ● ●

Foxes are farmed in some countries and have been domesticated for several decades. Almost all silver fox fur (also called blue) will be from farms with selectively bred animals. Wolves and coyotes are hunted from the wild to manage their numbers. They are mostly protected and so relatively stable in number. They are hunted mainly for their fur and the rest of the animal is of little use.

OSTRICH SKIN

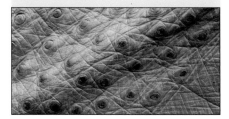

Description

Skin from farmed ostriches (*Struthio camelus*) produces exotic leather used to make high-value and luxurious items. It has distinctive markings as a result of the large quill follicles. The skins of adult birds (14 months or more) are 0.5 to 2 m² (5–20 ft²). The leg also produces usable leather and the shin is typically around 100 x 400 mm (4 x 16 in.).

Strength   • • • • •

The leather is strong with a hard-wearing surface. It is heavier than goat and sheep, but lighter than cow and calf. The properties are dependent on where the leather is derived from, because the sides are softer than the back and shins.

Resilience   • • • • •

Ostriches produce tough and pliable leather and the natural oils present in the skin prevent it from drying out and becoming stiff. This makes it long-lasting.

Quality   • • • • •

Markings produced by the quill follicles create a raised surface profile and the rest of the leather is smooth. The legs are scaled like a caiman's tail. It has good water repellency.

Colour   • • • • •

It is naturally light tan and often dyed. Like leather from cows and pigs, it may be produced with a variety of finishes including matt, glazed, metallic, two-tone and suede.

Versatility   • • • • •

Its high strength makes it useful for utility applications, while its desirability means that it is limited to more decorative items, such as boots, belts, bags and purses. It is also used in upholstery (page 404), such as for high-end car interiors and bicycle saddles. The feathers are used to make feather dusters.

Availability   • • • • •

The part of the leather marked with the distinctive quill follicles is relatively small and makes up only around one third of the total area. The rest is less valuable but still highly prized for its durability and resilience (the title image is ostrich shin, page 498). Ostriches are farmed and around half of the leather produced each year comes from South Africa. Many other countries have small populations.

Cost   • • • • •

It is exotic leather, takes many steps to produce to high standard and is available in relatively small quantities, all of which makes it very expensive. As a result of its high cost it is more common to find imitation ostrich made from cow or pig leather that has been embossed (page 220).

Environment   • • • • •

Ostriches are mainly farmed for their leather; meat and eggs are by-products, which is the reverse situation to farming cows, pigs and sheep. At around 14 months the leather is strong enough, without being too damaged, and the birds are slaughtered. Like all other leather, the skin has to pass through tanning and a series of processes to convert it from raw skin into durable leather. The choice of tanning technique will impact on the overall sustainability of the material.

FISH SKIN

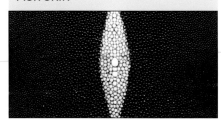

Description

Virtually any type of fish skin can be converted into leather. Stingray (genus *Dasyatis*) is native to Southeast Asia, whose skin is tanned to make exotic leather up to 750 mm (30 in.) long and 400 mm (16 in.) wide. The skin of salmon, perch (genus *Perca*), wolfish family (*Anarhichadidae*) and cod (genus *Gadus*), among others, is a by-product of meat production and is tanned to make thin, flexible leather. Perch can be quite big and up to 0.14 m² (1.5 ft²); cod is up to around 500 mm (20 in.) long and covers approximately 0.05 m² (77 in.²); wolfish are longer and narrower and cover up to 0.1 m2 (1 ft²).

Strength   • • • • •

The leather is strong considering its thinness, and retains its glossiness. Ray skin is particularly hard-wearing.

Resilience   • • • • •

Rays have tough skin covered with bony plates. Other types of fish used to make leather, such as salmon and wolfish, have soft flexible skins with good resilience. Atlantic Leather produces machine-washable salmon leather.

Quality   • • • • •

Each variety of stingray has distinctive markings. Saltwater stingrays have evolved a tough skin covered with bony plates and tanneries have methods for softening the hard-wearing hides. Fish have a variety of natural finishes, ranging from coarse to smooth. The scales are typically removed by the tanning process. Cod has finer scales than salmon and the texture may vary from coarse to smooth on the same piece. Perch skin has a rough surface. The wolfish has soft and smooth skin without scales. The finishing process ensures that the leather is waterproof and stain-repellent. Waxes and plastics may be sprayed on to increase durability, but this depends on the leather and on the finish required.

Colour   • • • • •

Cod has a fine uniform pattern, with a dark line down the centre (spine). The wolfish is covered with black dots, which will show through dyed colour, unless the dye is darker than the spots. All types of fish skin may be dyed. Stingray is not typically dyed and is used with its natural colour markings. The most common is black with white spots.

Versatility   • • • • •

There are many different types, which makes this reasonably versatile leather. It is used to make handbags, shoes, wallets and jewelry.

Availability   • • • • •

This depends on the location and on the skill of the tannery. Only certain leather-producing facilities have the knowhow to process fish skin effectively. Any type of fish skin can be converted into leather. In addition to the types already discussed, shark, groper and snapper, for example, are also used but in smaller quantities.

Cost   • • • • •

As a by-product of the fishing industry it is relatively low-cost.

Environment   • • • • •

The majority of fish skins are the by-products of meat production. Converting them into leather creates a high-value product from something that would otherwise be of little value. They are often converted into leather using vegetable tanning and dyeing.

REPTILE SKIN

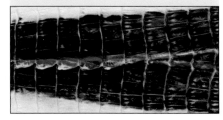

Description

Exotic leather is made from alligators (*Alligator mississippiensis*); members of the crocodile family (*Crocodilus*), such as the freshwater, saltwater and caiman; lizards such as the ring lizard (*Varanus salvator*) and Nile lizard (*Varanus niloticus*); and snakes such as the python (reticulated python, *Reticulatus*; short tail python, *Brongersmai*) and anaconda (*Eunectes notaeus*). They are endangered, or on the monitored species list, and so trapping in the wild is banned or strictly controlled. As a result, they are farmed for their skins, and producers must satisfy the criteria according to the International Trade in Endangered Species of Fauna and Flora (CITES). Size depends on the type, from small lizards to python 600 mm (2 ft) wide and crocodile skins up to 700 mm (2.3 ft) across.

Strength   • • • • •

The skins are strong, but this depends on the type, size and age of the animal.

Resilience   • • • • •

The skins are tough and hard-wearing. They are not very elastic, but can be surprisingly flexible.

Quality   • • • • •

They are prized mainly for their colour, pattern and texture. This varies, but is characterized by overlapping scales, the size of which depends on the species. They are bone-like and protect the animal during its life. Of the crocodiles, the freshwater variety from New Guinea is considered the most luxurious. Caiman is much cheaper, because the skins are usually inferior quality with imperfections that do not accept dye well.

Colour   • • • • •

Alligators, crocodiles and caiman are usually grey to dark green. Anaconda has larger scales than python, with a rounder pattern. They may be dyed and finished with a matt, glazed, suede or metallic finish. Python comes in the widest range of finishes, including pearlescent and multicoloured.

Versatility   • • • • •

They are used to make high-value items, such as wallets, handbags and boots. They are also used for some expensive upholstery and interiors.

Availability   • • • • •

They live mainly in tropical areas in southern parts of the USA and across South America, Africa and Asia. They are expensive and so restricted to high-end fashion and interior applications. Wild skins are rare and must have the relevant CITES paperwork. Certain species are entirely banned in other countries (even if they have the correct paperwork), such as the Siamese crocodile in the USA and Malaysian snakeskin in the EU.

Cost   • • • • •

Crocodile is the most expensive, followed by alligator, and caiman is several times less expensive for the same size piece. Snakeskin is expensive, but depends on the size, quality and species.

Environment   • • • • •

CITES, which was put in place to protect endangered and monitored species, requires that commercial farms demonstrate a viable second generation and so are not reliant on new stock from the wild.

Materials
Regenerated Fibres

These semi-synthetic materials are derived from renewable sources. Naturally occurring polymers, such as cellulose and protein, are extracted and converted into filament yarns. Cellulose may be derived from many different types of raw material, including wood, bamboo and cotton. The conventional techniques used to produce viscose are complex and unsustainable. A new method has been developed and commercialized – under the trade name Tencel – that utilizes renewable materials, is non-polluting and recycles chemicals in an almost closed-loop process. Protein-containing materials, such as soy, corn and milk, are converted into fibre using a similar technique to viscose, but this is much less widely practised and there are few commercially successful examples.

PROPERTIES OF REGENERATED FIBRES

Regenerated cellulose fibres are typically not very strong, but they have high water absorption and a lustrous surface. Their rich aesthetic qualities and drape are utilized in plain and fancy fabrics. The surface may be smooth or textured, and as a semi-synthetic they may be produced as continuous monofilaments or multifilaments, or chopped into staple fibres and twisted into yarn (page 56).

They were originally developed as an alternative to relatively expensive cotton (page 481), flax (page 481), wool (page 494) and silk (page 497). As such, they are often produced as a replacement, mimicking the visual qualities of these materials, or blended with them to reduce the cost of the fabric. However, the properties are quite different. Natural fibres are based on cellulose or protein, but they are not chemically regenerated. Therefore they contain other substances and retain the natural polymer structure. Regenerated fibres are semi-synthetic and their properties are more controllable. They

are more uniform and can be modified in many ways that are impossible or impractical for natural fibres. For example, when the fibre is being processed and still in solution (see Filament Spinning, page 50), it may be coloured (known as solution-dyed) and modified with additives to produce much higher-quality results than can be achieved once the filament has been spun. In addition, the diameter, length and surface profile may be changed during production.

Viscose was the first man-made fibre. It has been through several iterations, and more recent developments, such as high wet modulus (HWM) viscose and lyocell, overcome some of the shortfalls of the early types. Even so, they are one of the weakest groups of fibres, and their strength is reduced when wet. So, in addition to being used because they are low-cost, they are used for their visual appearance, such as in the production of fancy fabrics (see image opposite).

Cellulose fibres have a semi-crystalline molecular structure. Owing to the force of

attraction between the molecules areas of highly organized (crystalline) polymer chains are formed, joined together by less organized (amorphous) groups. The strengths and weaknesses of the fibre come from the properties of these two areas. While the crystalline areas are strong, with high wet strength, the amorphous areas are less so, and the sideways bonds between the polymer chains are weak.

With moisture and abrasion, the amorphous part of the fibres swells and breaks down into microfine hairs, which are less than 4 microns thick. Known as fibrillation, it can be both advantageous (creating a softer hand, known as peach skin) and problematic (pilling). Natural cellulose fibres, such as cotton and flax, also fibrillate, but to a lesser extent. In the production of lyocell it can be controlled and virtually eliminated by cross-linking the molecular structure in the amorphous areas.

Regenerated protein fibres, known as azlon, are manufactured using similar techniques to viscose and so share many of the same characteristics. Their inherent properties are quite different from natural protein fibres, such as wool and silk, but like viscose, they may be modified and manufactured to mimic a range of materials, including those.

Azlon exhibits some interesting properties, in addition to being sourced from renewable materials, that have appealed to many inventors and manufacturers over the years. For example, fabrics constructed with it have excellent vapour transmission combined with water absorption properties similar to cotton, which makes it a very comfortable material to wear close to the skin.

TYPES OF REGENERATED FIBRE

These fibres are all manufactured from naturally occurring polymers. Converting into fibre requires that the raw material is dissolved and separated, so that the polymer may be extracted and drawn into a filament.

Cellulose is obtained from many different types of raw material – such as wood, cotton and bamboo – which have varying concentrations of the natural polymer. Importantly, it is the method of extraction that defines the quality and characteristics of the fibre, and not so much the raw material. There are two principal techniques, which are wet and dry spinning.

Viscose is most commonly produced by the wet spinning method. In this process cellulose is chemically extracted from wood pulp and mixed with sodium hydroxide (caustic soda). After a suitable length of time to allow the material to mature, which can be up to 50 hours, carbon disulphide is introduced to form sodium cellulose xanthate. This produces a yellow crumb-like substance, which is dissolved in sodium hydroxide and further ripened into a viscous solution, which is forced through spinnerets. The high shear, caused by high pressure and high viscosity, result in alignment of the molecular structure lengthways. The filaments are drawn into a bath of dilute sulphuric acid, which decomposes the xanthate and regenerates the cellulose into a solid fibre.

HWM viscose has improved tenacity. By altering the chemical process, the structure of the fibre is improved. When at the cellulose xanthate stage, sodium hydroxide is introduced and the filament is stretched several hundred percent as it is squeezed through the spinneret. This maximizes the crystallinity and chain length in the finished fibre, resulting in a stronger material with reduced fibrillation.

Lyocell is a more recent development and commercial production began in the 1990s. The process was developed as an alternative to the conventional viscose, with the aim of reducing the environmental impact. Viscose production creates many by-products, such as sulphur, metal salts (copper and zinc) and ammonia, which are potentially harmful to people and the environment if not properly managed.

Lyocell overcomes this by using much less harmful chemicals in an almost closed-loop process. Production begins with the wood pulp, which is dissolved in a solution of N-methylmorpholine-N-oxide (NMMO) with water. The resulting dope is squeezed through spinnerets to form filaments with good molecular alignment. The filaments are then fed into diluted NMMO solution to set the fibre structure in place. The fibres are washed and wound and the NMMO solution is recovered and reused.

The modified wet spinning process produces a fibre without any by-products. The filament has a rounded profile and smoother longitudinal appearance than viscose. The physical properties are more comparable with cotton than viscose in many respects.

Wood pulp is converted into cellulose acetate by first swelling the fibres in acetic acid and using acetic anhydride to convert it into cellulose acetate. The resulting white powder – used in many other applications, such as spectacles and cellophane (see Plastic Film Extrusion, page 180) – is dissolved in acetone to produce a viscous solution suitable for spinning. The solution is pumped through the spinneret into warm air. The acetone evaporates to leave solidified filaments. This technique, known as dry spinning,

Devoré Viscose pile is digital printed (page 276) and selectively dissolved from the sheer ground fabric to produce a colourful patterned fabric. Devoré relies on the chemical incompatibility of two types of material, such as protein and cellulose, so that one can be dissolved while the other remains intact.

produces the most volatile organic compounds (VOCs). With modern control and recovery systems it is possible to capture and recycle most of the solvents, but not all.

There have been many attempts to make regenerated protein fibre from various raw materials, including soybean, milk and peanuts. Many have failed, including a widely publicized attempt by Henry Ford in the 1930s. Even so, the properties of regenerated protein and the abundance of some potential raw materials make this an attractive development idea. The most successful commercial developments so far extract protein from raw materials with a high concentration of the polymer. It is chemically modified and copolymerized with a synthetic polymer, such as polyvinyl alcohol (PVA) or acrylonitrile (among other ingredients), to make it suitable for spinning into filament yarn. It is converted by wet spinning in a process similar to viscose.

NOTES ON MANUFACTURING

As a group of materials, regenerated fibres have many of the same processing opportunities and limitations. However, each of the different types outlined above has some unique properties, which will affect how they are converted into textiles and apparel.

They may be modified at each stage in their conversion from natural material

Lyocell denim This twill fabric, manufactured from lyocell, is coloured and finished to mimic the natural variation and wearing characteristics of cotton. Lyocell produces a lighter and softer fabric with improved drape compared to regular denim.

Bamboo viscose Bamboo is used as raw material to make soft and silky viscose and lyocell. The sustainability of these yarns is affected more by the conversion process than the raw material.

to synthetic. At the dope phase they may be coloured or combined with additives to impart specific properties, such as increased strength or improved flame resistance. Chemical treatments may be introduced once the filament has been spun. This allows for greater flexibility, but may not yield such high-quality results, because the modification is applied to the surface as opposed to the raw material. They are often blended with fibres such as wool and cotton, and finished to mimic the natural appearance of these materials (see image above).

Fibres with low surface energy, such as lyocell, or smoother surface profile, such as HWM viscose, are more difficult to treat and dye (page 240). While it is possible to create the same range of colour as viscose, it will be more expensive and the colour may not be as long-lasting.

Produced as mono- or multifilament yarns, which may be cut into staples, the fibres may be used in all yarn formats, such as filament, staple (page 56) and plied (page 60). In addition, the surface and texture of the yarn can be modified during production.

The yarns are used to make woven (pages 76–105), knitted (pages 126–51) and nonwoven (page 152) fabrics. Cellulose types are utilized in many technical products, such as face wipes, pads, nappies, filters and bandages, because they can be produced to exact requirements, are safe for skin contact and are highly absorbent.

REGENERATED FIBRE DEVELOPMENTS

Bamboo is a fast-growing and renewable source of material used to produce viscose and lyocell, as well as natural plant fibre. The natural staple fibre (page 483), as well as fibres produced with the lyocell method may be sustainable. However, the majority of bamboo fibre is viscose (see image left) and therefore the yarn and fabric will have the same environmental impacts as all other types of material produced with the same method.

There have been many attempts to convert protein from natural materials into fibre. There are several types of protein, defined by the arrangement of amino acids. Examples include casein from milk and zein from corn (see also natural protein fibres, which use keratin and fibroin, page 490). Each material has different amounts of protein, which determine how efficient it is to convert into dope suitable for spinning.

The most exciting of these recent developments comes from China, where it is claimed they are using waste soybean paste from tofu production. Even though the process may not be closed-loop and sustainable as yet, there is room for development if the raw material is plentiful and renewable.

VISCOSE

Description
Also known as rayon and artificial silk, this fibre is manufactured from chemically extracted cellulose. It was the first man-made fibre and has been used in various forms and compositions since production began in the late 1880s.

Strength •••••
It is weaker than natural cellulose fibres (page 478), but has higher tenacity than wool (page 494). It loses much of its strength when wet. High wet modulus (HWM) viscose, which has improved molecular alignment and chain length, is stronger but still inferior to cotton and flax.

Resilience •••••
It has slightly better elasticity than many plant fibres, but is prone to creasing and losing its shape over time owing to its low wet strength. Creasing is reduced with wrinkle-resistant finishes (page 226).

Quality •••••
Viscose has a lustrous surface quality that can be controlled to some extent, and is the reason it is used to mimic natural fibres such as cotton, flax, wool and silk. It feels soft and drapes well. However, slippery types tend to lose their shape relatively easily, because the yarns move freely against one another. It is highly absorbent. Unlike synthetic fibres (page 510) and acetate, viscose does not melt and so is suitable for high-temperature pressing and ironing.

Colour •••••
It can be dyed a wide range of colours and has excellent fastness and resistance to ultraviolet light. Viscose dyed in solution, before spinning, has the best colour fastness. However, large volumes are required to justify dyeing an entire batch a single colour.

Versatility •••••
It is used mainly for its aesthetic and absorbent properties. Produced as a monofilament, multifilament or staple yarn, it is used in the construction of fabrics for fashion (ranging from plain to velvet) and interiors (tablecloths, upholstery and blankets). The yarn is used for embroidery (page 298). Technical applications include wipes, filters and medical supplies. It is used as a woven (pages 76–105) or nonwoven (page 152) and is often blended with natural fibres to bring their price down without compromising on aesthetics.

Availability •••••
Viscose is manufactured by several companies around the world that use different trade names (this is the only way to distinguish the different types). It ranges from solid to hollow and white to solution-dyed. It may be modified, such as with flame-retardant additives or to improve absorbency. It has been replaced by many less expensive and less polluting synthetics, including polyester (page 515) and polypropylene (page 515).

Cost •••••
Typically less than cotton, viscose is a relatively inexpensive fibre, but this depends on the source and quality.

Environment •••••
It is derived from potentially renewable sources, including softwood (page 487), bamboo and cotton. However, converting the raw material into a suitable yarn uses harmful chemicals and can be polluting. It is biodegradable at the end of its useful life.

LYOCELL

Description

This is a more recent development that has only been in public use since 1990s. It is often referred to under the trade name Tencel. It uses the same raw materials as viscose, such as wood and cotton, but the production uses fewer harmful chemicals and it is possible to contain everything in a closed-loop process to minimize pollution.

Strength   ● ● ● ○ ○

It is the strongest of the regenerated cellulose fibres, comparable with cotton. It retains more than three quarters of its strength when wet.

Resilience   ● ● ● ○ ○

It has moderate resilience, a little better than viscose, but still wrinkles relatively easily.

Quality   ● ● ● ● ○

The fibre has a rounded profile and smoother longitudinal appearance than viscose. The dense and homogenous fibre produces fabrics with a soft hand and good drape. Fibrillation occurs as a result of surface abrasion when wet.

Colour   ● ● ● ● ○

It can be dyed a wide range of vivid colours. However, the dyes are typically expensive, because the low surface energy of lyocell makes it difficult to colour. This increases the cost of the finished product.

Versatility   ● ● ● ● ○

Like viscose, it is available in a variety of forms and can be modified for applications in apparel and interiors. Examples include formal wear, casual wear (such as denim), bedding, baby clothes and facial wipes. It is often blended with natural fibres to produce a fabric with higher rates of absorption and improved drape.

Availability   ● ● ○ ○ ○

It is most commonly available as Tencel, which is the trade name of lyocell manufactured by Lenzing.

Cost   ● ● ● ● ○

It is more expensive than viscose owing to the additional processing cost.

Environment   ● ● ● ● ○

When using wood obtained from sustainably managed forests and producing the fibre in a closed-loop production process, recovering the majority of chemicals used, this material has lower environmental impacts than viscose. For example, Tencel is produced from the pulp of trees from sustainably managed forests certified by the Forest Stewardship Council (FSC). Raw materials that yield a high percentage of cellulose, such as eucalyptus trees, reduce waste and further improve sustainability. Once converted into yarn it uses the same chemical processes, such as dyeing, to convert it into finished goods. At the end of its life, lyocell can be recycled or composted.

CELLULOSE ACETATE

Description

Often simply referred to as acetate, it is derived, like viscose, from chemically extracted cellulose, but has some distinct properties. It softens at a relatively low temperature, like synthetics, and so is easily melted by pressing or ironing. It is readily identified, because it is the only fibre that is soluble in acetone.

Strength   ● ● ○ ○ ○

Like viscose, it has poor strength.

Resilience   ● ● ○ ○ ○

It has low elasticity and poor resistance to abrasion.

Quality   ● ● ● ○ ○

Acetate fabrics have good body and drape. However, they have a tendency to develop static electricity and so stick to the body. It is not very absorbent and so should be avoided in applications that press against the skin, especially for warm weather clothing.

Colour   ● ● ● ● ○

It has a lustrous surface and can be disperse dyed (page 420) or produced in a wide range of inherent colours. Fibres dyed in solution have excellent colour fastness, whereas yarn-dyed fabric is prone to fading over time and with exposure to the elements.

Versatility   ● ● ○ ○ ○

It is most commonly used to make nightwear, dresses and linings. Acetate is often blended with cotton and wool to provide added shape retention. With heat and pressure, acetate can be softened and formed, and so it is suitable for calendering, embossing (page 220), pleating and heat setting. It is modified with a range of additives, such as to improve flame-retardant properties, antibacterial properties and resistance to sunlight.

Availability   ● ● ● ○ ○

It is not as widely produced now since modern synthetics have replaced it in many applications.

Cost   ● ● ● ● ●

It is a low-cost fibre.

Environment   ● ● ○ ○ ○

It is comparable with viscose in many ways, because the raw material is chemically extracted in the same way. Dry spinning requires fewer steps and so reduces the use of chemicals and production of waste. However, this method produces the most volatile organic compound (VOC) emissions as a result of evaporation. The VOCs may be reclaimed and reused, but this does not always happen.

AZLON

Description

Certain types of protein-containing materials, such as soybean, peanut, corn and milk, are manufactured into regenerated fibre called azlon. It has a soft hand and is marketed as an alternative to cashmere (page 494). Although originally developed in the 1930s, it has only been produced commercially in the last decade. Regenerated protein from soybeans is relatively expensive and the raw material has low tensile strength. Synthetic materials such as polyester (page 515) have so far provided superior properties and are less expensive, making them the preferred choice for most applications.

Strength   ● ○ ○ ○ ○

On their own, regenerated protein fibres such as azlon have low tensile strength. They are chemically combined with a synthetic compound, such as vinyl alcohol or acrylonitrile, to improve strength. The resulting copolymer is stronger than viscose and has improved resistance to surface abrasion. It loses about one third of its strength when wet.

Resilience   ● ○ ○ ○ ○

It does not have good elastic recovery and are prone to creasing, so is treated with wrinkle-resistant finishes.

Quality   ● ● ● ● ○

Fabrics are soft and lightweight with excellent drape. It has a slippery surface and oval-shaped cross-section. It is comfortable to wear, because the fabric has excellent vapour transmission properties and similar absorption properties to cotton.

Colour   ● ● ● ● ○

It is naturally off-white and can be reactive dyed (page 240) a range of bright colours with good fastness. It has a smooth surface, which produces a natural lustre.

Versatility   ● ● ○ ○ ○

It is modified with additives, such as to improve flame-retardant and antibacterial properties. Applications include apparel and interior textiles. It is often combined with wool (page 494) and cotton (page 481) to add lustre. Because it has not been in commercial use for long, it is unlikely that the potential number of different applications has been fully explored.

Availability   ● ○ ○ ○ ○

A few specialized companies, mainly in China, manufacture regenerated protein fibre. Soybean has the highest concentration of protein suitable for yarn production.

Cost   ● ○ ○ ○ ○

Produced in low quantities, these fibres are relatively expensive.

Environment   ● ● ● ● ○

As with viscose, the process of extraction uses chemicals. However, depending on the technique used, these can be non-polluting and the majority can be reclaimed. A concern with these types of yarn is that protein comes from food crops, such as soya and corn. If it is more profitable to grow feedstock for yarn production (or other applications) than for food, then there is a likelihood that food production will be displaced, which is problematic in countries that have a food shortage. Ideally, the raw material will be a by-product of agriculture or food production, such as waste generated when making tofu.

Synthetic Materials

Epitomized by the commercialization of nylon by DuPont in 1938, which created a fashion revolution in high-quality, convenient and affordable garments, synthetic fibres have dramatically transformed textile design. Today, the application of synthetics in all their forms – fibres, films, foams, laminates, coatings and print – ranges from delicate lace to ballistic armour and from lightweight waterproofs to tensile structures. This diversity is the result of being able to modify the polymer structure, as well as the cross-sectional shape and diameter of the filament yarn, thus engineering it to meet specific demands as well as creating brand new opportunities for fashion, interiors and architecture. Many thousand different types are commercially available and new grades are continually being developed.

PROPERTIES OF SYNTHETIC MATERIALS

Like naturally occurring polymers, such as cellulose (page 478) and protein (page 490), man-made synthetics are made up of long chains of repeating units. These low-molecular-weight molecules, called monomers, consist of hydrogen, carbon and oxygen, as well as traces of other elements, such as fluorine, nitrogen and sulphur. They are combined through a process of polymerization. With adjustments in the process and by combining different monomers, a vast range of plastics is created.

Additives are used to add colour and enhance specific properties. Many

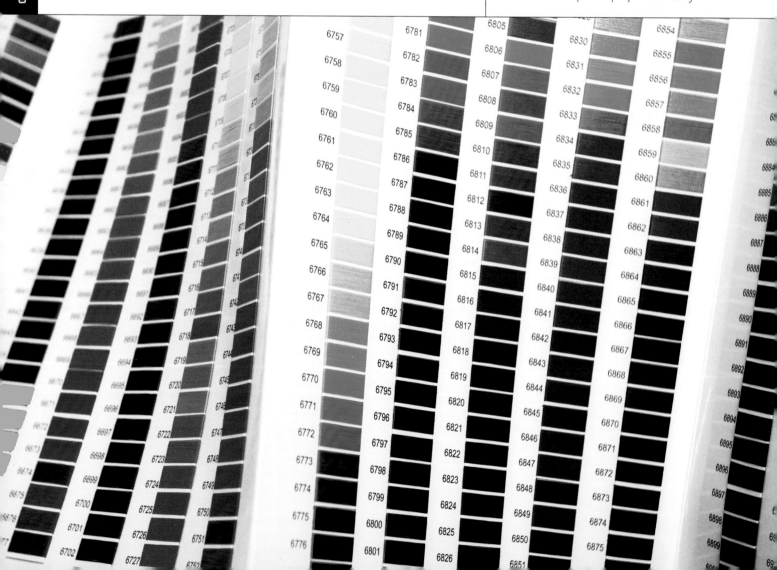

Right
Metal-coated polyester
Metal coatings are
applied for decorative
and functional
purposes. The finish
is bright and reflective.
A wide range of colours
is possible.

Right bottom
Artificial grass Strips
of polyethylene (PE) are
tufted (page 286) to
make resilient artificial
grass used for playing
football, hockey and
tennis.

different additives may be used in combination. For example, ultraviolet stabilizers are used to improve plastic's resistance to sunlight; flame retardants reduce flammability; plasticizers increase flexibility; and lubricants aid processing.

Synthetics range from amorphous to crystalline (that is, from randomly oriented polymer chains to a highly organized crystal structure). The amount of crystallinity affects the material properties: plastics with a high level of crystallinity have improved chemical resistance and lower creep (permanent deformation under load). At the other end of the spectrum, amorphous plastics soften over a wider temperature range, have improved ductility and can be transparent. Many types of fibres and films are produced from plastics that have a mix of amorphous and crystalline regions, known as semi-crystalline. Drawing out these fibres and films during production improves crystal orientation and thus improves mechanical performance. Examples of fibres processed in this way include polyethylene (PE), polypropylene (PP), nylon and polyester, all of which have an open (aliphatic) structure.

The same practices are used to improve the properties of synthetic films. Polyester is a widely used packaging material. When produced as film for high-performance applications it is drawn along its length and width to improve the crystal alignment. Commonly known under the trademark name of Mylar, these biaxial-oriented films are very strong for their weight. Coated with a thin film of metal, they are impermeable to liquid and gases. Application ranges from packaging to high altitude balloons. Slit into long strips, they are woven and knitted into fabrics for decoration (see image above top).

The length of the polymer chain, referred to as molecular weight, has a significant impact on the physical properties. For example, PE is a commodity plastic used to produce disposable packaging. It is chemically similar to PP and often grouped together under the name polyolefin. By increasing the length of the polymer chains, to produce ultra-high-molecular-weight PE (UHMWPE), the strength is greatly improved, because even through the molecular bonds remain unchanged the number of overlaps is increased, thus increasing the force of attraction. These fibres are commonly known under the trademark names Dyneema and Spectra.

They are applied alongside other high-performance fibres, such as aramid (page 524) and carbon (page 526), in demanding applications where strength to weight is paramount.

PP is one of the least expensive fibres. It is used in diverse applications ranging from single-use packaging to hardy artificial grass (see image above and page 292).

Plastic foams (page 172) provide insulation and cushioning. The choice of plastic depends on the application and many different types are suitable, including PP, polyurethane (PU) and polyvinyl chloride (PVC). They are foamed by mechanical action, or with the use of

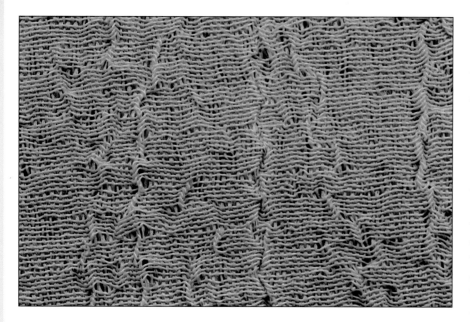

Textured nylon
Thermoplastics are
formed with heat and
pressure. Using two
types of material that
respond differently
to heating creates
dramatic three-
dimensional effects.

blowing agents. The density, colour and
hardness of foam can be specified to suit
the requirements of the application.

TYPES OF SYNTHETIC MATERIAL

Synthetic polymers are either thermoset
or thermoplastic. The principal
difference between these two is that
most thermoplastics can be melted and
shaped over and over again, whereas
thermosets cannot. This is due to the fact
that thermosets form permanent cross-
links between the polymer chains, which
will not allow the polymer to flow when
heated. These links also help to improve
the material's resistance to chemicals
and weathering.

Elastomers are flexible with rubber-like
properties. They may be thermoplastic
or thermoset, such as PVC and PU. They
are made up of an amorphous structure,
with permanent cross-links between the
polymer chains. This combination allows
the material to stretch to the limit of the
permanent bonds. Beyond that it fails.

The degree to which the polymer
chains are cross-linked affects the elastic
properties: increasing the cross-links
increases rigidity. Technogel is a unique
PU-based material that takes advantage
of this phenomenon. The polymer
structure, made up of a low number of
cross-links, creates a highly flexible and
resilient material that deforms readily
under load, spreading in all directions.
When the weight is removed the material

rebounds to its original shape. It is used
for padding, support and protection.

Each type of plastic is synthesized
differently. Some require more energy and
are more complex to produce than others.
High-energy inputs generally make
the material more expensive, a rule of
thumb that can be used to estimate the
embodied energy and thus the impact
its production has on the environment
(although this is not always the case).

The ingredients, molecular weight
and crystallinity define the properties of
thermoplastics. They become soft when
heated, because the polymer chains
are not permanently joined together
and so can slide over one another. This
has advantages and disadvantages. In
production, materials with lower melting
points are generally more efficient to
produce. For example, PP melts at 130°C
(266°F), whereas nylon remains rigid
up to 250°C (482°F). During its useful
life, the operating temperature (range
throughout which the plastic is usable)
affects applications and care. For example,

materials that become soft when heated
are suitable for thermoforming and
texturizing (see image left). And at the
end of their life, materials that can be
melted are much more efficient to recycle.
Acrylic is an exception to this rule, along
with special-purpose synthetics (page
520). These thermoplastic materials break
down before they melt.

Thermosetting plastics, often simply
referred to as thermosets, are used
to make dimensionally stable sheet
materials with high resistance to heat
and chemicals. PU, PVC and synthetic
rubber (although the first two may also
be thermoplastics) can be soft or rigid,
depending on the molecular structure.
The strength of the sideways bonds
between the polymers determines how
much they can move in relation to one
another before the material fractures. The
strong polymer structure of a thermoset
ensures a durable elastic material suitable
for very demanding applications. They
may be combined with thermoplastics,
such as PP, to produce elastomers that
can be melted and recycled, known as
thermoplastic elastomer (TPE). Similarly,
TPU, known as elastane or spandex, is
made up of rigid parts joined together by
flexible parts, which gives the material its
strength and elasticity.

The type of fibre determines whether
the spinning technique (page 50) uses
solvents or melting to form the polymer

Dyed colour
Polyester and acrylic yarns are package dyed with fluorescing colours. They are combined in various configurations to create a textured fabric that utilizes the vibrant and versatile properties of synthetics.

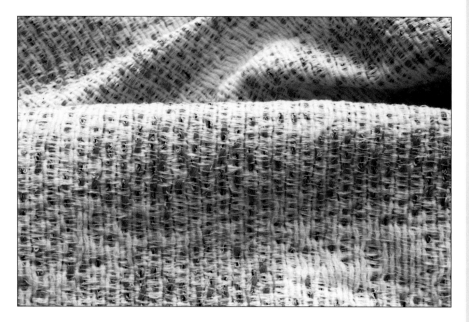

into filaments. Solvent spinning processes are either wet (into solution) or dry (into air). Used to form semi-synthetics such as viscose (page 508), they are also used to form acrylic and elastane filaments. Melt spinning is used to form thermoplastics including PP, PE, nylon and polyester. This technique does not generate volatile organic compounds (VOCs) like solvent spinning and is more energy-efficient. This means it is generally less polluting and more cost-effective.

The size and shape of filament can be adjusted from microdenier to large-diameter monofilaments. The shape is adjusted to provide different properties: flat, smooth filaments will reflect more light and so make colours appear more lustrous and vibrant. For example, trilobal nylon is triangular shaped and so reflects light to produce a shimmering surface on fabrics and flock (page 280).

Multiple materials may be extruded in combination to take advantage of the properties of two (bi-component filament), three (tri-component filament) or more different materials. Using two materials that respond to different temperatures means that when the yarn is heated it will twist into loops as only one side of the filament contracts.

NOTES ON MANUFACTURING
Synthetic plastics are extruded and spun into filament, which may then be combined into multifilament yarns, or cut into staples. Synthetics have in many cases been developed to mimic the qualities of natural fibres and are converted into yarns using the same techniques, such as spinning (page 56), twisting (page 60) and braiding (page 68). These are formed into fabrics by weaving (pages 76–105), knitting (pages 126–51), needle punching (page 152) and so on. The principal difference between thermoplastics and other types of fibre is that thermoplastics melt when heated and so can be combined with thermal energy as well as mechanical integration.

Synthetics are engineered: every ingredient has been selected for a specific purpose. Modifications are made with additives introduced during raw material production, spinning or later on during finishing. Colour is added before spinning – known as solution dyeing, dope dyeing or melt dyeing – or afterwards using conventional yarn and fabric dyeing techniques (page 240). Solution dyeing produces the most vibrant and colour-fast finish: mixed with the raw ingredients, the colour is throughout the fibre and forms a strong join with the polymer. If the colour is added afterwards it has to penetrate from the outside in. Therefore, hydrophobic materials (repel water) or those that are resistant to chemicals are difficult, if not impossible, to dye. These types of material – including PP and PE – are therefore commonly available in standard colours. Specifying a bespoke colour will require minimum order quantities of up to 10 tons. Other types of plastic, such as nylon, polyester and acrylic, can be dyed at any stage to produce vivid and saturated colour (see image above). This creates a more flexible solution for fashion textiles.

Polyester and nylon are suitable for disperse dyeing. As well as producing solid colour, this dye is utilized in transfer printing (page 272). This unique technique allows for a high level of flexibility in design and production. Dyes are applied to paper using conventional printing and later transferred to the fabric by sublimation to produce durable and long-lasting colour.

Synthetics are available in many formats, which may be combined in many different ways. The versatility of these materials as a result of being able to engineer specific properties, together with their cost-effectiveness, means they are used to mimic more expensive materials, as well as creating unique types. The high cost and desirability of natural materials such as silk (page 497), wool (page 494), fur and leather (page 498) makes them ideal for imitation by low-cost, easy-care and lightweight synthetics (see image overleaf).

SYNTHETIC MATERIAL DEVELOPMENTS
Several types of biobased and biodegradable synthetic material

Fake fur Knitted and napped polyester produces a dense, smooth and very soft pile that resembles fur. It is used to make apparel as well as to decorate interiors.

have emerged in recent years. There is significant development in this area thanks to the many advantages of building plastics from renewable ingredients that are capable of biodegrading at the end of their useful life. These types of material can be divided into three main areas: bioplastics, partially biobased plastics and biodegradable petroleum-derived plastics.

Bioplastics are derived from renewable biomass such as maize or potatoes. Their properties can be similar to petroleum-derived plastics, but they use 20 to 30% less energy to produce. They currently make up only a fraction of textile production worldwide. As an emerging technology, bioplastics have a great deal of potential for the future, but currently do not offer the convenience or versatility of petroleum-derived plastics. This is continually changing as developments in production bring more choice to designers.

Certain types of petroleum-derived plastics, including nylon, polyester and epoxy, may also contain a proportion of bio ingredients. These products are not always biodegradable and their environmental impact is not necessarily superior to that of conventional plastics if such factors as land and water use are taken into account.

Biodegradable petroleum-derived plastics are either compostable (partially biobased) or oxy-degradable (containing photoactive or thermoactive ingredients). Oxy-degradable plastics fragment into tiny particles, but their biodegradability is not scientifically proven. BASF Ecoflex and Ecovio are fully biodegradable plastic films. Ecoflex is partially biobased polyester: it has the properties of conventional PE, but is fully biodegradable under composting conditions. Ecovio consists of polylactic acid (PLA) and Ecoflex.

Perhaps the most exciting development in the area of composites is the discovery that certain thermoplastic fabrics may be consolidated into sheet material without the addition of a resin matrix. Called self-reinforced plastics (SRP), these high-strength materials have the potential to replace conventional composites in a range of applications. They are fabricated by heating and pressing woven thermoplastic fabric to form a consolidated sheet. The outsides of the fibres melt and join together when pressure is applied. The drawn fibre properties are maintained, because only the outside is melted, thus creating a sheet material with superior mechanical performance. PP and polyester have so far been developed for application in this way, used to make lightweight luggage and automotive panels. Other than the obvious mechanical advantages, this material is much more efficient to recycle than conventional composites.

Functional additives are incorporated into polymers to impart novel properties. They are sometimes referred to as 'smart', because their response to stimuli appears intelligent. A variety of functions may be incorporated by microencapsulation, such as adding fragrance (perfume release), thermochromatic properties (colour change in response to temperature) and phase-change (absorbing thermal energy when the outside temperature is warm and releasing it when it is cool).

Metal-coated yarns are being used to provide a range of high-performance functions, such as protection from intense radiant heat and electrical conductivity for wearable electronics. They may also be used to control static electricity and provide electromagnetic interference shielding. However, their durability is still a challenge, especially for military and aerospace applications, where their functionality has the greatest potential.

POLYPROPYLENE (PP)

Description

Polypropylene (PP) was first introduced to the market in the 1950s and is casually referred to as 'polyprop'. It is a polyolefin and chemically very similar to polyethylene (PE). Since improvements in the fibre increased the melting temperature it has seen continual growth. Production is likely to expand as new applications are found and it is progressively used to replace more expensive or less widely available materials, such as plant fibres (page 478). It is characterized by a slippery surface, as a result of its low coefficient of friction; high strength, thanks to the high proportion of crystallinity in the polymer structure; and low heat resistance.

Strength   • • • • ·

Tenacity is affected by the degree of crystallinity and orientation achieved by drawing. Thus, high-tenacity fibres are the result of increasing the amount of stretch applied after spinning. Its dimensional stability is not affected by moisture, making this a practical material for products that will be exposed to water, such as outdoors or on sailing boats. However, the strength of PP is affected by exposure to sunlight. Even with ultraviolet (UV) stabilizers, the fibre will eventually start to break down and become weaker.

Resilience   • • • · ·

PP has good resiliency (but lower than polyester and nylon).

It is flexible with good elasticity (up to 40% elongation before breaking). It is used as a homopolymer (containing a single type of monomer) or copolymerized with ethylene to improve flexibility. The strength begins to break down at the relatively low temperature of 130°C (266°F). However, it has excellent low-temperature properties and remains flexible down to -70°C (-94°F) or more. It has good resistance to abrasion.

Quality   • • • • ·

PP is the lightest yarn for its size at 0.9 g/cm³ (0.03 lb/in.³) and the lightest fibre overall (lowest specific gravity), being about one third lighter than polyester and one fifth lighter than nylon. It is less dense than water and so used to make ropes (it floats). It has very low thermal conductivity, more so than any other natural or synthetic fibre (a little below wool and around one third compared to cotton) and so retains heat for longer. Its low specific gravity helps to keep weight down while providing a high level of bulk and cover. It does not absorb moisture and so can feel uncomfortable if worn next to the skin. However, the benefit of low moisture absorption is that it is difficult to stain. It also has excellent resistance to chemicals, including acid, alkali and bleach. The surface of PP feels quite slippery as a result of its low coefficient of friction. This helps to make it resistant to microorganisms, but also affects the tactility. It also adversely affects printing (pages 256–79) and adhesive bonding (page 370), because very few chemicals will stick to the surface with sufficient strength. Therefore, to form a strong join, PP is welded (page 376).

Colour   • • • • ·

It may be tinted or opaque. Colour incorporated into the fibre during production, known as solution dyeing (see page 240), produces the most durable and saturated results. Additives help to reduce the effects of exposure to sunlight and the colour has good fastness. Colour applied to the fibre after production (yarn or fabric stage) will not be very durable, owing to the fibre's hydrophobic properties, the low number of active sites for chemical bonds, and the slippery surface that dyes will not stick to very well. Increasing the fibre's affinity for dyeing compromises other properties.

Versatility   • • • • ·

It is produced as fibre, film and sheet, and molded into products such as toys and packaging. As a widely used fibre it is available in a range of formats from microdenier to large-diameter monofilaments, including textured (although it does not texture as well as other types, such as polyester), flat and staple. The largest area of application is nonwovens (page 152), ranging from teabags and nappies (diapers) to geotextiles. It is also used to make carpet (nonwoven backing and woven pile, page 96), packaging, sacks, and fabrics for automotive interiors. It can mimic natural materials, such as to replace hemp rope on reconditioned ships. Applications are widespread beyond conventional textiles. For example, it is used as fibre reinforcement in concrete (to provide strength), as self-reinforced thermoplastic sheet and thermoplastic composite laminating (page 446). PP films are used for many textile applications too, including laminated fabrics for industrial and agricultural purposes. They are used as an alternative to PVC-coated fabric, because they are cost-effective and have lower environmental impacts. PP foams are used in laminated fabrics for insulation and packaging applications.

Availability   • • • • •

It is widely used fibre, film and sheet produced in large quantities. Specifying a unique solution-dyed colour or other property improvement added at the production stage will require a large minimum order to make it worthwhile for the manufacturer.

Cost   • • • • •

It is one of the cheapest plastics, but the price fluctuates according to global oil prices and demand for the material.

Environment   • • • · ·

PP is highly efficient to recycle and can be mixed with other similar materials during processing, such as PE. It is also incinerated to recover energy. PP film laminated with PP fabric can be readily recycled, unlike materials with similar properties but made up of a mix of different plastics.

POLYESTER

Description

Polyesters are thermosetting (unsaturated with alkyl resin backbone), semi-crystalline thermoplastic (saturated with plasticizer) or biobased. Polyester fibre and film used in textiles is mostly thermoplastic polyethylene terephthalate (PET) made up of copolymerized ethylene glycol and terephthalic acid. Polyester fibre and film were introduced to the market by DuPont in the 1950s under the trade names Dacron and Mylar. PCT is a widely used material, including for packaging and industrial applications. Other types include polytrimethylene terephthalate (PTT), polybutylene terephthalate (PBT) and polylactic acid (PLA) (see Bioplastic).

Strength   • • • • ·

Polyester fibre is strong, with a similar tenacity to that of polypropylene (PP). Drawing during spinning improves the alignment of the molecules and so increases strength. Film is also stretched during production to improve strength. Known as biaxial-oriented polyester (BOPET), thin films are very strong for their weight and used to make high altitude balloons, for example. Polyester is hydrophobic and so has the same dry and wet strength.

Resilience   • • • • ·

It has good resilience with good resistance to wrinkling and abrasion. However, after repeated bending the fibre properties start to break down.

Quality   • • • • ·

It has a relatively high melting temperature of up to 250°C (482°F), can be pressed during production and has good shape retention. This ability means that quilted (page 196) fabrics with a polyester filling will retain their shape well over time, as opposed to flattening out like natural fibres. This can be advantageous or uncomfortable, depending on the application. Hollow polyester fibres provide insulation by trapping air inside. Polyester is hydrophobic and so can feel clammy in warm weather; this same property means that it remains dry to the touch when used as a facing layer over an inner absorbent medium that becomes saturated. Pilling occurs on the surface as a result of abrasion.

Colour   • • • • •

It is naturally milky white and can be disperse dyed (page 240) a wide range of vivid colours with excellent fastness (see title image, page 510). Polyester is naturally lustrous and often used to imitate silk (page 497). This can be a problem – the sheen may be recognizable in low-cost clothing – and is reduced by adding titanium dioxide. Stretching the fibre, or film, results in a transparent material, which may have very good clarity.

Versatility   • • • • •

Thermoplastic polyester is available as fibre, film and molded plastic, making this a very versatile material. Fibres are produced as solid, hollow, textured and filament, and range from micro to large denier. As a staple fibre polyester is blended with natural fibres to improve wrinkle resistance and durability. The fibres are used to make every type of apparel including suits, sportswear, outdoor clothing, disposables and socks. It is available knitted (pages 126–51), woven (pages 76–105) and nonwoven (page 152) (although PP is overtaking in this area). Its high versatility and low cost have meant that it has replaced viscose (page 508) in many applications. It is used to make fabrics for interiors, including bedding, mattresses, curtains, upholstery and carpets. It is used to make yarn and thread for stitching (page 354) and embroidery (page 298). And it is twisted (page 60) and braided (page 68) to make rope for industrial and sailing applications.

BOPET film is used in a wide range of applications, ranging from a few microns up to 350 microns thick. It is metal coated for both technical and decorative applications, such as to create an impermeable barrier to gases (balloons and packaging), provide insulation, or create a light-reflective surface. Laminated with high-performance filaments, such as aramid (page 524) and carbon (page 526), it is used to make lightweight racing sails (page 374). It is also laminated onto paper to create a durable and weather-resistant surface, such as for packaging applications.

Availability   • • • • •

Polyester fibre is manufactured by many large-scale producers and sold under various trade names. Some notable examples include Coolmax, a high-performance yarn comprising polyester blended with wool to create warm, breathable and moisture-wicking fabrics; Thermocool, used to produce thermoregulating and moisture-wicking fabrics; and Dacron, a high-tenacity polyester used in all applications from woven garments to monofilament fishing line and braided high-performance rope. Films are also widely available and produced by several companies. The best-known example is Mylar by DuPont.

Cost   • • • • •

It is low-cost, similar to PP.

Environment   • • • · ·

Polyester can be recycled, for example by converting drinks bottles into fibre (a technique popularized by the success of Polartec recycled polar fleece, first produced in 1993), or broken down to the base monomer and repolymerized into new plastic.

NYLON

Description

Developed and commercialized by DuPont in the late 1930s, it was the first truly synthetic fibre. Amide bonds are produced from a mix of adipic acid and hexamethylene diamine, and the resulting material is known as polyamide (PA). Like polyester, it was developed as a low-cost alternative to natural fibres, such as silk (page 497) and cotton (page 481). As well as being one of the most important modern fibres, it is widely used to make molded engineering products such as for automotive applications. There are many different types available, numbered according to their molecular structure, including nylon 6, 12, 6/6, 6/10, 10/10 and 6/12. Fibres are produced from nylon 6 and 6/6, both of which are semi-crystalline.

Strength   • • • • •

Nylon is a light and very strong fibre. Cold drawing improves molecular alignment and so increases strength up to a point.

Resilience   • • • • •

It has very good elasticity (more than most fibres, except elastane and rubber), but has lower wrinkle resistance than polyester. Therefore, polyester (or polypropylene) is preferred for nonwovens (page 152), because it will retain its shape better. Nylon is very hard-wearing and used to make carpets and upholstery for high-traffic applications.

Quality   • • • • •

Nylon has good resistance to chemicals and abrasion. It is degraded by exposure to sunlight over long periods of time. It has a relatively high melting temperature of 250°C (482°F) or more, although nylon 6 is slightly lower at 220°C (428°F). Nylon is hydrophilic and so prone to absorbing moisture. Although a relatively small percentage compared to natural fibres, such as flax (page 481), this does not make it any more comfortable to wear. However, it can affect the dimensional stability of nylon and so cause problems during manufacture. Abrasion causes surface pilling.

Colour   • • • • •

Nylon is available in a wide range of solution-dyed colours, or it can be dyed (page 240) at a later stage. It is produced with a smooth lustrous surface – such as used to imitate silk or create a bright flocked colour (page 280) – or it is made with a dull surface finish to produce matt colour.

Versatility   • • • • •

It is available as monofilament, multifilament and staple, which may be fine, smooth or crimped. It is widely used in apparel (outdoor wear, sportswear, swimwear, dresses, suits and lingerie), interior (bedding, netting and curtains) and technical applications (fishing line, toothbrush bristles, sleeping bags, backpacks, tents, airbags, seatbelts, umbrellas, nets, webbing and racket strings). One of the first major applications was knitted hosiery: nylon was a great success in 1940s, because high-quality hosiery could be produce with cost-effective circular knitting (page 138) techniques combined with heat setting (see Boarding, page 416). Nowadays, owing to its relatively high cost compared to PP and polyester, it tends to be reserved for applications that demand its higher performance properties, ranging from lightweight and tightly woven windproof fabrics (outerwear and parachutes) to hardy nonwoven abrasive pads. It is equally used as sheer or pile (page 96). It is laminated (page 188) with a waterproof thermoplastic outer layer. Or it is combined with natural fibres, such as cotton or wool (page 494), to increase their strength and durability.

Availability   • • • • •

Nylon is manufactured by many companies and converted into a wide range of fabrics. Examples include Cordura, a high-performance textile used to make outdoor clothing, and ripstop (page 78), a lightweight fabric traditionally made from nylon that is resistant to ripping and tearing.

Cost   • • • • •

It is moderately expensive and higher priced than polyester and PP.

Environment   • • • • •

Nylon is recycled or incinerated to reclaim the embodied energy at the end of its useful life.

ACRYLIC

Description

Acrylic, also referred to by its chemical name, polyacrylonitrile (PAN), was introduced to the market in 1950s and is typically a copolymer comprising at least 85% acrylonitrile. Modifying with additional ingredients produces modacrylic (35–85% acrylonitrile). It comes in a variety of formats and the exact properties of the fibre depend on the production and ingredients. It is thermoplastic, but does not melt like the others. Instead, heating causes the fibre to break down. This helps to make it inherently flame-retardant. It is used in textile coating (page 226) and adhesive systems (page 370). Its inherent resistance to sunlight, heat and chemicals makes it durable and cost-effective for both indoor and outdoor applications.

Strength   • • • • •

Acrylic fibres have moderate strength, but this varies depending on the type.

Resilience   • • • • •

The fibre has good resilience and will not wrinkle easily. It has low elasticity, although this may be improved by adding plasticizer.

Quality   • • • • •

Acrylics have good resistance to staining and are extremely resistant to sunlight (depending on the type). The fibre is soft and lofty and often blended with wool (page 494) to take

advantage of the properties of both and produce fabrics that are light, durable, bulky and easy to care for. It is napped (page 216) to produce a soft hand. It is also blended with cotton (page 481) and silk (page 497) to improve durability and shape retention. Acrylic fibres are not good conductors of heat and so can feel warm. The fibre has very low water absorption. Moisture sticks to the surface and so migrates through fabrics, evaporating slowly. Soft fibres are prone to pilling, as a result of abrasion, which results in a softer and fuzzier surface. Modifications exist that increase resistance to pilling. Acrylic coatings and adhesives are durable, with good resistance to stiffening with age.

Colour   • • • • •

The fibre is dyed in solution or once formed into a fibre or fabric (see Dyeing, page 240). Coatings and adhesives are pigmented with colour. A wide range of vivid colours is possible, with good fastness.

Versatility   • • • • •

On their own or blended with wool, acrylic fibres are soft and easy to care for, making them useful in the production of jumpers, jackets, sportswear, socks and children's clothing. They are reasonably durable and flame-retardant and so are used for interior fabrics, such as blankets, curtains and upholstery. Modacrylics are distinguished by improved abrasion resistance, which makes them particularly useful in carpets. Acrylic has excellent resistance to sunlight and so is often used for upholstery and outdoor fabric applications, such as awnings and covers. Its lightness is utilized to make fleece and fake fur. Mixing fibres that are affected differently by heat treatment, such as varying levels of shrinking, produces fake fur with more natural-appearing variation. PAN is the chemical precursor to carbon fibre (page 526): oxidizing the resin at very high temperature causes tightly bonded carbon crystals to form.

Availability   • • • • •

As a fibre, it is not as widely used as polyester and PP, but nevertheless is available in a large array of different formats. Dralon is an example of acrylic fibre produced for many different applications, including interior, exterior and apparel.

Cost   • • • • •

These fibres are relatively low-cost, but this depends on the exact ingredients and method of production.

Environment   • • • • •

Acrylic requires more energy to convert into a fibre than polyester or PP, and acrylonitrile is highly toxic. Acrylic does not melt and so cannot be recycled as easily as other thermoplastics. As a coating it is most commonly used with polyester fabric, producing a composite that cannot be recycled by melting.

POLYETHYLENE (PE)

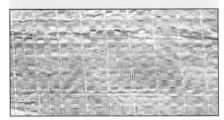

Description

Polyethylene (PE) is a type of polyolefin and chemically very similar to polypropylene (PP) . It is also sometimes known as polythene. It is fabricated from ethylene monomer and the polymer chain length determines the physical properties, from ultra-low molecular weight (ULMWPE) to ultra-high molecular weight PE (UHMWPE). The first commercial production of PE was in the 1950s, but it was not until much more recently that full-scale production of very high-strength PE fibre was established. The fibre is not as versatile as PP and so is not found in such a diversity of applications, but it can be produced with very high strength to weight, far superior to aramid (page 524). The film is used to make packaging (such as plastic shopping bags) and slit into thin strips woven into fabric (such as agricultural sacks).

Strength ●●●●●

PE's properties are defined by the molecular weight of the polymer (length of the polymer chain). Increasing chain length improves the strength and resilience, because even though the bonds between the polymer chains are not very strong, increasing chain length creates a denser material and this multiplies the number of overlaps. Fibres made from UHMWPE, also known as ultra-high-density polyethylene (UHDPE), have exceptional strength to weight (up to twice that of standard aramid). When placed under constant load, the fibres will deform, a drawback known as creep.

But this has been virtually eliminated in some recent fibre developments. It has a relatively low melting temperature of 130°C (266°F), but remains strong and flexible to below -150°C (-238°F).

Resilience ●●●●○

PE recovers well from wrinkling and can be heat set to improve wrinkle resistance. It is flexible and so exhibits excellent loop and knot strength, two essential properties for reliable high-strength rope (pages 60 and 68).

Quality ●●○○○

It has very low coefficient of friction, so the surface feels slippery (comparable with polytetrafluoroethylene). This means it is particularly tricky to bond with adhesives (page 370) or print (pages 256–79). Its slipperiness provides very high resistance to abrasion. It is non-toxic (as a molded plastic it is commonly used to make food packaging). It does not absorb much water and moisture will not stick to the surface, which helps to make it resistant to mold, bacteria and staining. It is lightweight and a little less dense than water, so like PP it will float. It has very good energy absorption, a quality utilized in armoured vehicles. It is degraded with exposure to sunlight. This problem is reduced, but not eliminated, with additives.

Colour ●●○○○

Owing to its water resilience, PE fibre is typically solution dyed. As such, a wide range of colour is possible, but minimum order quantities will apply for special colours. It is most commonly used in natural white.

Versatility ●●○○○

PE fibre in various forms is used to make low-cost apparel, including underwear, sportswear and swimwear. Like PP it is also utilized in high-wear applications, such as carpets, and for high-strength applications, such as industrial packaging. As a nonwoven (page 152), it is used in medical, cosmetic and geotextile products. However, it is not as widely used as PP for commodity applications. It is more commonly used for high-performance applications, such as yacht rigging and parachute cord. It is used to make slings and ropes for heavy lifting equipment. DuPont Tyvek is a high-strength and lightweight nonwoven used in protective clothing, such as disposable suits, as well as in tear-resistant envelopes, maps and other printed media. At the extreme end of the spectrum, it is constructed into reinforced laminates utilized in ballistic armour, cycle helmets and other safety equipment. Combined with a suitable resin matrix, thermoplastic composites (see Composite Press Forming, page 446) may be molded into three-dimensional parts, such as military helmets.

Availability ●●●●○

UHMWPE fibre is widely produced and more commonly referred to by its trademark names, such as Dyneema or Spectra. PE films are also common, but mainly used for non-textile applications such as packaging and industrial products.

Cost ●●●○○

For its high strength, PE is relatively inexpensive thanks to the straightforward manufacturing processes.

Environment ●●●●○

Like PP, it has lower embodied energy than other types of synthetic fibre. In other words, less energy is required in its production. On top of that, it is efficient to recycle (nowadays, PE packaging is one of the most widely recycled materials).

ELASTANE

Description

Elastane was developed as an alternative to rubber (page 168) in the 1950s. Also known as spandex (an anagram of 'expands'), it is thermoplastic polyurethane (TPU) (see also Polyurethane, page 518). The unique property of this elastomer, as a result of its randomly coiled polymer structure made up of a mix of stiff (diisocyanate) and flexible (macroglycol) segments, is that it can be stretched 500% or more without breaking or permanently deforming.

Strength ●●●●●

It is strong, and it has greater density than rubber but higher strength, and so can be used in lighter weights. It is generally more durable than rubber. But strength starts to break down at around 120°C (248°F).

Resilience ●●●●●

Its elastomeric properties are the result of flexible coiled parts of the structure straightening out as load is applied, and as soon as the tension is released the material instantly rebounds to its original shape. The amount of stretch ranges from 400% to 700% depending on the proportion of flexible and rigid parts.

Quality ●●●●○

The surface is soft and so conforms to other shapes, which makes it grip and stick. Therefore, when used in garments it is typically covered, or wrapped, with a more comfortable-feeling material, or concealed in a sleeve. Depending on the elasticity and proportion it can be engineered with varying degrees of tension (from body-hugging to compression). It is resistant to degradation by perspiration, oils and detergents. Stabilizers are added during production to improve the material's resistance to sunlight, heat and chlorine. As a thermoplastic yarn, it can be heat set into shaped profiles.

Colour ●●●●○

It is clear or opaque and suitable for dyeing a wide range of colours with disperse and acid techniques (page 240). It is prone to yellowing on exposure to sunlight and stabilizers are required to counter this problem.

Versatility ●●●●○

Elastane is most commonly blended with other fibres, such as cotton (page 481) or polyester. It is either combined by weaving (pages 76–105) or knitting (pages 126–51), or it is wrapped in another, nonelastic fibre, known as covered or core-spun. Up to around 40% produces a very stretchy garment, such as body-hugging sportswear, swimwear, undergarments, hosiery and leggings. It is used to make compression garments too, such as for medical, sports or cosmetic applications. A small amount of elastane (around 2%) combined with other fibres brings stretch and recovery to otherwise stiff woven garments, such as skinny jeans, while increasing the stretch and recovery of knitted constructions. It is also used to make fitted bedding and upholstery.

Availability ●●●●●

It is produced by several major manufactures and is widely available. The most well known is Lycra.

Cost ●●●●○

It is moderately expensive, but normally used in small amounts blended with other fabrics. In other words, it is cost-effective for its relatively high performance.

Environment ●●●●○

It is relatively complex to produce, but no more polluting than other synthetics. Thermoplastic materials are recycled or incinerated to recover their embodied energy.

POLYURETHANE (PU)

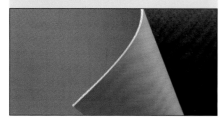

Description

Polyurethane (PU) is synthesized from diisocyanate and a polyester or polyether (polyol). It is used in many formats, including as a fibre (see Elastane, page 517), film and coating. It is either thermoset of thermoplastic. Thermosetting PU forms permanent cross-links between the polymer chains and for this reason cannot be melted and reprocessed. It is applied as a coating (page 226) or adhesive or molded to shape. Thermoplastic polyurethane (TPU) becomes soft when heated and so is suitable for laminating (page 188) and welding (page 376), as well as for applying as a single-component coating.

Strength ● ● ● ● ○

It is strong, lightweight and generally more durable than natural rubber (page 168). Whereas the strength of TPU starts to lessen at around 120°C (248°F), PU will not melt, owing to the thermoset polymer structure.

Resilience ● ● ● ● ●

It is flexible and elastic because of the polymer structure. It is made up of stiff cross-linked areas (diisocyanate) intersected with flexible segments (polyol). The flexible parts straighten as load is applied and rebound when released. The amount of stretch depends on the proportion of flexible and rigid parts.

Quality ● ● ● ● ○

PU coatings and laminates are performance materials with high flexibility and soft hand. PU is highly resistant to tearing and abrasion, which helps to keep material thickness to a minimum. Microporous PU is very durable and suitable for high-flex applications.

Colour ● ● ● ● ○

It is available in a wide range of vivid colours, including saturated and fluorescent. TPU is suitable for dyeing and PU is not. Therefore PU is coloured in solution.

Versatility ● ● ● ● ●

PU is very versatile. It is available as fibre, sheet or coating, and can be molded to shape. It may be homogenous or foamed. PU is used to make padding; as a coating (page 226), it is used to make synthetic leather and waterproof fabric. There are two main coating systems: one-part and two-part. One-part systems are suitable for PU and TPU resins. The polymer is dissolved in solution and cures (PU) or physically dries (TPU) after application. Two-part systems involve the reaction of two components to form cross-linked PU. TPU film is laminated with knitted and woven fabrics to make shoes, bags, inflatables and upholstery. It is molded into shoes and packaging. Because TPU can be softened with heat, it is suitable for welding, as well as adhesive bonding (page 370) and machine stitching (page 354). Combined with polyester knit by laminating, it is referred to as polyurethane fabric (PUL), and is hard-wearing and resistant to washing.

Availability ● ● ● ● ●

These are commonly produced and widely available.

Cost ● ● ● ● ○

PU is low-cost and TPU is more expensive.

Environment ● ● ○ ○ ○

Diisocyanate is toxic and must be handled carefully during production. At the end of its useful life, TPU is recycled or incinerated to recover the embodied energy. PU cannot be recycled and so is incinerated or disposed of.

SYNTHETIC RUBBER

Description

There are many types of synthetic sheet material used in place of natural rubber (page 168). First invented in 1909 and driven by the demand for transportation (tyres), synthetic rubbers in many different types and grades have been developed since. Some of the most common examples include styrene butadiene rubber (SBR), butadiene rubber (BR), polychloroprene rubber (CR), isoprene rubber (IR) and ethylene propylene rubber (EPM and EPDM). They have shape memory, which means that they will return to their original shape after deformation.

Strength ● ● ● ● ○

They are typically stronger and more resistant than natural rubber.

Resilience ● ● ● ● ●

They have excellent resilience. The level of flexibility depends on the ingredients and additives and ranges from highly compressible to semi-rigid.

Quality ● ● ● ○ ○

The properties of these materials vary: they may have outstanding properties in one or two areas, but are more commonly preferred for their balance of properties, such as good mechanical strength, resistance to chemicals and weathering, abrasion resistance and low flammability.

Colour ● ● ○ ○ ○

The colour depends on the base material, because while some (such as SBR) are available only in black, others (such as EPM and EPDM) can be produced in a range of inherent colours. CR wetsuit material is laminated (page 188) with knitted jersey (page 126) to add a layer of colour.

Versatility ● ● ○ ○ ○

They are mainly used as sheet materials, and the bulk of applications are in automotive tyres and industrial hoses. Fashion and textile items include swimsuits, footwear and other sports clothing. One of the best-known examples is neoprene (which is CR), developed by DuPont in 1930s and used to make a range of products including wetsuits and drinks coolers, among other things.

Availability ● ● ● ● ○

There are many different types, each with its own unique elastomeric properties.

Cost ● ● ○ ○ ○

They are moderate- to high-cost, depending on the properties and performance.

Environment ● ○ ○ ○ ○

The production techniques vary according to the base polymer, but may require high energy input. Owing to the cross-linked polymer structure, these materials cannot be recycled by melting and reprocessing. Therefore, they are typically disposed in landfill and will remain intact for thousands of years.

POLYVINYL CHLORIDE (PVC)

Description

Polyvinyl chloride (PVC) is thermoplastic or thermoset and ranges from stiff to flexible, depending on the proportion of plasticizer in the base material. It is versatile and low-cost, and thus widely used in apparel and industrial applications. PVC-coated fabrics are often simply referred to as vinyl.

Strength ● ● ● ● ○

PVC is predominantly used in combination with a woven (pages 76–105) or knitted (pages 126–51) base layer: the strength will be determined by the combination of materials. PVC itself is a strong polymer and combined with high-tenacity fibres makes a very durable composite.

Resilience ● ● ● ● ○

It is resilient and the amount of flexibility depends on the proportion of plasticizer and thickness of material.

Quality ● ● ● ● ○

It is inherently flame-retardant, dimensionally stable and resistant to sunlight, but not suitable for exposure to temperatures above 60°C (140°F) for prolonged periods. It can be gloss or textured, depending on the calender roll (page 220), and solid or foamed. It is not as high quality as polyurethane (PU), but considerably less expensive. It is waterproof and stain proof and so used to make protective clothing, bags and shoes. It is inert and so used for medical applications, such as blood storage bags. For particularly demanding applications, such as outdoors, it is coated with acrylic to increase resistance to the elements.

Colour ● ● ● ● ○

There is virtually no limitation on the range of colour that can be created in PVC. It is always fabricated in the finished colour, not dyed afterwards.

Versatility ● ● ● ● ○

Typically applied as a coating, PVC is used in a wide range of applications, from fashion textiles to industrial products and architecture. It is produced to mimic leather and rubber, with suitable surface texture and colour. These are used to make clothing, upholstery, water tanks, tablecloths, inflatables, banners, canopies, tarpaulin and industrial fabrics. Architectural applications include awnings, canopies, tents and walkways. PVC-coated fabrics are stitched (page 354) or welded (page 376). As well as pigments, it is often combined with additives to enhance performance, such as plasticizers, fillers, lubricants and flame-retardants. It is also used as the base resin for screen printing (pages 260–71) inks, known as plastisol.

Availability ● ● ● ● ●

It is widely produced. The source is important, because different grades have been developed that have reduced environmental impact.

Cost ● ● ● ● ○

PVC is low-cost, but the fabric it is applied onto affects cost, depending on the fibre, construction and density.

Environment ● ● ○ ○ ○

PVC plastics are often criticized for containing harmful ingredients. For example, phthalates and bisphenol A are subject to restrictions in many countries. Non-phthalate plasticizers are available for sensitive applications, such as toys and medical devices. Thermoplastic PVC can be readily recycled.

SILICONE

Description

Silicone is a high-performance coating (page 226) material applied to a variety of base fabrics. It is a polymer consisting of oxygen and silicon atoms, among others, and available in a wide range of molecular weights and viscosities.

Strength · · · · ·

Silicone is not very strong and tensile strength is therefore governed by the fabric layer. For this reason, coated fabrics are sometimes referred to as cloth-inserted silicone.

Resilience · · · · ·

It is resilient and resistant to abrasion. It does not absorb water and has a high level of flexibility.

Quality · · · · ·

It can be used over a very wide temperature range, from -100 to 300°C (-148–572°F), which is more than any other elastomeric material. It is chemically inert, resistant to sunlight and hydrophobic. It has low surface energy, so very few things will stick to it, which helps to keep it clean and free from bacteria and mildew. Produced with a smooth surface, it can provide a high level of grip as a result of conforming to shape and thus maximizing surface contact.

Colour · · · · ·

It can be produced in a range of vivid colours, from translucent to opaque.

Versatility · · · · ·

Thanks to its unique properties silicone is used for specialist and demanding applications, such as medical products, safety curtains, gaskets, high-pressure hoses and pumps. It is applied as a thin coating to protect fabrics used outdoors, such as tents. Its smooth surface, which helps it to grip without sticking, is used to prevent clothing from slipping. It is applied to fabrics to improve weathering, wrinkle resistance and tear strength, without detracting from the breathability. It may also be used to improve softness and handle or a range of woven (pages 76–105) and knitted (pages 126–51) fabrics. Silicone printing ink is presenting itself as a viable alternative to PVC-based inks, without the negative environmental implications.

Availability · · · · ·

Silicone is widely produced and each variant has its own unique properties.

Cost · · · · ·

It is an expensive material and its price typically dominates the cost of the coated fabrics.

Environment · · · · ·

Silicone is energy-intensive to produce and as a thermosetting resin it cannot be recycled.

BIOPLASTIC

Description

Starch contained in plants is converted into biobased monomers by bacterial fermentation or catalytic reduction. The monomers are polymerized into bioplastics, which are applied as fibre (page 50), film (page 180), coating (page 226) or prepreg (page 234) using conventional plastic manufacturing equipment. Depending on the type of polymer, the characteristics may be comparable with polypropylene (PP) and polyester.

Strength · · · · ·

Properties range from stiff and rigid to flexible. As a result of their biodegradability, the properties break down once the material has been hydrolysed (broken down through a chemical reaction with water). The point at which microbes can start to break down the polymer depends on which type is used and may be immediate or only after heating in a suitable humid environment.

Resilience · · · · ·

The resilience of the fibre or film depends on the type of polymer used, ranging from poor to good.

Quality · · · · ·

This depends entirely on the type of material. Thermoplastics types are classified as polyester, but their properties may be quite different, depending on the polymer structure. For example, polylactic acid (PLA) may be processed in much the same way and exhibits the same properties as conventional plastics, including high tenacity, good elastic recovery and high resistance to sunlight. Once hydrolysed at high temperature (60°C/140°F) the polymer becomes susceptible to microbial degradation. This means it is compostable.

Colour · · · · ·

This depends on the polymer and continues to be explored, because these materials are still relatively new. PLA has excellent colourability and can be dyed in the same way as polyester using disperse dyes (page 240).

Versatility · · · · ·

There is a range of thermoplastic types. PLA is applied as a fibre or a film. It can have similar properties to polyester or be quite different, depending on the polymer structure. It is produced as woven (pages 76–105), knitted (pages 126–51) and nonwoven (pages 152–57) fabric for garments, nappies (diapers), interiors, upholstery (page 404) and automotive, as well as fibre reinforcement in composite laminates. Polyhydroxyalkanoate (PHA) and poly-beta-hydroxybutyrate have similar properties to PP. Polyglycolic acid (PGA) and PLA are used in medical applications, such as temporary sutures (resorbable) and drug delivery systems. PGA is brittle and is degraded rapidly by hydrolysis and is therefore not suitable for many other applications. Polyfurfuryl alcohol (PFA) is a thermosetting resin derived from the catalytic reduction of crop waste such as sugar cane bagasse. It used as the resin matrix in prepreg biocomposites (page 239).

Availability · · · · ·

Even though they are more sustainable than synthetic materials, they currently make up a only minuscule fraction of global textile consumption. The production techniques used are a recent development and the technology is continually progressing through research and development. As a result they are not widely available, nor are they available in many different formats and finished. However, this is sure to change.

Cost · · · · ·

PHA and PLA are recent developments and so are expensive compared to synthetics such as PP and polyethylene (PE): PLA is twice the price and PHA is about four times more expensive, taking into account the density changes. The price is expected to continue to fall as production becomes increasingly efficient and demand increases.

Environment · · · · ·

Some are compostable, while others are biodegradable. Compostable means they fulfil US and EU standards (ASTM 6400 and EN 13432 respectively) for degrading in composting conditions (more than 90% must be converted into carbon dioxide, water and biomass within 90 days), whereas biodegradable simply means the material can be broken down into carbon dioxide, water and biomass by microorganisms within a reasonable length of time. The source of biomass is critical because the impact of growing the crops may outweigh the benefits – for instance, deforestation, genetic modification (GM), the use of petroleum-powered machinery for production and transportation or the displacement of local food production and increased food prices may all be taken into account.

Materials

Special Purpose Materials

These materials are defined by their unique properties and include certain synthetics, metals, minerals and carbon. They are converted into fibres, usually through complex and energy-intensive processes, to create fabrics with exceptional strength, stiffness or resistance. They are utilized in some everyday products, but are relatively expensive and so are mainly reserved for protective, technical, extreme sports, military and industrial applications. Metals are also used for their decorative and functional qualities. They are applied as coatings, such as in the production of bright and reflective packaging films, and as embroidery thread, which is used to embellish high-value fabric and leather; and they are foil calendered onto fabrics to produce a textured metallic surface.

MATERIAL PROPERTIES

Synthetic fibres are manufactured from long chains of monomers, joined through polymerization (see Synthetic Materials, page 510). Special purpose synthetics are some of the strongest materials available and are being utilized in the production of continually lighter and stronger structures. Super fibres, as they are sometimes called, are known mainly under the trademark names: aramid is known as Kevlar, Twaron and Nomex; polyphenylene benzobisoxazole (PBO) is known as Zylon; expanded polytetrafluoroethylene (ePTFE) is known as Tenara; and liquid crystal polymer (LCP) is known as Vectran.

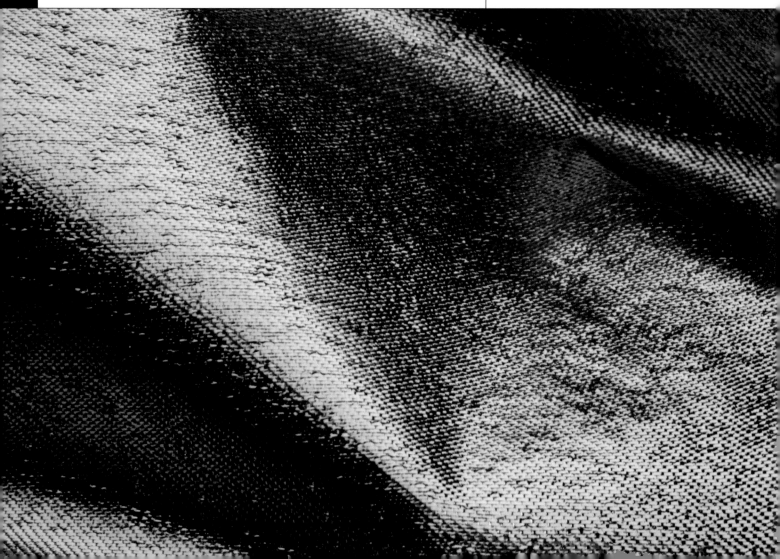

Nomex glove
Knitted meta-aramid fibre provides very good thermal insulation. It is applied in protective clothing, ranging from firefighters' outfits to barbecue gloves.

They are different from conventional synthetics, such as nylon (page 516) and polyester (page 515), as the result of modifications to the polymer structure to increase molecular weight and strength of bond. Combined with their inherent lightness, this makes them very desirable for niche and demanding applications.

Para-aramid is an aromatic polyamide, which means that unlike nylon (aliphatic polyamide) many of the amide bonds are cyclic (closed). The cyclic bonds ensure that the monomers join together into much longer unbroken chains (increased molecular weight), with highly oriented crystal structure. This results in a fibre with significantly higher tenacity, stiffness and chemical resistance. Its highly oriented structure, a phenomenon referred to as liquid crystal, means that full strength is achieved without needing to draw out the fibre. The same is true for LCP, a type of aromatic polyester.

Meta-aramid, often referred to by its trademark name Nomex, is based on a similar structure to para-aramid, but with a slight variation in the location of the bonds. This results in a material that is not as strong, but has exceptional heat resistance and flame-retardant properties. These unique qualities are utilized in protective clothing (see image above), as well as in composite laminates for aircraft and vehicles that have to operate within strict fire regulations.

Acrylic (page 516) and viscose (page 508) are both suitable as the chemical precursors to carbon fibre. Following controlled pyrolysis at very high temperature, the polymer is converted into a tightly packed crystal structure with very high-strength bonds between the molecules. It has similar composition to graphite, but the structure is oriented lengthways, resulting in a fibre with unmatched tensile strength and stiffness. These properties are used in ultra-lightweight composites that have revolutionized the design of aeroplanes, boats and cars alike.

Metals are elements, many essential for sustaining life on earth. They are extracted from their mineral ore, refined and processed to form combinations of metallic and other elements suitable for converting into yarn and coatings for textiles. This uses energy-intensive and environmentally damaging processes, so metals are typically more expensive than synthetics. However, they have many important characteristics unique to this group of materials: they are hard and highly resistant to abrasion; they can be bent without breaking to take on new shapes; they are dense and heavy; some are practically inert; and silver and copper are antimicrobial.

There are many different types of metal suitable for an unlimited range of applications. The types most commonly used as fibre include gold, silver, titanium, steel, aluminium and copper.

Metals are divided into two main groups according to whether they contain iron (ferrous) or not (non-ferrous). Ferrous metals, such as steel, are prone to surface corrosion. In the presence of oxygen and water the iron reacts to form a layer of iron oxide, commonly known as rust. This is a degradation process that gradually consumes metal that contains iron from the outside inward. Stainless steel contains chromium, which helps to protect the surface.

Some non-ferrous metals, such as gold and platinum, are practically inert. Others react with the atmosphere and produce a surface oxide layer that forms a protective barrier against further corrosion. The oxide layer is also very durable, such as anodized aluminium. The layer of oxide that is produced by anodizing (electrochemical process employed to build up the naturally occurring oxide) is among the hardest materials known.

Basalt and glass fibres are mined from rock with high silica content. The ingredients are melted at very high temperature and converted into fibres. They have high tensile strength – basalt is slightly stronger – and this property is utilized in a range of composite laminates, often together with other higher-strength fibres, such as carbon.

They are incombustible in application and so are very useful for providing protection and insulation. They are surprisingly flexible considering the brittleness of their raw ingredients. This is as a result of the high surface area to volume ratio. Asbestos is a mineral that was once used in large quantities for many of the same reasons. However, it was discovered that the fine fibres have a propensity to cause fatal illnesses when inhaled. As a result, asbestos is banned in most countries and special licences are required to handle the material.

Embroidery PET film is metallized to look like gold, cut into thin strips and wrapped around a filament core. It is embroidered onto fabrics for embellishment.

NOTES ON MANUFACTURING

Special purpose materials tend to be complex and costly to produce, owing to the difficultly of overcoming their inherently resistive properties. Glass and basalt are the slight exception, but still require heating up to 1,500°C (2,732°F) or more to melt the material. The price is a good indicator of the amount of energy required to extract and convert these materials into usable fibres.

Aramid, LCP and PBO are converted into fibre by solvent spinning (page 50). PTFE will not melt and is practically inert, so is spun in a temporary carrier and sintered to produce a consolidated fibre.

Basalt Fibres derived from organic rock are surprisingly flexible owing to the high surface area to volume ratio. They are woven into fabrics used for thermal insulation and composite laminating.

Metal wire is produced by drawing bulk metal through progressively smaller dies until the desired diameter is reached. It is typically round but can be square, triangular or another shape. Braided (page 68) and twisted (page 60) into cable, it is typically used for loadbearing, conductive and protective applications. In the case of textiles, metal yarn is also used for decorative applications, such as embroidery (page 298).

Glass, basalt and carbon are inherently stiff and brittle. This makes them trickier to process into fabrics. They are mostly woven (see image below) or unidirectional (see Spread Tow and Fabric Prepreg, page 234), because this requires the least amount of bend be applied to the fibre. Produced as filaments, they may also be chopped into shorter lengths and processed as staples (see Staple Yarn Spinning, page 56). This technique is used to produce more cost-effective fabrics, by increasing production output and material utilization. It is also used to convert recycled or waste material into uniform fabric. However, reducing the length of fibre reduces tensile strength.

As high-performance fibres used mainly in technical applications, colourability is limited. This is either due to their chemical inertness, which means they cannot be coloured, or because they are produced in such low quantities that dyeing is not considered economically worthwhile. Opaque materials, including

carbon and basalt, are naturally rich, dark colours. This gives them their distinctive appearance, a quality often mimicked in lower-cost materials to raise their perceived value.

Metals come in a variety of colours, depending on the ingredients. They are desirable as well as functional, and associated with luxury and long-lasting goods. They are expensive to apply to textiles as wire and foil, and difficult to work. Metal coating overcomes this by applying a thin metallic layer to more versatile materials, such as plastics and other metals. Cosmetic finishes are typically less than 6 microns thick, with a metal film of less than 1 micron. For functional coatings, thicknesses of 10 to 30 microns are built up. They are utilized as an impermeable barrier (balloons and packaging), reflective surface (sun shades and space suits) and conductive layer (shielding).

The highly reflective nature of aluminium is best exploited in coated plastic film. The plastic sheet, typically polyethylene terephthalate (PET), is coated and covered with a protective layer, which may be tinted with colour to give the appearance of gold or copper (see image above). Fabrics are constructed from strips of the film, which are woven (see title image, page 520) or knitted together.

Metallized films coated with an adhesive backing are applied as transfers. Calendered (page 220) to the surface of fabrics using heat and pressure, the foil is bonded in place to create a bright metallic finish. When applied to textured surfaces in this way – including leather, embroidery, lace (pages 106–19) and fancy woven fabrics (page 84) – the very thin foil adheres only to the high points (see image opposite top). Alternatively, the metallized film is laminated to the surface of the fabric under pressure. This provides very good coverage and is even suitable for applying metallic effect to four-way stretch fabrics (see image right bottom).

MATERIAL DEVELOPMENTS

These high-performance fibres have revolutionized applications where strength to weight is critical. For example, around half the primary structure of the Boeing 787 Dreamliner, including the fuselage and wings, is constructed from carbon-fibre-reinforced composite, helping to make it the most fuel-efficient passenger plane to date.

Process development, such as computer-guided tape placement, combined with material innovation, creates new application opportunities. Carbon and aramid are used in racing sails – encapsulated between two layers of PET film (see 3DL, page 457) or prepregged with resin and laminated (see 3Di, page 458) – that are 20% or so lighter than conventional equivalents. The fabrics are built around the properties of the materials, utilizing as much of their strength as possible. This gives teams that can afford to invest in new technology huge competitive advantage.

Shape memory polymers (SMP) are converted into fibres and films. These materials exhibit the unique property of being able to be mechanically deformed and then return to their original shape when heated to a specific temperature. This property is built into plastics by combining two or more materials that have different responses to thermal stimulation. The primary material has a softening point, referred to as glass transition temperature (Tg), above that of the secondary material. This combination means that the material can

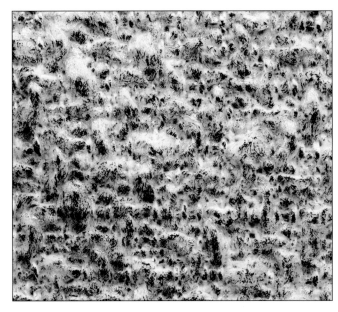

Gold foil Foil is bonded to the surface of woven fabric by calendering to produce a bright metallic finish reminiscent of gold leaf.

Gold finish This four-way stretch fabric comprises 80% nylon and 20% elastane. Gold-coloured film is applied by laminating to produce a bright metallic surface finish on one side.

be mechanically formed into new shapes – by bending, twisting and pressing – and held in place by the secondary material. Therefore, when heated to above the softening point of the secondary material, the force of the primary material overtakes and so reforms the material to its original shape. It may be thermoplastic or thermoset and can be engineered to respond to specific temperatures, or other stimuli, depending on the requirements of the application. There are many potential applications, particularly in space travel (collapsible), medical products and design for disassembly.

Shape memory alloys (SMA) are novel metallic materials that exhibit molecular rearrangement at specific temperatures. In other words, like SMP they can be deformed and then, when triggered at a specific temperature, return to their original shape. The temperature that triggers the rearrangement can be engineered to suit specific applications and can range from approximately -25°C to 100°C (-13°–212°F). An example is nitinol, used for example to make unbreakable spectacle frames, which is a nickel–titanium alloy. This property is referred to as superelasticity when the temperature for rearrangement is set lower than room temperature. The result is a material that constantly springs back to its original shape.

ARAMID

Description

Aramid is an aromatic polyamide. It is distinguished from aliphatic polymers (see Nylon, page 516) by the type of bonds formed in the polymer chains. At least 85% of the amide linkages are cyclic (benzene rings), and these help to ensure that the polymer forms into long unbroken chains. The two principal types, meta-aramids and para-aramids, are made from slightly different ingredients. As a result, they have different properties.

Strength   • • • • •

Aramid is strong and lightweight. Para-aramids, such as Kevlar and Twaron, have very high strength to weight and low creep (resistance to deformation under load). Meta-aramids, such as Nomex, are not as strong, but have excellent heat resistance and dimensional stability.

Resilience   • • • • •

They are relatively stiff, with moderate resilience. Repeated bending and compression cause kink bands and fibrillation (separation of the rod-like structure) to occur. This weakens the fibre – knot and loop strength is around one third of tensile strength – but the gradual degradation, as opposed to immediate failure, contributes to the high energy absorption of aramid.

Quality   • • • • •

Aramid is tough with good resistance to abrasion. It does not melt like other types of synthetic polymer, but will begin to degrade at around 500°C (932°F), depending on the exact type. However, it has poor resistance to ultraviolet (UV) light, making it vulnerable when used outdoors.

Colour   • • • • •

Para-aramid is naturally gold-coloured. Teijin produces a black filament, called Technora, coloured by solution dyeing. Meta-aramids are naturally off-white.

Versatility   • • • • •

They are produced as filaments and staples. The high strength and stiffness of para-aramids is utilized in rigid, lightweight composite laminates (page 446), nonwovens (page 152) and racing sails (pages 374 and 454). Para-aramids' high tear strength, cutting resistance and impact absorption are utilized in many demanding applications, ranging from reinforced tyres to bulletproof vests. The high heat resistance and flame-retardant properties of meta-aramids is utilized in heatproof clothing, helmets and gloves. Nomex is converted into a paper-like material used for heat resistance and electrical insulation, or as honeycomb core in composite constructions for aerospace and furniture.

Availability   • • • • •

Previously only available from DuPont (Kevlar and Nomex), they are now produced by several other manufactures, notably in China and Japan.

Cost   • • • • •

Owing to the complexity and cost of production, these are expensive fibres.

Environment   • • • • •

Production is energy-intensive and these materials cannot be melted and recycled like most other synthetic fabrics. However, their high strength to weight and excellent heat resistance are invaluable for reducing weight while increasing safety.

LIQUID CRYSTAL POLYMER (LCP)

Description

Liquid crystal polymer (LCP) is an aromatic polyester known under the trademark name Vectran. It is distinguished from aliphatic polyesters (page 515) by the presence of benzene rings in the polymer chain. This ensures a highly oriented crystal structure, similar to aramid, which results in exceptional tensile properties. Unlike aramid, LCP will soften at high temperature and so can be extruded and melt spun (page 50), and at the end of its useful life it can be recycled.

Strength   • • • • •

LCP is high strength with very low (virtually immeasurable) creep (non-recoverable stretch after prolonged load), which makes it reliable under load. The liquid crystal polymer structure is fixed straight from spinning and does not require further drawing to stretch out the chains.

Resilience   • • • • •

It has good flexibility compared to many other high-strength fibres, which provides good flex-fold characteristics.

Quality   • • • • •

LCP has good resistance to chemicals, fatigue and abrasion, and can withstand temperatures up to 330°C (625°F). It is inherently fire-retardant. It does not absorb moisture and has excellent vibration and impact absorption properties.

Colour   • • • • •

It can be solution dyed and is available in some standard colours including gold, black and red. It is opaque, so colours appear a little washed-out. The surface has a naturally bright and shiny appearance.

Versatility   • • • • •

LCP has a unique combination of properties that are utilized in demanding applications in aerospace, marine, sports and composite products. For example, it is used in the construction of loom-woven (page 76) airbags used by spaceships to assist in landing. As a braided (page 68) high-performance rope, it is strong and durable enough to be used without a jacket. This makes it quicker and easier to splice. However, it is degraded by exposure to sunlight and so requires a jacket if used outdoors. It has very good cut resistance and so is used in protective clothing. The benefits of LCP for such applications include its resistance to bleach (superior to aramid) and high-temperature performance (superior to UHMWPE).

Availability   • • • • •

It is a niche product, more expensive than aramid, and produced in relatively lower volumes.

Cost   • • • • •

To date, Vectran is the only commercially available melt-spun LCP. It is expensive.

Environment   • • • • •

Production is not particularly energy-intensive. It can be melted and recycled, but is unlikely to be processed in this way owing to the very low quantities and typical applications.

POLYPHENYLENE BENZOBISOXAZOLE

Description

Polyphenylene benzobisoxazole (PBO), known mainly under the trade name Zylon, has the highest strength and modulus of any synthetic fibre. Developed in the 1980s, it is an aromatic polymer with a rigid rod-like structure. It has almost zero creep (non-recoverable stretch after prolonged load) and very low elongation (2–4%), but poor resistance to abrasion and sunlight.

Strength ● ● ● ● ●

It is very strong (twice as strong for the same weight as aramid), but quickly loses its strength when exposed to sunlight, so is not recommended for use without protection. Knot and loop strength is around one third of tensile strength.

Resilience ● ● ○ ○ ○

It is stiff and more brittle than other types of synthetic fibre. Fibres are prone to kinking under compression.

Quality ● ● ○ ○ ○

Combined with very high strength, PBO has very little creep. It has very high flame resistance and good chemical resistance. However, it is degraded by humidity, sunlight (loses around one third of its strength within six months of exposure) and abrasion. These weaknesses were tragically highlighted by the failure of bulletproof vests made from PBO that had rapidly degraded (much faster than would be expected from UHMWPE, for example).

Colour ● ● ● ○ ○

It is available in standard colours including gold, brown and black, but can be produced in a wide range of vivid colours.

Versatility ● ● ● ● ●

It is used to make the core of rope used to rig racing boats. Owing to its vulnerability to sunlight, abrasion and salt water, it must be protected within a sealed jacket. Its high strength and lightness is utilized in safety gloves and protective clothing. As a continuous fibre it is used to reinforce composite material for high-performance and demanding safety applications, such as tethers used in Formula One racing cars and aerospace applications.

Availability ● ○ ○ ○ ○

It is manufactured by Toyobo in Japan.

Cost ● ● ○ ○ ○

It is expensive.

Environment ● ● ○ ○ ○

PBO is complex and energy-intensive to produce. It cannot be easily recycled.

POLYTETRAFLUOROETHYLENE (PTFE)

Description

Fluorinated polymer fibres are made up of long carbon chains surrounded by fluorine atoms. The bonds are very strong, resulting in a highly resistant material with a slippery surface. Polytetrafluoroethylene (PTFE) is often known under the DuPont trademark name Teflon.

Strength ● ● ● ○ ○

It maintains its strength up to 300°C (572°F) and can be used continuously between -250°C (-418°F) and 200°C (392°F), although at very low temperatures it becomes less ductile. PTFE fibres outperform films (page 180) of comparable PTFE materials, because they are processed with a higher degree of molecular orientation. Tenara fibre is manufactured from expanded PTFE (ePTFE). This architectural fibre is two to three times stronger than conventional PTFE.

Resilience ● ● ● ● ○

It has reasonably good resilience and flexibility thanks to a fibrillated (comprising multiple individual slender fibres) microstructure. It is prone to creeping (permanently deforming) under tension or compression. This is advantageous for some applications (seals that conform to shape), but disadvantageous for others (rope and cable).

Quality ● ● ● ● ○

PTFE is a very durable material and has high resistance to chemicals, the highest of any synthetic fibre. It has excellent resistance to sunlight and can be used outdoors without any protection. It is inherently flame-retardant and heat-resistant. It has a very low coefficient of friction, making the surface feel slippery, a quality referred to as self-lubricating. This is a valuable property for high-performance rigging, where the rope needs to slip around a winch drum, as well as gripping and easing smoothly. It is heavier than most other synthetics and around twice the density of nylon (page 516). It has inferior abrasion resistance to UHMWPE (page 517).

Colour ● ○ ○ ○ ○

It is naturally brown and can be bleached white, although this reduces its strength. It is hydrophobic and not possible to dye.

Versatility ● ● ○ ○ ○

PTFE fibres are woven (see Loom Weaving, page 76) and knitted (see Weft Knitting, page 126) to make lightweight fabrics and used in nonwoven needle felts (page 152) for filters. The filaments are used to make rope (pages 60 and 68) and seals. Staple fibres are even suitable for flocking (page 280). It is often blended with other fibres to impart some of the unique properties of PTFE while keeping the cost implication to a minimum. This is done with fibre construction techniques, or by using PTFE fibre to reinforce plastic composites. Owing to its high cost it is reserved for only the most demanding applications. PTFE films range from 25 microns to 3 mm (0.001–0.125 in.) thick. Thanks to its sintered construction, it can be stretched, to produce a microporous film. The billions of holes are smaller than water droplets, but large enough to let gases pass through. The best-known application of this property in textiles is Gore-Tex, which is waterproof but maintains its breathability thanks to the unique properties of ePTFE. Since the patents have expired – Gore-Tex was invented in 1960s – many alternatives and direct competitors have emerged.

Availability ● ○ ○ ○ ○

It is made by few companies and the principal producer is DuPont, which manufactures it under the trademark name of Teflon.

Cost ● ○ ○ ○ ○

It is the most expensive yarn.

Environment ● ○ ○ ○ ○

PTFE cannot be made by conventional synthetic forming processes as a result of its very high resistance to heat and chemicals. Therefore, it is fabricated using a sintering technique, which requires significant amounts of energy and creates waste. It cannot be melted for recycling.

CARBON FIBRE

Description

Carbon fibre (CF) is the product of the oxidizing of polyacrylonitrile (PAN) (see Acrylic, page 516) at between 1,000 and 3,000°C (1,832–5,432°F). This causes tightly bonded crystals to form, to produce a fibre with incredible strength and stiffness. Even though it had been known for some time, it was not until the 1960s that a high-quality fibre was produced suitable for the development of the strongest, stiffest and lightest composite materials. Known as carbon-fibre-reinforced plastics (CFRP), they are today mainly utilized in aerospace, automotive and marine applications to save weight and maximize efficiency.

Strength ● ● ● ● ●

Carbon fibre has exceptional strength for weight (and size), many times greater than steel. Materials produced from fibres have isotropic properties, in other words they are stronger in the direction of fibre. There are different grades of fibre, ranging from low to high modulus (indication of stiffness) and increasing in tensile strength, depending on the method of production. There are also different types of polymer precursor (other than PAN), such as viscose (page 508), but these are not so widely used.

Resilience ● ● ○ ○ ○

It is stiff and prone to breaking when bent. This is a result of the crystal structure, which is susceptible to cracking between the sheet-like formations. This property is advantageous in parts of racing cars designed to absorb impact, because energy is dissipated with each fracture. It suffers in more flexible applications, such as yacht sails, where it is rolled, folded and knocked against the rigging.

Quality ● ● ○ ○ ○

It has very good dimensional stability and the properties hardly change at all in hot or cold conditions. It has very good resistance to corrosion and will not deteriorate in normal atmospheric conditions. It is highly conductive, like graphite, and so is avoided in certain applications, because it will have the effect of blocking radio waves. The most common fabric construction is twill weave, because this provides the most versatile balance of stiffness and conformability. This gives composite laminates their typical diagonal-checked appearance.

Colour ● ○ ○ ○ ○

Made up of carbon crystals, it is available only in black. Colour is added by combining other fibres with carbon during weaving, such as polyester (page 515), or with a coating or laminate.

Versatility ● ● ○ ○ ○

Carbon fibre is used for high-performance applications where high stiffness and strength to weight are critical, such as aerospace, automotive and military. The fibres are typically around 6 microns or more in diameter. Twisted yarn (page 60) and tow (unidirectional bundle or tape) are measured according to the number of individual filaments. In other words, 3K, 6K and 12K contain 3,000, 6,000 and 12,000 individual filaments respectively. These are converted into various weaves (see Loom Weaving, page 76), braids (page 68) and unidirectional-type fabrics suitable for composite laminating (page 234). Alternatively, yarn or tow is laid down in a bespoke configuration for ultimate strength to weight, such as in the case of filament winding (page 460), thin ply technology (page 236) and 3D thermal laminating (page 454). The fibre is most commonly combined with thermosetting epoxy resin, or thermoplastic polyetheretherketone (PEEK), depending on the application, thanks to their superior mechanical properties.

Availability ● ● ○ ○ ○

Carbon fibre is readily available, but owing to its relatively high cost and the complexity of fabricating into high-performance materials, it is not widely used.

Cost ● ○ ○ ○ ○

It is energy-intensive to produce and so expensive. However, in recent years, owing to an increase in demand from the automotive industry, the price of carbon fibre has come down significantly.

Environment ● ○ ○ ○ ○

As a result of the high temperatures required, complexity of processing and large number of steps, the production of carbon fibre is harmful to the environment. Composite materials are very difficult, if not impossible, to recycle, although there has been some progress in the recycling of carbon-fibre composite.

GLASS FIBRE

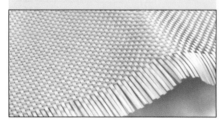

Description

Continuous glass fibres, first produced in the 1930s, are manufactured by melting silica together with other minerals at 1,500°C (2,732°F). The molten homogenous mass is coloured with additives, or a decolourant is added to make clear glass. The material is conditioned for 8 to 12 hours, which is long enough for all the bubbles to rise to the surface and dissipate, and cooled to 1,150°C (2,102°F). It is extruded and drawn into glass fibres (GF) starting from around 5 microns thick and coated with chemical sizing to prevent surface abrasion. There are many different types of glass, depending on the ingredients, including soda lime, borosilicate, alumino-borosilicate and aluminosilicate. They are all based on the same amorphous structure, which means they melt over a larger temperature range. Fibres are used to make insulation, optic cables, reinforced plastic (GFRP) and reinforced concrete.

Strength ● ● ● ○ ○

The molecular structure of high-strength glass is held under tension between the surface layer and the core. This is why when glass fails the cracks continue to develop even when the load is removed. The result is that glass has very good compressive strength. However, tensile strength is limited by its brittleness, because any defects in the surface layer, such as caused by abrasion, concentrate stress and readily propagate into cracks when the fibre is bent (they are coated during production to prevent this). Thus, the hardness and depth of surface layer determine the durability of the material in application.

Resilience ● ● ● ○ ○

Glass has excellent resiliency, but very poor elasticity.

Quality ● ● ● ○ ○

There are many different types of glass fibre, produced from different ingredients. E-glass is the most common and is considered a general-purpose material. The E stands for 'electrical', because it was originally used for electrical insulation. A-glass is similar to window glass, with high alkali content, S-glass is high-strength, C-glass has high chemical resistance, M-glass has high modulus (stiffness), AR-glass has good alkali resistance, D-glass has low dielectric constant for electrical applications, and ECR-glass has very high resistance to corrosion. All types of glass have high heat resistance, are incombustible and are inert to most liquids and chemicals. They have very good dimensional stability and stretch only around 5% before breaking, depending on the type.

Colour ● ○ ○ ○ ○

As a transparent material, glass can be produced in a range of bright, saturated colours, incorporated during raw material production. However, owing to the volume required to justify this, and the fineness of the fibres, they are typically produced clear and colour is incorporated in composite laminates through the resin matrix or coating system.

Versatility ● ● ● ○ ○

Glass fibres are used to make highly insulating materials, such as nonwoven 'glass wool', protective gloves and home furnishings. They are used for their high strength to weight to make composites, including thermosetting laminates, thermoplastic types (page 446) and filament winding (page 460). These materials are made from yarns converted into woven (typically plain, twill, basket or satin, pages 76–105), unidirectional (page 234) or nonwoven (page 152) fabrics. They are typically round, but special purpose types may be produced with hollow or triangular cross-section.

Availability ● ● ● ● ○

It is widely available in a range of formats.

Cost ● ● ● ● ○

As a fibre reinforcement, glass is considerably less expensive than carbon, aramid and polyphenylene benzobisoxazole (PBO), and has advantages of its own.

Environment ● ● ● ○ ○

Production is continuous and the temperature is maintained every day. Manufacturing directly from mixed raw materials eliminates inefficiencies associated with complex production with multiple steps. Unlike plastic, glass is not derived from petroleum, thereby eliminating all the associated environmental impacts (except of course in the production). A high proportion of glass is recycled, although when used in composite materials it is impractical, if not impossible, to extract the glass for recycling.

BASALT FIBRE

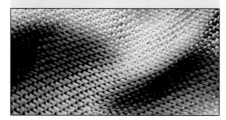

Description

Basalt fibre is derived from volcanic rock. The crushed rock is melted at around 1,500°C (2,732°F) and converted into fibres starting from around 10 microns in diameter. Even though like glass it consists of a high proportion of silica, basalt forms a crystalline structure, which is more ordered than the amorphous structure of glass. Originally developed for military application, basalt fibres have only recently been available for commercial development.

Strength   • • • • ○

It is stronger than E-glass (comparable with S-glass) and more elastic.

Resilience   • • ○ ○ ○

During production basalt fibres are chemically sized, like glass, to protect the surface and thus avoid breakages during fabrication.

Quality   • • ○ ○ ○

The surface is smooth, shiny and inert. The exact ingredients depend on the source, because no secondary materials are added. This means the quality is variable. It provides excellent heat insulation, several times more efficient than glass. It has high heat resistance, which means that it is suitable for applications from -250 to 650°C (-418–1,138°F).

Colour   • ○ ○ ○ ○

It is naturally dark metallic golden brown and is not available in any other colour.

Versatility   • • • ○ ○

Basalt fibres are converted into twisted (page 60) or unidirectional (page 234) yarn. Using conventional looms, the yarns are woven (see Loom Weaving, page 76), braided (page 68) or applied as unidirectional fabrics. It is converted into composites using conventional laminating and filament winding (page 460) techniques. As a result of its higher stiffness than E-glass, basalt is being used to replace or reinforce glass in critical applications, such as wind turbine blades and boat hulls, as well as providing improved strength to weight (at lower cost than carbon fibre) in camera tripods and sports equipment. As a result of its highly insulating properties, it is used in place of glass for automotive, construction and protective garment applications. Unlike carbon, it is non-conductive and so does not interfere with radio waves. This means it is useful for applications that require higher stiffness than glass can offer, without the electrical interference of carbon.

Availability   • ○ ○ ○ ○

Basalt rock is found around the world. But as a relative newcomer in composite manufacture, basalt fibres are not yet widely available. They are known under trademark names, such as Basfiber.

Cost   • • • ○ ○

The price varies considerably, but on the whole is comparable with glass fibres.

Environment   • • • ○ ○

Mined basalt is the only material that goes into the production of the fibres. In other words, unlike glass, no secondary materials are added. However, it does require a little more energy to produce a homogenous melt. It is naturally occurring in the earth's crust.

METALS

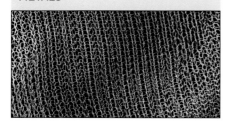

Description

Metals used in textiles include precious types – gold and silver – titanium, steel, aluminium and copper, among others. Metallic materials are characterized by a tightly packed cubic crystal structure, which means they are strong, hard and bright. They are highly conductive, of heat and electricity, and also malleable. The extent of their properties depends on the base elements and alloyed ingredients. This unique set of properties, unlike any of the previously described materials, is utilized in textiles for both functional and decorative purposes.

Strength   • • • • ○

Metals are very strong, because the bonds that make up the atomic structure do not break down easily. Aluminium is about twice the strength of steel for the same weight and titanium is stronger still. Copper is strong, but heavy. Gold and silver are soft and more malleable.

Resilience   • • • ○ ○

Metal holds its shape very well. Stranded wire, produced by twisting together smaller-gauge wire (page 68), is more flexible for the same density of solid wire (and more expensive). Increasing the number of individual strands improves flexibility and resistance to fatigue. Wire diameter starts from around 18 microns, depending on the type of metal. Fibres used in textile applications are prone to twisting and breaking if not managed carefully during fabrication.

Quality   • • • • ○

On the whole, metals are stiff and malleable compared to other types of yarn and sheet material. This means they do not drape well and crease readily. Some metallic elements are highly reactive, while others are practically inert. The iron content in ferrous metals reacts with water and oxygen to produce iron oxide (rust). Stainless steel is alloyed with chromium to help prevent this. Metals that do not contain iron (non-ferrous) react with oxygen to varying degrees. Gold is practically inert, whereas aluminium forms an oxide barrier layer on its surface that is very durable (aluminium oxide, or alumina, is one the hardest materials known to man). Copper is a bright reddish pink when first produced. This does not last long. The surface quickly develops a layer of dark-brown oxide, which gradually becomes green (verdigris) as it develops, depending on the air quality. Metals that react will need to be protected beneath an impermeable coating (plastic-coated wire is produced by extrusion, page 180), metal coating, or by enhancing the material's naturally protective surface layer.

Colour   • • • • ○

The natural colour depends on the base ingredients and additional metallic elements. Stainless steel is naturally bright grey; titanium is dark grey and when polished takes on a bright ceramic-looking finish; aluminium is available in many different grades, naturally light to mid-grey, and is one of the most reflective metals; copper ranges from gold to dark brown depending on the alloyed elements (brass and bronze are copper alloys); silver is bright and reflective but oxidizes readily and so has to be regularly polished or 'coloured over' to maintain its lustre; and pure gold is yellow, and alloyed with other elements can appear lighter, reddish or pink. Applying a tinted topcoat brings out the metallic quality of metal in a range of vivid colours.

Versatility   • • • • ○

Metals are surprisingly versatile. They are utilized for their strength in twisted rope and cable (page 60), for shape retention and protective clothing; and for their conductivity in shielding (against radiation and static for example) and in wearable electronics applications. They are woven (pages 76–105) and knitted (pages 126–51) into fabrics ranging from lightweight sheer and lace (pages 106–19) to heavyweight protective cloths. Decorative yarns, made from gold and silver, are produced from metal filaments twisted together to make them pliable, filament coiled around a core that is later removed (called purl), or flattened into sheets that are wrapped around a core to create a smooth surface (see Embroidery, page 298). Stainless steel is wrapped with natural fibres, such as silk (page 497) and wool (page 494), to create a very resilient material that can be fabricated and finished like more conventional fabrics, such as by dyeing and steam pressing. They may be used in knitting, warp and weft, depending on the strength. Gold is malleable enough to be beaten into very thin sheets that are applied as leaf. Thin layers of metal alloys are deposited onto plastics and other substrates as coatings to impart some of their beneficial properties. The bright and reflective qualities of aluminium are utilized in transfer films that are applied to textiles by calendering and embossing (page 220) to produce a metallic finish.

Availability   • • • • ○

All types of metal may be drawn into fibre and wire. Steel is commonly used for its strength and versatility, while copper is also common, because it is highly conductive. Other metals are used to a lesser extent and so are less widely available. Gold and silver are produced as leaf used for decorative applications. All metals can be applied as coatings, some more readily than others.

Cost   • • ○ ○ ○

Metal yarns tend to be quite expensive, while only certain specialists have the expertise (or willingness) to convert them into cable or fabric. This limits the possibilities somewhat.

Environment   • • ○ ○ ○

Mining has a significant impact on the environment and companies have to follow strict guidelines that help to minimize air, water and soil pollution, and loss of biodiversity. The impact depends on the type of metal. Steel is more efficient than aluminium and copper for example, while titanium is one of the most energy-intensive materials of all. On the whole, metals that require more energy to produce are more expensive, which helps to understand the embodied energy. This is not the case for precious metals. Recycling is much more efficient than mining and extracting virgin material. The advantage of recycling metal is that the properties are unaffected, as long as the added ingredients are controlled. In other words, if aluminium is recycled it is important that the different grades are separated so that the alloying elements do not reduce the overall quality (hence why aluminium cans are recycled separately).

Glossary and Abbreviations

Words in **semi-bold** refer to terms that have their own entry in the glossary.

3DL
Three-dimensional thermal laminating (page 454), a process developed by North Sails.

3Di
Three-dimensional **isotropic** (page 454), a process developed by North Sails.

acrylic
The common name for polyacrylonitrile (PAN) (page 516) used as a fibre. Not to be confused with polymethylmethacrylate (PMMA), a plastic sheet material also referred to as acrylic.

aliphatic
A **polymer** structure that comprises carbon atoms in open chains (not **aromatic**). They may be saturated or unsaturated. Examples include **nylon** (page 516) and **polyester** (page 514), two of the most common synthetic fibres.

anisotropic
A material whose properties are different widthways and lengthways. For example, loom-woven fabrics tend to be stronger in the direction of **warp** (machine direction), because the warp **yarn** has to be strong enough to withstand the loads applied during weaving. Likewise, wood has higher **tensile strength** along the **grain** (direction of growth) as opposed to across the grain. See also **isotropic**.

antimicrobial
A material or additive that kills or inhibits the growth of microbes.

aramid
Common name for **aromatic** polyamide (see Nylon, page 516). The two main types are meta- and para-aramids, typically referred to by their trademark names, such as Nomex and Kevlar.

aromatic
A **polymer** structure that contains unsaturated rings of atoms. In the case of synthetic plastics, this property means that during polymerization the **monomers** join together into longer unbroken chains (higher strength) with highly oriented crystal structure. Examples include **super fibres** (page 520) such as **aramid**

(page 524) and liquid crystal polymer (**LCP**) (page 524), which are aromatic polyamide and aromatic **polyester** respectively. See also **aliphatic**.

assist gas
Gas purged through the nozzle during laser cutting (page 328) to reduce scorching and remove debris.

balanced weave
A woven pattern made up of the same number of **warp** yarns and **weft** yarns per cm²/in². See also **unbalanced weave**.

basket weave
A fabric woven with multiple **warps** interlaced with multiple **wefts** to produce a repeating checked pattern reminiscent of a woven basket. See also **plain weave**.

batting
Cotton wadding.

beam
A large cylinder on which **yarns** are wound in preparation for dyeing (page 240), weaving (pages 76–119) or knitting (pages 126–51), for example.

bias
At 45° to the **warp** and **weft** in loom-woven fabric. Patterns cut on the bias will have more stretch (this property is utilized in dressmaking to produce body-hugging dresses from woven fabrics) than those cut with the **grain**.

biocomposite
High-performance composite materials fabricated from natural resins, such as **bioplastic**, reinforced with natural fibres, such as flax (page 481).

bioplastic
Polymers derived from biological sources (made without petrochemicals). Some are completely biodegradable and require less energy to produce than synthetic polymers. Examples include polylactic acid (page 000), polyhydroxyalkanoate (PHA) and poly-beta-hydroxybutyrate (PBH), polyglycolic acid (PGA) and polyfurfuryl alcohol (PFA) (see Bioplastic, page 519).

biopolymer
See **bioplastic**.

bisphenol A (BPA)
A carbon-based synthetic compound used as an ingredient in plastics –

such as polyvinyl chloride (PVC) and epoxy – applied as adhesive (page 370), coating (page 226) or laminate (page 188) and subject to restrictions in many countries owing to concerns about its safety.

blend
A mix of two or more types of fibre or **yarn**.

BOPET
Biaxial-oriented polyethylene terephthalate (PET), a high-strength plastic film also known by its trademark name Mylar (see Polyester, page 515).

BR
Butadiene rubber (page 518).

breathable
Used to described materials that allow atmosphere and vapour to pass through, but typically do not allow liquids to penetrate. This includes natural materials such as leather (page 498), and man-made coatings such as **silicone** (page 519) and polytetrafluoroethylene (**PTFE**) (page 515).

CA
Cellulose acetate, commonly referred to simply as acetate (page 509).

cable
Metal rope consisting of multiple laid (**twisted**) strands. The three most common configurations for marine and interior type applications are 1 x 19, 7 x 7 and 7 x 19. The first figure refers to the number of strands and the second to the number of wire **filaments** in each strand. So for example, 7 x 19 is constructed with seven twisted strands, each of which is made up of 19 wire filaments.

CAD
Computer-aided design, a general term used to cover computer programs that assist with the design and engineering of an item.

calendering
The process of compressing fabric between heavy metal rolls (page 220). Smooth-surfaced rolls produce a flat, polished or glazed finish on fabrics. Shallow-engraved rolls produce a textured finish, such as moiré and schreiner.

CAM
Computer-aided manufacturing, a term used to cover software required to drive computer-guided processes, such as CNC laser cutting (page 328) and CNC knife cutting (page 316).

carded
Mixed-length **staple fibres** that have been untangled and aligned on a toothed swift roller (page 23) in preparation either for **spinning** into **yarn** (page 56) or for fabricating a **nonwoven** (page 152). In the case of wool the same grade of fibre is referred to as **woollen**. See also **combed**.

CF
Carbon fibre (page 526).

CFRP
Carbon-fibre-reinforced plastic (see Composite Press Forming, page 446).

CMYK
In printing terms, the four process colours, cyan, magenta, yellow and key (black) are collectively abbreviated as CMYK.

CNC
Machine equipment operated by a computer is known as 'computer numerical control' (CNC). See also **CAM**.

combed
Long **staple fibres** that have been **carded** and combed to produce stronger, finer and more even **yarn**. In the case of wool the same grade of fibre is referred to as **worsted**.

computerized nesting
An automatic system used to arrange patterns on fabric in the most efficient manner (to reduce waste when cutting).

copolymer
A polymer made up of long chains of two repeating **monomers**: for example, ethylene propylene (EPM), which is a type of synthetic rubber (page 518). See also **polymer** and **homopolymer**.

cord
Flexible rope made from several **twisted yarns** or strands of material.

core-spun yarn
A **yarn** that has been wrapped within a separate yarn, usually to improve the look and feel. For example, sewing

thread consisting of a high-strength **polyester** filament wrapped in **staple yarn**, or latex wrapped with synthetic **filament**.

course
Horizontal row in a knitted structure (pages 126–51). See also **wale**.

CR
Polychloroprene rubber (page 518), commonly known as neoprene.

creep
A measurement of the permanent elongation (**stretch**) of a material after prolonged load. Low creep is particularly important for **twisted** (page 60) and braided (page 68) rope used in critical structural applications.

crimp
Non-linear deviations to the shape of natural and man-made fibres. It is applied to synthetic fibres to add texture and improve resilience. See also **texturing**.

cure
To polymerize rubber or resin (solid, coating, adhesive or composite matrix) to form a **thermosetting plastic**. The reaction may be stimulated using a catalyst or heat.

decitex (dtex)
A unit of weight used to indicate the fineness of single **yarns**, as well as **plied yarns**. It is equal to 1 gram per 10,000 metres. See also **tex** and **denier**.

denier
A unit of weight used to indicate the fineness of single **yarns**, as well as **plied yarns**. It is equal to 1 gram per 9,000 metres. See also **microdenier**, **tex** and **decitex**.

die
Device used to cut fabric to a given shape (see Die Cutting, page 336).

dobby
Small and typically geometric patterns on woven fabric, and the loom used to weave them (page 84).

doctor blade
Long, flat metal strip used to control the film thickness during laminating (page 188) and coating (page 226).

doubling
Combining two singles to make a **plied yarn** (see Plying and Twisting, page 60). See also **spinning**.

drawing
The process of stretching **filament yarns** during production (see Filament Spinning, page 50) to align the **polymer** chains and so increase strength.

Dyneema
Trademark name for ultra-high-molecular weight polyethylene (**UHMWPE**) (page 517). See also **Spectra**.

elastane
Common name for fibre produced from **thermoplastic polyurethane** (TPU). Also called spandex (North America). See page 517.

elastomer
A natural or man-made material that exhibits elastic properties and has the ability to deform under load and return to its original shape once the load has been removed. Examples include natural rubber (page 168), synthetic rubber (page 518) and elastane (page 517).

end
Single **yarn**, such as in the case of weaving (page 76–119) and knitting (page 126–51).

EPM
Ethylene propylene rubber (page 518).

exothermic reaction
A chemical reaction that generates heat, such as in the case of foam molding (page 410).

fancy
Used to describe **yarn** or fabric with a highly decorative appearance, for example intricately patterned **jacquard**-woven fabric. See also **novelty yarn**.

fashioned
A knitted garment with **courses** shortened or lengthened by transferring loops to create panels, or complete garments, with a body-shaped profile (see Weft Knitting, page 126). The same technique is used to produce seamless upholstery (page 404).

FEA
'Finite element analysis', a computer simulation technique that uses finite element method (FEM) to analyse designs and predict performance before manufacturing. It is used extensively in the design of racing sails (page 458) to help reduce weight.

fibrillate
To split or break apart the separate fibres or elements, such as plastic **yarn** used in tufted artificial grass (page 286) or the degradation of regenerated fibres (page 506).

figured
Artwork or patterns that represent human or animal forms, such as those produced by **jacquard** weaving (page 84) and embroidery (page 298).

filament
A continuous **yarn**. The only naturally occurring example is silk (page 497). Man-made fibres, such as **viscose** (page 508) and **polyester** (page 515) are spun as filament (see also **monofilament** and **multifilament**), which may subsequently be cut into short lengths, called **staples**.

filling yarn
Another term for **weft**.

float
A woven or knitted **yarn** that passes over several yarns for each one that it overlaps. This type of stitch (knit) or weave is used to create solid colour on the face of fabrics, such as **jacquard** patterns.

GFRP
Glass-fibre-reinforced plastic (see Composite Press Forming, page 446).

grain
This term is used to refer to woven textiles and leather (as well as wood). For textiles, it is the result of weaving with **warp** and **weft** yarns perpendicular to one another. The warp yarns tend to be stronger, because they have to be able to withstand the loads applied during weaving, so fabrics tend to stretch less along their length (machine direction, parallel to the **selvedge**). This is especially important for pattern cutting (page 312). See also **bias** and **on-grain**. In the case of leather, grain refers to the natural surface pattern, which is distinct for each species of animal.

greige
Fabric that has not been dyed or finished.

hand
Refers to the surface feel of a material, such as its softness.

handle
Refers to the way a material feels when handled, such as the way it drapes.

heat set
A **thermoplastic yarn** that has been heated to its softening point and set into the desired shape as it cools and resolidifies, such as to relieve the stress in woven fabric (see Stretching and Shrinking, page 210) or add permanent creases (pleats).

heddle
A fundamental part of a weaving loom. It is a length of wire or cord with an eyelet through which each **warp yarn** is fed. Attached to a harness, the heddles are moved up and down in groups, controlled by computer numerical control (**CNC**), or by depressing a foot pedal. Whether they are up or down determines whether the warp or the **weft** is visible on the surface of the fabric.

hide
Leather derived from larger animals, such as cows. See also **skin**.

homopolymer
A polymer made up of a single type of **monomer**, such as polypropylene (PP) (page 515). See also **polymer** and **copolymer**.

hydrophilic
Propensity to absorb or attract water: for example, natural plant fibres, such as cotton (page 481) and flax (page 481). The opposite of **hydrophobic**.

hydrophobic
Repels water and similar liquids: for example, **silicone** (page 519) and polytetrafluoroethylene (**PTFE**) (page 525), which are applied as coatings (page 226). The opposite of **hydrophilic**.

IR
Isoprene rubber (page 518).

isotropic
A material with uniform physical properties in all directions. It is referred to as quasi-isotropic when the properties are almost uniform. See also **anisotropic**.

jacquard
A system invented by Joseph M. Jacquard in the 19th century that uses punched cards to facilitate weaving intricate patterned fabrics, such as damask and brocade, using independently controlled **heddles** (see Fancy Loom Weaving, page 84). Originally developed for handlooms (page 90), jacquard looms nowadays are **CNC** and capable of producing any type of weave. Individual stitch selection in **weft** and **warp** knitting (page 126 and 144 respectively) is also referred to as jacquard.

jersey
Plain knitted fabric (page 126).

Kevlar
Trademark name for para-**aramid** (page 524). See also **Twaron**.

LCP
Liquid crystal polymer, also known by its trademark name Vectran (page 524).

loft
Refers to the bulkiness and thickness of insulating materials, such as **nonwovens** (page 152) and **batting** used in quilted fabrics (page 199).

Lycra
Trademark name for **elastane** (page 517).

lyocell
A type of viscose (page 508) fibre produced in an almost closed-loop manufacturing process. Also known by its trademark name Tencel.

mandrel
Shaped metal rod, such as used to form a cylinder over during **filament** winding (page 460).

medullated
The hollow structure of certain wool, hair and fur fibres (see Natural Protein Fibres, page 490), caused by the deterioration of a specific group of cells that make up the core. It is known as partially medullated if the hollow structure does not run the full length of the fibre.

microdenier
See **microfibre**.

microfibre
A fibre smaller than 1.1 **decitex** (1 **denier**). Used interchangeably with microdenier.

micron (µm)
A unit of measure equivalent to one thousandth of a millimetre (0.001 mm), or 0.00004 inches, used to determine the thickness of (among other things) fibres, such as wool (page 494) and cotton (page 481). Also called micrometre.

modulus
Also referred to as Young's modulus or the modulus of elasticity, it is the ratio of stress versus strain. It is a constant value (not to be confused with **tensile strength** or **Shore hardness**) and used to characterize materials: for example, whereas **aramid** (page 524) has high modulus (rigid), natural rubber (page 168) has low modulus (flexible).

moisture-wicking
The ability of certain fibres to draw moisture away from the wearer's skin to the garment's outer layer to facilitate faster drying.

molecular weight
A unit of measure of the length of **polymer** chain, such as in synthetic plastic. Higher molecular weight creates denser and higher-strength plastics, because there are more overlaps between the polymer chains. For example, ultra-high-molecular-weight polyethylene (UHMWPE) is several times stronger than commodity polyethylene (**PE**).

momme (mm)
A unit of weight used to indicate the fineness of silk (page 497) fabric 91.4 m (100 yd) long and 1.14 m (45 in.) wide. Thus, 15-momme silk is notionally taken from a piece of silk that would weigh 6.8 kg (15 lb) if measured 91.4 x 1.14 m. Heavier fabric would have a higher momme weight. See also **denier** and **decitex**.

monofilament
A single thread of silk or man-made fibre. See also **filament** and **multifilament**.

monomer
A small, simple compound with low molecular weight, which can be joined with other similar compounds to form **polymers**. Multiple identical monomers are joined to form a **homopolymer**, two different types form a **copolymer** and three form a terpolymer.

mordant
A substance combined with dye (see Dyeing, page 240) to cause the colour to fix to the fabric. Different dyes react to different mordants. In a technique known as space dyeing (see visual glossary, page 288) multiple colours are built up irregularly along a length of **yarn** in a single dyeing sequence by applying two or more different mordants beforehand.

multiaxial weave
A fabric consisting of more than two **yarns** interlaced at angles: for example triaxial and quadriaxial, which are used in upholstery (page 404) and molded composites (page 446).

multifilament
Multiple **filaments** extruded together during **spinning** to make a continuous **yarn**. See also **monofilament**.

Mylar
Trademark name for biaxial-oriented polyethylene terephthalate (**BOPET**) plastic film (see Polyester, page 514).

nappa
A soft full-**grain** leather, typically sheepskin, produced by tumbling.

neoprene
Common name for polychloroprene rubber (CR) (page 518).

Nomex
The trademark name for meta-**aramid** (page 524).

nonwoven
Fabric made up of randomly oriented fibres combined together by physical entanglement, welding or adhesive bonding.

novelty yarn
Yarns with a distinctive and decorative visual or tactile effect created by **plying** two or more different yarns together, or by finishing technique (such as metallizing). Also known as **fancy** or complex yarn.

nubuck
Leather that is mechanically buffed on the **grain** side to raise the surface and produce a soft finish. The grain remains visible.

nylon
Common name for polyamide (PA) (see Nylon, page 516).

olefin
Family of plastics that includes polypropylene (PP) (page 515) and polyethylene (PE) (page 517).

on-grain
Woven fabric that has been tentered so that the **warp** and the **weft** yarns are perpendicular to one another, straight and in parallel alignment lengthways and widthways respectively. See also **grain**.

PA
Polyamide, also known as nylon (page 516).

PAN
Polyacrylonitrile, also known as **acrylic** (page 516).

patent leather
Leather that has been coated (page 226) or laminated (page 188) with plastic film to produce a smooth, glossy finish.

patina
Surface layer, or pattern, developed over time with use, such as seen on leather (page 498).

pattern
A plan, design or artwork used by a tailor, shoemaker, dressmaker, seamster or embroiderer. See Manual Cutting (page 312).

PBO
Polyphenylene benzobisoxazole, also known by its trademark name Zylon (page 525).

PBT
Polybutylene terephthalate, a type of **thermoplastic polyester** (page 514).

peach
A soft and fuzzy appearance on fabric produced by napping (page 216), chemical treatment or laundering. See also Scouring and Mercerizing (page 202).

PE
Polyethylene (page 517). See also ultra-high-molecular weight polyethylene (UHMWPE).

PET
Polyethylene terephthalate, a type of **thermoplastic polyester** (page 514).

phthalates
A group of chemicals (esters of phthalic acid) used to make synthetic plastics such as polyvinyl chloride (PVC) (page 518) more flexible (in this application they are known as plasticizers). They are restricted in many countries owing to the potential harm they pose to people and the environment.

pick
A row of woven fabric formed by inserting one or more **weft yarns** into the **shed**, followed by beating up (pushing the weft to its final position) and securing the weft by alternating the raised and lowered **heddle** bars.

pile
The soft upright surface of fabric made up of projecting **yarns** produced by weaving (page 96), knitting (page 00), tufting (page 286) and napping (page 216). Examples include velvet (page 100), velveteen (page 98), corduroy (page 98) and carpet (page 102).

PLA
Polylactic acid, a type of **thermoplastic polyester** (page 514) produced from biological ingredients. See also **bioplastic**.

plain weave
The simplest weave pattern, whereby each **warp** goes over and under the adjacent **wefts**, and vice versa. Examples include lightweight sheer fabrics (such as chiffon and ninon), open-space fabrics (such as cheesecloth), **basket weave**, batiste (lightweight **balanced weave**), printed cloth (medium-weight fabrics such as chintz and calico), muslin, percale (bedding fabrics, printed or unprinted), ripstop (**filament**-yarn weave with slightly larger warp and weft **yarns** at intervals to stop ripping), hopsacking (coarse open weave) and gingham (balanced medium-weight fabric that uses different-coloured warps and wefts to create a checked pattern). See also **twill weave** and **satin weave**.

plated
In the case of knitting (page 126), two **yarns** are inter-looped side by side to produce a fabric with different properties on the face and the back, for example two different-coloured yarns or two different-feeling yarns (such as hard-wearing on the outside and soft on the inside). 'Plated' may also refer to a pile of fabric formed by folding a length back and forth, as opposed to winding onto a roll. The benefit of plating is that it takes up less space for transportation.

plied yarn, ply yarn
Yarn constructed from two or more individual yarns (singles) **twisted** together. They are called two-ply when two singles are combined, three-ply when three singles are combined and so on.

ply
Single layer of a laminated construction, such as in the case of laminated fabric (page 188), wood veneer (thin sheets) or high-performance composite (page 446).

polyester
The common name for polyethylene terephthalate (**PET**), polybutylene terephthalate (**PBT**), polytrimethylene terephthalate (**PTT**) and polylactic acid (**PLA**). See Polyester (page 514).

polymer
A natural or synthetic compound made up of long chains of **monomers**. Biological examples include cellulose (see Plant Fibres, page 478) and protein (see Natural Protein Fibres, page 484). See also **homopolymer** and **copolymer**.

PP
Polypropylene (page 515).

prepreg
Fabric that has been pre-impregnated with resin. For example, high-strength fibres are pre-impregnated with resin in preparation for composite press forming (page 446).

PTFE
Polytetrafluoroethylene (page 525), also known by its trademark name Teflon.

PTT
Polytrimethylene terephthalate, a type of **thermoplastic polyester** (page 514).

PU
Polyurethane (page 518).

PVA
Polyvinyl alcohol.

PVC
Polyvinyl chloride (page 518).

quasi-isotropic
See **isotropic**.

raschel
A **warp**-knitted (page 144) structure produced using two or more guide bars on a V-bed knitting loom. This technique is capable of producing double-faced, tubular and openwork structures. And by adding **jacquard** (independent guide bars), complex and intricate patterns may be introduced into the structure.

reed
A comb-like structure spanning the full width of a loom, made up of blunt blades that sit between each **warp** (see Loom Weaving, page 77). Its purpose is to beat up each newly inserted **weft** to its final position, thereby creating a fabric of the correct density.

resilience
The ability of a material to spring back to its original shape once load has been removed.

rib
A form of knitted fabric with raised **courses** (widthways) or **wales** (lengthways).

rope
A length of thick, flexible cord formed by **twisting** (page 60) or braiding (page 68) several yarns or strands together.

roving
A loosely **twisted** sliver of fibres that have been **carded** (and in some cases **combed**) in preparation for **spinning**.

satin weave
A weave pattern that produces fabric with a lustrous surface, owing to each **yarn floating** over four or more yarns for every under-lap. Fewer under-laps mean the yarns can be packed very tightly together. It is commonly produced from silk (page 497). Sateen is made using spun cotton (page 481) or flax (page 481). See also **plain weave** and **twill weave**.

SBR
Styrene butadiene rubber (page 518).

seersucker
A lightweight fabric with a relief surface created by applying chemicals in a linear pattern to produce a puckered finish, or with slack-tension weaving (page 97).

selvedge
The edge of woven fabric constructed in such a way as to avoid the **yarns** unravelling. Also spelt 'selvage' (North America).

semi-crystalline
A polymer structure that contains both crystalline (highly organized) and amorphous (random) regions, for example **polyester** (page 514).

semi-synthetic
A group of synthetic plastics manufactured from naturally occurring **polymers**, including cellulose and protein, and converted into fibre, for example viscose and acetate (see Regenerated Fibres, page 506).

shape memory
The ability of a material to be heavily manipulated and then return to its original shape, sometimes with the application of heat or electricity.

Materials that exhibit shape memory include natural rubber (page 168), synthetic rubber (page 518) and certain metal alloys.

shed
Space between the reed and the **warp yarns** (created by raising some warp yarns while lowering others, using **heddles**) on a weaving loom, through which the **weft** is inserted. See also **pick**.

Shore hardness
The hardness of a plastic, rubber or **elastomer** is measured by the depth of indentation of a shaped metal foot on a measuring instrument known as a durometer. The depth of indentation is measured on a scale of 0 to 100; higher numbers indicate harder (less flexible) materials. These tests are generally used to denote the flexibility of a material. The two most popular are Shore A and Shore D. Indenter feet with different profiles are used in each case, and so each process is suitable for different material hardnesses. For example, soft materials are measured on the Shore A scale, and hard materials are measured on the Shore D scale. There is not a strong correlation between different scales. Shore hardness is also known as durometer hardness.

silicone
A **thermosetting plastic** consisting of oxygen and silicon atoms, among others (page 519), and applied as a coating (page 226).

sinter
To coalesce a powder into a solid mass without the material passing through the liquid phase. For example, high-performance polytetrafluoroethylene (**PTFE**) **yarn** is produced by extruding the raw ingredients as a powder held in place by a temporary carrier, and with heat and pressure the PTFE particles are bonded together. This is the only practical way to form the yarn, because PTFE will not melt like most **thermoplastics**.

size
A chemical solution used to stiffen **yarn** and fabric, for example to increase the strength of **warp** yarns in preparation for loom weaving (page 76). Afterwards it is removed from the fabric by scouring (page 202).

skin
Leather derived from smaller animals, such as lamb, sheep and pig. See also **hide**.

SLA
Stereolithography (page 466).

sliver
Bundles of loose and untwisted natural or man-made fibres produced by **carding** in preparation for **spinning**.

SLS
Selective laser **sintering** (page 466).

spandex
Common name for fibre produced from **thermoplastic elastomer** (TPU) in North America. Also called elastane (page 517).

Spectra
Trademark name for ultra-high-molecular weight polyethylene (**UHMWPE**) (page 517). See also Dyneema.

spinning
A **yarn** formation process. See Filament Yarn Spinning (page 50) and Staple Yarn Spinning (page 56). When used to combine multiple individual yarns to make a **plied yarn** it is also called plying, **twisting** and **doubling**. See Plying and Twisting (page 60).

spread tow
Tape made up of high-performance **filaments** laid side by side, in parallel. See also **tow**.

strand yarn
See **strip yarn**.

staple
A cluster of wool (page 494) as it grows naturally.

staple fibre
Short lengths of fibre, which may occur naturally, such as cotton (page 481) and wool (page 494), or may be man-made **filaments** cut to length.

staple yarn
Continuous **yarn** produced by spinning **staple fibres** together. See also **filament**.

stitch
A loop of **yarn** relating to a single operation in a knitted (pages 126–51), embroidered (page 298) or sewn (pages 342–69) structure.

stretch
The amount a material deforms when tension is applied. It manifests as elastic behaviour (the material returns to its original shape once the load is removed, for example **elastane**, page 517), permanent (such as tentering, page 211) or **creep** (gradual elongation over time when under tension). Elastic fabrics are either two-way stretch (one direction) or four-way stretch (lengthways and widthways).

strike-off
A sample print used to test colour and design in preparation for printing.

strip yarn
Narrow width of plastic film or leather applied as a **yarn**. Also called strand yarn.

sublimation
The process of a substance going from solid to gas (when heated) without passing through the liquid phase. For example, sublimation dyes are utilized in transfer printing (page 272).

suede
Leather that has been mechanically buffed on the flesh side to raise the surface and produce a soft finish. Also called velour.

super fibre
A group of synthetic fibres differentiated from common types by their superior strength to weight (as a result of increased **molecular weight** and high-strength bonds between the **polymer** chains). Examples include polyphenylene benzobisoxazole (PBO), known as Zylon (page 525), and expanded polytetrafluoroethylene (ePTFE), known as Tenara (page 525).

Teflon
Trademark name for polytetrafluoroethylene (PTFE) (page 525).

tenacity
A measure of the strength of a fibre or **yarn** calculated as the breaking force required for a specific density (**denier** or **decitex**) of yarn. See also **tensile strength**.

Tencel
Trademark name for **lyocell** (page 508).

tensile strength
Also referred to as ultimate strength, it is a measure of the resistance of a material to break under tension (**tenacity**) and thus the force required to break it apart (breaking force). For example, **nylon** (page 516) has higher tensile strength than viscose (page 508), and carbon fibre (page 526) has higher tensile strength than glass fibre (page 526).

terry
A looped uncut **pile**, such as to aid cushioning in knitted socks (page 142) and absorbency in woven towels (page 98).

tex
A unit of weight used to indicate the fineness of single **yarns**, as well as **plied yarns**. It is equal to 1 gram per 1,000 metres. See also **decitex** and **denier**.

texturing
Also called bulking or crimping, this is the process of adding bulk and texture to **filament yarns** during production (page 50). It is used to increase cover (such as in garments and carpets) and improve **resilience**.

thermochromatic
A compound – such as additives used in **filament** spinning (page 50), coating (page 226) or printing ink (page 00) – that changes colour as its temperature goes up or down.

thermoplastic
A **polymer** that becomes soft and pliable when heated. Whereas amorphous types soften gradually, **semi-crystalline** thermoplastics have a more clearly defined melting point. They are melted and spun to make **filament** (page 50), extruded to make film and sheet (page 180) or converted into foam (page 172). See also **thermosetting plastics**.

thermoplastic elastomer
A general name used to describe **thermoplastics** that exhibit similar elastic properties to rubber, such as **elastane** (page 517).

thermosetting plastic
Commonly referred to as thermoset, this group of materials is formed by heating, catalysing or mixing two parts to trigger a one-way polymeric reaction. Unlike most **thermoplastics**, thermosets form cross-links between

the **polymer** chains that cannot be undone, which makes this material more complicated to recycle. However, this same property means that thermosets tend to have higher strength and resistance to fatigue and chemical attack than thermoplastics. They are predominantly used as adhesives (page 370) and coatings (page 226).

tow
A bundle of untwisted natural or man-made fibres. Also refers to bundles of flax (page 481) or hemp (page 482) that have been crushed and roughly **combed** in preparation for **spinning** into **yarn** (page 56). See also **sliver**.

TPE
Thermoplastic elastomer.

TPT
Thin Ply Technology, developed by North Sails (page 236). Increasing the number of plies in a composite laminate while optimizing filament placement allows greater design freedom and improved structural optimization. This is known as the 'thin ply effect'.

TPU
Thermoplastic polyurethane, known in fibre form as elastane (spandex in North America) (page 517).

transfer foil
A plastic film that has been pre-decorated with colour, graphics, metal foil or metallic pigment, applied to leather or textile with heat and pressure for decoration (see Calendering and Embossing, page 220).

tuft
A length of **yarn**, or bundle of yarns, pulled through fabric and projecting from the surface. It is either cut or left as a loop (see Tufting, page 286).

Twaron
The trademark name for para-**aramid** (page 524). See also **Kevlar**.

twill weave
Each **warp** or **weft** in this weave **floats** across two or more weft or warp **yarns** to create a diagonal pattern. The angle, from reclining to steep, is determined by the density of the weave and balance of the fabric, whereby regular twill is **balanced**. Examples include denim (yarn-dyed cotton), flannel, herringbone (twill reversed at regular intervals), chino (hard-wearing with a slight sheen) and houndstooth (even-sided twill with warp and weft in contrasting colours). See also **plain weave** and **satin weave**.

twist
A term to describe the number of turns for a given length of **yarn** and the direction of twist, which is either Z (right) or S (left) (see Staple Yarn Spinning, page 56, and Plying and Twisting, page 60).

unbalanced weave
A woven pattern made up of proportionally more of one **yarn** (**warp** or **weft**) than the other per cm²/in². See also **balanced weave**.

UHMWPE
Ultra-high-molecular weight polyethylene (page 517), commonly known by its trademark names, including Dyneema and Spectra.

unidirectional fabric
Fabric constructed with the **yarns** running in parallel (page 373).

urethane
Another name for polyurethane (**PU**) (page 518).

Vectran
The trademark name for liquid crystal polymer (LCP) (page 524).

Ventile
The trademark name for an uncoated woven cotton (page 481) fabric that is windproof and rainproof. Made from the highest-quality long-**staple** cotton using a dense Oxford weave (a type of **plain** or **basket weave**), Ventile utilizes the natural swelling of the fibres when wet to produce a weatherproof outer layer.

vinyl
See polyvinyl chloride (**PVC**).

viscose
The first man-made fibre, produced by chemically extracting cellulose from biological ingredients, such as wood and cotton, and converting it into fibre by **filament spinning**. See Viscose, page 508.

volatile organic compound (VOC)
Chemicals with a very low boiling point, which causes them to evaporate at room temperature. They may be harmful to people or the environment, in which case they are collected and recycled. For example, dry solvent spinning (see Filament Yarn Spinning, page 50) produces VOCs.

vulcanization
The process of **curing** natural rubber (page 168) with sulphur, heat and pressure in a one-way reaction to form a **thermosetting** material.

wadding
Soft insulating material, such as **nonwoven** (page 152), used to line garments and blankets for example.

wale
In the case of knitting (pages 126–51) it is a vertical row (machine direction) of inter-looped stitches (see also **course**). In the case of **pile** weaving (page 96), it is a vertical ridge (**warp** direction), such as found on corduroy.

warp
Yarn that runs the length of a piece of loom-woven fabric (machine direction and perpendicular to the **weft**, or filling yarn).

weft
Yarn that runs across the width of a piece of loom-woven fabric (perpendicular to the **warp**). Also called filling yarn.

woollen
Wool fibres (page 494) of mixed length that have been **carded** (but not **combed**). When spun they produce a bulky **yarn** that is not as strong as **worsted** wool (carded and combed).

worsted
Long wool fibres (page 494) that have been **carded** and **combed** (parallel alignment). It is spun to make high-quality, strong and fine yarn. See also **woollen**.

yarn
A continuous length of fibre. See also **staple yarn** and **filament yarn**.

Zylon
The trademark name for polyphenylene benzobisoxazole (PBO) (page 525).

Featured Companies

Agricultural Commodity Service Cooperative (ACSC) Bio-Farmer
Kyrgyzstan
www.organicfarming.kg

Anderson and Sheppard
London, United Kingdom
www.anderson-sheppard.co.uk

Area Rugs and Carpets
Mirfield, West Yorkshire
United Kingdom
www.arearugs.co.uk

Automated Cutting Solutions
Fareham, Hampshire
United Kingdom
www.automatedcutting.co.uk

Bridgedale Outdoor Ltd
Newtownards, N. Ireland
United Kingdom
www.bridgedale.com

British Wool Marketing Board
Bradford, United Kingdom
www.britishwool.org.uk

Calderdale Carpets
Dewsbury, United Kingdom
www.calderdalecarpets.com

Calder Dyeing
Dewsbury, United Kingdom
www.caldertextiles.co.uk

Camira Fabrics
Mirfield, West Yorkshire
United Kingdom
www.camirafabrics.com

Cavalier Carpets
Blackburn, Lancashire
United Kingdom
www.cavaliercarpets.co.uk

Cifra SpA
Verano Brianza (MB), Italy
www.cifra-spa.net

Cluny Lace Co. Ltd
Ilkeston, Derbyshire
United Kingdom
www.clunylace.com

Coakley & Cox
Attleborough, Norfolk
United Kingdom
www.coakleyandcox.co.uk

Composites Evolution
Chesterfield, United Kingdom
www.compositesevolution.com

Compositum Ltd
Cayley, Canada
www.compositum.ca

Dents
Warminster, Wiltshire
United Kingdom
www.dents.co.uk

Drake Extrusion
Bradford, United Kingdom
www.drakeuk.com

DTC Dutch Thermoplastic Components
Almere, The Netherlands
www.composites.nl

EKOTEX
Namysłów, Poland
www.ekotex.com.pl

Elisa Strozyk
Berlin, Germany
www.elisastrozyk.de

English Willow Baskets
Dereham, Norfolk
United Kingdom
www.robandjuliekingbasketmakers.co.uk

Felthams
March, Cambridgeshire
United Kingdom
www.cplfelthams.co.uk

Flexitec Structures
Fareham, Hampshire
United Kingdom
www.flexitec.com

Fybagrate
Liversedge , West Yorkshire
United Kingdom
www.fybagrate.co.uk

Hand & Lock
London, United Kingdom
www.handembroidery.com

Haworth Scouring Company
Bradford, United Kingdom
www.haworthscouring.co.uk

Heinen Leather
Wegberg, Germany
www.heinen-leather.de

Interfoam
Bedford, United Kingdom
www.interfoam.co.uk

Jean Leader
Glasgow, United Kingdom
www.jeanleader.net

John Holden
Bolton, United Kingdom
www.john-holden.co.uk

John Lobb
Northampton, United Kingdom
www.johnlobb.com

Johnstons of Elgin
Elgin, Moray, United Kingdom
www.johnstonscashmere.com

K2 Screen
London, United Kingdom
www.k2screen.co.uk

Mallalieus of Delph
Oldham, United Kingdom
www.mallalieus.com

Marlow Ropes
Hailsham, United Kingdom
www.marlowropes.com

Mazzucchelli
Castiglione Olona (VA), Italy
www.mazzucchelli1849.it

North Sails
Minden, Nevada
USA
www.northsails.com

North Thin Ply Technology
Penthalaz, Switzerland
www.thinplytechnology.com

OneSails GBR (East)
Ipswich, United Kingdom
www.onesails.co.uk

Palm Equipment International Ltd
Clevedon, Somerset
United Kingdom
www.palmequipmenteurope.com

PaperFoam
Barneveld, The Netherlands
www.paperfoam.com

Penang Batik
Penang, Malaysia
www.penangbatik.co.my

Philippa Brock
London, United Kingdom
www.theweaveshed.org

Philippine Fiber Industry Development Authority
Quezon City, Philippines
fida.da.gov.ph

PP3DP
Beijing, China
www.pp3dp.com

Precision Dippings Marketing
Bristol, United Kingdom
www.precisiondippings.co.uk

Prelle
Lyon, France
www.prelle.fr

Rig Magic
Ipswich, United Kingdom
www.rigmagic.co.uk

Rotor
Brussels, Belgium
www.rotordb.org

SCP
London, United Kingdom
www.scp.co.uk

H. Seal & Company Ltd
Coalville, Leicestershire
United Kingdom
www.hseal.co.uk

Seifert and Skinner & Associates
Berbroek, Belgium
www.seifert-skinner.com

Sharps
Leeds, United Kingdom
www.sharpsfabricprinters.co.uk

Silk & Burg
London, United Kingdom
www.silkandburg.com

Skyeskyns
Isle of Skye
United Kingdom
www.skyeskyns.co.uk

Solar Impulse SA
Lausanne, Switzerland
www.solarimpulse.com

Standfast & Barracks
Lancaster, United Kingdom
www.standfast-barracks.com

Studio Echelman
Brookline, Massachusetts
USA
www.echelman.com

The Fabrick Lab
Hong Kong
www.thefabricklab.com

The Grosvenor Wilton Co.
Kidderminster, Worcestershire,
United Kingdom
www.grosvenorwilton.co.uk

Thomas & Vines
London, United Kingdom
www.flocking.co.uk

Thornback & Peel
London, United Kingdom
www.thornbackandpeel.co.uk

Tianyuan
Beijing, China
www.3dscan.com.cn

TigerTurf
Hartlebury, Worcestershire,
United Kingdom
www.tigerturf.com

Turnbull Prints
Bury, Greater Manchester,
United Kingdom
www.turnbull-design.com

Umeco
Heanor, Derbyshire
United Kingdom
www.umeco.com

Vanners
Sudbury, Suffolk,
United Kingdom
www.vanners.com

Whiteley Fischer Limited
Luton, Bedfordshire,
United Kingdom
www.whiteley-hat.co.uk

Further Reading

Adanur, Sabit, *Handbook of Weaving* (Raton: CRC Press, 2001)

Black, Sandy, *Eco-Chic: The Fashion Paradox* (London: Black Dog Publishing, 2008)

Braddock, Sarah E., and Marie O'Mahony, *Techno Textiles: Revolutionary Fabrics for Fashion and Design*, (London: Thames & Hudson, 1998)

Briscoe, Lynden, *The Textile and Clothing Industries of the United Kingdom* (Manchester: Manchester University Press, 1971)

Collier, Ann M., *A Handbook of Textiles (Third edn)* (Oxford and New York: Pergamon Press., 1980)

Elsasser, Virginia Hencken, *Textiles: Concepts and Principles (Second edn)* (New York: Fairchild Publications, 2005)

Field, Anne, *Spinning Wool: Beyond the Basics (Revised edn)* (London: A & C Black, 2010)

Gordon, Beverly, *Textiles: The Whole Story* (London: Thames & Hudson, 2011)

Johnstons of Elgin, *Scottish Estate Tweeds* (Elgin: Johnstons of Elgin, 1995)

Kadolph, Sara J., *Textiles (Eleventh edn)* (New Jersey: Pearson Education, 2011)

Leydecker, Sylvia, *Nano Materials in Architecture, Interior Architecture and Design* (Basel: Birkhäuser, 2008)

Leitner, Christina, *Paper Textiles* (London: A&C Black, 2005)

Lupton, Ellen, *Skin: Surface, Substance + Design* (London: Laurence King Publishing and New York: Princeton Architectural Press, 2002)

McDonough, William, and Michael Braungart, *Cradle to Cradle: Remaking the Way We Make Things* (London: Jonathan Cape and New York: North Point Press, 2002)

McFadden, David Revere, *Radical Lace and Subversive Knitting* (New York: Museum of Arts & Design, 2007)

McQuaid, Matilda, *Extreme Textiles: Designing for High Performance* (London: Thames & Hudson and New York: Smithsonian Cooper-Hewitt, National Design Museum: Princeton Architectural Press, 2005)

Miller, Edward, *Textiles: Properties and Behaviour in Clothing Use* (London: B. T. Batsford, 1968)

Murphy, William S., *The Textile Industries: A Practical Guide to Fibres, Yarns & Fabrics in Every Branch of Textile Manufacture, Including Preparation of Fibres, Spinning, Doubling, Designing, Weaving, Bleaching, Printing,*

Dyeing and Finishing Vol. 8 (London: The Gresham Publishing Company, 1911)

Müssig, Jörg, *Industrial Application of Natural Fibres: Structure, Properties and Technical Applications* (Chichester: John Wiley & Sons, 2010)

O'Mahony, Marie, *Advanced Textiles for Health and Wellbeing* (London and New York: Thames & Hudson, 2011)

O'Mahony, Marie, and Sarah E. Braddock, *Sportstech: Revolutionary Fabrics, Fashion and Design* (London and New York: Thames & Hudson, 2002)

Potter, M. David, and Bernard P. Corbman, *Textiles: Fiber to Fabric (Fourth edn)* (New York: Gregg Division/McGraw-Hill Book Company, 1967)

Quinn, Bradley, *Fashion Futures* (London: Merrell, 2012)

Quinn, Bradley, *Textile Visionaries: Innovation and Sustainability in Textile Design* (London: Laurence King Publishing, 2013)

Richards, Ann, *Weaving Textiles that Shape Themselves* (Marlborough: Wiltshire, 2012)

Rossback, Ed, *Baskets as Textile Art* (London: Studio Vista, 1974)

Roulac, John W., *Hemp Horizons: The Comeback of the World's Most Promising Plant* (Vermont: Chelsea Green Publishing Company, 1997)

Sen, Ashish Kumar, *Coated Textiles: Principles and Applications (Second edn)* (New York: CRC Press, 2008)

Spencer, David J., *Knitting Technology: A Comprehensive Handbook and Practical Guide (Third edn)* (Cambridge: Woodhead Publishing, 2001)

Stattmann, Nicola, *Ultra Light, Super Strong: A New Generation of Design Materials* (Basel: Birkhäuser, 2003)

Tellier-Loumagne, Françoise, *The Art of Embroidery: Inspirational Stitches, Textures and Surfaces* (London and New York: Thames & Hudson, 2006)

Thompson, Rob, *Sustainable Materials, Processes and Production, The Manufacturing Guides* (London: Thames & Hudson, 2013)

Tyler, David J., *Carr & Latham's Technology of Clothing Manufacture (Fourth edn)* (Oxford: Blackwell Publishing, 2000)

Wright, Dorothy, *Baskets and Basketry* (London: B. T. Batsford Ltd., 1959)

Illustration Credits

Martin Thompson photographed the processes and materials in this book. The authors would like to thank the following for permission to reproduce their photographs and illustrations.

Introduction
Page 13 (super fibres): printed with the permission of Studio Echelman; photography by Peter Vanderwarker
Page 13 (paper fabric): Philippa Brock
Page 15 (seamless knitting): Cifra SpA
Page 16 (Thin Ply Technology): Solar Impulse
Page 18 (racing sail design): OneSails GBR (East)

Cotton
Page 28 (title image): photography by Murat Uzakovich Dzhergalbayev
Pages 32–33 (images 1–6 and 8–9): photography by Murat Uzakovich Dzhergalbayev
Page 33 (images 7 and 10): Agricultural Commodity Service Cooperative (ACSC) Bio-Farmer; photography by Zhanibek Borkoshev

Bast Fibres
Page 38 (images 2–3): EKOTEX
Page 39 (images 5–6): EKOTEX

Leaf Fibres
Page 40 (title image): Philippine Fiber Industry Development Authority
Pages 42–45 (all images): Philippine Fiber Industry Development Authority

Silk
Page 49 (images 1–7): Philippine Fiber Industry Development Authority

Weft Knitting
Page 126 (title image): Johnstons of Elgin

Warp Knitting
Page 150 (image 1): Cifra SpA

Foam
Page 175 (images 3–5): Caligen Foam
Page 176 (images 1 and 3): Caligen Foam

Plastic Film Extrusion
Page 187 (all images): Rotor

Laminating
Page 192 (three-dimensional fabric): The Fabrick Lab; photography by Elaine Ng Yan Ling

Spread Tow and Fabric Prepreg
Page 234 (title image): North Thin Ply Technology
Page 237 (all images): North Thin Ply Technology

Resist Dyeing
Page 248 (title image): Penang Batik
Page 254 (images 2–4): Penang Batik

Hand Block Printing
Page 259 (all images): Turnbull Prints

Tufting
Page 293 (image 13): TigerTurf

Slitting and Crosscutting
Page 327 (all images): Johnstons of Elgin

Machine Stitching
Page 362 (image 1): OneSails GBR (East)
Page 364 (image 1): Palm Equipment International Ltd
Page 365 (image 10): Palm Equipment International Ltd
Page 366 (image 1): Palm Equipment International Ltd
Page 369 (image 1): Palm Equipment International Ltd

Adhesive Bonding
Page 374 (image 1): OneSails GBR (East)
Page 375 (image 11): OneSails GBR (East)

Upholstery
Page 406 (image 1): SCP

3D Thermal Laminating
Page 454 (title image): North Sails
Pages 456–59 (all images): North Sails

Filament Winding
Page 463 (image 1): Compositum Ltd

Fibrous wood, grass amd leaves
Page 485 (wooden textile): Elisa Strozyk; photography by Sebastian Neeb

Acknowledgments

The technical detail and accuracy of the manufacturing case studies is the result of the extraordinary generosity of many individuals and organizations. Their knowledge of materials and processes, and in most cases their years of hands-on experience, were invaluable in understanding the opportunities for designers. The authors would like to give their personal thanks to the following: Chinara Babanova and Zhanybek Borkoshev at **Agricultural Commodity Service Cooperative (ACSC) Bio-Farmer**; Anda Rowland and John Hitchcock at **Anderson and Sheppard**; Andrew Warburton at **Area Rugs and Carpets**; James Mordant at **Automated Cutting Solutions**; Jim Campbell at **Bridgedale**; Olga Dorojevic, Jean Murphy, Mark Powell and Angela Marshall-Williams at **British Wool Marketing Board**; Roy Thornton, Greg Bedford and Steve Norris at **Calderdale Carpets**; Stephen Norris at **Calder Dyeing**; Tony Lay, Ian Burn, Darren Hill, Graham Berry and Catherine Watkins at **Camira Fabrics**; Hans Lowe and Russell Clark at **Cavalier Carpets**; Cesare Citterio, Mario Redaelli and Jenny Citterio at **Cifra SpA**; Charles Mason at **Cluny Lace Co. Ltd**; Tim Cox and Tim Hooker at **Coakley & Cox**; Brendon Weager at **Composites Evolution**; Maximilian Lundershausen at **Compositum Ltd**; Natalie Perret, Heather Ambrose, John Roberts and Deborah Moore at **Dents**; Tony Suddards at **Drake Extrusion**; Sjoerd Hooning, David Manten and Tom Stoil at **DTC Dutch Thermoplastic Components**; Marek Radwanski at **EKOTEX**; Rob King at **English Willow Baskets**; Colin Oswald at **Felthams**; David Banks at **Flexitec Structures**; Stephen Kershaw, Richard Leather and Andrew Leather at **Fybagrate**; June Denny and Prue Seal at **H. Seal & Company Ltd**; Scott Gordon Heron at **Hand & Lock**; David Gisbourne at **Haworth Scouring Company**; Thomas Heinen at **Heinen Leather**; Nick Reid at **Interfoam**; Jean Leader; Julie Thornhill at **John Holden**; Steve Johnson and Valerie Guerin-Goff at **John Lobb**; Jenny Stewart and Fiona Dalgleish at **Johnstons of Elgin**; Mark Jenkins at **K2 Screen**; David Mallalieu at **Mallalieus of Delph**; Paul Dyer at **Marlow Ropes**; Giovanni Mazzucchelli, Alessandro Fiocchi and Stefania Maffioli at **Mazzucchelli**; Bill Pearson at **North Sails**; John Parker at **OneSails GBR (East)**; Bob Slee at **Palm Equipment International Ltd**; Quah Chin Choon at **Penang Batik**; Petronilo B. Jabay and Nini P. Clemente at **Philippine Fiber Industry Development Authority**; the engineers at **PP3DP**; Chris Prickett at **Precision Dippings Marketing**; Guillaume Verzier and Oksana Vanoverberghe-Ortynska at **Prelle**; Nigel Theadom at **Rig Magic**; Kristof Weckx at **Seifert and Skinner & Associates**; Ross Thackery at **Sharps**; Clive Hartwell at **Skyeskyns**; Craig MacDonald and Stephen Thomas at **Standfast & Barracks**; Elaine Ng Yan Ling at **The Fabrick Lab**; Malcolm Foley at **The Grosvenor Wilton Co.**; Simon Thomas at **Thomas & Vines**; the engineers at **Tianyuan**; Eleanor Smith and Martine Hirrell at **TigerTurf**; Paul Turnbull at **Turnbull Prints**; David Tooth at **Vanners**; and Peter Whiteley at **Whiteley Fischer Limited**

The book's content would not have been so rich and colourful if it were not for the extraordinary generosity of individuals, organizations and professional photographers that have kindly supplied images. The authors would like to give personal thanks to the following: Sara Feinstein for her help setting up the photo shoot in Kyrgyzstan and Murat Uzakovich Dzhergalbayev for taking the lovely pictures there; Zhanybek Borkoshev (Agricultural Commodity Service Cooperative / ACSC Bio-Farmer) for taking photographs of the cotton ginning in Kyrgyzstan; Maximilian Lundershausen (Compositum Ltd.) for the ComposicaD screenshots; Philippa Brock for photographs of her beautiful fabrics; Melissa (Studio Echelman) for all her help; Solar Impulse SA for permission to use a picture of their magnificent solar-powered aircraft; Maarten Gielen (Rotor) for the extrusion blow molding photographs; Elaine Ng Yan Ling (The Fabrick Lab) for all her help, support and the lovely photographs; Paul Turnbull (Turnbull Prints) for the woodblock printing photographs; Cesare Citterio (Cifra SpA) for the examples of his company's advanced seamless knitting; John Parker (OneSails GBR / East) for all the photographs, screenshots and explanations; Marek Radwanski (EKOTEX) for the photographs of his farm; everyone at Philippine Fiber Industry Development Authority for their support and the lovely photographs; Jenny Stewart (Johnstons of Elgin) for the photographs of their textile mill; Angela Mistry (Caligen Foam); Eleanor Smith (TigerTurf); Bill Pearson (North Sails) for all the fantastic help, support and photographs of the great work that North Sails are doing; Bob Slee (Palm Equipment International Ltd); and Tim Cox (Coakley & Cox) for the photograph of the Oscar sofa.

Without the support, encouragement and input of colleagues, family and friends this book would not have happened. We would like to thank the designer Chris Perkins for his dedication to the project and for making another beautiful and informative book, the editor Kirsty Seymour-Ure for working through the text with incredible patience and attention to detail, and Thames & Hudson for again believing in such an ambitious project and for continuing to support our work.

Rob Thompson: I would particularly like to thank Martin, my Dad and the photographer on the project, who understands how to make beautiful pictures that communicate information better than anyone I know.

Martin Thompson: Special thanks to my wife Lynda, Rob's mother, and to his brothers for their suggestions and wise counsel. And thanks for all the friendly help I've received on factory visits: thank you to MDs and machinists alike.

We would like to give special thanks to Joe Hunter, of Vexed Design, for his valued contribution in the early stages of the book.

Index